Analogue and Digital Electronics

Other titles in Electrical and Electronic Engineering

B.W. Allen, *Analogue Electronics for Higher Studies*

G.J. Awcock and R. Thomas, *Applied Image Processing*

M. Beasley, *Reliability for Engineers*

P.V. Brennan, *Phase-Locked Loops — Principles and Practice*

Rodney F.W. Coates, *Underwater Acoustic Systems*

C.W. Davidson, *Transmission Lines for Communications, second edition*

J.D. Edwards, *Electrical Machines and Drives*

Peter J. Fish, *Electronic Noise and Low Noise Design*

M.E. Goodge, *Analog Electronics — Analysis and Design*

B.A. Gregory, *An Introduction to Electrical Instrumentation and Measurement Systems, second edition*

C.G. Guy, *Data Communications for Engineers*

Robin Holland, *Microcomputer Fault-finding and Design*

K. Jackson, *C Programming for Electronic Engineers*

Paul A. Lynn, *An Introduction to the Analysis and Processing of Signals, third edition*

Paul A. Lynn, *Digital Signals, Processors and Noise*

Paul A. Lynn, *Radar Systems*

R.C.V. Macario, *Cellular Radio — Principles and Design, second edition*

R.J. Mitchell, *Microprocessor Systems — An Introduction*

Noel M. Morris, *Electrical Circuit Analysis and Design*

M.S. Nixon, *Introductory Digital Design — A Programmable Approach*

F.J. Owens, *Signal Processing of Speech*

P.D. Picton, *Neural Network — An Introduction*

Dennis N. Pim, *Television and Teletext*

R.G. Powell, *Electromagnetism (Foundations of Engineering)*

M. Richharia, *Satellite Communications Systems — Design Principles*

Peter Rohner, *Automation with Programmable Logic Controllers*

R. Seals and G. Whapshott, *Programmable Logic: PLDs and FPGAs*

P.R. Shepherd, *Integrated Circuits — Design, Fabrication and Test*

M.J.N. Sibley, *Optical Communications, second edition*

P. Silvester, *Electric Circuits (Foundations of Engineering)*

A.J. Simmonds, *Data Communications and Transmission Principles*

P.M. Taylor, *Robotic Control*

T.J. Terrell and Lik-Kwan Shark, *Digital Signal Processing*

M.J. Usher and C.G. Guy, *Information and Communication for Engineers*

M.J. Usher and D.A. Keating, *Sensors and Transducers, second edition*

G.S. Virk, *Digital Computer Control Systems*

Lionel Warnes, *Electronic and Electrical Engineering*

Lionel Warnes, *Electronic Materials*

B.W. Williams, *Power Electronics — Devices, Drivers, Applications and Passive Components, second edition*

Analogue and Digital Electronics

Lionel Warnes

Department of Electronic and Electrical Engineering
Loughborough University of Technology

Dedicated to Vlad Oustimovitch

First published 1998 by
MACMILLAN PRESS LTD
Houndmills, Basingstoke, Hampshire RG21 6XS
and London
Companies and representatives
throughout the world

ISBN 0−333−65820−5

A catalogue record for this book is available
from the British Library.

This book is printed on paper suitable for recycling and made from fully managed and sustained forest sources.

10   9   8   7   6   5   4   3   2   1
07 06 05 04 03 02 01 00 99 98

Printed in Malaysia

Contents

Preface

EXACTLY one hundred years ago J J Thomson discovered the electron in Cambridge. During the next thirty years the nature of the new particle was progressively revealed by a succession of brilliant scientists. In parallel with this fundamental work applied scientists, engineers and inventors produced devices to control the flow of electrons which constitutes electric current and electronics was born. The earliest of these was Fleming's radio valve or vacuum-tube diode patented in 1905, though it played little part in the early development of radio, which mainly used the point-contact, crystal diode for detection. Lee de Forest's triode offered the possibility of amplification and sinewave generation as well as detection, and by the end of the First World War over a million had been produced. Electronically-based industries grew exponentially up to the invention of the bipolar junction transistor in 1947, assisted enormously by the pressing needs of the Second World War, and that invention enabled the rate of expansion in electronics to be maintained if not accelerated to the present day.

Today electronics pervades every technology so that the study of electronics is an essential part of nearly all engineering degree programmes during the first two years. Though the fundamentals of the subject change only slowly, the way in which electronics is applied and the means by which it is applied – the devices, circuits and systems – change almost daily, causing textbooks to become outdated rapidly unless frequently revised or rewritten. This book is designed to give non-specialist students a good foundation in analogue and digital electronics, and takes specialists in their second year of study to a higher level. The mathematics required to follow the text has been kept to A-level standard, but mostly rather below that. Wherever possible the text refers to up-to-date devices, integrated circuits and equipment, with many product specifications incorporated wherever appropriate. As many practical examples as possible are worked through at appropriate places in the text, with a selection of unworked examples given as problems at the end of each chapter.

In order to be as self-contained as possible, the circuit theory necessary for electronics – and only that – is contained in the first chapter, which can be skipped if the student has had or is taking a first course in it. The book also contains a small amount of semiconductor theory, just enough to understand the workings of semiconductor devices; and an appendix gives a brief outline of integrated circuit manufacture. There are also sections on transducers – a weak point in many electronic systems, not to say student projects – electromagnetic compatibility and measurements, both somewhat neglected too in electronics texts. The bulk of the book is divided between analogue electronics (more or less the first half) and digital electronics (most of the second half).

I have to thank my publisher, Malcolm Stewart, for being patient, and numerous students who have wittingly and unwittingly assisted in the preparation of this book: a generous portion of the contents comes from supervising and examining electronics projects over the course of two decades. Eugene Aw was especially helpful in providing me with data sheets from the Internet.

Lionel Warnes

Some useful constants

The magnetic constant or permeability of free space, μ_0 $4\pi \times 10^{-7}$ H/m

The electric constant or permittivity of free space, ε_0 8.85×10^{-12} F/m

The magnitude of the charge on the electron, q 1.6×10^{-19} C

Boltzmann's constant, k 1.38×10^{-23} J/K

The speed of light in a vacuum, c 3×10^8 m/s

Planck's constant, h 6.63×10^{-34} Js

The ice point, 0°C   273 K

Notes: $c = 1/\sqrt{(\mu_0\varepsilon_0)}$. The characteristic impedance of free space, $Z_0 = \sqrt{(\mu_0/\varepsilon_0)} = 377\ \Omega$. The constants listed are accurate, where approximated, to about 0.15% and are the values used in calculations in this book.

SI unit prefixes

prefix	name	factor
E	exa	10^{18}
P	peta	10^{15}
T	tera	10^{12}
G	giga	10^{9}
M	mega	10^{6}
k	kilo	10^{3}
m	milli	10^{-3}
μ	micro	10^{-6}
n	nano	10^{-9}
p	pico	10^{-12}
f	femto	10^{-15}
a	atto	10^{-18}

1 Electrical circuits

I N THIS chapter we shall introduce some ideal circuit elements — two types of power source and three types of component — and the laws governing their behaviour. These ideal elements will later be used to model non-ideal, practical devices, perhaps over restricted ranges of frequency, temperature, current and so forth.

1.1 Sign convention ·

Conventional current conveys positive charge, though the actual charge carriers in a length of copper wire are electrons flowing the other way. If a source of current is delivering current to a circuit as in figure 1.1a, the power sent from the source is VI when the current, I, flows out of the positive terminal. By the same convention the power consumed by the circuit or load is positive when the current flows into the positive terminal. The current must then leave the load and enter the source at the negative terminals. Power flows from the source to the circuit. If the current is reversed then power flows from the circuit to the source.

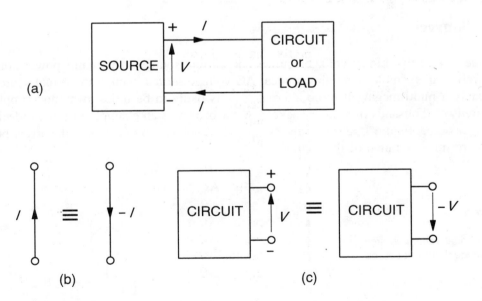

Figure 1.1 (a) Delivering power to a load circuit (b) current sign convention (c) voltage sign convention

In figure 1.1b, reversing the arrow showing the current's direction *and* changing the sign of the current to $-I$ does not affect the power sent from the source or the power

1

consumed by the load. A current of $+I$ flowing one way is exactly the same as a current of $-I$ flowing in the opposite direction. Reversing a voltage arrow and the sign of a voltage similarly leaves the voltage polarity unaltered, as in figure 1.1c.

1.2 Linear electric circuits

Consider the black box in figure 1.2, which contains some unspecified circuit elements, and let us take the input terminals to be those on the left and the output terminals to be those on the right. Suppose we excite the circuit with an input of $x_1(t)$ — which may be either a current or a voltage — and that a corresponding output of $y_1(t)$ is produced. Next we excite the circuit with an input of $x_2(t)$, obtain an output of $y_2(t)$. If the circuit produces a response of $y_1(t) + y_2(t)$ when the input is $x_1(t) + x_2(t)$, then it is said to be *linear*; alternatively one can say that it obeys the *principle of superposition*. Power is a product of voltage and current so that it cannot be taken as an input or output variable without violating the linearity rule. Input and output variables in linear electric circuits are confined to voltage and current.

Figure 1.2

A linear circuit

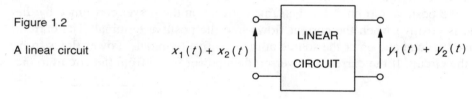

1.3 Sources

Sources are of two kinds: voltage sources and current sources. The mains power point in the wall is an example of an almost ideal AC voltage source, though the actual source — probably a multi-megawatt turbo-alternator — is likely to be miles away and supplying hundreds of thousands of similar sockets. A car battery is an example of a near-ideal DC voltage source. Figure 1.3a shows the symbol for a voltage source, where the arrow points to the positive terminal of the source.

Figure 1.3

(a) An ideal voltage source
(b) An ideal current source

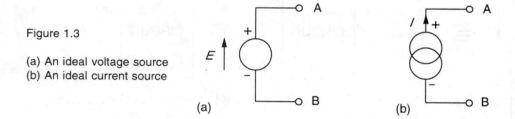

The polarity signs are normally omitted. Frequently a practical voltage source is said to have an *electromotive force* or e.m.f., denoted by the letter E, so that for example a car battery is said to have an e.m.f. of 12 V, or a turbo-alternator an e.m.f. of 6.9 kV. The voltage appearing between terminals AB is always E no matter what the load. The current from the voltage source depends on the load and flows out of A and returns through B. The power supplied is *EI* if the current out of the positive terminal is *I*.

Current sources are given the symbol shown in figure 1.3b, and again the polarity signs are usually omitted, though it is sometimes important to remember that there is a voltage across a current source. Transistor amplifiers are often effective current sources. No matter what load is connected across AB, the current flowing through it is always *I*. The voltage across AB depends on the load. If the voltage across the source is *E*, then the power it supplies is *EI* when the current flows out of the positive terminal.

If we were to short circuit an ideal voltage source the short-circuit current would be infinite; similarly, were we to open circuit an ideal current source, the open-circuit voltage would be infinite. Neither event occurs in practice because practical sources have *internal resistances* which restrict the current flowing from the voltage source and the open-circuit voltage of the current source. The practical sources are shown in figure 1.4, from which we see that the practical voltage source has a series internal resistance and the practical current source has a parallel internal resistance.

Figure 1.4

(a) A practical voltage source
(b) A practical current source

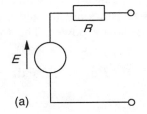
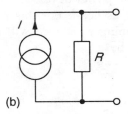

1.4 Ideal devices

Devices are of many kinds, but are principally divided into passive devices and active devices. Passive devices are in essence those such as resistors, capacitors, inductors and diodes that do not require an external source of power: they always either consume power (resistors and diodes) or store energy (inductors and capacitors). Active devices such as operational amplifiers and transistors require a source of external power before they can perform their function within a circuit. Passive devices are subdivided further into linear devices — resistors, capacitors and inductors — and non-linear devices — mostly diodes.

1.4.1 Resistors: Ohm's law[1]

The operation of a passive device within a circuit can be summarised by the relationship between the current through and the voltage across it. We consider here the linear members among ideal passive devices; diodes are given a chapter to themselves.

---

[1] Georg Simon Ohm (1787-1854) published at least three versions of his law until in 1827 he produced one which is in essence that of equation 1.1; however, it was only gradually accepted and Ohm endured considerable obloquy before he eventually became Professor of mathematics and physics at Munich only in 1849.

An ideal resistor possesses only resistance[2], the size of which is given by Ohm's law

$$V = IR \qquad (1.1)$$

V being the potential across the resistance in volts (V), I the current through it in amps (A) and R, its resistance, is then in ohms (Ω).

Figure 1.5

The circuit symbol and Ohm's law for resistance

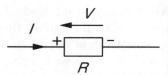

In figure 1.5, the voltage across the resistance, V, has the polarity shown by the voltage arrow and polarity signs (which are usually omitted) when the current flows through the resistance in the direction indicated by the current arrow. If V is in volts (V) and I is in amperes (A), then R is in ohms (Ω).

The power developed in the resistance is VI, which is I^2R or V^2/R by Ohm's law, and is lost as heat[3]. For this reason resistors are termed dissipative. Any non-ideal element containing resistance will also dissipate power. The total energy lost in a resistance is given by

$$W = \int P \mathrm{d}t = \int I^2 R \mathrm{d}t \qquad (1.2)$$

1.4.2 Capacitors

A capacitor is a device that separates charge. The process of charging a capacitor is one in which one electrode or 'plate' collects a positive charge and the other electrode collects a negative charge of equal magnitude, the overall charge being zero. Ideal capacitors possess only capacitance whose magnitude is given by the ratio of the positive charge to the potential difference between the plates.

Figure 1.6

The circuit symbol for capacitance, showing the charging current and resulting voltage and charge polarities

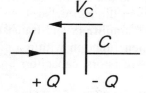

In figure 1.6, the charge on the positive plate is Q coulombs (C) and the voltage across the capacitance is V_C volts, so that the capacitance in Farads (F) is

$$C = Q/V_C \qquad (1.3)$$

---

[2] In circuit diagrams, the elements are taken as ideal and are spoken of as resistances, inductances or capacitances rather than resistors etc. since it is assumed that the terms resistor etc. refer to practical, non-ideal devices.

[3] It appears that James Prescott Joule (1818-99) first discovered this relation in 1841. Joule's work on the mechanical equivalent of heat was rejected by scientific journals and only appeared in a Manchester newspaper.

Now suppose the current flowing into the ideal capacitor is I amps, then as current is the rate of flow of charge past a point in a circuit, a current of 1 A being a rate of charge movement equal to 1 C/s, we can write

$$I = dQ/dt \tag{1.4}$$

But from equation 1.3 we know that $Q = CV_C$, and substituting for Q in equation 1.4 yields

$$I = \frac{d(CV_C)}{dt} = C\frac{dV_C}{dt} \tag{1.5}$$

given that C is constant. This equation shows that *the voltage across a capacitor cannot change instantaneously*, since then the current would be infinite. Though equation 1.5 is a perfectly adequate current-voltage relationship for a capacitor it is usually rearranged and integrated to give

$$dV = \frac{I\,dt}{C} \quad \rightarrow \quad \int dV = \frac{1}{C}\int I\,dt \tag{1.6}$$

If the capacitor is initially uncharged before being charged to a potential, V_C, then

$$V_C = \frac{1}{C}\int I\,dt \tag{1.7}$$

This is the customary form of current-voltage relation for a capacitor; from it the voltage across a capacitor can be found if the current is known as a function of time. As the integral is the area under the I–t graph scaled by $1/C$, calculating that area is the easiest way of finding V_C.

The power delivered to the capacitor is $V_C I$ in watts (W) and since I is given in equation 1.5 we can write this as

$$P = V_C I = V_C C\frac{dV_C}{dt} \tag{1.8}$$

The power is not lost as in a resistor but stored as energy in the electric field between the capacitor's plates. From equation 1.7 we can find the energy in joules (J) by integration

$$W = \int P\,dt = C\int V_C\,dV_C = \tfrac{1}{2}CV_C^2 \tag{1.9}$$

The energy stored is proportional to the square of the voltage across the capacitor. When the capacitor is discharged, the energy is given back. In an ideal capacitor, no energy is lost, it is just stored.

1.4.3 Inductors

Inductors ideally possess only inductance. Though capacitors and resistors may often be considered ideal without serious error, it is in practice usually impossible to ignore the resistance associated with real inductors.

However, for an ideal inductor of figure 1.7, whose inductance is L Henrys (H), the voltage across it, V_L, is related to the current through it, I, by the equation

$$V_L = L\frac{dI}{dt} \tag{1.10}$$

This equation shows that *the current through an inductor cannot change instantaneously*, since then the voltage across it would be infinite.

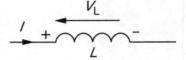

Figure 1.7 The circuit symbol for inductance

Note that if the current into the inductor in the sense shown is increasing, then the voltage across the inductor is increasing as dI/dt is positive. By the same token, if this current is decreasing then the voltage is negative, that is the voltage arrow and polarity signs on the inductor are reversed.

The power delivered to the inductor is $V_L I$, and since $V_L = LdI/dt$, it follows that

$$W = \int P dt = \int LI dI = \tfrac{1}{2}LI^2 \tag{1.11}$$

The energy stored by an inductor is proportional to the square of the current it carries, and is released during the fall of the current. In an ideal inductor energy is not lost, just stored.

Example 1.1

In the circuit of figure 1.8a, what are the voltages across the inductance, V_L, the capacitance, V_C, and the whole circuit[4], V_{AB}, given the voltage across the resistance, V_R, is as shown in figure 1.8b?

Notice firstly that the current through all the components is the same and is given by Ohm's law: $I = V_R/R$, so that the current waveform is the same as the voltage across the resistance scaled down by a factor of 100. Thus at $t = 1$ ms, $I = 20/100 = 0.2$ A and so on. The I–t graph is shown in figure 1.9a. We can then find the voltages across the other components from their $V - I$ relations since the current flowing through them is I also.

For the first millisecond the voltage across the inductance is given by

$$V_L = L\frac{dI}{dt} = 0.15 \times 200 = 30 \text{ V}$$

since the rate of change in current is 0.2 A in 1 ms or +200 A/s and $L = 0.15$ H. After this time the current remains constant at 0.2 A, so that dI/dt is zero and so too is V_L. At $t = 2$ ms the current ramps down to zero in 2 ms, which means that

$$\frac{dI}{dt} = \frac{-0.2}{0.002} = -100 \text{ A/s}$$

---

[4] Here we use *double-suffix notation*, in which the first suffix, A, indicates the terminal to which the voltage arrow points, normally the positive terminal, while the second suffix, B, indicates the negative terminal. By this convention, $V_{AB} = -V_{BA}$.

The sign shows that V_L is now negative:

$$V_L = L\frac{dI}{dt} = 0.15 \times -100 = -15 \text{ V}$$

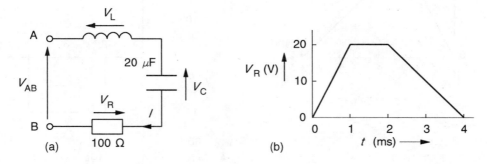

Figure 1.8 (a) The circuit for example 1.1 (b) The voltage across the resistance as a function of time

The graph of V_L against time is shown in figure 1.9b. Note that the changes in V_L that occur at 1 and at 3 ms are instantaneous: just before $t = 1$ ms, or $t = 1-$, $V_L = 30$ V and just after $t = 1$ ms, or $t = 1+$, $V_L = 0$ V.

To find the voltage across the capacitance we use equation 1.7, which is the same as saying that the change in V_C is the area under the I–t graph scaled by $1/C$. Rather than carrying out the integration, it is easier to find this area. Because the current is initially a linear function of time (or a ramp function), integrating it to find V_C must give a quadratic function of time. However, after 1 ms, the area under the I–t graph is

$$\tfrac{1}{2} \times 0.2 \times 10^{-3} = 10^{-4} \text{ As}$$

making

$$V_C = \text{area} \div C = 10^{-4} \div 20 \times 10^{-6} = 5 \text{ V}$$

From 1 to 2 ms the current is constant at 0.2 A and the area under the I–t graph is 2×10^{-4} As. The area under the graph from $t = 0$ to $t = 2$ ms is therefore 3×10^{-4} As and the voltage must be

$$V_C = 3 \times 10^{-4} \div 20 \times 10^{-6} = 15 \text{ V}$$

Between 1 ms and 2 ms the voltage increases linearly from 5 V to 15 V. After 2 ms the current ramps down to zero at 4 ms. Since the current is still positive the area under the I–t graph is still positive and the capacitor voltage continues to increase, though at a slower and slower rate. The area under the I–t graph from $t = 2$ ms to $t = 4$ ms is

$$\tfrac{1}{2} \times 0.2 \times 0.002 = 2 \times 10^{-4} \text{ As}$$

giving a total area from 0 to 4 ms of 5×10^{-4} As and a final voltage of

$$V_C = 5 \times 10^{-4} \div 20 \times 10^{-6} = 25 \text{ V}$$

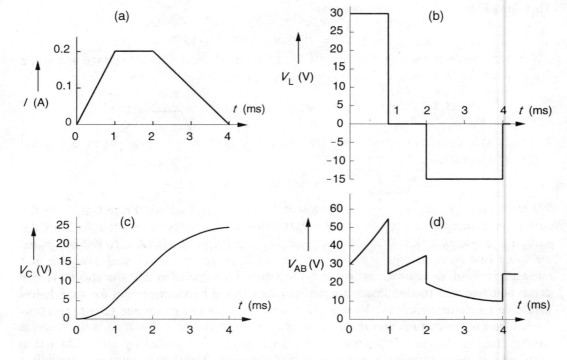

Figure 1.9 Current and voltage waveforms for example 1.1

We can see that at $t = 3$ ms, $I = 0.1$ A and the area under the $I–t$ graph is the total area less the area of the triangle between 3 and 4 ms. This triangular area is $0.5 \times 0.1 \times 0.001 = 0.5 \times 10^{-4}$ As, so the rest of the area is 4.5×10^{-4} As and $V_C = 22.5$ V. The graph of V_C against t is shown in figure 1.9c.

Finding V_{AB} requires V_R, V_L and V_C to be added, as in the table below:

t (ms)	0	1−	1+	1.5	2−	2+	3	4−	4+
V_R (V)	0	20	20	20	20	20	10	0	0
V_L (V)	30	30	0	0	0	−15	−15	−15	0
V_C (V)	0	5	5	10	15	15	22.5	25	25
V_{AB} (V)	30	55	25	30	35	20	17.5	10	25

Figure 1.9d shows the resultant graph of V_{AB} against time.

Example 1.2

In the circuit of example 1.1, what is the maximum energy stored in the inductance? And in the capacitance? What average power and total energy are dissipated in the resistance during the 4 ms that the current flows?

The energy stored in the inductance is $\frac{1}{2}LI^2$, so we need only find the maximum current, 0.2 A, to find the maximum energy:

$$W_{\mathrm{Lm}} = \tfrac{1}{2}LI_{\mathrm{m}}^2 = 0.5 \times 0.15 \times 0.2^2 = 3 \text{ mJ}$$

The energy stored in the capacitance is $\tfrac{1}{2}CV^2$, so we need the maximum voltage across the capacitance, 25 V, to find the maximum energy stored in it

$$W_{\mathrm{Cm}} = \tfrac{1}{2}CV_{\mathrm{m}}^2 = 0.5 \times 20 \times 10^{-6} \times 25^2 = 6.25 \text{ mJ}$$

The resistance dissipates power whenever current flows in it. The total energy dissipated is given by equation 1.2

$$W_{\mathrm{R}} = \int P\,\mathrm{d}t = \int I^2 R\,\mathrm{d}t$$

We must therefore express I as a function of t if we are to find W_{R}. From 0 to 1 ms the current is a linear function of time: $I = 200t$ when t is in seconds. From 1 to 2 ms the current is constant at 0.2 A and for the last 2 ms it is falling at 100 A/s. In the last phase we could find an equation for I in terms of t, but it would be complicated to square and integrate it. Were the current to start at 0 A when $t = 0$ and rise to 0.2 A at the same rate that it actually fell, 100 A/s, the energy dissipated would be the same but the calculation would be much simpler. Thus we can put $I = 100t$ for the last phase and take t from 0 to 2 ms. The energy lost is then given by

$$W_{\mathrm{R}} = \int_{0}^{1\,\mathrm{ms}} (200t)^2 R\,\mathrm{d}t + \int_{1\,\mathrm{ms}}^{2\,\mathrm{ms}} 0.2^2 R\,\mathrm{d}t + \int_{0}^{2\,\mathrm{ms}} (100t)^2 R\,\mathrm{d}t$$

Carrying out the integration yields

$$W_{\mathrm{R}} = R\left(200^2 \left[\tfrac{1}{3}t^3\right]_0^{0.001} + 0.2^2\left[\,t\,\right]_{0.001}^{0.002} + 100^2\left[\tfrac{1}{3}t^3\right]_0^{0.002}\right)$$

$$= R(200^2 \times 10^{-9}/3 + 0.2^2 \times 10^{-3} + 100^2 \times 8 \times 10^{-9}/3)$$

$$= 100 \times (13.3 + 40 + 26.7) \times 10^{-6} = 8 \text{ mJ}$$

So that the average power is

$$P_{\mathrm{R}} = W_{\mathrm{R}}/T = (8 \times 10^{-3})/(4 \times 10^{-3}) = 2 \text{ W}$$

1.5 Kirchhoff's laws

The currents and voltages in all electrical circuits are governed by two laws formulated by G R Kirchhoff (1824-87) in 1848. No analysis of any circuit is possible without their implicit or explicit use, and many circuit theorems are based on them. Kirchhoff's laws and the voltage-current (V–I) relationships for circuit elements are sufficient for finding all the voltages and currents in any circuit.

1.5.1 Kirchhoff's voltage law

Kirchhoff's voltage law is used so often in circuit analysis that it is customarily abbreviated to KVL. It states that

The algebraical sum of voltages around any closed loop in a circuit is zero

'Algebraical' simply means 'with due regard to sign', clockwise pointing voltages being taken as opposite in sign to anticlockwise. The law may be written symbolically as

$$\sum V = 0 \qquad (1.12)$$

Another way of putting it is to say that around any loop the sum of the clockwise voltages is equal to the sum of the anticlockwise voltages.

Figure 1.10

Illustrating the application of Kirchhoff's voltage law

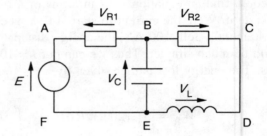

The use of KVL is illustrated with reference to figure 1.10, where the voltages may be summed round the left-hand loop ABEFA, taking clockwise voltages and e.m.f.s as positive and the anticlockwise ones as negative. This sum is

$$E - V_{R1} - V_C = 0 \qquad (1.13)$$

The voltages across the passive components are found from the appropriate *V–I* relation, such as Ohm's law for resistances. Summing the voltage around the right-hand loop BCDEB yields

$$V_C + V_{R2} - V_L = 0 \qquad (1.14)$$

Equations 1.13 and 1.14 can be added to give

$$E - V_{R1} + V_{R2} - V_L = 0$$

which is precisely the same as obtained from using KVL around the large loop ABCDEFA. A network of two meshes (loops containing no smaller loops within them), such as that of figure 1.10, yields only two independent equations form KVL.

1.5.2 *Kirchhoff's current law*

This law is also used so often that it is commonly abbreviated to KCL. It may be stated

The algebraical sum of the currents into a node is zero

The currents out of a node are taken to be opposite in sign to those into it. In symbolic form KCL can be written

$$\sum I = 0 \qquad (1.15)$$

Figure 1.11

Illustrating the application of Kirchhoff's current law

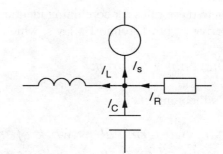

An alternative way of stating KCL is to say the sum of the currents into a node is equal to the sum of the currents out of it. Figure 1.11 illustrates the use of KCL. Calling currents into the node positive and those out negative, then we see that

$$I_R + I_C - I_L - I_s = 0$$

1.5.3 *Consequences of Kirchhoff's laws*

Using Kirchhoff's laws one can prove that resistances in series and conductances[5] in parallel can be added to give respectively the equivalent resistance or conductance. Thus if resistances R_1 and R_2 are in parallel, the equivalent conductance is

$$G_{eq} = G_1 + G_2 = \frac{1}{R_1} + \frac{1}{R_2} = \frac{R_1 + R_2}{R_1 R_2}$$

from which we find the equivalent resistance by taking the reciprocal

$$R_{eq} = \frac{1}{G_{eq}} = \frac{R_1 R_2}{R_1 + R_2}$$

This result is sometimes known as the *product-over-sum rule*.

---

[5] Conductance, G, is the reciprocal of resistance: $G = 1/R$. Its units are siemens (S), which are the same as Ω^{-1}.

One can also show that inductances in series and parallel combine in the same way as resistances, but that capacitances combine like conductances; that is parallel capacitances must be added to give the equivalent capacitance: $C_{eq} = C_1 + C_2 + \ldots$. And series capacitances are combined by taking the reciprocal of the individual reciprocals

$$\frac{1}{C_{eq}} = \frac{1}{C_1} + \frac{1}{C_2} + \ldots$$

1.6 Circuit theorems

In addition to these rules for combining identical components, a number of useful theorems may be derived from Kirchhoff's laws, which we shall examine next.

1.6.1 Thévenin's theorem

Thévenin's theorem[6] states

Any linear, two-terminal network of independent sources and resistances may be replaced by a single voltage source in series with a resistance

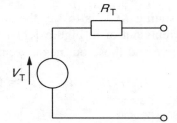

Figure 1.12

Thévenin's equivalent circuit

The value of the voltage source is the open-circuit voltage at the network's terminals. The value of the resistance is the resistance measured at the terminals of the network after all the sources have been replaced by their internal resistances.

Figure 1.12 shows Thévenin's equivalent circuit, which is identical to the practical voltage source of figure 1.4a. An example is the best way to demonstrate its use.

Example 1.3

Replace by its Thévenin equivalent the circuit to the left of terminals A and B in figure 1.13 and hence find the current in, voltage across and power lost in the 5Ω resistance.

We proceed first by removing the 5Ω resistance and obtain the circuit of figure 1.14. Then we require the open-circuit voltage, V_{AB}, which is $3I$, the voltage across the 3Ω resistance, by Ohm's law. The current, I, passes through all three resistances and is found

---

[6] L Thévenin (1857-1926) was a French telegraphic engineer; he formulated his theorem in 1883 to help solve the problems thrown up by long-distance telegraphy.

by using KVL round the loop

$$12 - I - 3I - 2I = 0$$

Whence $I = 2$ A and then the voltage across AB is $3 \times 2 = 6$ V.

Figure 1.13

The circuit for example 1.3

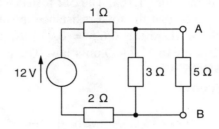

This illustrates the *voltage-divider rule*: since the current is the same through series resistances the voltage across all of them is $V_s = IR_1 + IR_2 + IR_3 = I\Sigma R$, and the voltage across the kth one is IR_k. But the current is

$$I = \frac{V_s}{\Sigma R}; \quad \text{hence} \quad V_k = IR_k = V_s \times \frac{R_k}{\Sigma R}$$

The voltage across resistances in series divides in proportion to the resistance. Here, $\Sigma R = 6$ Ω, $R_k = 3$ Ω and $V_s = 12$ V, and thus $V_k = 12 \times 3/6 = 6$ V, as found before.

Figure 1.14

The circuit after removing the 5Ω resistance. V_{AB} is the voltage of the Thévenin source

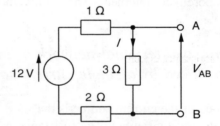

The second part of the Thévenin circuit is the series resistance which is found by replacing the voltage source by its 'internal resistance'. Examination of figure 1.13 reveals two resistances in series with the source, either or neither of which could be its internal resistance; but if the source were ideal, then its internal resistance would be zero, that is a short circuit.

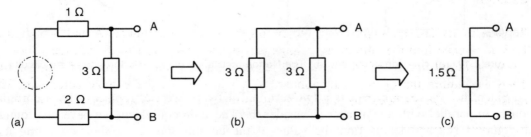

Figure 1.15 (a) The voltage source is shorted (b) The series resistances are added (c) Final resistance

The circuit whose resistance is to be found is then that of figure 1.15a. The series resistances can be added to give the circuit of figure 1.15b, then the two resistances in parallel are combined by the product-over-sum rule to give the final Thévenin resistance of $(3 \times 3)/(3 + 3) = 9/6 = 1.5\ \Omega$. The Thévenin circuit is therefore a voltage source of 6 V in series with $1.5\ \Omega$ as in figure 1.16a. The 5Ω resistance is then connected to AB as in figure 1.16b and we deduce that the total resistance in series with the source is $5 + 1.5 = 6.5\ \Omega$. The current, I, is therefore $6/6.5 = 0.923$ A, and the voltage across the 5Ω resistance is $5 \times 0.923 = 4.615$ V, by Ohm's law. The power developed in the resistance is $VI = I^2R = V^2/R = 4.26$ W.

Figure 1.16

(a) The Thévenin equivalent
(b) The 5Ω resistance is reconnected across AB

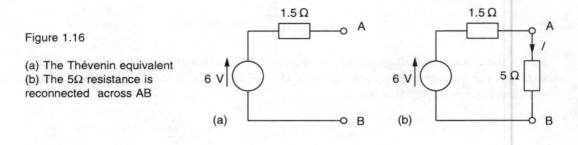

1.6.2 Norton's theorem

Norton's theorem complements Thévenin's, though it was propounded much later in 1926; it states

Any linear, two-terminal network of independent sources and resistances can be replaced by a single current source in parallel with a resistance.

The value of the current source is the short circuit current flowing when the network's terminals are short circuited. The resistance is found exactly as in Thévenin's theorem.

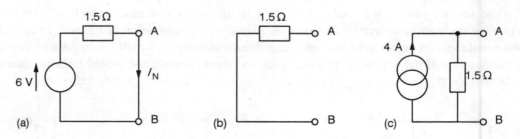

Figure 1.17 (a) The Norton current (b) The Norton resistance (c) The Norton equivalent circuit

Since the Thévenin circuit is a two-terminal network we can replace it by a current source in parallel with a resistance. Consider the Thévenin equivalent we have just found in figure 1.16a and let us find the value of the current source by short-circuiting its terminals as in figure 1.17a. The current flowing is $I_N = 6/1.5 = 4$ A by Ohm's law, and this is the Norton equivalent source current. The Norton resistance in parallel with this is found by replacing the voltage source by its internal resistance, which is zero as it is ideal, that is a short circuit as in figure 1.17b. We see then that *the Norton resistance is the same as the Thévenin resistance* and is the ratio of the open-circuit (or Thévenin) voltage to the short-circuit (or Norton) current:

$$R_N = R_T = \frac{V_{oc}}{I_{sc}} = \frac{V_T}{I_N} \qquad (1.16)$$

The Norton equivalent of the Thévenin equivalent circuit is therefore that of figure 1.17c.

This is also of course the Norton equivalent of the circuit to the left of AB in figure 1.15a, and we can find the current and voltage through the 5Ω resistance by connecting it across the terminals of the Norton circuit as in figure 1.18a. To find the current flowing through the 5Ω resistance, we combine the parallel resistances by the product-over-sum rule: $(5 \times 1.5)/(5 + 1.5) = 1.153$ Ω as in figure 1.18b. All the source current must pass through this equivalent resistance and then the voltage across it can be found by Ohm's law to be $V = 4 \times 1.154 = 4.616$ V. The voltage is the same (apart from a rounding error) as before and leads to the same current though the 5Ω resistance: $4.616/5 = 0.923$ A.

Figure 1.18

(a) The 5Ω resistance is attached to the Norton equivalent circuit
(b) The parallel resistances are combined

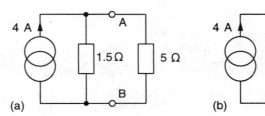

Thévenin's and Norton's theorems can be used for network reduction among other things, as the next example shows.

Example 1.4

Find the Thévenin equivalent of the circuit of figure 1.19a between terminals A and B by source transformations using Norton's and Thévenin's theorems.

The usual procedure is to start at the end of the network furthest from the terminals, that is at the 6V source, which is readily transformed into a current source with Norton's theorem. The short-circuit current from a 6V source with a 6Ω series resistance will be $6/6 = 1$ A, while the resistance stays the same, but is placed in parallel as in figure 1.19b. Note that the current arrow in the transformed source must point to the same terminal as the original voltage arrow of the voltage source.

The rules for combining sources are

- *Current sources can be combined only when in parallel*
- *Voltage sources can be combined only when in series*

We can therefore combine the two current sources in figure 1.19b by adding the source currents algebraically as in figure 1.19c. The source resistances of 6 Ω and 3 Ω are combined next by the product-over-sum rule to give the circuit of figure 1.19d.

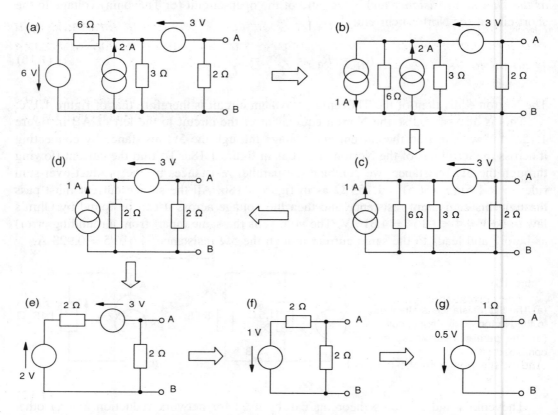

Figure 1.19 (a) The original circuit (b) The 6V source is transformed (c) Current sources are combined (d) Parallel resistances are combined (e) The current source is transformed (f) The voltage sources are combined (g) The Thévenin equivalent circuit

Transformation of the current source into a voltage source by Thévenin's theorem yields the circuit of figure 1.19e, wherein the two series voltage source may be combined into one, the voltages adding algebraically to −2 + 3 = 1 V, taking anticlockwise voltages as positive. The circuit of 1.19f is then transformed into the final Thévenin circuit of figure 1.19g, since the terminal voltage will be half the source voltage by the voltage-divider rule and the network resistance (found by short-circuiting the voltage source) is

$$2 \ \Omega \ // \ 2 \ \Omega \ = \ \frac{2 \times 2}{2 + 2} \ = \ 1 \ \Omega$$

The next theorem concerns power rather than currents or voltages and is used when it is important to get as much useful power as possible from a source or network.

1.6.3 The maximum power transfer theorem

The maximum power transfer theorem states that

Maximum power is developed in a load connected to the terminals of a two-terminal network of sources and resistances when the load resistance is equal to the Thévenin resistance of the network

The theorem follows readily from Thévenin's theorem, since the network can be reduced to a single voltage source in series with a resistance and we then connect our load to this as in figure 1.20.

Figure 1.20

Maximum power is transferred from the network's terminals when $R_T = R_L$

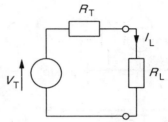

The current through the load is given by

$$I_L = \frac{V_T}{R_T + R_L}$$

And so the power in the load is

$$P_L = I_L{}^2 R_L = \left(\frac{V_T}{R_T + R_L}\right)^2 R_L \tag{1.17}$$

We find the maximum value of this by differentiating equation 1.17 with respect to R_L and setting the result equal to zero, as follows

$$\frac{dP_L}{dR_L} = \frac{V_T{}^2(R_T + R_L)^2 - 2(R_T + R_L)V_T{}^2 R_L}{D^2} = 0$$

where $D = (R_T + R_L)^2$. This expression is zero when the numerator is zero, which occurs when $R_L = R_T$, corresponding to a maximum in the load power, P_L. Substituting R_T for R_L in equation 1.17 leads to

$$P_L(\text{max}) = \left(\frac{V_T}{2R_T}\right)^2 R_T = \frac{V_T{}^2}{4R_T} \tag{1.18}$$

Example 1.5

What resistance attached across AB in the circuit of figure 1.21a will give maximum power transfer and what is the power transferred? What power is produced by the sources?

We will find the Thévenin equivalent circuit this time by using the superposition principle mentioned in section 1.2, which applies only to voltages and currents, not power. The principle enables us to consider the effects of each source separately and to add these up to give the combined effect. In the circuit of figure 1.21a we want to find the voltage at the terminals, which is the Thévenin voltage. Taking the 4V source first we can find its effect on the open-circuit voltage by replacing the current source by its internal resistance. An ideal current source has infinite internal resistance, so we just replace it with an open circuit, as in figure 1.21b. Then as the 2Ω and 6Ω resistances are in series, the open-circuit voltage, V_1, due to the 4V source is $4 \times 6/8 = 3$ V by the voltage-divider rule.

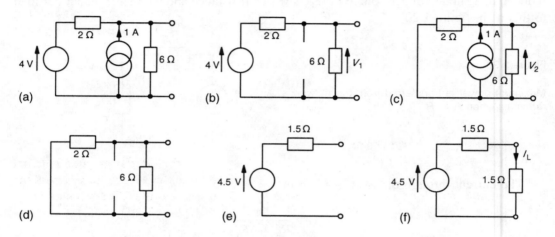

Figure 1.21 (a) The circuit of example 1.5 (b) Effect of the 4V source (c) Effect of the 1A source (d) Circuit for the Thévenin resistance (e) Thévenin circuit (f) The load for maximum power is connected

The effect of the 1A current source is found by replacing the ideal 4V source by a short circuit as in figure 1.21c, then finding the voltage across the 6Ω resistance from the current through it. We can use the *current-divider rule* for this purpose. The rule states that the currents through parallel conductances divide in proportion to the conductance; since the conductances are $\frac{1}{2}$ S and $\frac{1}{6}$ S, so the current through the $\frac{1}{6}$ S conductance is

$$I_{6\Omega} = \frac{1/6}{1/6 + 1/2} \times 1 = \frac{1}{1 + 3} \times 1 = 0.25 \text{ A}$$

From Ohm's law $V_2 = 6I_{6\Omega} = 1.5$ V, making the Thévenin voltage

$$V_T = V_1 + V_2 = 3 + 1.5 = 4.5 \text{ V}$$

The Thévenin resistance is found by replacing both sources by their internal resistances, leaving the 2Ω and 6Ω resistances in parallel as in figure 1.21d. Hence

$$R_T = 2 \text{ }\Omega \text{ // } 6 \text{ }\Omega = \frac{2 \times 6}{2 + 6} = 1.5 \text{ }\Omega$$

and the Thévenin equivalent circuit is that shown in figure 1.21e.

Maximum power is transferred when a 1.5Ω load is placed across the terminals of the Thévenin circuit, as in figure 1.21f, producing a total resistance of $3\ \Omega$ in series with the 4.5V Thévenin source. The load current flowing is therefore $4.5/3 - 1.5$ A, and the power developed in the load is

$$P_{\text{Lmax}} = I_{\text{L}}^2 R_{\text{L}} = 1.5^2 \times 1.5 = 3.375 \text{ W}$$

To find the power supplied by the sources we must return to the full circuit of figure 1.21a and then place a 1.5Ω load across the terminals as in figure 1.22: *we cannot use the Thévenin circuit for this*. We find power from a source by multiplying the current it supplies by the voltage across it. The 1A source is in parallel with the load so the voltage across it is

$$V_{1\text{A}} = V_{\text{L}} = I_{\text{L}}R_{\text{L}} = 1.5 \times 1.5 = 2.25 \text{ V}$$

The power supplied by the 1A source is therefore

$$P_{1\text{A}} = (VI)_{1\text{A}} = 2.25 \times 1 = 2.25 \text{ W}$$

and since voltage and current arrow point the same way, this is power supplied, not consumed. The current from the 4V source can be found by finding the voltage across the 2Ω resistance and using Ohm's law. By KVL the voltage across the 2Ω resistance is

$$V_{2\Omega} = 4 - V_{\text{L}} = 4 - 2.25 = 1.75 \text{ V}$$

and the current through it is then $1.75/2 = 0.875$ A. The power from the 4V source becomes

$$P_{4\text{V}} = (VI)_{4\text{V}} = 4 \times 0.875 = 3.5 \text{ W}$$

The total power supplied is $2.25 + 3.5 = 5.75$ W, of which 3.375 W is consumed by the load, implying that 2.375 W is the power consumed by the rest of the circuit.

Figure 1.22

The circuit for calculating the power delivered by the sources

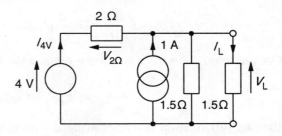

Had we replaced the current source by its Thévenin equivalent before we calculated the power, a different answer would have been obtained (see problem 1.10), because power is not a linear quantity and circuit equivalents only hold for voltages and currents.

1.7 Phasors

The theorems we have used previously in this chapter can be used to analyse sinusoidally-excited circuits ('AC circuits') once the voltage and currents in these have been converted into phasor notation[7]. In sinusoidally excited circuits the voltages and currents are of the form $A \sin(\omega t + \phi)$, where A is the magnitude, or peak value of the quantity, ω is its angular frequency, t is the time and ϕ is its phase angle with respect to a chosen or reference sinewave. The angular frequency, ω, in radians/s, is related to the frequency in Hz by

$$\omega = 2\pi f = 1/T$$

where T is the period of the waveform, the time between successive maxima or minima.

Look at the circuit of figure 1.23, in which the current through R, L and C is $I_m \sin \omega t$: what are the voltages across the components and across the whole circuit? We proceed exactly as with example 1.1 by using the v–i relationships for each of the components[8].

Figure 1.23

A series RLC circuit excited by a sinusoidal current source

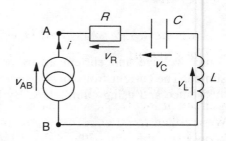

For the resistance we use Ohm's law:

$$v_R = iR = I_m \sin \omega t \times R = I_m R \sin \omega t$$

The voltage across the resistance is simply a replica of the current waveform with a scale factor of R. Note particularly that the voltage and current are *in phase*.

For the inductance we use

$$v_L = L \frac{di}{dt} = L \frac{d}{dt}(I_m \sin \omega t) = \omega L I_m \cos \omega t$$

But $\cos \omega t = \sin(\omega t + 90°)$ and if we replace ωL by X_L we can rewrite this equation as

$$v_L = I_m X_L \sin(\omega t + 90°)$$

Examining this relation and comparing it with that for the resistance, we see that

---

[7] The English telegraphic engineer, physicist, mathematician and eccentric, Oliver Heaviside (1850-1925), first used complex numbers in AC circuit analysis, in about 1880.

[8] By convention DC values and magnitudes of complex quantities are given italic capitals: *V, I, Z*. Instantaneous values are written as lower case italics: *v, i*. Phasors and complex quantities are given bold capitals: **V, I, Z**.

- *An inductance acts as if it had a resistance of ωL ohms*
- *The current through an inductance lags the voltage by 90°*

The apparent resistance of an inductance, ωL, is known as its *reactance* and, because this is proportional to the frequency, at low frequencies an inductance acts as a very low resistance or short circuit. Conversely at high frequencies it acts as an open circuit.

The capacitor's voltage-current relationship yields

$$v_C = \frac{1}{C}\int i\,dt = \frac{1}{C}\int I_m \sin \omega t\,dt = \frac{I_m}{\omega C}(-\cos \omega t)$$

Replacing $1/\omega C$ by X_C and $-\cos \omega t$ by $\sin(\omega t - 90°)$ gives

$$v_C = I_m X_C \sin(\omega t - 90°)$$

And we see that

- *A capacitance acts as if it had a resistance of 1/ωC ohms*
- *The current through a capacitance leads the voltage by 90°*

Because the capacitance has a reactance which is proportional to $1/\omega$ or $1/f$, it acts as an open circuit at low frequencies and a short circuit at high.

We are then faced with finding the overall voltage across the circuit, which by KVL we know to be the sum of the voltages across the components

$$v_{AB} = I_m R \sin \omega t + I_m X_L \sin(\omega t + 90°) + I_m X_C \sin(\omega t - 90°)$$

$$= I_m[R \sin \omega t + (X_L - X_C)\cos \omega t]$$

The terms can be combined by putting $R = Z\cos\phi$ and $(X_L - X_C) = X = Z\sin\phi$, where $Z = \sqrt{(R^2 + X^2)}$, and $\tan\phi = X/R$. The much simpler relationship

$$v_{AB} = I_m Z[\cos\phi \sin \omega t + \sin\phi \cos \omega t] = I_m Z \sin(\omega t + \phi)$$

is thereby found, which could be written as

$$\mathbf{V} = \mathbf{IZ}$$

The form of this relationship is very similar to $V = IR$, that is Ohm's law with R replaced by $\mathbf{Z}$, the *impedance* of the circuit, while $\mathbf{V}$ and $\mathbf{I}$ have become *phasors*.

Manipulating the trigonometric forms of current and voltage is tedious, and can be simplified if we drop the time-dependent part, ωt, altogether since it is unchanged throughout the calculations. A further simplification follows if we introduce a *j*-operator to produce phase changes, where multiplication by $\pm j$ has the effect of shifting the phase of current or voltage by ±90°. Currents and voltages are then written as a magnitude and phase such as $5\angle60°$ A or $9\angle-150°$ V and are called *phasors*. They can be plotted on an Argand diagram which has a horizontal axis for the *real* part and a vertical axis for the *imaginary* or *j*-part, as shown in figure 1.24a. The real part of a phasor is its projection on the horizontal axis, $I\cos\phi$ or $V\cos\phi$ and its imaginary part is its projection on the vertical axis, $I\sin\phi$ or $V\sin\phi$. Hence the real part of $5\angle60°$ A is $5\cos60° = 2.5$ A and the imaginary part is $j5\sin60° = j4.33$ V.

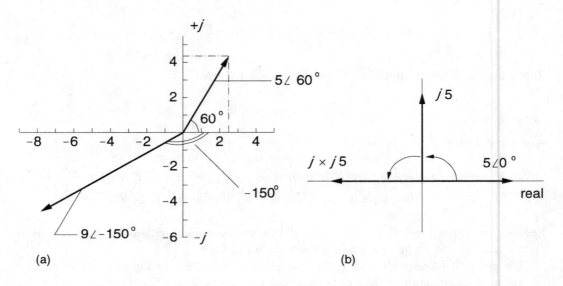

Figure 1.24 (a) The phasors $5\angle60°$ and $9\angle-150°$ (b) The effect of multiplying $5\angle0°$ by *j* twice

We can see that multiplying a phasor such as $5\angle0°$ (wholly real) by *j* will produce the phasor $j5$ or $5\angle90°$, a rotation of 90° in an anticlockwise sense, as shown in figure 1.24b. A further multiplication by *j* will rotate it clockwise another 90° to a phase angle of 180°, which is a phasor directed along the real axis in the opposite direction to the original phasor, that is

$$j \times j5 = 1\angle90° \times 5\angle90° = 5\angle180° = -5$$

Thus $j \times j = j^2 = -1$, so that $j = \sqrt{(-1)}$. Multiplying by $-j$ will rotate a phasor $-90°$, or 90° in an anticlockwise sense.

1.7.1 Root-mean-square (r.m.s.) values

A sinusoidal current such as $i = 4\sin\omega t$ A has a peak value of 4 A and a peak-to-peak value of 8 A (really it should be called the peak-to-trough value). When it passes through a 10Ω resistance the power dissipation is $i^2R = 160\sin^2\omega t$ W. The time dependence is of little account and we can instead say that the average power over one cycle is

$$P(\text{av}) = \frac{1}{T}\int_0^T 160\sin^2\omega t\,dt = \frac{1}{T}\int_0^T 80(1 - \cos2\omega t)\,dt = 80 \text{ W}$$

since the cosine part integrates to zero over 1 cycle. If we were to represent i as a phasor we would like the average power to remain I^2R as for DC current. That is we require the effective value of I to satisfy

$$I^2R = 10I^2 = 80 \text{ W} \quad \Rightarrow \quad I = 2\sqrt{2} \text{ A}$$

The effective value of I for power calculations is $2\sqrt{2} = 4/\sqrt{2}$ A $= I_m/\sqrt{2}$, and this is the r.m.s. value:

$$I_{rms} = I_m/\sqrt{2} = I_{pp}/2\sqrt{2}$$

where I_{pp} is the peak-to-peak value or $2I_m$. Phasor magnitudes are always taken as the r.m.s. value of the current or voltage in order to retain the power relationships found for direct voltages and current, namely

$$P = VI = I^2R = V^2/R$$

1.7.2 Complex arithmetic

Complex numbers obey rather different rules to real numbers. We have seen that the phasor $5\angle60°$ can we written as a real part, 2.5, and an imaginary part, $j4.33$, or

$$5\angle60° = 5\cos60° + j5\sin60° = 2.5 + j4.33$$

When we add or subtract two complex numbers we must add or subtract the real and the imaginary parts separately, as follows

$$5\angle60° + 4\angle-120° = 2.5 + j4.33 + 4\cos(-120°) + j4\sin(-120°)$$

$$= 2.5 + j4.33 + -2 + j(-3.46) = 0.5 + j0.87$$

Recombining these real and imaginary parts (the so-called *rectangular* form) into a magnitude and phase (the *polar* form) can be accomplished by noting that

$$(V\cos\phi)^2 + (V\sin\phi)^2 = V^2 = Re^2 + Im^2$$

Where Re = real part and Im = imaginary part. Thus $V = \sqrt{(Re^2 + Im^2)}$ and $\tan\phi = Im/Re$. Hence $0.5 + j0.87$ in polar form is

$$0.5 + j0.87 = \sqrt{0.5^2 + 0.87^2}\angle\tan^{-1}(0.87/0.5) = 1\angle60°$$

Care is needed in determining the phase angle when $|\phi| > 90°$, since the function $\tan^{-1}$ is limited to $\pm90°$. For example the phase angle of $(-1 + j3)$ is $180° - \tan^{-1}(3) = 108.4°$ and that of $(-1 - j3)$ is $-180° + \tan^{-1}(3) = -108.4°$.

Multiplication of phasors is achieved by using the Euler relation:

$$V\cos\phi + jV\sin\phi = V\exp(j\phi)$$

The multiplication of polar forms such as $2\angle30° \times 3\angle40°$ is then simplified because only the magnitudes are multiplied while the phases are added, as in

$$2\angle30° \times 3\angle40° = 2\exp(j30°) \times 3\exp(j40°)$$

$$= 6\exp[j(30° + 40°)] = 6\exp(j70°) = 6\angle70°$$

The exponential forms are usually omitted. In general

$$A\angle\theta \times B\angle\phi = AB\angle(\theta + \phi)$$

Division follows the same form:

$$3\angle40° \div 2\angle30° = 3\exp(j40°) \div 2\exp(j30°)$$

$$= (3/2)\exp[j(40° - 30°)] = 1.5\angle10°$$

Magnitudes are divided and the phase of the divisor is subtracted from the dividend. In general

$$A\angle\theta \div B\angle\phi = (A/B)\angle(\theta - \phi)$$

1.7.3 Impedance

Though not a phasor — because it is independent of time — impedance has, like a phasor, both magnitude and phase. The impedance of the series RLC circuit of figure 1.25a may be expressed in complex form as

$$\mathbf{Z} = R + j(X_L - X_C) = R + jX$$

that is it has magnitude, $Z = \sqrt{(R^2 + X^2)}$ and phase angle $\phi = \tan^{-1}(X/R)$. Being complex, impedances can be plotted on an Argand diagram like phasors too.

Example 1.6

In the circuit of figure 1.25a the excitation frequency is 50 Hz and the excitation voltage is $v_s = 339\sin\omega t$ V. Find the current flowing and the voltages across the components. Draw the phasor diagram.

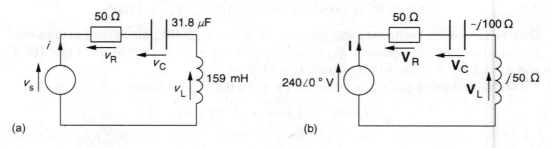

(a)   (b)

Figure 1.25 (a) The original circuit in the time domain (b) The circuit in the phasor domain

The first requirement is to put everything in phasor form. The exciting voltage has a peak value of 339 V, so its r.m.s. value is 339/√2 = 240 V. As its phase angle is zero we can write it in phasor form

$$\mathbf{V_s} = 240 \angle 0° \text{ V}$$

Then we need the reactances of the capacitance and inductance. The capacitance's reactance is

$$X_C = \frac{1}{\omega C} = \frac{1}{2\pi f C} = \frac{1}{2\pi \times 50 \times 31.8 \times 10^{-6}} = 100 \ \Omega$$

Capacitance causes the current to lead the voltage by 90°, or the voltage to lag the current by 90°:

$$\mathbf{V_C} = \mathbf{I}X_C \angle -90° = \mathbf{I} \times -jX_C$$

so we must write the capacitive reactance as $100\angle-90°$ or $-j100 \ \Omega$ in phasor form.

The inductive reactance is given by

$$X_L = \omega L = 2\pi f L = 2\pi \times 50 \times 159 \times 10^{-3} = 50 \ \Omega$$

Now the current lags the voltage or the voltage leads the current by 90°:

$$\mathbf{V_L} = \mathbf{I}X_L \angle 90° = \mathbf{I} \times jX_L$$

And the inductive reactance is $50\angle90° = j50 \ \Omega$. In phasor form the circuit is as figure 1.25b. We find the impedance of the circuit by adding up the resistances and reactances just like resistances in series:

$$\mathbf{Z} = 50 + -j100 + j50 = 50 - j50 \ \Omega$$

The magnitude of $\mathbf{Z}$ is $Z = \sqrt{(R^2 + X^2)} = \sqrt{(50^2 + 50^2)} = 70.7 \ \Omega$, while its phase angle is

$$\phi = \tan^{-1}(X/R) = \tan^{-1}(-50/50) = \tan^{-1}(-1) = -45°$$

In polar form $\mathbf{Z} = 70.7\angle-45° \ \Omega$.

The current through the circuit follows from Ohm's law

$$\mathbf{I} = \frac{\mathbf{V}}{\mathbf{Z}} = \frac{240\angle0°}{70.7\angle-45°} = 3.39\angle45° \text{ A}$$

The voltages across the components are

(1) $\mathbf{V_R} = \mathbf{I}R = 3.39\angle45° \times 50 = 169.5\angle45° \text{ V}$
(2) $\mathbf{V_C} = \mathbf{I}X_C = 3.39\angle45° \times 100\angle-90° = 339\angle-45° \text{ V}$
(3) $\mathbf{V_L} = \mathbf{I}X_L = 3.39\angle45° \times 50\angle90° = 169.5\angle135° \text{ V}$

It is left as an exercise to show that these add up to $240\angle0°$ V. The phasor diagram showing these voltages is that of figure 1.26.

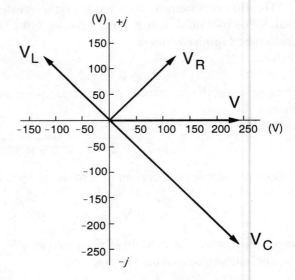

Figure 1.26

The phasor diagram for the circuit of figure 1.25b

1.7.4 Circuit theorems for AC

Kirchhoff's laws hold for phasors as well as direct currents and voltages and so do Thévenin's and Norton's theorems, except that 'resistance' is replaced by impedance. The same method is used to find the equivalent impedance of a circuit and the open-circuit voltage and short-circuit current. The maximum power transfer theorem is slightly modified in that instead of replacing '$R_L = R_T$' by '$Z_L = Z_T$' we use the *complex conjugate* of $\mathbf{Z_T}$. This is denoted by $\mathbf{Z_T}^*$, where if $\mathbf{Z_T} = R_T + jX_T$, $\mathbf{Z_T}^* = \mathbf{Z_L} = R_T - jX_T$: the sign of the imaginary part is changed to form the complex conjugate. The power in the load is the same as in equation 1.18,

$$P_L(\text{max}) = V_T^2/4R_T$$

since the imaginary components of load and source cancel out.

1.8 Power in AC circuits

In any circuit containing inductances, resistances and capacitances, power is only consumed by the resistances: inductors and capacitors store energy in one half cycle and give it back in the next. Consider the circuit of figure 1.26, in which an impedance $\mathbf{Z} = R + jX$ is connected to a voltage source, $\mathbf{V}$. The current which flows through $\mathbf{Z}$ is $\mathbf{I} = \mathbf{V}/\mathbf{Z}$ by Ohm's law, so the power consumed is

$$P = I^2R = (V/Z)IR = (V/Z)IZ\cos\phi = VI\cos\phi$$

since the magnitude of $\mathbf{I}$ is $I = V/Z$ and $R = Z\cos\phi$, where $\phi = \tan^{-1}(X/R)$ is the phase angle of the load.

The *apparent power* consumed by the circuit is $S = VI$, so that $P = S\cos\phi$, is the power actually consumed. $\cos\phi$ is known as the *power factor* of the load. The power factor is said to be leading when ϕ is negative (capacitive load) and lagging when ϕ is positive (inductive load).

We can define the imaginary part of the apparent power as

$$Q = S\sin\phi = VI\sin\phi = I^2X$$

The units of S are volt-amperes, VA, and the units of Q are reactive volt-amperes or var. From the definitions we see that

$$S^2 = P^2 + Q^2$$

Figure 1.27

The power triangle for an inductive load with a lagging p.f.

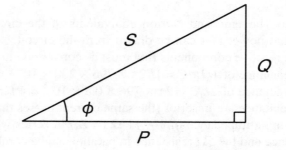

In terms of complex numbers we can write the complex power as

$$\mathbf{S} = \mathbf{VI}^* = VI\cos\phi + jVI\sin\phi = P + jQ$$

The complex conjugate of $\mathbf{I}$ is chosen so that in circuits with a lagging power factor (p.f.) Q is positive and represents vars supplied to the load. When the p.f. is leading the load supplies vars to the source and Q is negative. When a variety of loads is connected to the supply then we can add the power consumed by each load to find the total power consumption (ΣP), and the algebraical sum of the vars (ΣQ). The overall p.f. is then found from

$$\tan\phi = \frac{\Sigma P}{\Sigma Q}$$

The power triangle shown in figure 1.27 is a convenient graphical representation of the relation between P, Q and S.

Example 1.7

Find the apparent power, the reactive power and the dissipated power in the circuit of figure 1.25b.

The power consumption of the circuit is entirely due to the resistance in it since neither capacitance nor inductance consumes power: in one half cycle they store energy and in the next they release it to the circuit again. Therefore the power consumption is

$$P = I^2R = 3.39^2 \times 50 = 575 \text{ W}$$

The reactive power taken is

$$Q = I^2 X = I^2(X_L - X_C) = 3.39^2(50 - 100) = -575 \text{ var}$$

The negative sign indicates that the circuit supplies reactive volt-amperes to the source rather than drawing them — normal for a capacitive load (one with a leading phase angle). The apparent power is

$$S = VI = 240 \times 3.39 = 814 \text{ VA}$$

which can be shown to be $\sqrt{(P^2 + Q^2)}$.

Example 1.8

Find the Thévenin and Norton equivalents of the circuit of figure 1.28a and hence the maximum power that can be drawn from the circuit.

The reactive components first must be converted into reactances: the 200μF capacitance has a reactance of $1/2\pi fC = 1/(2\pi \times 796 \times 200 \times 10^{-6}) = 1\ \Omega$, while the 0.8mH inductance has a reactance of $2\pi fL = 2\pi \times 796 \times 0.8 \times 10^{-3} = 4\ \Omega$. Noting that the 1Ω resistance and the capacitance are in series (the same current passes through each), we can combine them into a single impedance of $1 - j1\ \Omega$ ($= \mathbf{Z_1}$) in series with the voltage source. The 0.3mH inductance and the 5Ω resistance in parallel can be combined too, this time by the product-over-sum rule to give an impedance of

$$\mathbf{Z_2} = \frac{5 \times j4}{5 + j4} = \frac{20\angle90°}{6.4\angle38.66°} = 3.123\angle51.34°\ \Omega$$

The simplified circuit in the phasor domain is that shown in figure 1.28b.

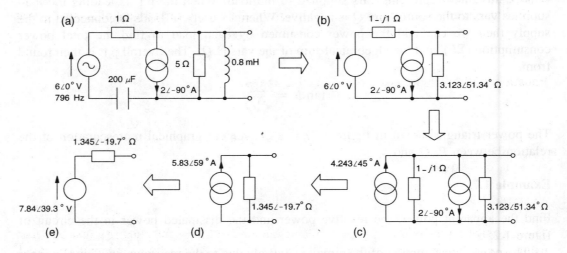

Figure 1.28 (a) The circuit for example 1.8 (b) The circuit in the phasor domain (c) After transforming the current source (d) The Norton equivalent (e) The Thévenin equivalent

By using Norton's theorem the voltage source can be transformed into a current source, whose value is

$$I = \frac{V}{Z_1} = \frac{6\angle 0°}{1 - j1} = \frac{6\angle 0°}{1.414\angle -45°} = 4.243\angle 45° \text{ A}$$

This is in parallel with Z_1 and the transformed circuit is that of figure 1.28c.

Since the current sources are now in parallel they may be combined by adding them algebraically, currents pointing up the page being taken to be positive and those pointing down negative, to give a resultant current source of

$$I_N = 4.243\angle 45° - 2\angle -90° = 3 + j3 - (0 - j2) = 3 + j5 = 5.83\angle -59° \text{ A}$$

The impedances, Z_1 and Z_2, are in parallel too and are combined by the product-over-sum rule:

$$Z_N = \frac{Z_1 \times Z_2}{Z_1 + Z_2} = \frac{1.414\angle -45° \times 3.123\angle 51.34°}{1 - j1 + 1.95 + j2.44} = \frac{4.416\angle 6.34°}{3.283\angle 26°} = 1.345\angle -19.7° \text{ } \Omega$$

and the final Norton equivalent circuit is that of figure 1.29d. From this we can find the voltage of the Thévenin equivalent circuit, using a rearranged form of equation 1.16, in which R_T is replaced by Z_T ($= Z_N$)

$$V_T = I_N Z_N = 5.83\angle 59° \times 1.345\angle -19.7° = 7.84\angle 39.3° \text{ V}$$

The Thévenin and Norton impedances are identical, so that the final Thévenin circuit is as in figure 1.28e.

For maximum power transfer the load must be the complex conjugate of Z_T or Z_N, that is

$$Z_L = Z_T^* = 1.345\angle 19.7° \text{ } \Omega$$

Equation 1.18 gives the power developed in the load

$$P_L(\text{max}) = \frac{V_T^2}{4R_T} = \frac{5.83^2}{4 \times 1.345\cos 19.7°} = 6.71 \text{ W}$$

since $R_T = Z_T \cos\phi$.

Example 1.9

The circuit of figure 1.29a represents an antenna which is to be matched to a resistive load using a capacitance in parallel with the source and an inductance in series with the load. Find the reactances of the components and the maximum power that can be transferred to the load. What are the component values if the frequency is 300 kHz?

The circuit with the matching components is as shown in figure 1.29b, and in effect we have to find the Thévenin equivalent between A and B and then match the series RL load to it. To do this we need only find the Thévenin impedance which is 200 Ω in parallel with $-jX_C$ and that is

$$Z_T = \frac{200 \times -jX_C}{200 - jX_C} = \frac{-j200X_C}{200 - jX_C}$$

by the product-over-sum rule. Thus the matching impedance for maximum power transfer is

$$Z_L = Z_T^* = \frac{j200X_C}{200 + jX_C} = \frac{j200X_C(200 - jX_C)}{200^2 + X_C^2} = \frac{200X_C^2}{200^2 + X_C^2} + \frac{j200^2X_C}{200^2 + X_C^2}$$

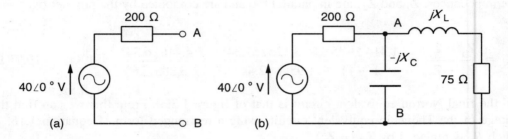

Figure 1.29 (a) The source which is to be matched (b) The source with matching network and 75Ω load

But we can see from figure 1.29b that Z_L is jX_L in series with 75 Ω, which implies that

$$Z_L = 75 + jX_L = \frac{200X_C^2}{200^2 + X_C^2} + \frac{j200^2X_C}{200^2 + X_C^2} \tag{1.19}$$

Now two complex quantities can be equal only if both real and imaginary parts are separately equal. Equating the real parts in equation 1.19 gives

$$75 = \frac{200X_C^2}{200^2 + X_C^2} \tag{1.20}$$

From this we can find X_C

$$75(200^2 + X_C^2) = 200X_C^2 \quad \Rightarrow \quad X_C = \sqrt{\frac{75 \times 200^2}{200 - 75}} = 155 \ \Omega$$

But $X_C = 1/\omega C$, so that $C = 1/\omega X_C = (2\pi \times 300 \times 10^3 \times 155)^{-1} = 3.42$ nF.

Equating the imaginary parts of equation 1.19 gives

$$X_L = \frac{200^2X_C}{200^2 + X_C^2} \tag{1.21}$$

Dividing equation 1.20 by equation 1.21 yields

$$75/X_L = X_C/200 \quad \Rightarrow \quad X_L = 75 \times 200/X_C = 75 \times 200/155 = 97 \ \Omega$$

And as $X_L = \omega L$, then $L = X_L/\omega = 97/2\pi \times 300 \times 10^3 = 51.5 \ \mu H$.

With this matching network the load resistance is effectively 200 Ω and so the load current is $I_L = V/2R_T = 40/400 = 0.1$ A and the load power is $I_L^2 R_T = 0.1^2 \times 200 = 2$ W. Confirmation by calculation from the circuit of figure 1.29b can be achieved by using KCL at node A to find V_L. The current through the capacitance is $I_C = V_L/-j155$. The current through the load is $I_L = V_L/(75 + j97)$ and the current through the 200Ω resistance is $I_s = (40\angle 0° - V_L)/400$. Thus

$$I_s = I_C + I_L = \frac{40 - V_L}{200} = \frac{V_L}{-j155} + \frac{V_L}{75 + j97} = \frac{V_L}{155\angle -90°} + \frac{V_L}{122.6\angle 52.3°}$$

Collecting terms in V_L leads to

$$V_L\left(\frac{1}{200} + \frac{1}{155\angle -90°} + \frac{1}{122.6\angle 52.3°}\right) = V_L\left(\frac{5 + 6.45\angle 90° + 8.16\angle -52.3°}{1000}\right) = \frac{40}{200}$$

$$\Rightarrow \quad V_L\left(\frac{5 + j6.45 + 5 - j6.45}{1000}\right) = 0.2 \quad \Rightarrow \quad V_L = 20\angle 0° \ V$$

Hence the load current

$$I_L = \frac{20\angle 0°}{122.6\angle 52.3°} = 0.163\angle -52.3° \ A$$

And the power consumed by the load is

$$P_L = I_L^2 R_L = 0.163^2 \times 75 = 2 \ W$$

as only the resistive part dissipates energy. The previous answer is laboriously confirmed.

1.9 Transients

Though in many cases circuits can be analysed using phasors quite satisfactorily, in some cases we are interested in what happens when circumstances are changed and not in the steady-state condition. These conditions exist for only a limited time before the circuit settles down and so they are often called *transients*. Transients only occur because circuits invariably store energy — usually in inductors and capacitors, though any circuit will always contain some inductance and capacitance — which takes time to transfer to or from the circuit.

1.9.1 The RC circuit's step response

Consider the series RC circuit of figure 1.30a, which is connected at time $t = 0$ to a constant-voltage source, of value V_s.

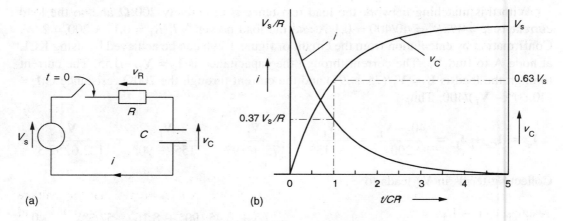

(a)   (b)

Figure 1.30 (a) The series RC circuit (b) Its response to a step-voltage input

The voltages around the circuit sum to zero by KVL so that

$$V_s - v_R - v_C = 0 \qquad (1.22)$$

But we know $v_R = iR$ and $v_C = C^{-1}\int i\,\mathrm{d}t$ since these are the v–i relationships for these elements, and substituting into equation 1.22 yields

$$V_s - iR - \frac{1}{C}\int i\,\mathrm{d}t = 0$$

Differentiating with respect to t produces

$$\frac{\mathrm{d}V_s}{\mathrm{d}t} - R\frac{\mathrm{d}i}{\mathrm{d}t} - \frac{i}{C} = 0 \qquad (1.23)$$

But the source voltage is constant, so that $\mathrm{d}V_s/\mathrm{d}t = 0$ and equation 1.23 reduces to

$$\frac{\mathrm{d}i}{\mathrm{d}t} = \frac{-i}{CR} \quad \Rightarrow \quad \frac{\mathrm{d}i}{i} = \frac{-\mathrm{d}t}{CR}$$

And this we can integrate

$$\ln i = \frac{-t}{CR} + \alpha \quad \Rightarrow \quad i = \beta\exp(-t/CR) \qquad (1.24)$$

where α is a constant of integration and $\beta = \exp(\alpha)$ is another constant, which can be found by examining the initial condition of the circuit. At $t = 0+$, the instant after the switch is closed, the capacitor is uncharged and its voltage is 0, hence equation 1.22 is just $V_s = v_R = iR$, that is $i = i(0+) = V_s/R$. But according to equation 1.24, $i(0+) = \beta\exp(0) =$

β, so we have found $\beta = V_s/R$. Equation 1.24 then becomes

$$i = \frac{V_s}{R}\exp(-t/CR) \qquad (1.25)$$

It is wise to check that this gives the expected answer when $t = \infty$. Now when the capacitor is fully charged to the source voltage we expect the current to be zero, and substituting $t = \infty$ into equation 1.25 yields $i = 0$ as required.

We can find the capacitor voltage from equation 1.25 by integration

$$v_C = \frac{1}{C}\int i\,dt = \frac{1}{C}\int_0^t \frac{V_s}{R}\exp(-t/CR)\,dt = \frac{V_s}{CR}\left[-CR\exp(-t/RC)\right]_0^t \qquad (1.26)$$

$$= V_s[-\exp(-t/CR) - (-1)] = V_s[1 - \exp(-t/CR)]$$

The reader may confirm that this leads to the correct results at $t = 0$ and $t = \infty$.

A graph of v_C against t has the form of figure 1.30b, where we see that the voltage reaches $V_s(1 - 1/e) = 0.632V_s$ when $t = CR$. For $t = 3CR$, $v_C = V_s(1 - 1/e^3) = 0.95V_s$ and by the time $t = 5CR$, $V_C = 0.993V_s$ — for practical purposes the transient is over when $t = 3CR$.

The product CR has dimensions of seconds and is known as the *time constant*, τ, of the circuit because it determines the circuit's response rate to changes. We can see that if $C = 1$ F and $R = 10$ kΩ, the response will be slow as $CR = \tau = 10^4$ s, roughly 3 hours. On the other hand, if $C = 10$ pF (a typical capacitance for a small semiconductor device) and $R = 1$ Ω, then $\tau = 10$ ps, an extremely small time, indicating a very rapid response.

1.9.2 The RL circuit's step response

We can analyse the response of the series RL circuit of figure 1.31a to a step-voltage input in the same way as above.

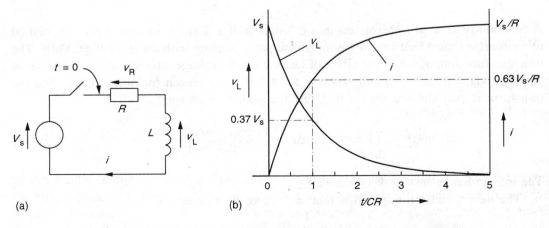

Figure 1.31 (a) The series RL circuit (b) Its response to a step voltage

The use of KVL and the v–i relations for L and R gives the *first-order* differential equation

$$V_s = iR + L\,\mathrm{d}i/\mathrm{d}t \quad \Rightarrow \quad \mathrm{d}i = (V_s/L - iR/L)\mathrm{d}t = (V_s/R - i)(R/L)\mathrm{d}t$$

which can be integrated, yielding

$$\int \frac{\mathrm{d}i}{i - V_s/R} = -\frac{R}{L}\int \mathrm{d}t \quad \Rightarrow \quad \ln(i - V_s/R) = -Rt/L + \alpha$$

where α is a constant of integration. Taking antilogs of the above produces

$$i - V_s/R = \beta\exp(-Rt/L) \quad \Rightarrow \quad i = V_s/R + \beta\exp(-Rt/L)$$

β ($= \exp\alpha$) is a constant found by examining the initial conditions, which are that $i = 0$ at $t = 0$, making $\beta = -V_s/R$. The final result is

$$i = (V_s/R)[1 - \exp(-Rt/L)] = (V_s/R)[1 - \exp(t/\tau)]$$

Here, the time constant, τ, is seen to be L/R compared to CR for the RC circuit: in an RL circuit R acts on τ in the opposite way than in an RC circuit. If L is large and R is small, τ will be large; for example when $L = 10$ H and $R = 2$ mΩ, then $\tau = L/R = 5000$ s.

1.9.3 The Laplace transform

When the circuit contains more than one reactive element the solution of the differential equations — in fact, even correctly formulating them — becomes considerably more difficult. The usual method for solving these cases was pioneered by Heaviside, though the method is now recognised as Laplace's. It can be used to transform the circuit elements to produce transformed current-voltage relationships which are then detransformed into the required time-dependent form using tables.

The Laplace transform is defined by

$$\mathscr{L}[f(t)] = \int_0^\infty f(t)\exp(-st)\mathrm{d}t = F(s)$$

A function of time is transformed into a function of s. Before we can apply the method to a circuit we must first find some transform pairs, starting with a step voltage, $Vu(t)$. The unit step function, $u(t)$, has the effect of turning on the voltage, $v(t) = V$ at $t = 0$, and saves writing '$v(t) = 0$ for $t \le 0$ and $v(t) = V$ for $t \ge 0$'. The reason for doing this is that the transform is only defined for $t \ge 0$. Using the definition above

$$\mathscr{L}[Vu(t)] = \int_0^\infty V\exp(-st)\mathrm{d}t = \left[\frac{V\exp(-st)}{-s}\right]_0^\infty = \frac{V}{s} \qquad \text{(LT1)}$$

The integration is easy as V is constant.

The next transform required is that of $f'(t)$, or $\mathrm{d}f(t)/\mathrm{d}t$, which is

$$\mathcal{L}[f'(t)] = \int_0^\infty f'(t)\exp(-st)\,dt = [f(t)\exp(-st)]_0^\infty - \int_0^\infty f(t)(-s)\exp(-st)\,dt$$

We have used integration by parts ($\int u\,dv = uv - \int v\,du$). This expression becomes

$$\mathcal{L}[f'(t)] = -f(0) + s\int_0^\infty f(t)\exp(-st)\,dt = -f(0) + sF(s) \qquad \text{(LT2)}$$

where $F(s)$ stands for $\mathcal{L}[f(t)]$ and $f(0)$ is $f(t)$ when $t = 0$.

The third transform is that of $\exp(-at)$, which is

$$\mathcal{L}[\exp(-at)] = \int_0^\infty \exp(-at)\exp(-st)\,dt = \int_0^\infty \exp[-(s+a)t]\,dt$$

$$\text{(LT3)}$$

$$= \left[\frac{\exp[-(s+a)t]}{-(s+a)}\right]_0^\infty = \frac{1}{s+a}$$

The fourth transform required is that of $\int f(t)\,dt = g(t)$, which is

$$\mathcal{L}\left[\int f(t)\,dt\right] = \mathcal{L}[g(t)] = G(s) \qquad (1.27)$$

But $f(t) = g'(t)$, so that

$$F(s) = \mathcal{L}[f(t)] = \mathcal{L}[g'(t)] = sG(s) - g(0) = sG(s) \qquad (1.28)$$

since

$$g(t) = \int_0^t g(\tau)\,d\tau \;\Rightarrow\; g(0) = \int_0^0 g(\tau)\,d\tau = 0$$

Comparing equations 1.27 and 1.28 we see that the required transform is

$$\mathcal{L}\left[\int f(t)\,dt\right] = G(s) = s^{-1}\mathcal{L}[f(t)] = \frac{F(s)}{s} \qquad \text{(LT4)}$$

Table 1.1 gives these four transform pairs and four others that will be of use.

The step response of a series RC circuit

We can now apply the transform to circuits, starting with the first-order RC circuit excited by a step voltage, $Vu(t)$, as in figure 1.32. The capacitor is initially uncharged. The current flowing is $i(t)$ and the use of KVL leads to the equation

$$Vu(t) = iR + C^{-1}\int i\,dt$$

The whole of this equation is transformed term by term to give

$$\mathcal{L}[Vu(t)] = \mathcal{L}[iR] + \mathcal{L}[C^{-1}\int i\,dt]$$

In this equation R and C are independent of t, so that $\mathcal{L}[iR] = R\mathcal{L}[i(t)] = RI(s) = R\mathbf{i}$. The transformed variable is written in boldfaced lower case. We can transform $\int i\,dt$ with LT4 and then the equation for the circuit in the s-domain, as it is called, becomes

$$V/s = R\mathbf{i} + \mathbf{i}/Cs \quad \Rightarrow \quad \mathbf{i} = \frac{V}{Rs + 1/C} = \frac{V/R}{s + 1/RC}$$

Figure 1.32

Step excitation of a series RC circuit

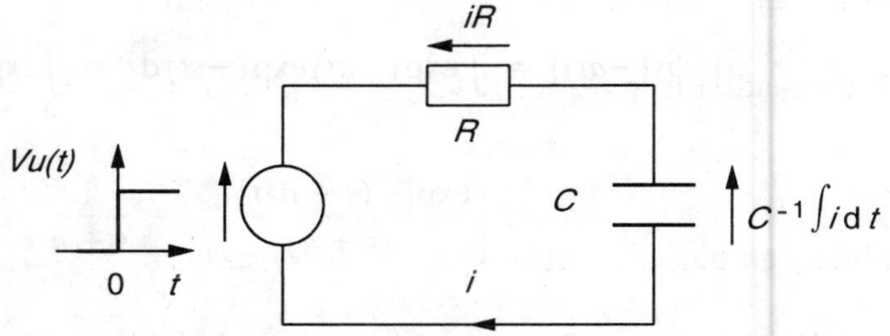

Table 1.1 *Laplace transform pairs*

$f(t)$	$F(s)$	
$Vu(t)$	$\dfrac{V}{s}$	(LT1)
$f'(t)$	$sF(s) - f(0)$	(LT2)
$\exp(-at)u(t)$	$\dfrac{1}{s + a}$	(LT3)
$\displaystyle\int_0^t f(\tau)\,d\tau$	$\dfrac{F(s)}{s}$	(LT4)
$Vt\exp(-at)u(t)$	$\dfrac{V}{(s + a)^2}$	(LT5)
$f(t)\exp(-at)u(t)$	$F(s + a)$	(LT6)
$\sin(\omega t)u(t)$	$\dfrac{\omega}{s^2 + \omega^2}$	(LT7)
$\cos(\omega t)u(t)$	$\dfrac{s}{s^2 + a^2}$	(LT8)

To find $i(t)$ we have to *detransform* **i** back into the time-domain, using the Laplace transforms we have derived already. And we can see, **i** is of the form $A/(s + a)$, with $A \equiv V/R$ and $a \equiv 1/RC$, which is LT3. The detransformation goes

$$\mathscr{L}^{-1}\mathbf{i} = i(t) = \mathscr{L}^{-1}[A/(s + a)] = A\exp(-at)u(t)$$

Substituting for A and a gives the result we have seen previously in section 1.9.1:

$$i(t) = \frac{V}{R}\exp(-t/RC)u(t)$$

The step response of a series RL circuit

The equation from using KVL on this circuit, shown in figure 1.33, is

$$Vu(t) = iR + L\,di/dt$$

which can be transformed into the s-domain using LT1 and LT2:

$$V/s = Ri + Lsi - i(0) = (R + Ls)\mathbf{i}$$

since $i(0)$, the initial current though the inductance, is zero. Solving for **i** gives

$$\mathbf{i} = \frac{V/s}{R + Ls} = \frac{V/L}{s^2 + Rs/L} = \frac{V/L}{s(s + R/L)} \qquad (1.29)$$

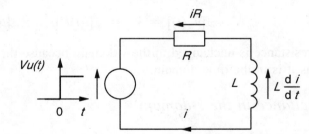

Figure 1.33

Step excitation of a series
RL circuit

Equation 1.29 is not in one of the four transforms we have derived. However, the expression for **i** can be expanded into partial fractions, using

$$\frac{V/L}{s(s + R/L)} \equiv \frac{A}{s} + \frac{B}{s + R/L}$$

where A and B are constants to be determined. The expression on the right-hand side can be written

$$\frac{A}{s} + \frac{B}{s + R/L} = \frac{A(s + R/L) + Bs}{s(s + R/L)} = \frac{(A + B)s + AR/L}{s(s + R/L)}$$

Comparing the numerator of this to the numerator of the right-hand side of equation 1.29 we see that $A + B = 0$ (coefficient of s) and that $AR/L = V/L$ (constant term), that is $A = V/R$ and $B = -V/R$. Thus **i** can be expressed as

$$\mathbf{i} = \frac{A}{s} + \frac{B}{s + R/L} = \frac{V/R}{s} - \frac{V/R}{s + R/L}$$

We recognise the forms on the right as LT1 and LT3, so that the detransformed current is

$$i(t) = (V/R)u(t) - (V/R)\exp(-Rt/L)u(t) = (V/R)[1 - \exp(-Rt/L)]u(t)$$

which is the same result obtained in section 1.9.2.

1.9.4 Circuit elements in the s-domain

It is not necessary to apply circuit laws and theorems firstly to the time-domain circuit to obtain a differential equation, and then use the Laplace transformation; instead it is better to transform the circuit into the s-domain and apply the laws and circuit theorems to the transformed circuit to obtain an s-domain expression for the voltage or current that is needed. This expression can then be detransformed into a time-domain form. The circuit elements are transformed by transforming their $V-I$ relationships. In the case of resistance, the relation is Ohm's law:

$$v_R = iR$$

which can be transformed into

$$\mathscr{L}[v_R(t)] = \mathbf{v_R} = \mathscr{L}[i(t)R] = R\mathscr{L}[i(t)] = \mathbf{i}R$$

Thus resistance is unchanged in the s-domain because the V-I relationship is of the same form as it is in the time domain.

Capacitance in the s-domain

The $V-I$ relation for a capacitance, initially charged up to a potential, V_0, and carrying a current $i(t)$ is

$$v_C(t) = V_0 + C^{-1}\int i\,dt$$

which can be transformed into

$$\mathbf{v_C} = \frac{V_0}{s} + \frac{\mathbf{i}}{Cs} \tag{1.30}$$

using LT1 and LT4. Thus the s-domain form of the capacitance is that shown in figure 1.34a, comprising a voltage source of value V_0/s in series with a reactance of $1/Cs$, since the application of Ohm's law to the reactance carrying a current of **i** must give a voltage of $\mathbf{i}/Cs$ for equation 1.30 to be true. The initial voltage, V_0, is transformed as a constant.

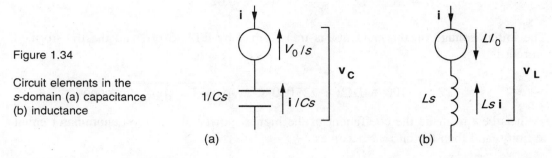

Figure 1.34

Circuit elements in the
s-domain (a) capacitance
(b) inductance

Inductance in the s-domain

The $V-I$ relationship for an inductance in the time domain is

$$v_L(t) = L\,di/dt$$

and we can transform this into the s-domain using LT2:

$$\mathbf{v_L} = L(s\mathbf{i} - I_0) = Ls\mathbf{i} - LI_0$$

taking the initial current through the inductance as I_0 in the same direction as $i(t)$. Thus the reactance of the inductance is Ls in the s-domain, and it must be in series with a voltage source of value LI_0, whose arrow points in the opposite direction to the voltage across the inductance, $Ls\mathbf{i}$. The s-domain circuit model of an inductance is therefore that shown in figure 1.34b.

Example 1.10

In the circuit of figure 1.35a, find the current through the inductance and the voltage across it as a function of time after the switch is closed at $t = 0$. What is the maximum current?

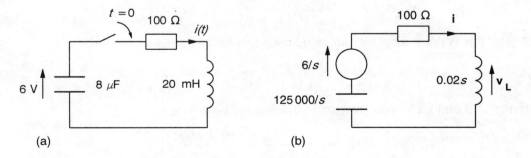

Figure 1.35 (a) The circuit for example 1.10 in the time domain (b) in the s-domain

The circuit is transformed into the s-domain first, as in figure 1.35b, then we solve for **i** using Ohm's law. This is quite easy as the circuit elements are all in series, so we can add them to find the total impedance in the s-domain:

$$Z = R + Ls + \frac{1}{Cs} = 100 + 0.02s + \frac{125\,000}{s}$$

The initial voltage on the capacitor is transformed by LT1 to $6/s$, and the transformed current is

$$i = \frac{v}{Z} = \frac{6/s}{100 + 0.02s + 125\,000/s} = \frac{6}{0.02s^2 + 100s + 125\,000}$$

It is best to make the coefficient of the highest power of s in the denominator equal to unity, and then the current becomes

$$i = \frac{300}{s^2 + 5\,000s + 6.25 \times 10^6} = \frac{300}{(s + 2500)^2}$$

This is not one of the transforms we have derived, but it can be shown that

$$\mathscr{L}[t\exp(-at)] = \frac{1}{(s + a)^2} \tag{LT5}$$

and so the detransformed current is

$$i(t) = 300t\exp(-2500t) \text{ A} \tag{1.31}$$

The maximum current is found by differentiating equation 1.31 with respect to time and setting the differential coefficient equal to zero:

$$di/dt = 300\exp(-2500t) - 2500 \times 300t\exp(-2500t) = 0$$

whence $t = 1/2500 = 0.4$ ms, for maximum i. Substituting this value for t into equation 1.31 yields $i_{max} = 44$ mA.

The voltage across the inductor can be found by using $v = Ldi/dt$, or by finding the transformed voltage and then detransforming it. Using the latter approach we obtain

$$v_L = Lsi = 0.02s \times \frac{300}{(s + 2\,500)^2} = \frac{6s}{(s + 2\,500)^2}$$

Again, this is not a form we can recognise immediately, but we note that

$$\frac{s}{(s + a)^2} = \frac{(s + a) - a}{(s + a)^2} = \frac{1}{s + a} - \frac{a}{(s + a)^2}$$

Using LT3 and LT5 gives the partial fraction expansion for v_L as

$$v_L = \frac{6}{s + 2500} - \frac{6 \times 2500}{(s + 2500)^2}$$

Whence the detransformed voltage is seen to be

$$v_L(t) = 6\exp(-2500t) - 15\,000t\exp(-2500t) \text{ V} \tag{1.32}$$

It is always worth checking that the answer corresponds with the known facts. First, what will $v_L(t)$ be when the switch is closed at $t = 0+$? No current will flow at first, so the voltage across the resistance will be zero and then the voltage across the inductance willbe 6 V. Equation 1.32 gives $v_L(0) = 6$ V, as it should; and when $t = \infty$, $v_L = 0$ V, which is also as required.

1.10 Resonance

In a circuit containing both inductance and capacitance energy can be transferred from one to the other — indefinitely if there is no resistance present. The energy is dissipated by resistance and unless it is replaced by some means current will eventually cease to flow. When an RLC circuit like this is connected to a sinusoidal source there will be a certain frequency, the *resonant frequency*, at which the energy stored in the circuit is a maximum.

Take the series RLC circuit of figure 1.36 which is excited by a sinusoidal voltage source of r.m.s. value, V, and angular frequency, ω. The reactances are in series, giving a total reactance of $j(\omega L - 1/\omega C)$ and an impedance of $\mathbf{Z} = R + j(\omega L - 1/\omega C)$.

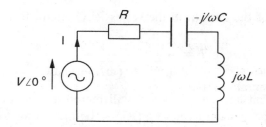

Figure 1.36

The series RLC circuit

Thus by Ohm's law the current flowing is

$$\mathbf{I} = \frac{\mathbf{V}}{\mathbf{Z}} = \frac{V \angle 0°}{R + j(\omega L - 1/\omega C)} \tag{1.33}$$

The magnitude of $\mathbf{I}$ is clearly a maximum when $\omega L = 1/\omega C$, making the imaginary part zero and then $\omega^2 = 1/LC$, or

$$\omega = \omega_0 = \frac{1}{\sqrt{LC}} \tag{1.34}$$

ω_0 is the *series resonant frequency*. For example if $L = 203$ mH and $C = 50$ µF, $\omega_0 = 314$ rad/s and $f_0 = \omega_0/2\pi = 50$ Hz. Components such as resistors, inductors or capacitors cannot be made without some of the other two circuit elements: thus a resistor will have some capacitance and inductance, an inductor some resistance and capacitance, and a capacitor some resistance and inductance. There is therefore always a frequency at which components become self-resonant and their use is normally restricted to frequencies well below this. If a circuit containing inductance and capacitance is subjected to a sudden change in excitation — such as turning the power on or off — and if the resistance is low enough, then oscillations in current will occur with a frequency which is approximately f_0.

1.10.1 The Q-factor

The Q-factor of the series RLC circuit at resonance is given by

$$Q = \frac{\omega_0 L}{R} \tag{1.35}$$

and since $\omega_0^2 = 1/LC$ by equation 1.34, we can also write $L = 1/\omega_0^2 C$ and substitution into equation 1.35 gives

$$Q = \frac{\omega_0 L}{R} = \frac{1}{\omega_0 CR}$$

At resonance, the current in the series RLC circuit, by equation 1.33, is simply V/R, and so the voltage across the inductance is $IX_L = I\omega L = V\omega L/R = VQ$.

1.10.2 Bandwidth

The magnitude of the current in the series RLC circuit from equation 1.33 is

$$I = \frac{V}{\sqrt{R^2 + (\omega L - 1/\omega C)^2}} \tag{1.36}$$

A graph of current against frequency will therefore start off at zero when $\omega = 0$, since $1/\omega C = \infty$ — the capacitor blocks DC — before rising to a maximum of V/R at the resonant frequency. It will then decline steadily to zero when $\omega = \infty$ as the inductance will block high frequencies (when $\omega L \to \infty$). At two frequencies, one above and one below ω_0, the power consumed by the resistance will be half what it would be at resonance. Since the power is proportional to the current, the current at half maximum power will be $I_{max}/\sqrt{2}$ = $V/R\sqrt{2}$. Substituting this into equation 1.36 leads to

$$\frac{V}{R\sqrt{2}} = \frac{V}{\sqrt{R^2 + (\omega_{1/2}L - 1/\omega_{1/2}C)^2}}$$

Squaring both sides of this leads to $(\omega_{1/2}L - 1/\omega_{1/2}C)^2 = R^2$, and the quadratic

$$\omega_{1/2}^2 \pm \frac{R}{L}\omega_{1/2} - \frac{1}{LC} = 0$$

which can be written

$$\omega_{1/2}^2 \pm \beta\omega_{1/2} - \omega_0^2 = 0$$

where $\beta \equiv R/L$ and $\omega_0 \equiv 1/\sqrt{(LC)}$. The solutions are then $\omega_{+1/2} = \beta/2 + \sqrt{(\beta^2/4 + \omega_0^2)}$ and $\omega_{-1/2} = -\beta/2 + \sqrt{(\beta^2/4 + \omega_0^2)}$. The difference between these two half-power frequencies is

$$\omega_{+1/2} - \omega_{-1/2} = \beta = R/L = \text{bandwidth}$$

Since $Q = \omega_0 L/R = \omega_0/\beta$, we see that the bandwidth is given by

$$\text{bandwidth} = \beta = \frac{\omega_0}{Q}$$

When Q is 'large' (which in practice means $Q \geq 5$), then $\omega_0{}^2 \gg \beta^2/4$ and the half-power angular frequencies are

$$\omega_{\pm\frac{1}{2}} \approx \omega_0 \pm \beta/2$$

Example 1.11

A series RLC circuit is required with a Q-factor of 10 and a bandwidth of 20 kHz. It must operate with a supply voltage of 12 V to dissipate 8 W at resonance. What should the component values be? What are the upper and lower half-power frequencies?

The power dissipation at resonance is $V^2/R = 8$, so that

$$R = V^2/8 = 144/8 = 18 \ \Omega$$

Expressing the bandwidth in krad/s leads to the inductance in mH:

$$\beta = 20 \times 2\pi = 40\pi = R/L \quad \Rightarrow \quad L = R/\beta = 18/40\pi = 0.143 \ \text{mH}$$

The circuit's resonant frequency, $\omega_0 = Q\beta = 10 \times 20 \times 2\pi = 1.257$ Mrad/s. Now $\omega_0 = 1/\sqrt{(LC)}$, making

$$C = \frac{1}{\omega_0^2 L} = \frac{1}{(1.257 \times 10^6)^2 \times 0.143 \times 10^{-3}} = 4.4 \ \text{nF}$$

The upper half-power frequency is $f_{+\frac{1}{2}} \approx f_0 + \frac{1}{2}\Delta f = 200 + 10 = 210$ kHz, where $\Delta f = \beta/2\pi$; and the lower is approximately at $200 - 10 = 190$ kHz. (The exact half-power points are at 210.25 kHz and 190.25 kHz, but in practice this difference would never be noticed.)

Suggestions for further reading

There are numerous textbooks on circuit theory. The reader is advised to seek one which they find easiest to understand. The few listed below have been tried and trusted.

Electrical circuit analysis and design by Noel M. Morris (Macmillan 1993).
Electric circuits by J W Nilsson and S A Riedel (Addison-Wesley 1996). A lengthy text in the American fashion.
Electric circuit fundamentals by S Franco (Saunders College Publishing 1995). Lengthy but not difficult.
SPICE: A guide to circuit simulation and analysis using PSPICE by P Tuinenga (Prentice-Hall, 1988). SPICE (Simulation Program with Integrated Circuit Emphasis) can be used to analyse any circuit. This book is a commendably compact guide to all the extensive facilities of PSPICE, a freely-available subset of SPICE.

Problems

1 What is the power supplied by (+) or (−) to each of the sources of figure P1.1?
[(a) +8 W (b) +24 µW (c) +8 kW (d) −12 mW (e) −99 MW (f) −800 W]
2 What power is dissipated in each resistance in the circuits of figure P1.2? What power is supplied by each source?
[(a) 60 Ω: 66.7 kW, 30Ω: 33.3 kW (b) 20 Ω: 50 kW, 40 Ω: 4 kW, 600V: 30 kW, 400 V: 24 kW (c) 5 MΩ: 8 kW, 10 MΩ: 4 kW (d) 40 Ω: 7.34 W, 30 Ω: 74.1 W, 2 A: 94.3 W, −30 V: −12.85 W]

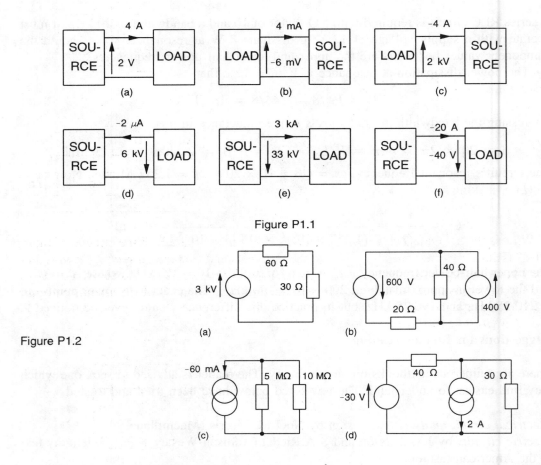

Figure P1.1

Figure P1.2

3 A capacitance of 20 000 µF is charged from 0 V with a current of $2\exp(-3t)$ A from $t = 0$. What is the voltage after charging? What energy is stored? What energy is dissipated during charging if it is from a voltage source through a constant resistance?
[33.3 V, 11.1 J, 11.1 J]
4 A current of $600t\exp(-3t)$ A flows through an inductor of inductance 100 mH and resistance 0.6 Ω from $t = 0$. What is the most energy stored? What energy is dissipated?
[271 J, 2 kJ]
5 A capacitance of 40 000 µF is charged to 20 kV and then discharged through a 100µH

inductance. If the resistance is negligible, what is the maximum current flowing? *[400 kA]*

6 What heat does a current of $4 \exp(-2t)u(t)$ mA produce when it flows through a resistance of 5.6 kΩ? *[22.4 mJ]*

7 If a constant current starts to flow through a capacitance of 0.02 F and an inductance of 60 mH at $t = 0$, at what time is the energy stored in each component the same if neither component stored energy before the current flow commenced? *[t = 34.6 ms]*

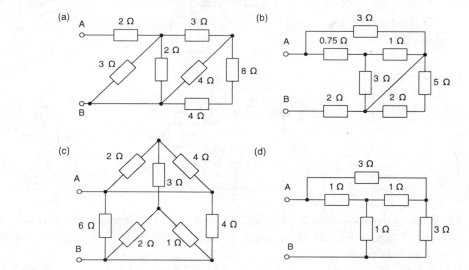

Figure P1.8

8 What is the equivalent resistance between the terminals A and B of the circuits of figure P1.8? (Hint: in P1.8d apply a 1V p.d. to the terminals and find the current, I, flowing into the circuit by Kirchhoff's laws; then $R_{eq} = 1/I$) *[(a) 3 Ω (b) 3 Ω (c) 7.4 Ω (d) 1.5 Ω]*

9 Find the equivalent capacitance between the terminals A and B of the circuits of figure P1.9. (Hint: use the method of P1.8d for P1.9c,d)
[(a) 1 F (b) 0.416 nF (c) 1.18 μF (d) 1.4C]

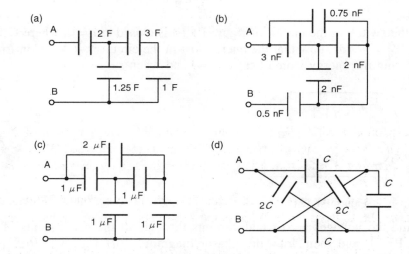

Figure P1.9

10 What are the Thévenin and Norton equivalents between terminals A, B of the circuits of figure P1.10? *[(a) V_T = 5 V, R_T = 2.5 Ω, I_N = 2 A (b) V_T = 3.57 V, R_T = 4.43 Ω, I_N = 0.806 A (c) V_T = 12 kV, R_T = 3 kΩ, I_N = 4 A (d) V_T = 25.5 V, R_T = 4.5 Ω, I_N = 5.67 A]*

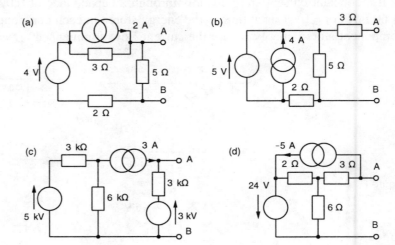

Figure P1.10

11 What is the most power that can be transferred to a load attached to AB in each of the circuits of figure P1.10? How much power is supplied to the circuit and load by each source in this case?

[(a) 2.5 W, P_{4V} = 6 W, P_{2A} = 3 W (b) 0.72 W, P_{5V} = −15 W, P_{4A} = 20 W (c) 12 kW, P_{5kV} = 12.8 kW, P_{3A} = 26 kW, P_{3kV} = −3 kW (d) 1.125 W, P_{24V} = 33 W, P_{5A} = 131.25 W]

12 In the circuit of figure P1.12, what must R be to make the power taken from each source identical? What is this power? What power is developed in each resistance then? *[0.48 Ω. 23.3 W. R, 2.6 W; 1 Ω, 4.5 W; 2 Ω, 39.5 W]*

13 Show that the current flowing is

$$i = \frac{V_1}{R}\exp(-t/RC)$$

when the switch in the circuit of figure P1.13 is closed at $t = 0$, where $C^{-1} = C_1^{-1} + C_2^{-1}$. Using $W_R = \int i^2 R dt$ show that the energy lost in the circuit is $\frac{1}{2}CV^2$. Confirm this result by considering the energy stored before $t = 0$ and when $t = \infty$.

Figure P1.12   Figure P1.13

14 Find the voltages across and currents through all the components of the circuits of figure P1.14, and then draw the phasor diagrams. *[(a) $V_{10Ω}$ = 3.16∠116.6° V, $I_{10Ω}$ = 0.316∠116.6° A, $V_{20Ω}$ = 4.47∠71.6° V, $I_{20Ω}$ = 0.224∠71.6° A, I_C = 0.224∠161.6° A (b)*

$V_{200\Omega} = 6.96\angle{-35.4°}$ V, $I_{200\Omega} = 34.8\angle{-35.4°}$ mA, $V_{1k\Omega} = 24.7\angle{9.4°}$ V, $I_{1k\Omega} = 24.7\angle{9.4°}$ mA, $I_L = 24.5\angle{-80.6°}$ mA (c) $V_C = 0.29\angle{62.2°}$ V, $I_C = 36\angle{152.2°}$ mA, $V_{5\Omega} = 0.16\angle{5.7°}$ V, $I_{5\Omega} = 32\angle{5.7°}$ mA, $V_L = 0.24\angle{95.7°}$ V (d) $V_R = 20.5\angle{-89.6°}$ V, $I_R = 6.22\angle{-89.6°}$ mA, $V_C = 3\angle{0.4°}$ kV, $I_C = 0.47\angle{90.4°}$ A, $I_L = 0.48\angle{-89.6°}$ A]

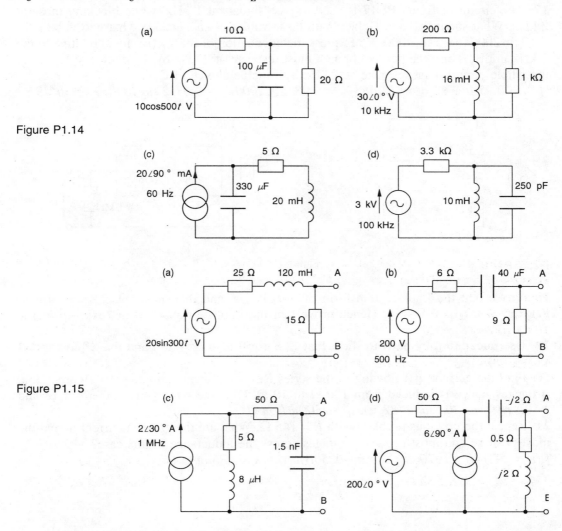

Figure P1.14

Figure P1.15

15 Find the Thévenin and Norton equivalents between the terminals A, B of the circuits of figure 1.15 and hence find the maximum power that can be developed in a load connected across A, B. What then is the reactive power supplied to this load?
[(a) $Z_T = 12.22\angle{13.23°}$ Ω, $V_T = 3.94\angle{-42°}$ V, $I_N = 0.323\angle{-55.22°}$ A, $P_{max} = 0.327$ W, $Q = -0.077$ var (b) $Z_T = 5.28\angle{-25.04°}$ Ω, $V_T = 106\angle{27.95°}$ V, $I_N = 20.1\angle{52.99°}$ A, $P_{max} = 587$ W, $Q = +274$ var (c) $Z_T = 100.9\angle{-2.15°}$ Ω, $V_T = 136.8\angle{69.75°}$ V, $I_N = 1.356\angle{-71.9°}$ A, $P_{max} = 46.4$ W, $Q = +1.74$ var (d) $Z_T = 2.043\angle{73.67°}$ Ω, $V_T = 14.72\angle{132.27°}$ V, $I_N = 7.21\angle{58.6°}$ A, $P_{max} = 94.3$ W, $Q = -322$ var]

16 The circuit of figure P1.16 is called Wien's bridge. The bridge is balanced when no current flows through the galvanometer, G, that is when A and B are at the same potential.

Show that the bridge-balancing condition is $Z_1/Z_2 = Z_3/Z_4$ where $Z_1 = R_1 - j/\omega C_1$, $Z_2 = R_2$, $Z_3 = R_3 \ // -j/\omega C_3$ and $Z_4 = R_4$. From this show, by equating real and imaginary parts separately, that the bridge can only balance when $C_3/C_1 = R_2/R_4 - R_1/R_3$ and then only at a single frequency given by $\omega^2 = 1/C_1 C_3 R_1 R_3$.

17 The circuit of figure P1.17 is used to pass signals at 1 kHz, while blocking those at 2 kHz. What should C_1 and C_2 be? With these values of capacitance what will v_o be if v_{in} = $10 \sin 2000\pi t + 2 \sin 4000t$ V? Express the attenuation of the 2 kHz signal relative to the 1 kHz in dB. (Hint: the parallel branch must resonate at 2 kHz by the choice of C_2. Then the whole circuit must resonate at 1 kHz by suitable choice of C_1.)

[$C_1 = 0.95 \ \mu F$, $C_2 = 0.317 \ \mu F$; $v_o = 9.982 \sin 2000\pi t + 0.0312 \sin 4000\pi t$ V; 36 dB]

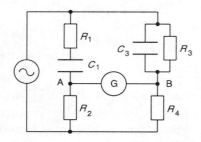

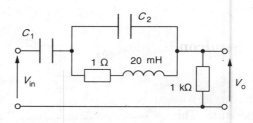

Figure P1.16   Figure P1.17

18 Prove that the Laplace transform of $tu(t)$ is $1/s^2$ and that the Laplace transform of $f(t)\exp(-at)u(t)$ is $F(s + a)$. Hence prove that the Laplace transform of $t\exp(-at)u(t)$ is $1/(s + a)^2$.

19 Use integration by parts to show that $\mathcal{L}[\sin(\omega t)]$ is $\omega/(s^2 + \omega^2)$ and that $\mathcal{L}[\cos(\omega t)]$ is $s/(s^2 + \omega^2)$.

20 Find the current, $i(t)$ flowing in the series RLC circuit of example 1.10, figure 1.35a, if the resistance is reduced from 100 Ω to 10 Ω. (Use the results of problems 1.18 and 1.19.) *[$i(t) = 120.6\sin(2487.5t)\exp(-250t)u(t)$ mA]*

21 Repeat the previous problem with $R = 145 \ \Omega$. What are the greatest current in and the most positive and most negative voltages across the inductance in this case?
[$i(t) = 57.14[\exp(-1000t) - \exp(-6250t)]u(t)$ mA, 33.86 mA, +6 V, −0.554 V]

2 Diodes

BEFORE discussing some of the numerous applications of diodes we shall first examine some essential aspects of the theory of semiconductors which govern their operation. This theoretical background will also be valuable when transistors are discussed in chapters three and four. We have not the space for even a partial treatment of the quantum theory of electronic behaviour: what follows is the barest outline.

2.1 Semiconductors

Electrical materials are usually classified as conductors, semiconductors or insulators according to what is perceived to be their intrinsic ability to conduct electricity at room temperature. This classification overcomes the difficulty of categorising superconductors which for the present operate only at lower temperatures, but it is really an oversimplification. Looking at the Periodic Table, of which table 2.1 is a part, we can see that most elements are metals which are characterised by malleability, ductility and high electrical conductivity. Only eleven elements in the region of the top right-hand corner are not metals and solid at room temperature: boron (B), carbon (C), silicon (Si), phosphorus (P), sulphur (S), germanium (Ge), arsenic (As), selenium (Se), antimony (Sb), tellurium (Te) and iodine (I). Five of these — boron, silicon, germanium, selenium and tellurium — are generally considered to be semiconductors and only silicon is much used in electronic devices. However, many uses are found for compound III-V semiconductors[1] such as gallium arsenide (GaAs), particularly in optoelectronics.

Table 2.1   *The Periodic Table of the elements*

GROUP	I	II	III	IV	V	VI	VII	0
	H							He
	Li	Be	**B**	**C**	N	O	F	Ne
	Na	Mg	Al	**Si**	*P*	*S*	Cl	Ar
	K	Ca	Ga	**Ge**	*As*†	**Se**	*Br*!	Kr
	Rb	Sr	In	Sn	*Sb*†	**Te**	*I*	Xe
	Cs	Ba	Tl	Pb	*Bi* †	Po*	At*	Rn*

Notes: A-subgroup elements only. All B-subgroup elements are metals
Typefaces: Metals — Li. **Semiconductors — Si**. *Non-metals — P*.
Gases — He. ! liquid. * radioactive. † semi-metals

---

[1] So called because one element is drawn from group III and one from group V of the Periodic Table

2.1.1 Electrical conductivity

If we consider a rectangular bar of material of length, l, and cross-sectional area, A, which has a resistance, R, then its electrical conductivity, σ, is

$$\sigma = \frac{l}{RA} = \frac{1}{\rho} \tag{2.1}$$

The resistivity, ρ, is the reciprocal of the conductivity. If l is in m, A in m^2 and R in Ω, then ρ is in Ωm and σ in S/m.

In metals the outer electrons are not attached to any particular atom but are free to move under the influence of an applied electric field, $\mathscr{E}$. If a potential drop along a wire of length, l, is V, then the field is $\mathscr{E} = V/l$ and its units are V/m. In this field the free electrons rapidly accelerate but soon collide with ions of the lattice and acquire an average speed known as the *drift velocity, v_d.* The ratio of the drift velocity to the field is constant for a given material at a given temperature and is known as the *mobility*, μ. Thus

$$\mu = v_d / \mathscr{E} \tag{2.2}$$

If the charge on an electron is q coulombs (C) and there are n electrons/m^3 all moving at a speed of v_d, the charge passing a point will be $nqAv_d$ per unit time. Now charge/time is current, that is

$$I = nqAv_d = nqA\mu\mathscr{E} = nq\mu AV/l$$

where we have replaced v_d by $\mu\mathscr{E}$ and $\mathscr{E}$ by V/l. Rearranging this equation we find

$$\frac{Il}{VA} = \frac{l}{RA} = \sigma = nq\mu \tag{2.3}$$

since $I/V = 1/R$ by Ohm's law. The conductivity is the product of the concentration of free electrons per m^3, their charge and their mobility.

Example 2.1

A round copper wire of length 1 km and diameter 1 mm is found to have a resistance of 22 Ω. What are the conductivity and resistivity of copper? If there is one free electron for each atom of copper what is the mobility of the electrons in copper? If the voltage drop along the wire is 35 V, what is the electron's drift velocity? [The density of copper is 8.9 tonnes/m^3 and its relative atomic mass is 63.54.]

Using equation 2.1 we find

$$\sigma = \frac{l}{RA} = \frac{l}{R \times \pi r^2} = \frac{1000}{22 \times \pi \times (0.5 \times 10^{-3})^2} = 57.9 \text{ MS/m}$$

where r is the wire's radius in cross-section. Copper's conductivity is high. Its resistivity is $\rho = 1/\sigma = (57.9 \times 10^6)^{-1} = 17.3$ nΩm, which is very small.

To find the mobility we use equation 2.3, which means we must calculate n first. Now the r.a.m. of copper is 63.54 and 1 atomic mass unit is 1.66×10^{-27} kg, so an atom of copper weighs $63.54 \times 1.66 \times 10^{-27}$ kg. Thus in 1 m³ (= 8900 kg) the number of atoms will be

$$n = \frac{8\,900}{63.54 \times 1.66 \times 10^{-27}} = 8.44 \times 10^{28}$$

If there is one free electron/atom, then this is also the free electron concentration per m³. The electronic charge is 1.6×10^{-19} C, therefore

$$\sigma = nq\mu = 8.44 \times 10^{28} \times 1.6 \times 10^{-19}\mu = 57.9 \times 10^{6} \text{ S/m}$$

Solving, we find $\mu = 4.29 \times 10^{-4}$ m²V⁻¹s⁻¹.

By equation 2.2 the drift velocity is

$$v_{\mathrm{d}} = \mu\mathscr{E} = \frac{\mu V}{l} = \frac{4.29 \times 10^{-4} \times 35}{1000} = 1.5 \times 10^{-5} \text{ m/s}$$

The drift velocity is very small and the electron concentration very large.

2.1.2 Intrinsic conduction in semiconductors

In a pure (or *intrinsic*) semiconductor at a temperature of 0 K there are no free electrons and its conductivity is therefore zero: it is an insulator. As the temperature is increased the valence electrons are able to acquire thermal energy which breaks some of the covalent bonds, freeing electrons for conduction. In place of the liberated electron a *hole* is created which can also move from bond to bond and from atom to atom, as shown in figure 2.1. The movement of the hole is really the movement of an electron the other way, that is the hole acts like a positively charged electron. Production of a free electron is always accompanied by the creation of a hole and therefore of two charge carriers of opposite sign, both of which will contribute to electrical conduction.

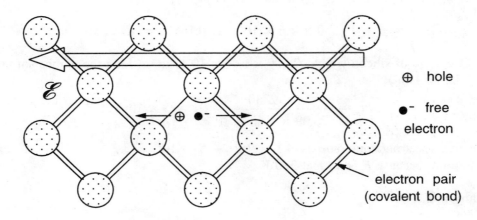

Figure 2.1 Creation of an electron-hole pair in a semiconductor crystal. These charge carriers can then move in opposite directions in the electric field, $\mathscr{E}$.

The hole mobility, μ_p, will in general be different from the electron mobility, μ_n, and the conductivity will be the sum of the two contributions

$$\sigma = nq\mu_p + pq\mu_p = nq(\mu_n + \mu_p) \tag{2.4}$$

where p_i, the intrinsic hole concentration, is equal to n_i, the intrinsic free electron concentration. The concentration of hole-electron pairs that forms in this way depends on the energy required to free the electron from its valence bond. The valence electrons occupy a large number of discrete, quantised energy levels or states which are very close together and are said to form the *valence band*. Similarly the free electrons occupy energy levels in the *conduction band* which are of higher energy than those in the valence band. Between valence and conduction bands is an energy gap of size E_g, known as the *gap energy* or just the *bandgap*. The larger the bandgap, the smaller the number of electron-hole pairs that can form at a given temperature. This number is given by

$$n_i = p_i = N\exp(-E_g/2kT) \tag{2.5}$$

where N is a constant, k is Boltzmann's constant ($= 1.38 \times 10^{-23}$ J/K) and T is the temperature in K. This expression is what one might expect for a classical particle which has to surmount an energy barrier of $E_g/2$, the factor of two is a consequence of the non-classical, quantum behaviour of electrons. We can combine equations 2.4 and 2.5 to give

$$\sigma = q(\mu_n + \mu_p)N\exp(-E_g/2kT) \tag{2.6}$$

Example 2.2

A sample of pure silicon at 300 K is found to have a resistivity of 950 Ωm. If the electron-to-hole mobility ratio is 3:1 and the electron mobility is 0.12 m^2/V/s what are the intrinsic hole and electron concentrations? If the density of silicon is 2330 kg/m^3 and its atomic weight is 28.09 what is the ratio of free electrons to atoms?

The hole mobility, $\mu_p = \mu_n/3 = 0.04$ m^2/V/s; then using equation 2.4 yields

$$\sigma = 1/\rho = 1/950 = n_i q(\mu_n + \mu_p) = n_i \times 1.6 \times 10^{-19} \times (0.12 + 0.04)$$

Hence

$$n_i = p_i = (950 \times 1.6 \times 10^{-19} \times 0.16)^{-1} = 4.1 \times 10^{16} \text{ m}^{-3}$$

One atom of silicon weighs $28.09 \times 1.66 \times 10^{-27}$ kg, making the number of atoms in 2330 kg ($= 1$ m^3)

$$n_a = \frac{2330}{28.09 \times 1.66 \times 10^{-27}} = 5 \times 10^{28} \text{ m}^{-3}$$

The free electron/atom ratio is $4.1 \times 10^{16} \div 5 \times 10^{28} = 8.2 \times 10^{-13}$ — a very tiny proportion, because $E_g \ll kT$ at 300 K.

Example 2.3

The conductivity of an intrinsic silicon sample is found to be 1.02 mS/m at 297.2 K and 2.15 mS at 307.9 K. What is the bandgap energy in silicon?

Equation 2.6 indicates that if the pre-exponential term $q(\mu_n + \mu_p)N$ is independent of temperature then to find E_g all we need is two measurements of σ at different temperatures. Let these be σ_1 at T_1 and σ_2 at T_2 and from equation 2.6 we find

$$\sigma = A \times \exp(-E_g/2kT) \quad \Rightarrow \quad \ln\sigma = B - E_g/2kT$$

where A and B are constants. Substituting σ_1, T_1 and then σ_2, T_2 into this and subtracting leads to

$$\ln\sigma_1 - \ln\sigma_2 = -E_g/2kT_1 + E_g/2kT_2 = (E_g/2k)(1/T_2 - 1/T_1)$$

hence

$$E_g = \frac{2k\ln(\sigma_1/\sigma_2)}{1/T_2 - 1/T_1} = \frac{2 \times 1.38 \times 10^{-23} \times \ln(1.03/2.15)}{1/307.9 - 1/297.2} = 1.74 \times 10^{-19} \text{ J}$$

Gap energies are always given in eV where 1 eV = 1.6×10^{-19} J, so that 1.74×10^{-19} J becomes 1.09 eV.

2.1.3 Extrinsic conduction in semiconductors

Devices are generally made from semiconductor material whose conductivity results from electrically-active impurities. These are introduced in a controlled manner during manufacture, a process known as *doping*, and the resulting conductivity is then *extrinsic*. Silicon is in group IV of the Periodic Table and has four valence electrons. Each atom in the solid has four nearest neighbours and form four covalent, shared-electron-pair bonds as shown schematically in figure 2.2a.

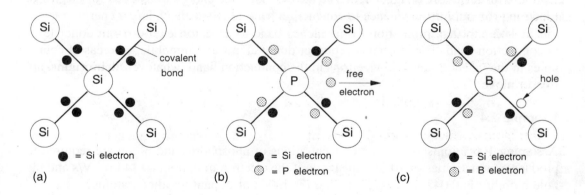

Figure 2.2 Bonding arrangements in silicon (a) Pure, undoped Si (intrinsic) (b) Phosphorus-doped Si (n-type) (c) Boron-doped Si (p-type)

If we now remove the central silicon atom (as in figure 2.2b) and replace it with a phosphorus atom from group V, which has five valence electrons, then one electron is surplus when the four bonds are formed. This electron is readily detached from its parent and enters the conduction band as a free electron available for conduction. The ionised phosphorus atom is immobile and the resulting conductivity is solely due to electrons from the dopant if we add enough phosphorus. The conductivity is then said to be *n-type* and phosphorus is said to be a *donor*.

Suppose we were to replace the silicon atom with a boron atom from group III as in figure 2.2c, then as the boron atom has only three outer valence electrons it leaves a hole in the valence band where an electron is missing. The hole is free to move in an electric field; the conductivity, due solely to these holes if we add enough boron, is called *p-type* and boron is said to be an *acceptor*, since it accepts electrons when the hole moves away. Figure 2.3 shows the various energy levels in a semiconductor.

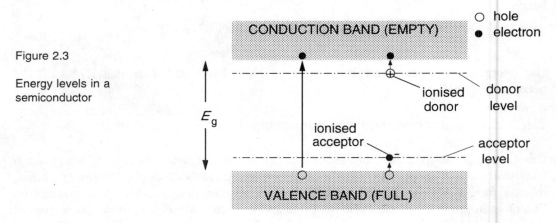

Figure 2.3

Energy levels in a semiconductor

The donor level is only about 0.01 eV from the bottom of the conduction band (the band edge) and the acceptor level is very close to the top of the valence band. In figure 2.3 these distances are exaggerated. The arrows show the energy transitions for electrons. Electron promotion from valence to conduction band involves much larger energy transfer (about 1–2 eV) than promotion from valence band to acceptor level or from donor level to conduction band. The ionised acceptor or donor atoms are immobile, whereas of course holes in the valence band and electrons in the conduction band, whatever their origins, are free to move.

Example 2.4

Assuming 100% ionisation, what concentrations of phosphorus and boron are required to produce conductivities of 10 S/m in silicon if the electron mobility is 0.144 m²/V/s and the hole mobility is 0.047 m²/V/s? What are the ratios of dopant to silicon atoms?

We can use equation 2.3, $\sigma = nq\mu$, in the form

$$n = \frac{\sigma}{q\mu_n} = \frac{10}{1.6 \times 10^{-19} \times 0.144} = 4.34 \times 10^{20} \text{ m}^{-3}$$

And the boron concentration is

$$p = \frac{\sigma}{q\mu_p} = \frac{10}{1.6 \times 10^{-19} \times 0.037} = 1.69 \times 10^{21} \text{ m}^{-3}$$

From example 2.2 we know that $n_a = 5 \times 10^{28}$ m^{-3} making the dopant/silicon atomic ratios

$$\frac{n}{n_a} = \frac{4.34 \times 10^{20}}{5 \times 10^{28}} = 8.7 \times 10^{-9}$$

and

$$\frac{p}{n_a} = \frac{1.69 \times 10^{21}}{5 \times 10^{28}} = 3.4 \times 10^{-8}$$

The dopant concentrations are of the order of 10 parts per billion.

The law of mass action for semiconductors

In an intrinsic semiconductor $p_i = n_i = N\exp(-E_g/2kT) = $ a constant, at a given temperature. Thus the product, $p_i n_i$, is also constant. If one then dopes the material with say, boron, the holes produced by the dopant will be in equilibrium with the electrons produced by intrinsic thermal generation and the product of hole and electron concentration remains constant. This is the law of mass action for semiconductors. In any semiconductor, doped or not

$$pn = p_i n_i = p_i^2 = n_i^2 = \text{constant} \tag{2.7}$$

Example 2.5

In example 2.4, what are the concentrations of holes and electrons in the two doped samples of silicon if the intrinsic carrier concentration is 4.1×10^{16} m^{-3}?

Using equation 2.7 with $n = 4.34 \times 10^{21}$ and $n_i = 4.1 \times 10^{16}$ we have in the n-type material

$$p = \frac{n_i^2}{n} = \frac{(4.1 \times 10^{16})^2}{4.34 \times 10^{21}} = 3.87 \times 10^{11} \text{ m}^{-3}$$

The hole (the *minority carrier*) concentration is far less than the electron (the *majority carrier*) concentration in n-type material. In the p-type sample, if $p = 1.69 \times 10^{21}$ m^{-3},

$$n = \frac{n_i^2}{p} = \frac{(4.1 \times 10^{16})^2}{1.69 \times 10^{21}} = 10^{12} \text{ m}^{-3}$$

Once more the majority carrier (this time the hole) concentration is far greater than the minority carrier (now the electron) concentration.

When the intrinsic carrier concentration is not negligible compared to the dopant concentration, then we cannot assume that the majority carrier concentration is the same as the dopant's. In order to find the actual carrier concentrations as well as the law of mass action, we must use the condition of overall charge neutrality, that is that in a p-doped sample, $p = n + N_A$, where N_A is the number of ionised acceptor atoms; and in an n-doped sample, $n = p + N_D$, where N_D is the number of ionised donor atoms.

Example 2.6

If the semiconductor in example 2.5 is doped with 6×10^{16} boron atoms/m^3, what will the electron and hole concentrations be if all the boron atoms are ionised?

We have $np = n_i^2 = (4.1 \times 10^{16})^2$ and $p = n + 6 \times 10^{16}$, hence

$$n(n + N_A) = n_i^2 \quad \Rightarrow \quad n(n + 6 \times 10^{16}) = (4.1 \times 10^{16})^2$$

which leads to the quadratic

$$n^2 + 6 \times 10^{16}n - 1.68 \times 10^{33} = 0 \quad \Rightarrow \quad n = 2.08 \times 10^{16} \text{ m}^{-3}$$

and hence $p = 8.08 \times 10^{16}$ m^{-3}.

2.1.4 The p-n junction and the rectifier equation

Having seen that it is possible to produce two conductivity types in a semiconductor we can now look at what happens when the two types are brought into contact in a p-n junction, shown in figure 2.4.

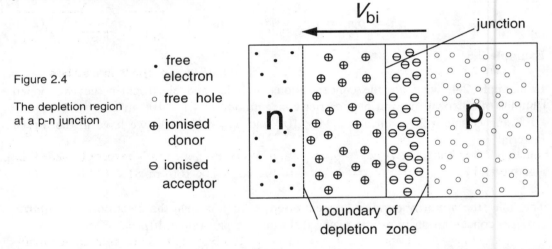

Figure 2.4

The depletion region at a p-n junction

- free electron
- ○ free hole
- ⊕ ionised donor
- ⊖ ionised acceptor

boundary of depletion zone

As soon as contact is made between the two types electrons from the n-type side diffuse across the junction into the p-type side and there combine with holes and disappear. Similarly holes from the p-type material will diffuse across the junction and combine with electrons on the n-type side. This process sets up a potential barrier across the junction, V_{bi}, the built-in potential, which opposes the movement of charge carriers by diffusion. In the region of the junction there are very few mobile charge carriers, only the fixed charges of the ionised dopants which produce V_{bi}. This thin region (of the order of 1 μm thick) is therefore variously known as the *depletion region* (or zone) or the *space-charge region* (or zone). Outside of this region the carrier concentrations are 'normal'. The process of thermal generation of minority carriers in the depletion region continues still and produces a current called the *generation current*, I_g, which is equal and opposite to the diffusion or *recombination current*, I_r. Both I_g and I_r are very small.

Now suppose we apply *reverse bias* to the junction by making the n-side higher in potential than the p-side. As can be seen in figure 2.5b, the bias potential reinforces the built-in potential, thus widening the depletion region and further reducing I_r by a factor of $\exp(-qV/kT)$, which is about 1.6×10^{-17} at 300 K at a reverse bias of 1 V. Then the current flowing is effectively only I_g which is known as the *reverse saturation current*, I_s. In a silicon signal diode, I_s is of the order of 10^{-14} A at 300 K, a vanishingly small current.

If *forward bias* is applied to the junction by making the p-side more positive than the n-side, the recombination current is now facing a smaller potential barrier and a reduced depletion region width as in figure 2.5c. Thus I_r is increased by a factor of $\exp(qV/kT)$ and overwhelms I_g even at small forward bias voltages.

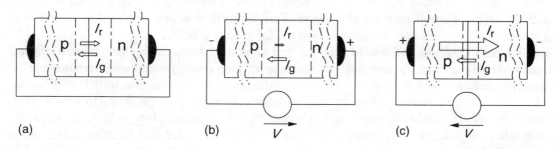

Figure 2.5 A p-n junction (a) with no bias (b) with reverse bias and (c) with forward bias

The current is given by

$$I = I_s[\exp(qV/kT) - 1] \tag{2.8}$$

Equation 2.8 is known as the *rectifier* equation. When the junction is reverse biased, V is negative and $I \approx -I_s \approx 0$. When it is forward biased, V is positive and

$$I \approx I_s\exp(qV/kT) \approx I_s\exp(40V) \quad \text{at 290 K} \tag{2.9}$$

A p-n junction can therefore be used to control the flow of current in a circuit according to the bias applied to it. In reverse bias it stops the current flow and in forward bias it allows current to pass, that is it *rectifies*. The exponential form of the rectifier equation means that the forward voltage drop across the junction is effectively constant. Rectifying devices made from p-n junctions are called p-n junction diodes.

Example 2.7

In a small germanium diode if I_s is 50 nA at 290 K, what forward bias will result in current flows of 1 mA and 10 mA? Repeat the calculations for a silicon diode having I_s = 0.01 pA.

The solution involves substitution in equation 2.9, leading to

$$10^{-3} = 50 \times 10^{-9} \times \exp(40V)$$

$$\Rightarrow V = \left(\frac{1}{40}\right)\ln\left(\frac{10^{-3}}{5 \times 10^{-8}}\right) = 0.25 \text{ V}$$

When the current is increased to 10 mA, $V = 0.3$ V, a small increase in voltage for a ten-fold increase in current.

For the silicon diode, with $I_s = 0.01$ pA and $I = 1$ mA, we find

$$10^{-3} = 0.01 \times 10^{-12} \exp(40V) \quad \Rightarrow \quad V = \frac{\ln 10^{11}}{40} = 0.63 \text{ V}$$

The voltage is $2\frac{1}{2}$ times that of the germanium diode. Increasing I to 10 mA leads to $V = 0.69$ V, again a small increase in voltage. This increase is nearly constant at 60 mV for each ten-fold (or *decade*) increase in current for silicon diodes.

The small changes in forward bias required to produce large changes in forward current leads one to say that the forward voltage drop of gemanium diodes when conducting is nearly constant at 0.3 V, and that for silicon diodes is nearly constant at 0.7 V. We can also see that I_s effectively determines the diode's forward voltage drop. In turn I_s is a function of the thermally-generated charge carrier concentration, which is strongly dependent on the bandgap energy, E_g. The forward drop of a p-n junction diode with 'normal' doping levels is roughly $E_g - 0.4$ V with E_g in eV. In gallium arsenide, GaAs, the bandgap is 1.4 eV, so we should expect a diode drop of about 1 V and in gallium phosphide, GaP, the bandgap is 2.2 eV and the expected forward voltage is 1.8 V. Figure 2.6 shows plots of equation 2.9 for diodes made of various materials.

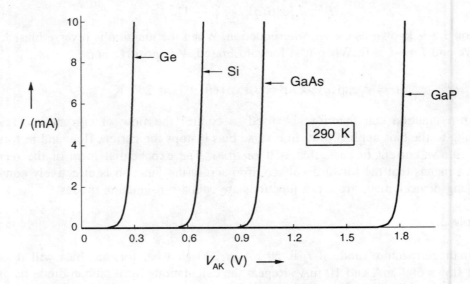

Figure 2.6 *I-V* characteristics at room temperature for p-n junctions in germanium, silicon, gallium arsenide and gallium phosphide according to equation 2.9

The effect of temperature

Equation 2.8 hides the fact that the reverse saturation current, I_s, is not a constant but also varies with temperature:

$$I_s \propto n_i^2 \propto \exp(-E_g/kT) \tag{2.10}$$

where E_g is the bandgap energy in joules. Thus the current-voltage relation for a p-n junction becomes

$$I \approx A\exp(-E_g/kT)\exp(qV/\eta kT)$$

where η is the *ideality factor* (≈ 1.2 for silicon) and A is a constant. Taking logs gives

$$\ln I = \ln A + qV/\eta kT - E_g/kT \Rightarrow V = \eta E_g/q + (\eta kT/q)\ln(I/A)$$

Differentiating V with respect to T yields

$$dV/dT = (\eta k/q)\ln(I/A) = (V - \eta E_g/q)/T$$

Since $V - \eta E_g/q$ is about -0.7 V in silicon, we can see that at 290 K the change of voltage with temperature is about -2 mV/K, that is, if the temperature *in*creases by 1°C, the voltage for a given forward current *de*creases by 2 mV. The implication is that if a p-n junction in forward bias is to be used as a voltage reference, then its temperature must be kept constant.

Taking logs of equation 2.10 and differentiating with respect to temperature gives

$$\delta I_s/I_s = E_g\delta T/kT^2 = 0.14\delta T \quad \text{(for Si at 300 K)}$$

The reverse saturation current (or leakage) current of a silicon p-n junction increases 14% for every 1 K rise in temperature, doubling every 7°C. This can cause problems in devices when the increase in temperature becomes excessive.

2.2 Semiconductor p-n junction diodes

The diodes used in most electronic circuits are based on p-n junctions in silicon. They are therefore not ideal rectifiers, as we have seen, mainly because of the significant forward voltage drop of about 0.7 V. The p-type side of the diode is known as the *anode* (A), while the n-type side is called the *cathode* (K). Most small discrete diodes are packaged as a cylindrical shape with a line around the cathodic end as in figure 2.7, which also shows the circuit symbol for a diode. In forward bias V_{AK} will be about 0.7 V for a silicon diode passing rated current. We shall deal with ordinary diodes first while later sections will be concerned with special types such as Zener diodes and optical diodes.

If a large forward current has to be passed the junction area is made larger in proportion to the current. Signal diodes operate with forward currents of about 10 mA maximum and the junction area is about 0.2 mm² or 2×10^{-7} m², so the operating current density is around 50 kA/m². In power rectifiers the current density is about 1 MA/m², and a device passing a current of 100 A will be about 25 mm in diameter.

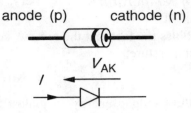

Figure 2.7

A small, signal diode. The package is
shown above and the circuit symbol below

2.2.1 The dynamic resistance

The slope of the diode's *I-V* characteristic can be found from the rectifier equation, equation 2.9; it has dimensions of AV^{-1} or S (Ω^{-1}) and is the reciprocal of the *dynamic resistance*, r_d. Let us differentiate equation 2.9, which is $I \approx I_s \exp(40V)$,

$$dI/dV = 1/r_d \approx 40 I_s \exp(40V) = 40I$$

The approximation is very close as I_s is very small and so we can write

$$r_d \approx 0.025/I \qquad\qquad (2.11)$$

The dynamic resistance of a p-n junction diode in forward bias is quite small: for example if $I = 10$ mA, $r_d = 0.025/10 \times 10^{-3} = 2.5\ \Omega$. But in bipolar junction transistors, where the current is small, it becomes much larger: about 1 kΩ.

2.2.2 Junction capacitance

The p-n junction acts like a parallel plate capacitor, whose capacitance is given by

$$C_j = \varepsilon_r \varepsilon_0 A/d \qquad\qquad (2.12)$$

where ε_r is the relative permittivity of the semiconductor (about 12 for silicon), ε_0 is the electric constant ('permittivity of free space') whose value is 8.85×10^{-12} F/m, A is the area of the junction normal to the current flow and d is the width of the depletion region. For a small silicon signal diode, $A \approx 0.2$ mm^2, $d \approx 1$ µm and then $C_j \approx 21$ pF. The width of the depletion region, however, depends on the voltage across the junction, being considerably larger when reverse biased than when forward biased so that C_j is voltage-dependent. *Varactor* diodes or *varicap* diodes put this property to good use as voltage-variable capacitors in tuning circuits. Junction capacitance can be a problem at high frequencies and in this case *point-contact* diodes are used which do not contain p-n junctions, but instead rely on the rectifying properties of a metal-semiconductor contact.

In parallel with the junction capacitance there is a capacitance caused by the change in concentrations of minority carriers, called the *diffusion capacitance*, C_d. The charge stored in a forward biased p-n junction carrying a current, I, is given by

$$C_d = \frac{dQ}{dV} = \frac{dQ}{dI}\frac{dI}{dV} = \frac{dQ}{dI}\frac{1}{r_d} = \frac{q_s}{r_d} = \frac{q_s I}{0.025} = 40 I q_s$$

where q_s is the stored charge per amp and r_d is the dynamic resistance, $\approx 0.025/I$ at 290 K. In a small signal diode, $q_s \approx 4$ nC/A and so $C_d \approx 160$ nF/A or 160 pF/mA. Thus in a conducting diode, $C_d \gg C_j$.

Example 2.8

A silicon p+/n junction (p+ indicates heavy p-type doping) of area 1 mm^2 is subjected to a reverse bias of 10 V. If the donor atom concentration is 10^{21} m^{-2}, what is its capacitance? At what voltage will its capacitance be halved?

The depletion region in a p-n junction in silicon has a width given by

$$w = \sqrt{2\varepsilon_r\varepsilon_0 V/qN} \tag{2.13}$$

where ε_r = relative permittivity = 12 in silicon, N = dopant concentration on the more lightly-doped (n) side = 10^{21} and V is the reverse bias, 10 V. Thus we find

$$w = \sqrt{\frac{2 \times 12 \times 8.85 \times 10^{-12} \times 10}{1.6 \times 10^{-19} \times 10^{21}}} = 3.6 \ \mu\text{m}$$

and so

$$C_j = \frac{\varepsilon_r\varepsilon_0 A}{w} = \frac{12 \times 8.85 \times 10^{-12} \times 10^{-6}}{3.6 \times 10^{-6}} = 29.5 \ \text{pF}$$

The capacitance will clearly be halved when the depletion zone is twice as wide, that is when the reverse bias is quadrupled to 40 V.

2.2.3 Switching time

When the diode goes from forward bias to reverse bias the current flow does not cease abruptly, but carries on for a short time in the reverse direction as there is significant excess minority carrier storage either side of the junction. For example in a p+/n junction the current is largely due to hole conduction. At the moment of bias reversal the holes on the n-side are in excess of the equilibrium concentration for zero bias and constitute stored charge which must be removed by the flow of reverse current, I_R. Generally $I_R \approx I_F$, the forward current prior to bias reversal, the magnitudes of both currents being dependent on the circuit external to the diode. The switching time is proportional to the minority carrier lifetime in the vicinity of the junction and can be greatly reduced by doping with an element such as gold, which forms a strong *trapping centre* in the middle of the bandgap, thereby quickly removing the excess minority carriers. Without such doping the lifetime may be up to 1 ms and the diode can then only be used for DC or very low frequencies.

Manufacturers of switching and fast-recovery diodes generally specify the *reverse recovery time*, t_{rr}, which is approximately the minority carrier lifetime. The upper frequency limit for these diodes is much lower than $1/t_{rr}$, by a factor of from 100 to 1000.

2.2.4 Junction breakdown

A p-n junction will pass only a very small ('leakage') current in reverse bias until quite suddenly a very large current flows which will destroy the diode unless limited in some way; the voltage at which this occurs is the *breakdown voltage* of the junction. Its value depends on the doping concentrations and concentration gradients at the junction. Normally the breakdown process takes place by *avalanche multiplication*, wherein an electron gains sufficient energy from the bias field that it can remove a valence electron from a lattice atom when it collides with one and create an electron-hole pair. The newly-created carriers can then be accelerated in turn and produce more electron-hole pairs on collision so that an avalanche of current is created. For an abrupt junction (that is with no dopant concentration gradient) the breakdown voltage is inversely proportional to the dopant concentration on the more lightly-doped side. In silicon the relation is

$$V_B = 1.5 \times 10^{17} N^{-0.7} \tag{2.14}$$

where N is the dopant concentration on the lightly-doped side. Since it is difficult to reduce impurity concentrations much under 10^{20} m^{-3}, we can calculate that breakdown voltages above 1.5 kV are hard to achieve in silicon.

The breakdown voltage is directly proportional to the bandgap and is therefore about 30% higher in gallium arsenide $E_g = 1.42$ eV than in silicon ($E_g = 1.1$ eV). Breakdown voltage also depends on junction geometry and can be greatly reduced if this has small radii of curvature. When the doping is very heavy another mechanism known as *tunnelling* or *Zener breakdown* can occur, but this is only important at breakdown voltages below about 6 V in silicon.

2.2.5 Circuit models for diodes

In a circuit a diode's behaviour is largely governed by equation 2.8, the rectifier equation, which relates the current through the diode to the voltage across it, as shown in figure 2.6. When we come to analyse a circuit containing a diode it is convenient to model this behaviour, or idealise it in some way to simplify the analysis.

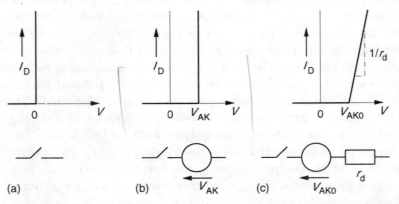

Figure 2.8 Diode models (a) as an ideal switch (b) as a switch in series with a voltage source (c) as a switch, voltage source and resistance in series

If the voltages and currents in a circuit are relatively large a reasonable approximation to a diode is an ideal switch as shown in figure 2.8a. However, when the diode's forward voltage drop is significant we can model it as a constant voltage source, V_{AK}, in series with a switch as in figure 2.8b. Finally, if the dynamic resistance is significant we can add a resistance, r_d, in series with the voltage source as in figure 2.8c, so that the diode's voltage varies with the current and the voltage drop at zero current is V_{AK0}.

Example 2.9

Calculate the currents flowing through each branch in the circuit of figure 2.9, given that the diodes are identical with a forward voltage drop of 1 V and zero dynamic resistance.

Figure 2.9

The circuit for example 2.9

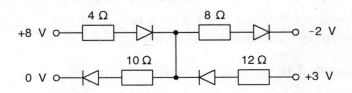

The diode model used is that of figure 2.8b with $V_{AK} = 1$ V. We will assume that all the diodes are forward biased and carry forward current. Examination of the circuit reveals that if we knew the potential at the node, then the currents could be found. Let us call this potential V, then we can find the branch currents in terms of V using KVL and Ohm's law.

Figure 2.10

The circuit of figure 2.9 split into branches and annotated with the voltages

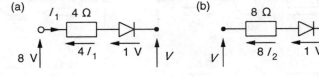

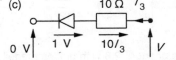
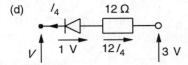

In the first branch, shown with appropriate voltage arrows in figure 2.10a, the potential across the 4Ω resistance is $4I_1$ by Ohm's law. Thus by KVL

$$8 = 4I_1 + 1 + V \quad \Rightarrow \quad I_1 = (7 - V)/4$$

Then in the second branch of figure 2.10b we can deduce similarly that

$$V = 8I_2 + 1 + (-2) \quad \Rightarrow \quad I_2 = (V + 1)/8$$

And in the third branch (figure 2.10c) we find

$$V = 10I_3 + 1 + 0 \quad \Rightarrow \quad I_3 = (V - 1)/10$$

Finally in the fourth branch (figure 2.10d) we have

$$3 = 12I_4 + 1 + V \quad \Rightarrow \quad I_4 = (2 - V)/12$$

we know that $I_1 + I_4 = I_2 + I_3$, giving

$$(7 - V)/4 + (2 - V)/12 = (V + 1)/8 + (V - 1)/10$$

whence $V = 3.39$ V.

Substituting this value of V into the equations for the currents leads to $I_1 = 0.903$ A, $I_2 = 0.549$ A, $I_3 = 0.239$ A and $I_4 = -0.116$ A. Now although this solution satisfies the equations generated by Kirchhoff's laws, it is incorrect because I_4 is negative an impossibility in an ideal diode. We must therefore rework our solution taking $I_4 = 0$. The equations for the three other currents produced by using KVL are still valid, but the KCL equation now reads $I_1 = I_2 + I_3$, giving

$$(7 - V)/4 = (V + 1)/5 + (V - 1)/10$$

whence $V = 3$ V and $I_1 = 1$ A, $I_2 = 0.8$ A and $I_3 = 0.2$ A: this time all the currents are positive and the solution is therefore legitimate.

Example 2.10

In the circuit of figure 2.8 the diodes are all identical having a dynamic resistance of 0.5 Ω and $V_{AK} = 1$ V at a forward current of 1 A. Find all the currents and the voltages across the diodes.

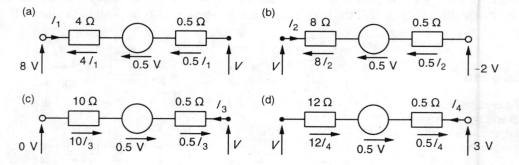

Figure 2.11 The circuit branches of figure 2.8 with diodes modelled as in figure 2.7c

The diode model used is that of figure 2.7c, for which we must calculate V_{AK0}, the voltage drop of the diode at zero current. With a dynamic resistance of 0.5 Ω and a current of 1 A, the voltage drop across r_d is 0.5 V and so $V_{AK0} = 1 - 0.5 = 0.5$ V. Thus the branch circuits of figure 2.11 result and we use KVL to find the currents as before.

In the branch of figure 2.11a we obtain

$$8 = 4I_1 + 0.5 + 0.5I_1 + V \quad \Rightarrow \quad I_1 = (7.5 - V)/4.5$$

From figure 2.11b

$$V = 8I_2 + 0.5 + 0.5I_2 + (-2) \quad \Rightarrow \quad I_2 = (V + 1.5)/8.5$$

Then from figure 2.11c

$$V = 0.5I_3 + 0.5 + 10I_3 + 0 \quad \Rightarrow \quad I_3 = (V - 0.5)/10.5$$

And finally from figure 2.11d

$$3 = 0.5I_4 + 0.5 + 12I_4 + V \quad \Rightarrow \quad I_4 = (2.5 - V)/12.5$$

KCL gives $I_1 + I_4 = I_2 + I_3$ as before, so that

$$(7.5 - V)/4.5 + (2.5 - V)/12.5 = (V - 0.5)/10.5 + (V + 1.5)/8.5$$

whence $V = 3.374$ V and again I_4 is negative, an impossibility. Setting $I_4 = 0$ and repeating the analysis, KCL now becomes $I_1 = I_2 + I_3$, the currents being given by the same expressions as before, so that

$$(7.5 - V)/4.5 = (V - 0.5)/10.5 + (V + 1.5)/8.5$$

Solving, $V = 3.534$ V, hence $I_1 = 0.881$ A, $I_2 = 0.289$ A and $I_3 = 0.592$ A.

The voltages across the diodes can now be calculated from $V_{AK} = V_{AK0} + I_D r_d = 0.5 + 0.5 I_D$. This makes $V_1 = 0.5 + 0.5 \times 0.881 = 0.9405$ V, $V_2 = 0.5 + 0.5 \times 0.289 = 0.6445$ V and $V_3 = 0.5 + 0.5 \times 0.592 = 0.796$ V. We cannot find V_4 in this way because it is reverse biased, the current is zero and the diode is an open switch. But by KVL, $3 = V_4 + V$, whence $V_4 = 3 - V = 3 - 3.534 = -0.534$ V.

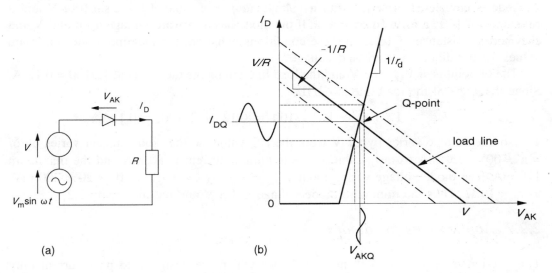

Figure 2.12 (a) A biased diode with an AC source (b) Load line, Q-point and waveforms for (a)

2.2.6 *Load lines and DC bias*

If we consider the circuit of figure 2.12a, in which an AC voltage source is biased by a DC voltage source, so that the diode always conducts, then we can solve graphically for the diode current and voltage. Ignoring the AC part for the moment, $V = V_{AK} + IR$, by KVL, which can be written as

$$I = (-1/R) V_{AK} + V/R$$

$$(y = m \ x \ + \ c)$$

Plotting this line on a graph of I against V_{AK} gives a straight line of slope $-1/R$ and intercept V/R on the I-axis, as shown in figure 2.12b. This is the locus of all possible values of I and V_{AK}, no matter what diode is used, and is known as the *load line*. When the diode's *I-V* characteristic is plotted on the same graph, its intercept with the load line defines the value of V_{AK} and I_D for the DC bias with no AC component. These values are known as the *quiescent* conditions of the circuit and are given an additional Q subscript: V_{AKQ} and I_{DQ}. The load line's intercept with the diode characteristic is called the *Q-point*.

When AC is superimposed on the DC, then the maximum supply voltage is $V + V_m$ and the minimum $V - V_m$. These voltages correspond to two lines either side of the load line, shown dot-dashed in figure 2.12b, and from their intercepts on the diode characteristic we can deduce the maximum and minimum diode current and voltage. If the AC excursions from the Q-point are small, so that the diode characteristic is approximately a straight line over this range, then the resulting diode current and voltage waveforms will be sinusoidal, but with a large DC component.

Example 2.11

A diode is connected in series with a voltage source of value $12 + 2 \sin 100t$ V and a resistance of 100 Ω, as in figure 2.12a. If the diode has a forward voltage drop of 2 V and a dynamic resistance of 10 Ω in these conditions, what are the maximum and minimum values for the diode's current and voltage?

The Q-point is at $V_{AKQ} = 2$ V and the load line intercepts the I-axis at $12/100 = 0.12$ A. Since the slope of the load line is $-1/100$,

$$I_{DQ} = 12/100 - V_{AKQ}/100 = 0.12 - 2/100 = 0.1 \text{ A}$$

The source voltage varies by ± 2 V from the DC value, so the diode current varies by $\delta I = \pm 2/100 = \pm 20$ mA from I_{DQ}, that is the minimum current is 80 mA and the maximum 120 mA. The diode voltage varies about $V_{AKQ} = 2$ V by $\delta V = r_d \delta I = 10 \times 20 = 200$ mV, as $r_d = 10$ Ω. Thus the minimum diode voltage is 1.8 V and the maximum 2.2 V.

2.2.7 Applications for diodes

This section is concerned with normal diodes which are designed to pass current only when forward biased. Only the most important applications will be decribed here: voltage reference and stabilisation — for which mostly special types of diode are used — are described in the section on Zener diodes.

The half-wave rectifier

In its most basic form the half-wave rectifier consists of an AC voltage source in series with the diode and a resistive load as in figure 2.13a. In this circuit the diode is forward biased during the positive half wave from the source, and so allows current to pass. The diode is also said to have turned on or to be conducting.

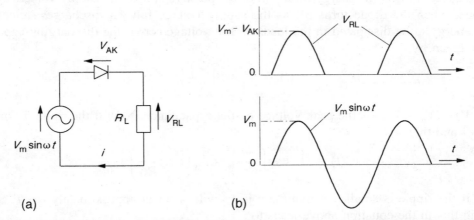

(a)

(b)

Figure 2.13 (a) The unsmoothed half-wave rectifier (b) Load and source voltage waveforms

The voltage across the load is the same as that from the supply apart from the voltage drop, V_{AK}, across the diode. Unless V_s is small, V_{AK} can be ignored. Then during the negative half cycle the diode is reverse biased, does not conduct and $V_{AK} = -V_s$. The peak-to-peak *ripple* in the load voltage (and current) is $V_m - 0.7$ V. Defining the relative peak-to-peak ripple by

$$v_{rpp} = \frac{V_{max} - V_{min}}{V_{max} + V_{min}} \times 100\%$$

we can see that in this case the ripple is almost 100%. Figure 2.13b shows the voltage waveforms. Generally ripple is not wanted and it must be reduced to tolerable levels by storing energy for release when the diode is not conducting. In circuits consuming relatively little power, capacitors are used for this purpose, but in high-power circuits inductors are preferred. When a capacitor is used for energy storage it must be placed in parallel with the load whose voltage is to be smoothed as in figure 2.14a.

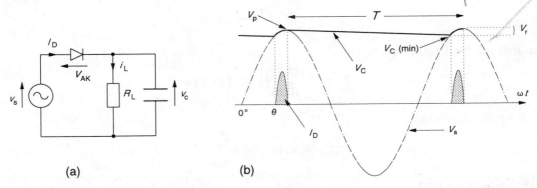

(a)

(b)

Figure 2.14 (a) A capacitively-smoothed half-wave rectifier (b) The voltage and current waveforms

The capacitor absorbs energy for a part of the positive half of the supply voltage cycle and then, when the diode turns off as the supply voltage falls, it discharges, releasing stored energy to be dissipated in the load, R. The voltage across the discharging capacitance is given by

$$v_C(t) = V_p \exp\left(\frac{-t}{RC}\right)$$

where V_p $(= V_m - V_{AK})$ is the peak voltage on the capacitance. Now if the ripple is small, $RC \gg t$ and then

$$v_C(t) = V_p \exp\left(\frac{-t}{RC}\right) \approx V_p\left(1 - \frac{t}{RC}\right)$$

Again if the ripple is small, the minimum in v_C will occur at approximately $t = T$. Substituting this in the equation above leads to

$$v_C(\text{min}) \approx V_p\left(1 - \frac{T}{RC}\right) = V_p\left(1 - \frac{1}{RCf}\right)$$

But the ripple is $V_p - v_C(\text{min})$, that is

$$V_r = V_p - v_C(\text{min}) \approx V_p - V_p\left(1 - \frac{1}{RCf}\right) = \frac{V_p}{RCf}$$

Example 2.12

A capacitively-smoothed half-wave rectifier operates from a 240V, 50Hz supply which delivers 1 kW to a resistive load. What is the load resistance neglecting the ripple and diode voltage drop? If the peak-to-peak ripple is to be 4%, what is the value of smoothing capacitance required? What are the maximum and minimum of the energy stored in the capacitor? What is the peak current through the diode? What is the maximum power dissipated in the diode if the diode's forward voltage drop is 0.8 V?

The effective voltage at the load is $240\sqrt{2}$ V, the peak of the supply, if we neglect the ripple and the diode drop. Thus the power in the load is

$$P = \frac{(240\sqrt{2})^2}{R} = 1000$$

so that the load resistance is

$$R = \frac{(240\sqrt{2})^2}{1000} = 115 \ \Omega$$

4% ripple means that

$$V_r = 0.4V_p = \frac{V_p}{RCf}$$

from which we find

$$C = \frac{1}{0.04Rf} = \frac{1}{0.04 \times 115 \times 50} = 4350 \ \mu F$$

The maximum energy stored in the capacitor is $\frac{1}{2}CV_p^2$, where

$$V_p = V_m - V_{AK} = 240\sqrt{2} - 0.8 = 338.6 \text{ V}$$

Thus the maximum stored energy is

$$0.5 \times 4.35 \times 10^{-3} \times 338.6^2 = 258 \text{ J}$$

The minimum voltage is $V_p - V_r = 0.96V_p$ and so the minimum energy stored is $0.96^2 \times 258 = 238$ J. The energy loss per cycle is then 20 J and the load power must be $20 \times 50 = 1000$ W since this energy is lost in the load f times per second, which is the correct value.

The diode conducts at an angle

$$\theta = \sin^{-1}(V_{Cmin}/V_p) = \sin^{-1}(1 - V_r/V_p) = \sin^{-1}0.96 = 73.7°$$

as shown in figure 2.14b. It switches off just after the peak voltage is attained (because the exponential decay of capacitor voltage is slower at first than the fall of the supply voltage) that is at $\theta \approx 90°$. Thus the diode conducts for $90° - 73.7° = 16.3°$ out of a $360°$ cycle, which is $0.045T$. If the diode and load currents are constant, then the diode current must be $I_L/0.045 = 22I_L$. And as

$$P_L = I_L^2R \quad \Rightarrow \quad I_L = \sqrt{(P_L/R)} = \sqrt{(1000/115)} = 2.95 \text{ A}$$

and while conducting the average diode current is $22 \times 2.95 = 65$ A. The average diode power loss during conduction is $V_{AK}I_D = 0.8 \times 65 = 48$ W. The diode current is not really constant during conduction and the maximum current and maximum power are about 50% larger. The peak power loss in the diode will then be $1.5 \times 48 = 72$ W. However, the average diode power over a cycle is 22 times less than the average during conduction, or $48/22 = 2.2$ W. The diode's power rating need be little more than this.

The full-wave rectifier

Making use of only the positive half cycle of the supply frequency means that the ripple and diode current rating are both double what they could be were both halves of the cycle as in the full-wave rectifier of figure 2.15.

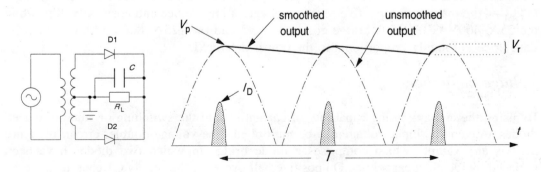

Figure 2.15 A smoothed, full-wave rectifier using a centre-tapped transformer, and the voltage and current waveforms. I_D is shown only for the capacitively-smoothed circuit

In this particular type a transformer with a centre-tapped secondary is used. The load is connected to the centre of the secondary and grounded at that end. For smoothing a capacitor is placed in parallel with the load to give the waveform shown also in this figure. Because the capacitor now discharges for only half a cycle instead of a whole cycle the ripple is halved for a given capacitance, as is the average current through the diodes; however, two diodes are now required. Full-wave rectification is often done with a diode bridge using four diodes, discussed next.

The diode bridge rectifier

The diode bridge, shown in figure 2.16, obviates the need for a centre-tapped transformer by allowing current to flow alternately through the positive and negative supply terminals.

Figure 2.16

The diode bridge rectifier. Here it is shown without smoothing. A smoothing capacitor would be placed across AB

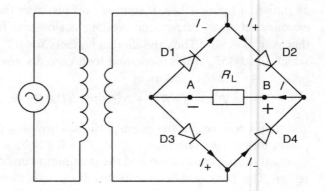

When the supply goes positive, I_+ flows through D2 and D3 while D1 and D4 are reverse biased and so turned off. Then on the next half cycle the supply polarity reverses, and I_- flows through D4 and D1 while D2 and D3 turn off. The current flows through the load, R_L, in the same direction throughout so it is DC. Terminal B is always positive while A is always negative with respect to B. The ripple can be reduced to any desired level using a parallel capacitance and is half that of the half-wave rectifier for the same value of capacitance. Now, however, there are two diode drops, not one and $V_p = V_m - 2V_{AK}$. Bridge rectifiers are available as single packages, for example the DF04M is a 1A, 400V (V_{RRM} — the maximum repetitive reverse voltage rating) device and costs only 40p, while the 25A, 400V GBPC2504 bridge costs about £3 and the 125A, 800V PSB125/08 costs £50. Price is roughly proportional to the power delivered.

Voltage multipliers

By using the energy-storing capability of capacitors and the switching of suitably placed diodes a given AC input voltage can be multiplied. The voltage-doubler circuit of figure 2.17 is an example which comprises a diode bridge in which two diodes have been replaced by identical capacitors. On positive half cycles of the supply C1 charges up to V_C ($= V_m - V_{AK}$) and then C2 charges up on negative half cycles so that the load voltage is $2V_C \approx 2V_m$. The ripple is controlled by making the capacitance sufficiently large.

Figure 2.17

A voltage doubler which converts AC to
DC of approximately twice the peak AC
voltage

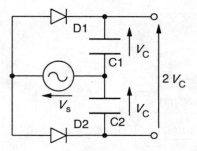

Circuit protection

There are two types of circuit protection diodes: those which limit the voltage at a point
to one or more diode drops and those which allow stored energy to discharge harmlessly.
The former use is illustrated in figure 2.18a which restricts the input voltage to $\pm V_{AK}$.

Figure 2.18b shows a circuit-protection diode which allows the energy stored in the
inductance to discharge through the diode when the circuit is broken. Were the diode not
provided, the voltages produced by a sudden open circuit could cause damaging arcing at
the switch, or might produce insulation breakdown between adjacent turns of the coil. The
induced voltage, V_L, is LdI/dt and this is equal to V_{AK} if there is no resistance in the circuit.

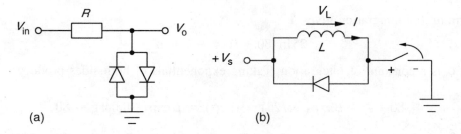

(a)   (b)

Figure 2.18 (a) A diode limiter. V_o is restricted to about ± 0.7 V with silicon diodes (b) A catch diode.
When the switch opens the current discharges harmlessly through the diode

Example 2.13

Find an expression for the change of current with time when an inductance of 250 mH,
carrying an initial current of 40 A, discharges through a diode with a constant forward
voltage drop of 0.8 V and a series resistance of 10 mΩ, as shown in figure 2.19a.

Note that the voltage on the inductor, v_L, (like that on the resistance) must point in the
opposite direction to the current, though eventually v_L will turn out to be negative. The
diode can be modelled as a constant voltage source which opposes the flow of current.
KVL gives

$$v_L + v_{AK} + iR = 0$$

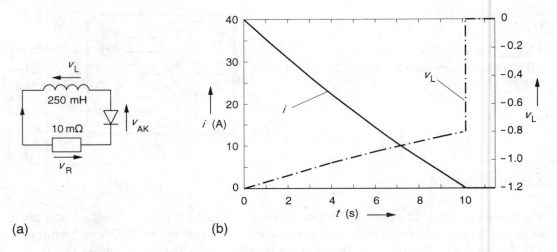

(a)   (b)

Figure 2.19 (a) The circuit for example 2.13 (b) i-t and v_L-t graphs

But $v_L = L di/dt = 0.25 di/dt$, $v_{AK} = 0.8$ V and $iR = 0.01i$. Substituting these into the equation above leads to

$$0.25 di/dt + 0.8 + 0.01i = 0 \quad \Rightarrow \quad \int \frac{0.25 di}{0.8 + 0.01i} = \int \frac{25 di}{80 + i} = -\int dt$$

Performing the integration gives

$$25 \ln (80 + i) = -t + \alpha$$

where α is a constant of integration. Taking exponentials of both sides produces

$$80 + i = \beta \exp(-t/25) \quad \Rightarrow \quad i(t) = \beta \exp(-0.04t) - 80$$

where β, another constant, must be found from the initial condition that $i(0) = 40$ A, which leads to $\beta = 120$ A and the final equation for the current

$$i(t) = 120 \exp(-0.04t) - 80 \text{ A}$$

Now we should check that the final conditions are satisfied, which usually means that when $t = \infty$, $i = 0$; however, this time that is not the case, since the current falls at an ever-decreasing rate until $v_L = -0.8$ V $= -v_{AK}$. At this point the current is zero and the time can be found by setting

$$i(t) = 120 \exp(-0.04t) - 80 = 0$$

Solving this gives $t(i = 0) = 10.14$ s.

The voltage across the inductance is found from

$$v_L = 0.25 \frac{di}{dt} = -0.04 \times 0.25 \times 120 \exp(-0.04t) = -1.2 \exp(-0.04t) \text{ V}$$

v_L is negative as the current is decaying. Graphs of $i(t)$ and $v_L(t)$ are shown in figure 2.19b.

DC restoration

In a circuit such as that of figure 2.20a, the DC level of the original signal is lost after the capacitor, but can be restored by a diode placed as shown. The restoration is incomplete because of the forward voltage drop across the diode; full restoration needs two diodes with DC bias as shown in figure 2.20b.

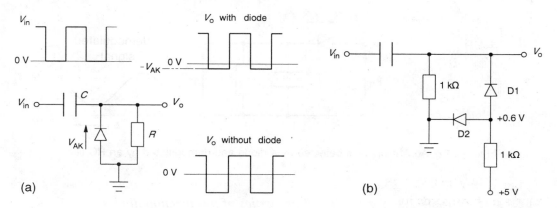

Figure 2.20 (a) Partial DC restoration when DC is lost after a capacitor (b) Full restoration requires a further compensation diode and a DC supply

Logarithmic conversion

This makes use of the rectifier equation so that in the circuit shown in figure 2.21a, the voltage across the diode has a component proportional to the logarithm of the resistor voltage. If the signal, V_{in}, is smaller than 0.6 V, then a compensation diode and a DC supply is required as in figure 2.21b. In general, since V_{in} might be negative, DC bias is required either at the input or on D2. In practice log converters are difficult to make so that accurate conversion is maintained over a wide range of inputs and they therefore require much more complicated circuitry than figure 2.21 indicates.

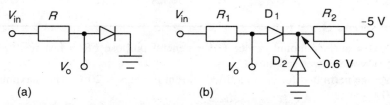

Figure 2.21 (a) A prototype log converter (b) A log converter with compensation for diode drop

Demodulation and signal detection

When an AM (amplitude modulated) signal is received it must be demodulated so that the signal impressed on the carrier wave is recovered as cleanly as possible. The incoming signal from the receiving antenna of figure 2.22 is passed through a detector diode which charges a capacitor-resistor combination. The time constant of this is adjusted so that $f_m \approx 1/2\pi RC \ll f_c$, then the carrier frequency, f_c, is suppressed and the output is only at the modulating frequency, f_m.

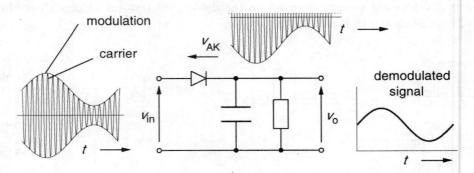

Figure 2.22 An AM signal is detected by a diode and demodulated by an RC filter

Table 2.2 *Some selected properties of p-n junction diodes*

Number	Type[A]	$I_F(AV)$[B]	V_{RRM}[C]	$I_r(max)$[D]	t_{rr}[E]	$C(0\ V)$[F]	Dimensions[G]	Price[H]
1N4148	signal	0.075	75	25 nA	8 ns	4	4×1.6ϕ	£0.02
LL4148	SMD	0.15	100	25 nA	8 ns	4	3.5×1.5ϕ	£0.05
BAS16W	SMD	0.25	75	25 nA	6 ns	4	2×1.25×1	£0.05
1N4004	GP	1	400	10 μA	-	-	5×3ϕ	£0.05
HER103	FR	1	200	5 μA	50 ns	-	5×2.7ϕ	£0.13
BA155	GP	0.1	150	1 μA	300 ns	2	4×2ϕ	£0.16
FR303	FR	3	200	5 μA	150 ns	-	9.5×5ϕ	£0.32
UF5404	SMPS	3	400	5 μA	50 ns	-	9.5×5.3ϕ	£0.51
RURP3060	SMPS	30	600	-	55 ns	-	TO220	£3.35
D798N	power	800	1200	-	-	-	14×50ϕ	£90.00
BB212	varicap	-	12	50 nA	-	550	TO92	£2.34
						17 (8 V)		

Notes: [A] SMD = surface-mount device GP = general purpose FR = fast recovery SMPS = for switch-mode power supplies

[B] Average permitted forward rectified current in amps at 20°C. The maximum constant DC is 50% greater.

[C] Repetitive maximum reverse voltage. Breakdown voltage is about 50% greater.

[D] Maximum reverse current at 20°C and −10 V.

[E] Typical reverse recovery times under various test conditions

[F] Typical junction capacitance in pF

[G] For rectangular packages L × W × H in mm. ϕ = cylinder diameter.

[H] Price for single quantities.

2.2.8 *Practical diodes*

The requirements for a diode depend on its purpose of course, but generally speaking the price increases with the maximum allowable forward rectified current, $I_F(AV)$, and the maximum allowable repetitive reverse voltage, V_{RRM}, or peak inverse voltage (PIV). The product of these two, $V_{RRM}I_F(AV)$, is proportional to the power-handling capability of the diode. Table 2.2 gives some data for a random selection of diodes and rectifiers, the latter term being reserved for diodes used in high-power applications. Most of the data refer to operation at 20°C and significant degradation in performance may occur at higher temperatures. The maximum operating temperature is about 150°C: all are silicon diodes.

2.3 Zener diodes

A Zener diode is made just like a 'normal' diode from a p-n junction in a semiconductor, usually silicon. Unlike ordinary diodes, however, Zener diodes are designed to conduct at a specified reverse bias. The symbol for a Zener diode is shown in figure 2.23.

The backward-directed line on the cathode indicates that a Zener diode conducts in reverse bias, but it will also conduct in forward bias just like any p-n junction, as shown by the current-voltage characteristic. At high doping levels the junction's breakdown mechanism is one of quantum mechanical tunnelling, but at moderate to low doping levels it is one of avalanche multiplication, the usual breakdown process in reverse-biased junctions. Unless the reverse current is restricted — by a series resistor for example — the diode will be destroyed by heat. In small Zeners the power rating may be less than $\frac{1}{2}$W, which implies a fairly small current, especially if the Zener voltage is high.

Figure 2.23

A Zener diode and its current-voltage characteristic. Note that in normal use the current flow is the reverse of that shown

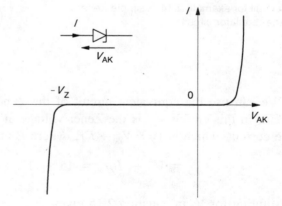

Figure 2.24b shows a model for a Zener diode which comprises a voltage source, V_{Z0}, in series with a resistance, r_z, representing the diode's dynamic resistance in reverse bias. The dynamic resistance is the reciprocal of the slope of the V-I characteristic, $(dI/dV)^{-1}$, (shown in figure 2.24a) and is usually given at a specified current. Depending on r_z, Zeners can be fairly good voltage regulators when the power requirements are small, as will be seen in the next example.

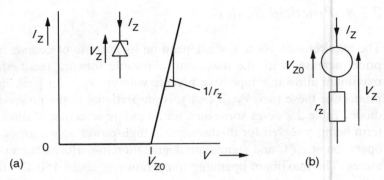

Figure 2.24

(a) Linearised *I–V* graph for
a Zener diode (b) The circuit
model for (a)

Example 2.14

In the circuit of figure 2.25 the Zener diode has a maximum power rating of 650 mW and
has a Zener voltage of 15 V at a reverse current rating of 15 mA. The load current, I_L,
varies from 10 to 25 mA, while the supply voltage varies independently from 25-30 V.
Calculate a suitable value for R_s, and hence find the maximum and minimum voltages
across the load if the diode's dynamic resistance, r_z, is 12 Ω at the rated current.

We must first find an expression for the Zener current, since from this we can
determine the load voltage and the value required for R_s. By KVL in figure 2.25

$$V_s = V_R + V_Z = I_s R_s + V_Z = (I_Z + I_L)R_s + V_Z \qquad (2.15)$$

since by KCL, $I_s = I_Z + I_L$.

Figure 2.25

The circuit for example 2.14, a simple Zener
voltage-regulator circuit

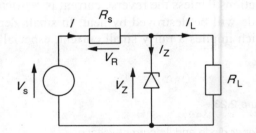

We can use the equivalent circuit of the Zener as in figure 2.24 to find the Zener
current. In this circuit V_{Z0} is the Zener voltage at zero current, not the Zener voltage at
rated current, which is $V_Z = V_{Z0} + I_Z r_z$, where I_Z is the rated Zener current, 15 mA. Thus

$$V_{Z0} = V_Z - I_Z r_z = 15 - 15 \times 10^{-3} \times 12 = 14.82 \text{ V}$$

Substituting for V_Z in equation 2.15 gives

$$V_s = (I_Z + I_L)R_s + V_{Z0} + I_Z r_z = I_L R_s + V_{Z0} + I_Z(R_s + r_z)$$

From this we can determine I_Z and thence R_s and V_Z ($= V_L$):

$$I_Z = \frac{V_s - V_{Z0} - I_L R_s}{R_s + r_z} = \frac{V_s - 14.82 - I_L R_s}{R_s + 12} \tag{2.16}$$

We can now see that I_Z will be a maximum when V_s is a maximum (30 V) and I_L a minimum (10 mA). Substituting these values into equation 2.16 gives

$$I_Z(\text{max}) = \frac{30 - 14.82 - 10 \times 10^{-3} \times R_s}{R_s + 12} = \frac{15.18 - 0.01 R_s}{R_s + 12} \tag{2.17}$$

Now the maximum permissible Zener current can be found from $P_Z(\text{max}) = V_Z \times I_Z(\text{max})$, taking $V_Z = 15$ V, which gives $I_Z(\text{max}) = 650/15 = 43$ mA. It would be unwise to design a circuit to operate at the maximum Zener rating: a suitable maximum current might be 30 mA, allowing a 40% safety margin (we need to know the operating temperature to be sure that this margin was enough). R_s is now found by putting $I_Z = 30$ mA into equation 2.17

$$\frac{(15.18 - 0.01 R_s)}{(R_s + 12)} = 30 \times 10^{-3} \quad \Rightarrow \quad R_s = 370 \ \Omega$$

Having decided on a value for R_s we can find the minimum Zener current, which occurs when V_s is a minimum (25 V) and I_L a maximum (25 mA):

$$I_Z(\text{min}) = \frac{V_s(\text{min}) - V_{Z0} - I_L(\text{max})R_s}{R_s + r_z} = \frac{25 - 14.82 - 0.025 \times 370}{382} = 2.4 \ \text{mA}$$

We then calculate the maximum and minimum Zener voltages:

$$V_Z(\text{max}) = V_{Z0} + I_z r_z = 14.82 + 30 \times 10^{-3} \times 12 = 15.18 \ \text{V}$$

And the minimum is $14.82 + 2.4 \times 10^{-3} \times 12 = 14.85$ V. The load voltage is regulated to about $\pm 1\%$: quite a good performance from a device costing only 5p.

Besides the variation in Zener voltage with current, there is also a variation in Zener voltage with temperature, which is a complicated function of nominal breakdown voltage (determined by the doping levels either side of the junction) and the actual current carried. Roughly speaking, the temperature coefficient, θ_Z, is about +0.1% of the nominal Zener voltage per °C for breakdown voltages above 9 V. Below this, the temperature coefficient falls to zero for a 5V Zener (but is current-dependent) and is negative below this. However, it is possible to achieve a temperature coefficient which is practically zero by placing an ordinary diode in series with the Zener. The 1N827 Zener, for example, has a quoted θ_Z of +0.001%/°C, but costs about £4 in single quantities.

A bigger practical problem with Zeners than temperature drift is the variation from the nominal voltage, which is usually 5%, that is a 10V Zener could actually be a 9.5V or a 10.5V Zener. An example will show how far from the expected voltage the Zener can be, even though it operates within specifications.

Example 2.15

A $\frac{1}{2}$W Zener diode has a nominal breakdown voltage of 12 V, at a current of 5 mA and a junction temperature of 25°C, with a manufacturing tolerance of ±5%. The diode has a dynamic resistance of 25 Ω and a voltage temperature coefficient (tempco) of +0.1%/°C. It is connected into a circuit which produces a thermal resistance of 0.3°C/mW, while the ambient temperature is 30°C. If the operating current is from 2-20 mA, what is the maximum range of operating voltages?

The minimum operating voltage will be for a Zener at the low limit from the nominal voltage, $0.95 \times 12 = 11.4$ V, which carries the minimum current of 2 mA. We can find the Zener voltage at zero current, V_{Z0}, which is $V_Z - I_Z r_z = 11.4 - 5 \times 10^{-3} \times 25 = 11.25$ V. This gives a Zener voltage with a current of 2 mA of

$$V_Z(\text{min}) = V_{Z0} + I_Z r_z = 11.25 + 0.002 \times 25 = 11.275 \text{ V}$$

The Zener's power dissipation is

$$P_Z = V_Z I_Z = 11.275 \times 2 = 22.55 \text{ mW}$$

so the junction temperature increase from ambient is $0.3 \times 22.55 = 6.8$°C, that is the junction temperature is 36.8°C, which is 11.8°C above the rated temperature of 25°C. Thus the increase in Zener voltage is

$$\Delta V_Z = 11.8 \times 0.1\% = 1.18\% = 0.0118 \times 12 = 0.14 \text{ V}$$

taking a percentage of the nominal Zener voltage. The eventual minimum Zener voltage is therefore $11.3 + 0.14 = 11.44$ V.

The maximum Zener voltage will be found with maximum current in a Zener which is 5% above nominal value, that is $V_Z = 1.05 \times 12 = 12.6$ V. For this diode $V_{Z0} = 12.6 - 0.005 \times 25 = 12.475$ V and the operating voltage is

$$V_Z(\text{max}) = V_{Z0} + I_Z r_z = 12.475 + 0.02 \times 25 = 12.975 \text{ V}$$

But now the power dissipation is $12.975 \times 20 = 259.5$ mW, so the junction temperature is $259.5 \times 0.3 = 77.85$°C above ambient, or 82.85°C above the rated temperature, giving a Zener voltage increase of

$$\Delta V_Z = 82.85 \times 0.1\% = 8.285\% = 0.08285 \times 12 = 0.994 \text{ V}$$

The actual maximum Zener voltage then becomes $12.975 + 0.994 = 13.97$ V. The total range of Zener voltage is from 11.44 V to 13.97 V, the maximum deviation from nominal being +1.97 V or 16%. This is made up of 0.6 V from the manufacturing tolerance (5%), 0.375 V (3.1%) from the dynamic resistance and 0.99 V (8.3%) from the voltage-temperature coefficient.

Although it is possible to buy diodes with very low voltage-temperature coefficients, it is hard to find one with a low manufacturing tolerance. However, there is a newer class of voltage-reference devices know as *bandgap references*, some of which have closely-controlled breakdown voltages. They are not discrete devices but integrated circuits and are discussed later in section 3.10.7; they cost from £1 to £3 each.

2.4 Schottky diodes

Schottky, or Schottky barrier, diodes[2] make use of the rectifying properties of a metal-semiconductor contact — usually aluminium and silicon — in order to achieve a relatively low forward-voltage drop at modest forward currents. The penalty paid for this is usually a greater reverse current flow than in p-n junction diodes. This is not a great handicap except for applications with stringent conditions on the leakage current. Schottky diodes are given a special symbol shown in figure 2.26. In addition to, and perhaps more importantly than, their low forward voltage drops, Schottky diodes have a very fast reverse recovery time, t_{rr}, since they do not store minority carriers like p-n junction diodes. They are therefore used for fast switching, heavy current duties such as switch-mode power supplies (SMPS), as well as in fast logic circuits (see chapter 9).

Table 2.3 gives a few data for some randomly-chosen Schottky diodes.

Table 2.3 *Schottky diodes*

Device	$I_F(AV)^A$	V_{RRM}	V_F^B	Package^C	Price^D
1N5711	0.01	70	0.41	$4.3 \times 1.9\phi$	£0.40
1N5817	1	20	0.38	$4.3 \times 1.9\phi$	£0.50
SB130	1	30	0.5	$5.2 \times 2.7\phi$	£0.50
BAT254^E	0.2	30	0.45	SOD110^F	£50
SS32^E	3	20	0.5	$4.5 \times 2.8 \times 2.3$	£1.00
SB530	5	30	0.45	$6 \times 3.5\phi$	£2.30
MBR1035	10	35	0.57	TO220	£3
MBR3535	35	35	0.64	$10 \times 10\phi S$	£9
240NQ045	240	45	0.55	$19 \times 19 \times 13$	£19

Notes: ^{A} Average forward current in A. ^{B} Forward voltage drop at $I_{F(AV)}$
^{C} ϕ Cylindrical, length $\times$ diameter in mm. S stud mounting. The
TO220 package is shown in figure 3.59 ^{D} For single quantities
^{E} Surface-mount technology (SMT) device ^{F} ceramic flat package

A metal-semiconductor contact has a voltage-current relationship which obeys the Richardson equation,

$$I = A^* A_c T^2 \exp\left[\frac{q(V - \phi)}{kT}\right] \tag{2.18}$$

where A^* = effective Richardson constant ≈ 1 MA/m^2/K^2, A_c = cross-sectional area of contact and $\phi = 0.7$ V for an aluminium or tungsten contact to silicon. Figure 2.26 shows a plot of equation 2.18 for a small, silicon, Schottky diode. For a small diode with $A_c = 0.25$ mm^2 equation 2.18 gives $V = \phi - 0.4$ V at 300 K and $I = 10$ mA, that is $V = 0.3$ V, roughly the same as a germanium diode. If lower voltage drops are required for a given forward current then one may increase A_c, hence the device size and cost.

---

[2] Schottky published his theory of metal-semiconductor contacts in 1938.

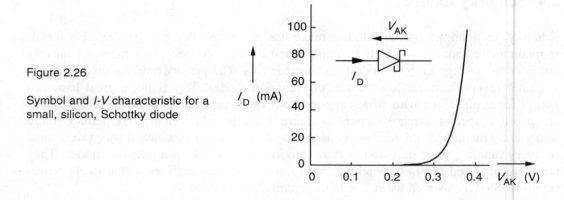

Figure 2.26

Symbol and *I-V* characteristic for a
small, silicon, Schottky diode

Example 2.16

Estimate the contact area of the SBR530 Schottky-barrier rectifier in table 2.3, assuming
it has an aluminium-silicon contact with $\phi = 0.7$ V and $A^* = 300$ kA/m^2/K^2. The data are
for 300 K.

We use equation 2.18 and from table 2.3 we find $I = 5$ A when $V = 0.45$ V. Thus

$$A_c = \left(\frac{I}{A^*T^2}\right)\exp\left[\frac{q(\phi - V)}{kT}\right]$$

$$= \frac{5}{3 \times 10^5 \times 300^2}\exp\left(\frac{1.6 \times 10^{-19}(0.7 - 0.45)}{1.38 \times 10^{-23} \times 300}\right) = 2.9 \times 10^{-6}\ \text{m}^2$$

This would correspond to a circular contact diameter of about 1.9 mm, and fits in with a
package diameter of 3.5 mm.

2.5 Optical diodes

Optical diodes can be divided into several groups as shown in table 2.4. Light-emitting
diodes or LEDs are designed to emit electromagnetic radiation in the wavelength range
from about 1.6 µm down to 0.45 µm, that is from the near infra-red region down through
the visible region from red (0.7 µm) to blue (0.45 µm). The infra-red emitters are used in
television and other remote, short-distance controllers, while the optical emitters are used
in indicator lamps and displays. Fibre-optic transmitters are usually laser diodes as these
have much higher intensities and efficiencies than LEDs, but they are considerably dearer.
Many laser diodes are used in CD players as they can be made relatively cheaply (about
£3 in large quantities), are very small (typical dimensions in mm are $1 \times 2 \times 5$) and very
reliable.

Table 2.4 *Optical diodes*

Device	Type	Purpose	Colour	λ (nm)[A]	Material	Cost[B]
BPW41D	PIN	IR detector	IR	820-1040	Si	£0.60
RS194-098	solar cell	battery charger	daylight	400-1000	Si	£10
LD274	LED	IR emitter	IR	950	GaAs	£0.19
HFBR-1402	laser	fibre-optic Tx	IR	820	GaAs	£30
HLMP-D150	LED	lamp	red	637	(Ga,Al)As	£0.30
HLMA-CL00	LED	lamp	amber	592	(In,Ga,Al)P	£0.67
HLMP-4740	LED	lamp	green	569	(Ga,Al)As	£0.24
HDSP-5601	LED	7-segment display	green	555	Ga(As,P)	£1.50
HLMP-DB15	LED	lamp	blue	480	SiC	£0.68
HLMP-4000	LED	bicolour lamp	green/red	569/626	(Ga,Al)As	£0.96
RS589-187	LED	5×7 dot-matrix	red	626	(Ga,Al)As	£3.99
RS211-755[C]	LED	message display	tricolour	569-626	(Ga,Al)As	£399

Notes: [A] wavelength range or that of peak response [B] Single quantities [C] 14 characters, programmable 5 × 7 red/yellow/green dot matrix, supplied with IR remote keyboard input

2.5.1 *Principles of operation*

In a LED the current flowing across a p-n junction causes electrons from the n-side to cross the junction to the p-side, where they recombine with holes and in doing so cause the emission of a photon. Holes from the p-side likewise cross to the n-side, there to recombine with electrons and produce a photon. The process is illustrated in the energy diagram of figure 2.27.

If the hole or electron crosses the whole of the bandgap before recombining then the photon's energy is simply E_g, the bandgap energy. If they can combine with ionised impurity centres located within the bandgap (and produced by suitable doping) then the photon's energy will be less than E_g. Now the energy of a photon is related to its frequency by

$$E_P = hf = hc/\lambda \tag{2.19}$$

where h = Planck's constant, 6.626×10^{-34} Js, c = speed of light in vacuo = 3×10^8 m/s and λ is the wavelength in vacuo. Thus if $E_P = E_g$ we find that the wavelength of the emitted light in nm is

$$\lambda = hc/E_g = 1242/E_g \quad \text{(nm)} \tag{2.20}$$

when E_g is in eV (1 eV = 1.6×10^{-19} J). For GaP with $E_g = 2.24$ eV, the wavelength for a photon coming from a transition across the whole of the bandgap is therefore 1242/2.24 = 554 nm, corresponding to the colour green when absorbed by the human retina.

Producing a photon is no guarantee that it will emerge from the LED: in fact a high proportion of photons do not, and the overall efficiency is very low — of the order of 0.01% (a tungsten filament lamp has an efficiency of about 2%). If the efficiency were 100% the light output of a LED would be 683 lumens/W, but in practice it is only about 0.1 lumens/W. (The lumen has replaced *candle power*, and is effectively a unit of power.)

Figure 2.27

An electron falling from the conduction band
to the valence band and recombining there
with a hole. The photon emitted has an
energy of $E_g = hf = hc/\lambda$

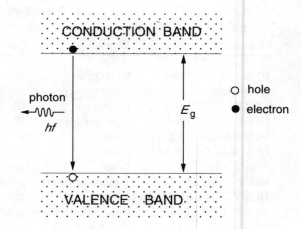

Example 2.17

Photons are emitted from a LED in a cone of semi-angle 12° as shown in figure 2.28. It
is found that the intensity of the emitted light is 240 mcd when the diode's current is
20 mA and its forward voltage drop is 2.2 V. What is the efficiency of the LED
approximately and what is the minimum wavelength of the emitted light? [1 candela (cd)
= 1 lumen (lm) per steradian (sr) = 1.464 mW/sr.]

Figure 2.28

The cone of light coming from a LED has an
area normal to the direction of propagation
of $\pi(r \sin \theta)^2$ at a distance, r

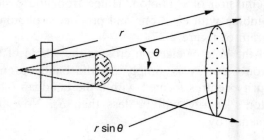

The candela is a unit of source intensity which must be multiplied by the solid angle
over which it emanates (assuming uniform intensity distribution) to find the light power
(known as the *luminous flux*) from the source. The solid angle in steradians subtended by
a cone of semi-angle θ, if θ is small, is $\pi(r \sin \theta)^2 \div 4\pi r^2 = \frac{1}{4}\sin^2\theta \approx \frac{1}{4}\theta^2$, where θ is in
radians. If $\theta = 12°$, that is $12\pi/180 = 0.2094$ rad, then the light is emitted in a solid angle
of 0.011 sr. If the intensity is 240 mcd, then the luminous flux emitted is 240×0.011 mlm
= 2.64 mlm, which converts to $2.64 \times 10^{-3} \times 1.464 - 10^{-3} = 3.9$ µW. But the input power
is $VI = 2.2 \times 20 = 44$ mW, making the overall efficiency

$$\eta = \frac{\text{light power out}}{\text{power in}} = \frac{3.9 \times 10^{-6} \times 100}{44 \times 10^{-3}} = 0.009\%.$$

The voltage drop is 2.2 V, which equates to a bandgap of about 1.8 eV, if we can take the voltage drop of a forward-biased diode to be $E_g/q - 0.4$ V, which is usual for silicon and germanium p-n junction diodes, depending on the doping levels and other factors, but will not be far out for LEDs. Thus the minimum wavelength is 1242/1.8 = 690 nm: it is likely to be a red LED.

2.5.2 Laser diodes

Laser action requires that the light emitted is coherent, that is the wavetrains emitted have definite phase relationships, which can only be obtained by stimulated emission. The stimulus is provided by photons produced in the same way as in LEDs, and though there is not much difference in principle between LEDs and laser diodes, a far higher current density is required for laser action. This presents certain practical problems, as does the requirement for a stable, concentrated beam of photons; in consequence the construction of laser diodes is more complicated and more costly than that of LEDs.

2.5.3 PIN diodes

The detection of photons requires their absorption by a semiconductor to create hole-electron pairs, which is accompanied by an increase in the current flowing across a reverse-biased p-n junction in PIN diodes. (The letters P, I, N refer to the doping of the layers used in the diode: P — p-type, I — intrinsic and N — n-type.) The photon can only be absorbed if the gap energy is smaller than the photon energy. Thus silicon, with a bandgap of 1.1 eV, cannot be used to detect infra-red photons below about 1242/1.1 = 1130 nm (using equation 2.19), or 1.13 µm. Using germanium ($E_g = 0.67$ eV) will extend the range to 1.85 µm. By using silicon-germanium 'alloys' (only metals form alloys and neither silicon nor germanium is a metal of course) it is possible to cover all this range with reasonable efficiency. Specially manufactured photodiode arrays are now used as sensors in solid-state spectrophotometers and other instruments. Photodiode detector arrays also find use as position sensors of very high accuracy.

2.5.4 Solar cells

Solar cells, like PIN diodes, must also absorb a photon to create a hole-electron pair and deliver energy to an external circuit. No junction bias is required for operation and, if left open circuit, a voltage will appear at the terminals which is approximately $E_g/q - 0.6$ V, about 0.5 V for silicon. Photons with more than the bandgap energy may be absorbed to produce a hole-electron pair, but any excess energy is lost as heat. Photons with less than the bandgap energy cannot produce hole-electron pairs and cannot deliver electrical power to an external circuit. It follows that solar cells have to be made of a material which is well-matched to the solar spectrum, and it turns out that the ideal bandgap is about the same as that in amorphous silicon (about 1.3 eV). Solar cells have an overall efficiency of about 10-15%. Given that the solar flux on a cloudless summer's day is about 1 kW/m^2, then it is clear that a small solar-powered calculator, operating from a solar cell whose area is about 1 cm × 3 cm, or 3×10^{-4} m^2, can only be charged at most at 0.3 mW, or at about 0.1 mA, given a charging voltage of 3 V. (See also problem 2.15.)

Suggestions for further reading

Semiconductor devices by S M Sze (John Wiley 1995)
Electronic engineering semiconductors and devices by J Allison (McGraw-Hill 1990)
Electronic materials by L A A Warnes (Macmillan 1994)

Problems

1 Silver has one free electron/atom and these have a mobility of 6.7×10^{-4} m^2/V/s at 300 K. If the r.a.m. of silver is 107.9 and its density is 10500 kg/m^3, what is the conductivity of silver at 300 K? Suppose a kg of silver costs the same as a kg of copper, which would you use for electrical power-transmission lines? *[62.8 MS/m]*

2 A tungsten filament in an electric light bulb is 1 m long and has a resistivity of 1 μΩm at its operating temperature. If it is to consume 100 W with a 240V supply, what is the diameter of the filament if it is of circular coss-section? Given that the free electron concentration in tungsten is 2×10^{29} m^{-3}, what is the drift velocity of the electrons in the filament? *[r = 23.5 μm, v_d = 7.5 mm/s]*

3 In a certain intrinsic semiconductor the electron/hole mobility ratio is 2:1 and it has a resistivity of 0.9 Ωm at 0°C. If the electron mobility is 0.38 m^2/V/s, what are the concentrations of electrons and holes at 0°C? If the bandgap energy is 0.7 eV, what will these be at 100°C, assuming unchanged mobilities? What is the intrinsic resistivity at 100°C? *[1.22 × 10^{19} m^{-3} at 0°C. 6.56 × 10^{20} m^{-3} at 100°C. 16.7 mΩm]*

4 The semiconductor material in problem 2.3 is doped p-type with 5×10^{20} acceptor atoms/m^3, all of which are fully ionised. What are the carrier concentrations at 0°C and at 100°C? What will the resistivity be at 0°C and at 100°C? (Use the law of mass action. Note that *n* cannot be neglected and that charge neutrality must be satisfied.)
[n = 2.98 × 10^{17} m^{-3}, p = 5 × 10$^{20·}$ m^{-3}. n = 4.52 × 10^{20} m^{-3}, p = 9.52 × 10^{20} m^{-3}. ρ_0 = 0.066 Ωm, ρ_{100} = 0.018 Ωm]

5 For the circuit of figure P2.5, find I_1 and I_2, given that the forward-biased diodes have a voltage drop of 0.7 V at all currents. *[I_1 = 0.14 A, I_2 = −0.39 A]*

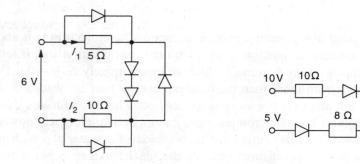

Figure P2.5   Figure P2.8

6 Repeat the previous problem but with the source polarity reversed.
[I_1 = −0.92 A, I_2 = 0.07 A]

7 Repeat problem 2.5, but now consider the diodes all to have dynamic resistances of 2 Ω and $V_{AK0} = 0.7$ V. $[I_1 = 0.176$ A, $I_2 = -0.266$ A$]$

8 In figure P2.8, what is the smallest value of R for which I_2 is zero? What are the currents then? What are the currents when $R = \infty$ and when $R = 0$? At what value of R will $I_3 = I_4$? (Take the forward voltage drops of all diodes to be 1 V and their dynamic resistances to be zero.)

$[30\ \Omega.\ I_1 = 0.5$ A, $I_2 = 0$, $I_3 = 0.4$ A, $I_4 = 0.1$ A. $I_1 = I_3 = 0.467$ A, $I_2 = I_4 = 0.$ $I_1 = 0.8$ A, $I_2 = 0.375$ A, $I_3 = 0.25$ A, $I_4 = 0.925$ A. $5.87\ \Omega]$

9 Repeat problem 2.8 for diodes with $V_{AK0} = 1$ V and dynamic resistances of 0.5 Ω.

$[R = 34.4\ \Omega.\ I_1 = 0.476$ A, $I_2 = 0$, $I_3 = 0.39$ A, $I_4 = 0.086$ A. $I_1 = I_3 = 0.42$ A, $I_2 = I_4 = 0.$ $I_1 = 0.725$ A, $I_2 = 0.308$ A, $I_3 = 0.263$ A, $I_4 = 0.77$ A. $5.4\ \Omega]$

10 In the circuit of figure P2.10, the diode has a characteristic given by $I = I_s \exp(33V)$ in forward bias, where $I_s = 6 \times 10^{-20}$ A. Replace the circuit to the left of the diode by its Thevenin equivalent. Use the result to draw the diode characteristic and load line. Hence determine the quiescent current and voltage across the diode. Calculate the dynamic resistance at the Q-point and determine the maximum and minimum diode currents and voltages (a) approximately, using the dynamic resistance and (b) exactly.

$[I_{DQ} = 11.95$ mA, $V_{AKQ} = 1.207$ V. $r_d = 2.53\ \Omega$ (a) 13.28, 10.62 mA; 1.2104, 1.2036 V (b) 13.264, 10.644 mA; 1.21022, 1.20355 V$]$

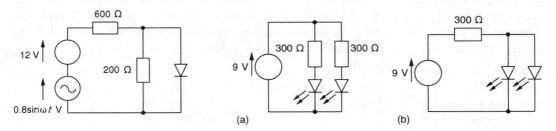

Figure P2.10   Figure P2.12

11 The *V-I* characteristic of the diode in problem 2.10 is dependent on temperature and can be written $I = I_s \exp(10^4 V/T)$. The reverse saturation current is given by $I_s = I_0 \exp(-E_g/kT)$, where $I_0 = \text{const.} = 23.15$ MA, $E_g = 1.6$ eV, k is Boltzmann's constant and T is the temperature in K. The diode's effective temperature is given by $T = T_A + P/1.5$, where T_A is the ambient temperature in K and P is the power in mW, dissipated at the Q-point. Calculate the maximum and minimum diode currents and voltages for $T_A = 10°C$ and $T_A = 40°C$. (Take the ice point to be 273 K.)

$[1.2257$ V, 1.2323 V; 10.47 mA, 13.13 mA. 1.1625 V, 1.1695 V; 10.91 mA, 13.58 mA$]$

12 In the circuits of figure P2.12, D1 and D2 are LEDs whose *I-V* relationship is $I = I_s \exp(40V)$, where I_s is 4×10^{-30} A for D1 and 5×10^{-31} A for D2. What are the brightness ratios of the diodes in each circuit if the brightness is proportional to the current?

$[(a)\ 1.007:1,\ (b)\ 8:1]$

13 In the circuit of figure P2.13, the first Zener has $V_{Z1} = 8$ V at $I_{Z1} = 50$ mA, and $r_{z1} = 10\ \Omega$. The second Zener has $V_{Z2} = 5$ V at $I_{Z2} = 50$ mA and $r_{z2} = 20\ \Omega$. When forward biased the Zeners both have $V_{AK} = 0.7$ V and negligible dynamic resistance. The load current varies from 0 to ±50 mA, while V_s is a square wave of ±12 V. The power dissipation limit for each of the Zeners is 750 mW. Choose the minimum value of R_s that will just avoid overloading either Zener. (Take into account the dynamic resistances of the Zeners for the power calculation.) With this value of R_s what are the load voltage variations? What then are the maximum power dissipations in the Zeners when reverse biased? *[R_s = 41.9 Ω. V_L = +8.61 V (I_L = 50 mA), = +8.93 V (I_L = 0); V_L = −6.38 V (I_L = −50 mA), = −7.06 V (I_L = 0). $P_{Z1\ max}$ = 628 mW, $P_{Z2\ max}$ = 750 mW]*

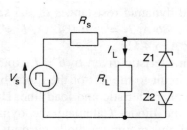

Figure P2.13

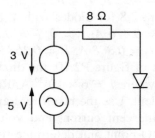

Figure P2.14

14 Find the phase angle at which the diode conducts and that at which it turns off in the circuit of figure P2.14, given the diode has a forward voltage drop of 0.8 V and zero dynamic resistance. Hence find the average power dissipation in the 8Ω load and in the diode. (The AC source is sinusoidal and its r.m.s. voltage is given.)
[32.5°, 147.5°. 310 mW, 88 mW]

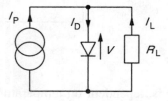

Figure P2.15

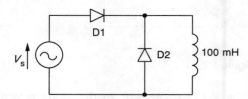

Figure P2.16

15 The equivalent circuit of a solar cell is shown in figure P2.15, from which we see by KCL that $I_L = I_D - I_P$. If the diode's *I-V* relation is $I_D = I_s \exp(35V)$, where I_s = 1 pA and I_P, the photocurrent, is 1 mA at a certain level of illumination, find the open-circuit voltage of the cell. Calculate the maximum load power by differentiating the expression $P = -VI_L = -V(I_D - I_P)$ with respect to V and setting the result equal to zero. What is the load resistance for maximum power? What power would be developed in a 700Ω load? Repeat these calculations for the case where the illumination is halved.
[0.592 V; 0.481 mW; 537 Ω; 0.431 mW. 0.49V; 0.231 mW; 1036 Ω; 0.175 mW]

16 The circuit of figure P2.16 illustrates the use of a free-wheeling diode (D2), which allows some of the current stored in the inductor to discharge when D1 is not conducting. If $V_s = 5 \sin 100\pi t$ V, and the diodes both have forward voltage drops of 1 V and zero dynamic resistance, deduce an expression for the current in the inductor for $0 \le t \le 0.02$ s, if it was zero when the supply was turned on. By how much does the current in the inductor increase for each cycle of the supply?

[i(t) = 0, $0 \le t \le 0.64$ ms; i(t) = $-159.2\cos100\pi t - 10^4t + 162.4$ mA, $0.64 \le t \le 9.36$ ms; i(t) = 224.8 - 10t mA, $9.36 \le t \le 20.64$ ms]

17 If V_s is a square wave of ±5 V and the inductor has a resistance of 1 Ω in the circuit of figure P2.16, the diodes being the same, what will the current in the inductor be in the steady state (a) immediately after D1 has turned off and (b) immediately after D2 has turned off? Estimate the power supplied to the circuit by the source and the power lost in the diodes and the inductance in the steady state.

[(a) 1.625 A (b) 1.325 A. Power supplied is 3.69 W, power lost in D1 = 0.74 W, in D2 = 0.74 W and in the inductor = 2.18 W]

18 Derive from equation 2.18 an expression for the temperature coefficient of a Schottky diode's forward voltage drop. If $A^*A_c = 1$ A/K², $\phi = 0.65$ V, $I_F = 10$ mA and $T = 300$ K, find the percentage change in voltage when the temperature increases by 10°C.

[−6.6%]

3 Bipolar junction transistors

THE BIPOLAR junction transistor, or BJT for short, is a three-terminal device comprising two p-n junctions placed back-to-back, as shown in the notional diagrams of figure 3.1, and looks therefore like two diodes joined together. Yet the BJT has profoundly different properties to those of a diode, particularly in its ability to amplify and to control. The BJT came from a long process of research and discovery, particularly in semiconducting materials; Bell Telephone Laboratories, where the transistor's invention was proclaimed by Shockley, Brattain and Bardeen in 1948, had undertaken a programme of fundamental materials research as early as 1930. Once made, the transistor, being a solid-state device, was rapidly adopted and improved: improved in speed, in power and voltage capability and in reliability. And it led to the revolutionary concept of the integrated circuit or IC, in which hundreds, then thousands and millions of transistors could be made simultaneously. We shall be examining many transistor circuits in due course, but first we should look at the behaviour of electrons and holes in a BJT to see how it works.

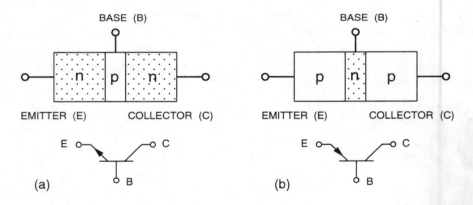

Figure 3.1 (a) An npn bipolar transistor and its circuit symbol (b) A pnp transistor and symbol

3.1 The physics of BJTs

The BJT has two junctions and three terminals to the three different regions of the device, which are called the *emitter*, *base* and *collector*. The base region sandwiched between collector and emitter is very thin, usually between 1 and 5 μm. The doping sequence in these layers can be either n-p-n or p-n-p as shown in figure 3.1, which also shows the circuit symbols of the two types of BJT. To distinguish between them an arrow is placed on the emitter connection which points out of the emitter for an npn device and into the emitter for a pnp, indicating the direction of conventional (positive) current. The two transistor types work in the same way, but from opposite supply polarities.

Figure 3.2 shows a pnp BJT connected to two voltage sources with the base input common to both, so that the resulting amplifier is known as a *common-base* amplifier. The emitter-base junction is a p-n junction which is forward biased by V_{EE}, the emitter DC supply. This junction therefore conducts current quite readily if the voltage across it, V_{EB}, is greater than about 0.7 V in silicon. In the emitter, holes are the majority carriers and the emitter current, I_E, is carried by holes. Electrons from the base can also cross the base-emitter junction and so at the junction I_E is due to both types of charge carrier. It is for this reason that the device is known as bipolar. However, the emitter is normally doped much more heavily than the base (it is really a p+/n junction) and the hole current is much larger than the electron current.

Once in the base the holes from the emitter are minority carriers which should recombine with majority-carrier electrons. This recombination current forms the base current, I_B. As previously said, the base is very thin, so holes crossing into it from the emitter do not have to go far to reach the base-collector junction, and only a small fraction recombine before doing so. The base-collector junction is reverse biased, which means that the n-type base is more positive than the p-type collector. With no forward bias of the emitter-base junction there would be very few holes in the base and little current would flow across the base-collector junction. But with forward bias of the emitter-base junction there are now plenty of injected holes in the base, that readily cross the base-collector junction, falling down the potential gradient into the collector so to form the collector current, I_C. There are very few electrons in the p-type collector to go the other way, so the current across the base-collector junction is all due to injected holes from the emitter. In an npn transistor the bias voltages would be reversed in polarity and most of the emitter current would be carried by electrons rather than holes.

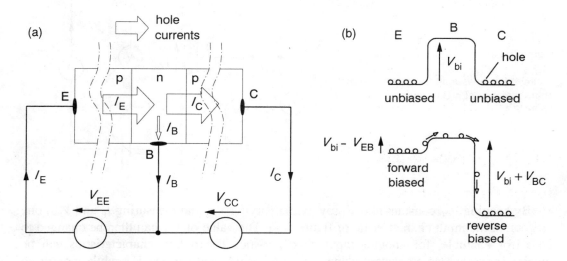

Figure 3.2 (a) Current flow in a biased pnp BJT in the common-base connection. In an npn transistor most of the current would be carried by electrons (b) Energy diagram for holes in a pnp BJT

What has been achieved by this process? With DC alone: nothing. But if an AC signal is superimposed on V_{EE}, it will appear in an amplified form as a component of the collector-base voltage, V_{CB}. The current gain is actually less than unity: if the fraction of holes that recombine in the base is ε, then $I_B = \varepsilon I_E$, and by KCL

$$I_E = I_C + I_B = I_C + \varepsilon I_E$$

$$\Rightarrow \quad I_C/I_E = 1 - \varepsilon = \alpha$$

And

$$I_C/I_B = \beta = \alpha/\varepsilon = \alpha/(1 - \alpha)$$

Now ε is small, normally between 0.02 and 0.002, making α from 0.98-0.998 and $\beta \approx 1/\varepsilon$ is from 50-500. As α is nearly unity the approximation $I_C \approx I_E$ is close.

Though the current gain is a little less than one, the voltage gain may be considerable, depending on the ratio of bias voltages, so that the power gain can be a hundredfold or more. Large current gain can be achieved by connecting the transistor differently to the biasing and signal voltages, in what is called the *common-emitter* configuration, and this is our next consideration.

3.2 The common-emitter (CE) amplifier

Figure 3.3 shows a circuit which can be used to determine the input and output characteristics of a common-emitter amplifier. In this connection the emitter is common to both input and output.

Figure 3.3

A circuit for measuring the common-emitter characteristics of a BJT, such as those of figure 3.4

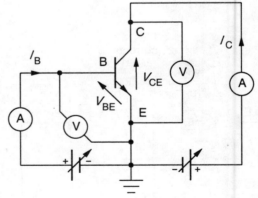

By keeping V_{CE} constant at 6 V, say, while varying V_{BB} and measuring I_B and V_{BE}, one can obtain the input characteristic of figure 3.4a. The value of V_{CE} can then be changed to 24 V for example, for another input characteristic. The output characteristics can be obtained by selecting a constant value of I_B, say 20 μA, and varying V_{CE} while measuring it and I_C. These can be plotted as in figure 3.4b, where they have been plotted for 20μA increments in I_B.

The input characteristic of figure 3.4a is seen to be exactly the same as the *I-V* relationship of a forward biased p-n junction, that is $I_B \approx I_s \exp(40 V_{BE})$ at room temperature. Since the input to the device is the base-emitter junction, which is a forward biased p-n junction, this might be expected. The important point to note in figure 3.4a is that V_{BE} is essentially constant, being virtually independent of the base current and the voltage of the output, V_{CE}. The slope of the input characteristic is dI_B/dV_{BE}, is $1/h_{ie}$ and h_{ie} is the same as the dynamic resistance of a diode, r_d, that is

$$h_{ie} \approx 1/40 I_B = 0.025/I_B \tag{3.1}$$

Sometimes this is written as $h_{ie} = V_{th}/I_B$, where V_{th} (= $kT/q \approx 25$ mV at 300 K) is the *thermal voltage*. h_{ie} is the small-signal input resistance[1] of the transistor and it partly controls the current and voltage gain. We can see that if $I_B = 25$ μA, a typical value, h_{ie} = 1 kΩ.

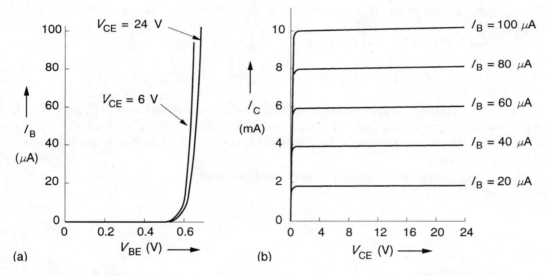

Figure 3.4 (a) The CE amplifier's input characteristics (b) The CE amplifier's output characteristics

The output characteristics of figure 3.4b have a more interesting appearance than the input characteristics. When the voltage across the transistor, V_{CE}, is small the collector current rises steeply with a slope determined essentially by the collector resistance. But, when V_{CE} is greater than about 0.3 V, the collector current becomes almost constant at a value determined by I_B. The ratio of I_C to I_B here, in the *operating region*, is nearly constant and is equal to β

$$I_C/I_B = \beta \tag{3.2}$$

From figure 3.4b it can be seen that when $I_C = 10$ mA, $I_B = 100$ μA, so that for this

---

[1] h_{ie} is one of the four *hybrid* or h-parameters, which describe certain properties of 2-port networks and are used later in modelling the CE amplifier.

transistor $\beta = 10/0.1 = 100$, so that $\varepsilon \approx 1/\beta = 0.01$ and $\alpha = 1 - \varepsilon = 0.99$; about 1% of the injected minority carriers recombine in the base, the great majority reaching the collector. Equation 3.2 is the key to understanding BJT behaviour in all amplifier circuits: if the base current is determined, so is the collector current.

β is identical to h_{FE} which is virtually the same as h_{fe} ($= \delta I_C/\delta I_B$, the small-signal current gain). In practical work β, h_{fe} and h_{FE} are used interchangeably.

3.2.1 Biasing the transistor: Q-point and load line

The circuit of figure 3.3 has two power supplies, one of which is redundant. The base supply can be derived from the collector supply, V_{CC}, by incorporating a suitable value of resistor, R_B, to produce whatever base current — and therefore collector current — we desire. In order to vary V_{CE} we must also include a resistor, R_C, in series with the collector, so ending up with the base-bias (sometimes called the fixed bias circuit since it 'fixes' the base current) circuit of figure 3.5, in which all the currents and voltages can be found if we know β.

In figure 3.5, by KVL and Ohm's law

$$V_{CC} = V_{RC} + V_{CE} = I_C R_C + V_{CE} \qquad (3.3)$$

Since V_{CC} is a constant supply voltage, varying I_C will cause a variation in V_{CE}. And as $I_C = \beta I_B$, varying I_B will also cause a corresponding variation in V_{CE}. By applying a signal to the base, we can thereby cause an amplified version of it to appear at the transistor's collector terminal.

Equation 3.3 can be rearranged to give I_C in terms of V_{CE}:

$$I_C = \left(\frac{-1}{R_C}\right)V_{CE} + V_{CC}$$

$$(\ y\ =\ m\ x\ +\ c\ ) \qquad (3.4)$$

This equation is a straight line, known as the *load line*. It is the locus of all possible values of V_{CE} and I_C. When we have chosen a value for I_B by choosing R_B, we will have thereby fixed I_C and, by equation 3.4, V_{CE} also.

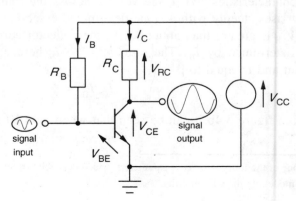

Figure 3.5

The fixed-bias CE amplifier. In this circuit a common supply is used to provide current to both base and collector

With R_B and R_C determined all the voltages and currents in the circuit will be constant, and nothing more will happen until we apply a signal to the base. For this reason the operating point of the transistor is called its *quiescent point* or Q-point, and the operational values of the voltages and current are given the subscript Q, as I_{BQ}, V_{CEQ}, I_{CQ} and so on. An example will show better how the quiescent conditions are found.

Example 3.1

In the circuit of figure 3.5, the supply voltage, V_{CC} = +12 V and the transistor has a beta of 100. It is desired to operate with V_{CEQ} = 5 V and I_{CQ} = 5 mA. Assuming V_{BEQ} = 0.6 V, what values of R_C and R_B are required? If the output characteristics are those of figure 3.4b, plot the load line and Q-point for this circuit on them.

Figure 3.6

(a) The output (collector) side of the circuit of figure 3.5
(b) The input (base) side

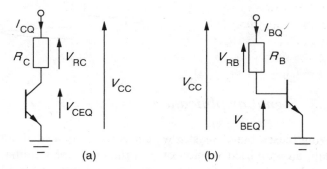

On the output side of figure 3.5, redrawn as figure 3.6a, we can see that by KVL

$$V_{CC} = V_{RC} + V_{CEQ} = I_{CQ}R_C + V_{CEQ}$$

$$\Rightarrow\ 12 = 5R_C + 5$$

$$\Rightarrow\ R_C = 7/5 = 1.4\ \text{k}\Omega$$

where we have worked in V, mA and kΩ, as is customary. To achieve I_{CQ} = 5 mA with β = 100 requires that $I_{BQ} = I_{CQ}/\beta$ = 5/100 = 0.05 mA (= 50 μA), and a value of R_B must be calculated that will give this.

Examining the input side of the circuit, figure 3.6b, we can see by KVL that

$$V_{CC} = I_{BQ}R_B + V_{BEQ}$$

$$\Rightarrow\ 12 = 0.05R_B + 0.6$$

$$\Rightarrow\ R_B = 11.4/0.05 = 228\ \text{k}\Omega$$

again working in V, mA and kΩ. The load line and Q-point are shown in figure 3.7, from which we can see that since the Q-point is not in the middle of the load line, the maximum output voltage swing without clipping will be ±5 V.

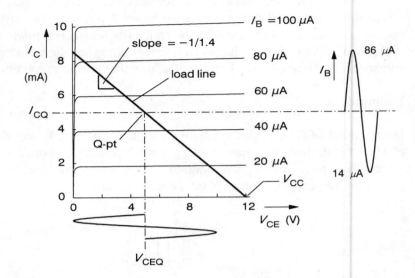

Figure 3.7

The Q-point and load line for the circuit of figure 3.5, with V_{CC} = 12 V, R_C = 1.4 kΩ and R_B = 228 kΩ

3.2.2 Signal amplification

Having biased our transistor we can now apply a small alternating voltage to the base and obtain an amplified version of it at the collector terminal. Let us suppose that the output is not loaded, but is fed into a device or other circuit which has infinite impedance and draws no current. What sort of gain will we get? We can see from figure 3.7 that the maximum available output voltage occurs when V_{CE} = 12 V, that is when I_C = 0, while the minimum is V_{CE} = 0 V, which occurs when I_C = 12/1.4 = 8.6 mA. In turn these values for I_C correspond to values of I_B of 0 and 86 μA, respectively. In the standard nomenclature for describing currents and voltages with alternating and direct components we write

$$i_B = I_{BQ} + i_b = I_{BQ} + I_{bm} \sin \omega t \qquad (3.5)$$

$$i_C = \beta i_B = I_{CQ} + i_c = I_{CQ} + I_{cm} \sin \omega t \qquad (3.6)$$

$$v_{CE} = V_{CC} - i_C R_C = V_{CEQ} + v_{ce} = V_{CEQ} + V_{cem} \sin \omega t \qquad (3.7)$$

The total current or voltage, that is the sum of DC and AC parts is given a lower-case symbol with an upper-case suffix: v_{CE}, i_B. DC components are given upper-case symbols and upper-case suffixes: I_{CQ}, V_{BEQ}. Peak AC values are given upper-case symbols with lower-case suffixes ending with an 'm': V_{bem}, I_{cm}.

We now can write for the quiescent (DC) base current, I_{BQ} = 50 μA, while the maximum possible value of the AC part is I_{bm} = 36 μA, since $i_B(\max) = I_{BQ} + I_{bm}$ = 86 μA. Thus $i_b = I_{bm} \sin \omega t = 36 \sin \omega t$ μA and i_B = 50 + 36 sin ωt μA. None of this, however, tells us what the input voltage is, for which we need the input resistance of the device, h_{ie}, given by equation 3.1 as $25/I_{BQ}$ = 25/0.05 = 500 Ω, with I_{BQ} in mA. Then

$$v_{\text{IN}} = v_{\text{BE}} = V_{\text{BEQ}} + v_{\text{be}} = i_{\text{B}}h_{\text{ie}}$$

$$= (50 + 36\sin \omega t) \times 10^{-6} \times 500 = 0.025 + 0.018\sin \omega t \ \text{V}$$

or $v_{\text{IN}} = 25 + 18\sin \omega t$ mV. The *signal* part of this is $v_{\text{be}} = 18\sin \omega t$ mV. Notice how small this is; and yet it is actually too large, since driving the amplifier to its absolute limits leads inevitably to distortion of the signal: we should expect to get no more than 80% of the maximum possible with reasonable fidelity. This implies that the input signal should be at most 14 mV peak or $14/\sqrt{2} = 10$ mV r.m.s. The output voltage would be correspondingly reduced by 20% from 5 V peak to 4 V peak or $4/\sqrt{2} = 2.8$ V r.m.s. Then the no-load voltage gain is $2.8/0.01 = 280$, a typical value for a CE amplifier.

3.2.3 Signal distortion

As mentioned above, even if the Q-point is well chosen, distortion of the amplified signal must occur if the input signal is too large, because the AC output amplitude is necessarily restricted to $\frac{1}{2}V_{\text{CC}}$ when $V_{\text{CEQ}} = \frac{1}{2}V_{\text{CC}}$, and less if any other Q-point is chosen. One should expect the distorted signal to show clipping at both its maximum and minimum when $V_{\text{CEQ}} = \frac{1}{2}V_{\text{CC}}$. However, when V_{CEQ} is less than $\frac{1}{2}V_{\text{CC}}$, then the negative-going part of the output is clipped before the positive, as in figure 3.8a. Likewise, when V_{CEQ} is too high, the positive part of the output is clipped before the negative as in figure 3.8b.

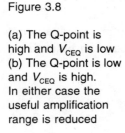

Figure 3.8

(a) The Q-point is high and V_{CEQ} is low
(b) The Q-point is low and V_{CEQ} is high.
In either case the useful amplification range is reduced

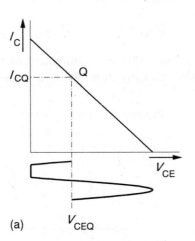

(a)

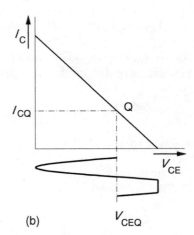

(b)

3.2.4 The practical CE amplifier

In a practical amplifier it would be undesirable to have DC at either the input or output, because it might cause problems with later amplifier stages, or there might be a DC component of the source which would affect the biasing of the transistor. For this reason decoupling capacitors are connected in series with the input and output as figure 3.9 shows. The capacitors are chosen to have negligible reactances at the lowest operating frequency: for example if 100 Ω at 100 Hz is 'negligible', then $C_1 = C_2 = 1/2\pi fR = 1/2\pi \times 100 \times 100 = 16$ μF. Values of about 10-100 μF are usually chosen.

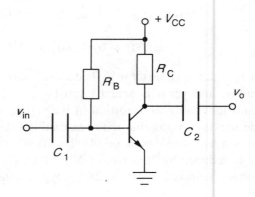

Figure 3.9

A practical CE amplifier, with decoupling
capacitors to block DC components

3.2.5 The small-signal equivalent circuit

The analysis of the CE amplifier can be assisted considerably if we use an equivalent
circuit for it. There are numerous such, but the one suitable for many purposes is the
small-signal equivalent circuit of figure 3.10, which models the transistor as a resistance
on the input side and a current source in parallel with R_C on the output side. The model
requires only two of the four hybrid parameters: the input resistance, h_{ie}, and the current
ratio h_{fe} ($= i_c/i_b \approx \beta$). On the output side the current source has a value of $-h_{fe}i_b$, which is
$-i_c$. The negative sign arises since from equation 3.7 it can be seen that

$$v_{CE} = V_{CC} - i_c R_C = V_{CC} - (I_{CQ} + i_c)R_C = V_{CEQ} - i_c R_C = V_{CEQ} + v_{ce}$$

that is $v_{ce} = -i_c R_C$. The model is valid if this AC component of the output voltage is not
significantly distorted. An example will show how the equivalent circuit can be used.

transistor

Figure 3.10

The small-signal equivalent circuit
for the CE amplifier

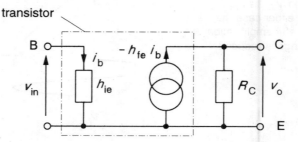

Example 3.2

Use the small-signal equivalent circuit to find the small-signal current and voltage gains
of a CE amplifier which has a load resistance, R_L, and whose input is a sinusoidal signal
source, v_s, having an internal resistance, R_s. Ignore the base-bias resistor, R_B. Hence
calculate voltage, current and power gains when $R_C = 1.4$ kΩ, $R_L = 1$ kΩ, $R_s = 75$ Ω, h_{fe}
$= 100$ and $I_{BQ} = 50$ μA.

Figure 3.11

The circuit for example 3.2

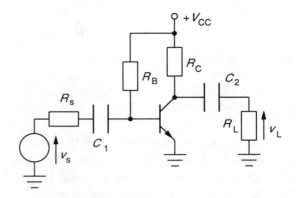

The circuit to be analysed is that of figure 3.11, and its small-signal equivalent is that of figure 3.12, where the 228kΩ base bias resistance has been ignored.

Figure 3.12

The small-signal equivalent of the circuit in figure 3.11

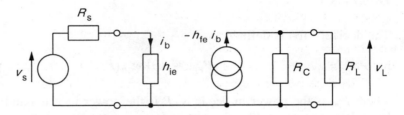

The load is placed in parallel with R_C because the output voltage at the collector goes to ground through R_C and V_{CC} as well as through the load. (To see this, just draw V_{CC} as a voltage source with its positive terminal connected to the terminal marked $+V_{CC}$, and with its negative terminal connected to ground.) Thus the output resistance is $R_o = R_C /\!/ R_L$. The output voltage is then $-h_{fe}i_b R_o$, while i_b is given by

$$i_b = v_s/(h_{ie} + R_s)$$

The overall voltage gain is

$$A_{vL} = v_o/v_s = -h_{fe}i_b R_o/v_s$$

Substituting for i_b gives

$$A_{vL} = \frac{-h_{fe}R_o}{h_{ie} + R_s} = \frac{-h_{fe}R_C R_L}{(h_{ie} + R_s)(R_C + R_L)} \tag{3.8}$$

If there is no load ($R_L = \infty$) and $R_s = 0$, the no-load voltage gain is

$$A_{v0} = -h_{fe}R_C/h_{ie} \tag{3.9}$$

The current gain is

$$A_{iL} = i_L/i_s = i_L/i_b$$

We see from figure 3.12 that

$$i_L = \frac{R_C}{R_C + R_L} \times -h_{fe}i_b$$

by the current-divider rule, and so

$$A_{iL} = \frac{-h_{fe}R_C}{R_C + R_L} \tag{3.10}$$

When $I_{BQ} = 50$ μA, $h_{ie} = 0.025/0.05 = 0.5$ kΩ, and substituting this and the other values into equation 3.8 leads to $A_{vL} = -101.4$. The minus sign merely indicates that the output is 180° out of phase with the input, which is not usually of much significance.

Substitution for h_{ie}, R_C and R_L into equation 3.10 gives $A_{iL} = -41.7$, and then the power gain is $A_{pL} = A_{vL}A_{iL} = 101.4 \times 41.7 = 4228$, omitting the cancelling minus signs.

Decibels

The decibel (dB) is defined as

$$P_{dB} = 10\log_{10}(P/P_0) \tag{3.11}$$

where P_{dB} is the power in decibels, P is the power in watts and P_0 is a reference power level. For example if the reference power level is 1 W and the power, P, is 25 W, then the power in dB is 13.8 dB. Sometimes the reference power level is 1 mW and then the power is written as so many dBm, as in 12 dBm, which is 15.8 mW. In acoustics the reference level is sometimes expressed as //μW, as in −30 dB//μW, which would be 1 nW.

Very often voltages and currents are expressed in dB using the relations

$$P_{dB} = 20\log_{10}V \quad \text{or} \quad P_{dB} = 20\log_{10}I \tag{3.12}$$

the reference power being 1 W and the voltage or current is assumed to be across or through a 1Ω resistance, so that $P/P_0 = V^2$ or $P/P_0 = I^2$ and application of equation 3.11 leads to 3.12 immediately.

The use of decibels is further extended to voltage and current gains using

$$A_{dB} = 20\log_{10}|A_v| \quad \text{or} \quad A_{dB} = 20\log_{10}|A_i| \tag{3.13}$$

Thus a voltage gain of −101.4 becomes one of 40.1 dB and a current gain of −41.7 becomes 32.4 dB. However, when the power gain is desired, these two cannot be added! The power gain is $101.4 \times 41.7 = 4228$, and this in dB is $10\log 4228 = 36.3$ dB, half the sum of the voltage and current gains in dB. The problem arises because of the artificial way in which voltages and currents are assumed to be on or in a 1Ω load when gains are converted to dB.

3.2.6 *A better way of biasing: the voltage-divider bias circuit*

The fixed-bias circuit, which keeps the base current constant by means of a single resistor connected between V_{CC} and the base, is not the ideal biasing circuit since it relies on the constancy of the transistor's β (or h_{fe}) for maintaining the quiescent collector current. Unfortunately β is not a reliable parameter. As we have seen $\beta \approx 1/\varepsilon$, where ε is the fraction of the emitter current that recombines in the base. This fraction is small and critically dependent on manufacturing variables over which control is limited. Thus in practice β varies widely even for transistors made in the same batch, and the batch-to-batch variation is even wider. Examination of transistor specification sheets will confirm this; for a general-purpose, low-power device β typically ranges from 50 to 300. Not only does β vary from one transistor to another, but it also changes with temperature, base current and collector-emitter voltage. The consequences are serious for the fixed-bias circuit.

Example 3.3

In the circuit of figure 3.5, example 3.1, $V_{CC} = 12$ V, $R_C = 1.4$ kΩ and $R_B = 228$ kΩ as before, but the transistor is replaced by one with $\beta = 200$ instead of $\beta = 100$. What are the quiescent conditions now? What are the quiescent conditions if $\beta = 50$?

I_{BQ}, as we have seen, is constant at its former value, 50 μA, so

$$I_{CQ} = \beta I_{BQ} = 200 \times 50 = 10\,000\ \mu\text{A} = 10\ \text{mA}$$

But we see from the load line in figure 3.7 that with this circuit the most I_C can be is 8.6 mA. The transistor cannot be supplied with enough collector current and is said to be *saturated* or to be *in saturation*. The Q-point lies at $V_{CEsat} = 0.2$ V and the undistorted amplification will be nil.

Figure 3.13

The effect of changing β on the Q-point of the fixed-bias (= constant base current) circuit of figure 3.5. The base currents of the characteristics must be scaled by a factor of $\beta/100$ as β changes

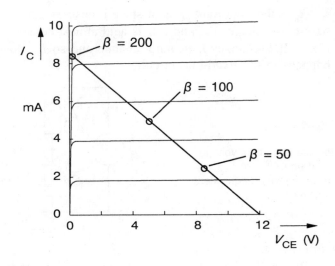

When $\beta = 50$,

$$I_{CQ} = 50 I_{BQ} = 50 \times 50\ \mu\text{A} = 2.5\ \text{mA}.$$

The quiescent voltage across R_C is $I_{CQ}R_C = 2.5 \times 1.4 = 3.5$ V, making

$$V_{CEQ} = V_{CC} - I_{CQ}R_C = 12 - 3.5 = 8.5 \text{ V}$$

Thus the Q-point is moved down nearer *cut-off* ($I_C = 0$) and the available amplification is ± 3.5 V at most. Figure 3.13 shows these Q-points on the output characteristics and load line.

What is needed is a quiescent collector current that is almost independent of β. The voltage-divider (sometimes called the emitter-bias) circuit of figure 3.14 does this by including a resistor, R_E, in the emitter circuit and a further resistor, R_2, in the base-bias circuit. The base-biasing resistors, R_1 and R_2, form a voltage-divider network which maintains the base-biasing voltage, V_B, at a constant value, provided the base current is small compared to I_1. Because there is now an emitter resistor, R_E, the base quiescent voltage is no longer V_{BEQ} but $V_B = V_{BEQ} + V_{RE} = V_{BEQ} + I_{EQ}R_E \approx V_{BEQ} + I_{CQ}R_C$, as $I_{EQ} \approx I_{CQ}$.

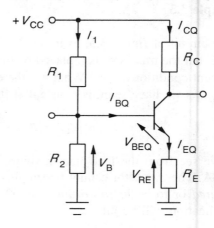

Figure 3.14

The voltage-divider bias circuit for the CE amplifier. By keeping V_B almost constant, I_{CQ} is kept constant despite large variations in the transistor's β

R_E is the important resistor since it ensures that any changes in I_C are fed back to the base-bias voltage, V_B, and this is held almost constant by the voltage divider, R_1 and R_2. Hence if β changes I_{CQ} cannot change and so I_{BQ} does instead — the converse of what happens in the fixed-bias circuit.

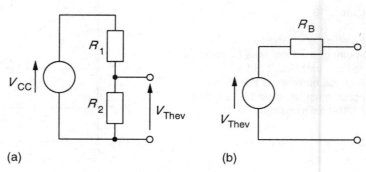

Figure 3.15

(a) The base-bias voltage divider
(b) Its Thévenin equivalent circuit

We shall analyse the circuit of figure 3.14 assuming that V_{BEQ} is constant. Looking at the input side as shown in figure 3.15a, we can see that it can be replaced by its Thévenin equivalent of figure 3.15b, in which $R_B = R_1 // R_2 = R_1 R_2/(R_1 + R_2)$ by the product-over-sum rule and

$$V_{Thev} = \frac{R_2 V_{CC}}{R_1 + R_2} \qquad (3.14)$$

by the voltage-divider rule.

Now examining the lower part of figure 3.14 as in figure 3.16a, we see that

$$V_B = V_{BEQ} + V_{RE} \approx V_{BEQ} + I_{CQ}R_E = V_{BEQ} + \beta I_{BQ}R_E \qquad (3.15)$$

Because the quiescent current through R_E is approximately I_{CQ}, we can replace it by a resistance of βR_E carrying a current of I_{BQ} in the base circuit. Thus we can attach to the terminals of the Thévenin circuit of figure 3.15b a voltage source, V_{BEQ}, and a resistance, βR_E, as in figure 3.16b. And hence the quiescent base current is

$$I_{BQ} = \frac{V_{Thev} - V_{BEQ}}{R_B + \beta R_E} \qquad (3.16)$$

Hence

$$I_{CQ} = \beta I_{BQ} = \frac{\beta(V_{Thev} - V_{BEQ})}{R_B + \beta R_E} = \frac{V_{Thev} - V_{BEQ}}{R_E + R_B/\beta} \qquad (3.17)$$

The resistors should be such that $R_E \gg R_B/\beta$, which is the same as saying $I_1 \gg I_{BQ}$, and then equation 3.17 becomes $I_{CQ} \approx (V_{Thev} - V_{BEQ})/R_E$. We shall see later that a further requirement is that $R_E \ll R_C$ to keep the voltage gain as great as possible.

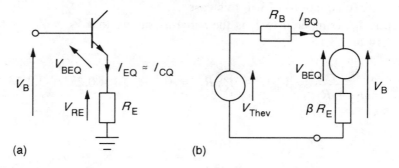

(a) (b)

Figure 3.16 (a) Base-emitter bias: $V_B = V_{BEQ} + V_{RE}$ (b) The equivalent of this part of the circuit is attached to the Thévenin circuit of figure 3.15b

Example 3.4

Calculate suitable values for R_1, R_2, R_C and R_E in the circuit of figure 3.14 to give $I_{CQ} = 10$ mA and $V_{CEQ} = 6$ V when $V_{CC} = 16$ V. Firstly by assuming that I_{BQ} is negligible compared to I_1 and secondly by using the equivalent circuit of figure 3.16b. Bear in mind that R_B should be as large as possible, but much less than βR_E and that $R_E \ll R_C$. Take $V_{BEQ} = 0.7$ V and $\beta \approx 100$.

The output side of figure 3.14 is shown in figure 3.17.

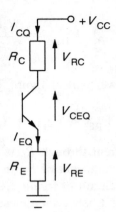

Figure 3.17

The output side of figure 3.14, in which KVL
is used to find $R_C + R_E$

In this circuit we use KVL to find

$$V_{CC} = V_{RC} + V_{CEQ} + V_{RE} = I_{CQ}R_C + V_{CEQ} + I_{EQ}R_E$$

As $I_{EQ} \approx I_{CQ}$, the above equation yields

$$R_C + R_E = (V_{CC} - V_{CEQ})/I_{CQ} = (16 - 6)/10 = 1 \text{ k}\Omega \qquad (3.18)$$

The requirements that $\beta R_E \gg R_B$ and that $R_E \ll R_C$ are mutually exclusive and a compromise is needed. As a rule we take $\frac{1}{4}R_C < R_E < \frac{1}{2}R_C$, say $R_E = 0.25$ kΩ and $R_C = 0.75$ kΩ to satisfy equation 3.18 in this case.

Neglecting I_{BQ} compared to I_1 is the same as saying that $V_B = V_{Thev}$ and is given by equations 3.14 and 3.15

$$V_B = \frac{R_2 V_{CC}}{R_1 + R_2} = V_{BEQ} + I_{CQ}R_E = 0.7 + 10 \times 0.25 = 3.2 \text{ V} \qquad (3.19)$$

Hence

$$\frac{R_2 V_{CC}}{R_1 + R_2} = \frac{16R_2}{R_1 + R_2} = 3.2 \quad \Rightarrow \quad R_1 = 4R_2$$

It is customary to take $R_B \approx 0.1\beta R_E = 0.1 \times 100 \times 0.25 = 2.5$ kΩ, since R_B should be as large as possible while still being small compared to βR_E. This compromise is the one that works best in practice with beta values of 50-300. Then

$$R_B = \frac{R_1 R_2}{R_1 + R_2} = \frac{4R_2^2}{4R_2 + R_2} = 2.5 \text{ k}\Omega$$

From which we find $R_2 = 3.125$ kΩ and $R_1 = 12.5$ kΩ.
These values can be substituted into equation 3.14 to find V_{Thev}

$$V_{\text{Thev}} = \frac{R_2 V_{\text{CC}}}{R_1 + R_2} = \frac{3.125 \times 16}{3.125 + 12.5} = 3.2 \text{ V}$$

And from equation 3.17

$$I_{\text{CQ}} = \frac{V_{\text{Thev}} - V_{\text{BEQ}}}{R_{\text{E}} + R_{\text{B}}/\beta} = \frac{3.2 - 0.7}{0.25 + 2.5/100} = 9.1 \text{ mA}$$

For all practical purposes this would be close enough to the desired value of 10 mA. In addition $V_{\text{CEQ}} = V_{\text{CC}} - I_{\text{CQ}}(R_{\text{C}} + R_{\text{E}}) = 16 - 9.1 \times 1 = 6.9$ V; close enough to the 6 V desired too. However, we shall calculate values for R_1 and R_2 that will give a collector current of exactly 10 mA.

Assuming $R_{\text{E}} = 0.25$ kΩ, $R_{\text{C}} = 0.75$ kΩ and $R_{\text{B}} = 2.5$ kΩ, we can adjust R_1 and R_2 again using equation 3.17 to find V_{Thev}

$$I_{\text{CQ}} = 10 = \frac{V_{\text{Thev}} - V_{\text{BEQ}}}{R_{\text{E}} + R_{\text{B}}/\beta} = \frac{V_{\text{Thev}} - 0.7}{0.25 + 2.5/100} \quad \Rightarrow \quad V_{\text{Thev}} = 3.45 \text{ V}$$

But from equation 3.14

$$V_{\text{Thev}} = \frac{R_2 V_{\text{CC}}}{R_1 + R_2} = 3.45 = \frac{16 R_2}{R_1 + R_2} \quad \Rightarrow \quad R_1 = 3.638 R_2$$

As before, $R_{\text{B}} = R_1 \;//\; R_2 = 3.638 R_2/4.638 = 2.5$ kΩ, so that $R_2 = 3.187$ kΩ and $R_1 = 12.75$ kΩ. It is highly unlikely that any of these resistor values can be found in in-house stores, which normally will stock preferred values[2] such as 3.3 kΩ and 12 kΩ.

Example 3.5

Recalculate the resistance values of example 3.4 using preferred resistor values such as given in footnote 2.

The first values to find are R_{C} and R_{E}, which must add up to 1 kΩ by equation 3.18. The preferred values are 820 Ω and 220 Ω or 680 Ω and 330 Ω respectively. Though both pairs have resistance ratios of about the right magnitude ($\frac{1}{4} < R_{\text{E}}/R_{\text{C}} < \frac{1}{2}$) the former pair is preferable because R_{E} is smaller. There is no point in doing an 'exact' analysis since preferred values are to be used; we just take $V_{\text{B}} = V_{\text{Thev}}$ and use equations 3.14 and 3.15:

$$V_{\text{B}} = \frac{R_2 V_{\text{CC}}}{R_1 + R_2} \approx V_{\text{BEQ}} + I_{\text{CQ}} R_{\text{E}} = 0.7 + 10 \times 0.22 = 2.9 \text{ V}$$

Therefore

$$\frac{16 R_2}{R_1 + R_2} = 2.9 \quad \Rightarrow \quad \frac{R_2}{R_1 + R_2} = 0.18$$

---

[2] Preferred values are in geometric progression and cover a decade of resistance with a dozen values such as 10, 12, 15, 18, 22, 27, 33, 39, 47, 56, 68, 82 and 100 kΩ. The tolerance for resistors is usually ±1%.

Again we will make R_B ($= R_1 \mathbin{/\!/} R_2$) $= 0.1\beta R_E = 0.1 \times 100 \times 0.22 = 2.2$ kΩ, and so $R_1 = R_B/0.18 = 12.2$ kΩ, the nearest preferred value being 12 kΩ. Then $R_2 = 0.18 \times 12/0.82 = 2.6$ kΩ, the nearest preferred value being 2.7 kΩ. One can perform the 'exact' analysis to find the quiescent conditions obtained with these values and one obtains $I_{CQ} = 9.25$ mA from equation 3.17 and hence $V_{CEQ} = V_{CC} - I_C R_C - I_E R_E = 16 - 9.25 \times 0.82 - 1.01 \times 9.25 \times 0.22 = 6.36$ V.

The results of the analyses of examples 3.4 and 3.5 are summarised below:

Table 3.1 *Effect of resistance changes on the Q point*

	R_C	R_E	R_1	R_2	I_{CQ}	V_{CEQ}
approximate	0.75	0.25	12.5	3.125	9.1	6.9
'exact'	0.75	0.25	12.75	3.187	10.0	6.0
preferred values	0.82	0.22	12	2.7	9.25	6.36

Note: Resistances are in kΩ, currents in mA and voltages in V

The effect of variations in β on the Q-point

Having made the biasing more complicated to stabilise the amplifier against variations in the transistor's β, it is reasonable to inquire into the degree of constancy in the Q-point. For this we can use equation 3.17, which is

$$I_{CQ} = \frac{V_{Thev} - V_{BEQ}}{R_E + R_B/\beta} \tag{3.17}$$

Taking logs and differentiating with respect to β (the only varying quantity):

$$\ln I_{CQ} = \ln(V_{Thev} - V_{BEQ}) - \ln(R_E + R_B/\beta)$$

$$\frac{1}{I_{CQ}} \frac{\mathrm{d}I_{CQ}}{\mathrm{d}\beta} = \frac{R_B/\beta^2}{R_E + R_B/\beta} = \frac{1}{\beta} \frac{1}{(1 + \beta R_E/R_B)}$$

And so

$$\frac{\delta I_{CQ}}{I_{CQ}} = \frac{\delta\beta}{\beta} \frac{1}{(1 + \beta R_E/R_B)} = \frac{\delta\beta}{11\beta}$$

if $R_B = 0.1\beta R_E$. We see that an 11% change in beta will produce only a 1% change in quiescent collector current. Using equation 3.17 again, with $V_{Thev} - V_{BEQ} = 2.5$ V, $R_E = 250$ Ω and $R_B = 2.5$ kΩ, we find for $\beta = 50$, $I_{CQ} = 8.3$ mA, $I_{BQ} = 166$ μA; for $\beta = 100$, $I_{CQ} = 9.1$ mA, $I_{BQ} = 91$ μA and when $\beta = 200$, $I_{CQ} = 9.5$ mA, $I_{BQ} = 47.5$ μA: all quite acceptable values. The collector current changes by a little, the base current by a lot.

3.2.7 Gain degeneration and avoidance: the emitter bypass capacitor

Placing R_E in series with the emitter results in an increase in the input resistance of the transistor. An alternating current in the base of i_b becomes an emitter current of i_e and so the small-signal input voltage to the base (see figure 3.18a) is not v_{be} $(= h_{ie}i_b)$ but

$$v_b = v_{be} + i_e R_E \approx v_{be} + h_{fe}i_b R_E = i_b(h_{ie} + h_{fe}R_E)$$

where we have taken $i_e \approx i_c = h_{fe}i_b$ $(\approx \beta i_b)$. Accordingly in the equivalent circuit of figure 3.10 we must replace h_{ie} by $h_{ie} + h_{fe}R_E$ as in figure 3.18b.

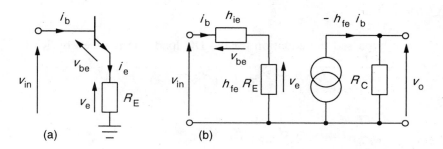

Figure 3.18 (a) The small-signal voltages and currents from input to ground
(b) The corresponding equivalent circuit for calculating amplifier gain

Then using this circuit, the no-load voltage gain given by equation 3.9 is found

$$A_{v0} = \frac{-h_{fe}R_C}{h_{ie} + h_{fe}R_E} \approx \frac{-R_C}{R_E} \approx -4$$

since

$$h_{ie} \approx 0.025/I_{BQ} = 0.25 \text{ k}\Omega \ll h_{fe}R_E = 100 \times 0.25 = 2.5 \text{ k}\Omega$$

using the values in example 3.4. Without R_E the no-load voltage gain would have been about -300. Stabilizing the quiescent conditions has caused most of the gain of the amplifier to be lost.

The position is remediable to a large extent by using a large capacitor to bypass R_E as in figure 3.19. The capacitor blocks DC and so the quiescent current, I_{EQ}, and resistor voltage are unaffected. However, i_e now passes through the capacitor, which shorts out R_E to AC, causing the voltage and current gains to be restored to the values given in equations 3.8, 3.9 and 3.10. The reactance of C_E must be small at the lowest frequency to be amplified. Thus if $X_{CE} = 10 \ \Omega$ at 10 Hz, then

$$C_E = \frac{1}{\omega X_{CE}} = \frac{1}{2\pi \times 10 \times 10} = 1600 \ \mu\text{F}$$

There is still a small penalty to be paid for R_E (as well as the additional expense of it and its by-pass capacitor) and that is the loss of amplification frequency range.

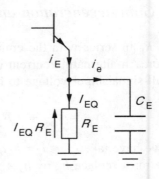

Figure 3.19

The emitter-bypass capacitor. C_E must have a small reactance compared to R_E and h_{ie} at the lowest frequencies to be amplified

The AC load line

When R_E is not bypassed the equation for the DC load line is given by KVL:

$$V_{CC} = I_C R_C + V_{CE} + I_E R_E \approx V_{CE} + I_C(R_C + R_E)$$

$$\Rightarrow \quad I_C \approx \frac{-V_{CE}}{R_C + R_E} + \frac{V_{CC}}{R_C + R_E}$$

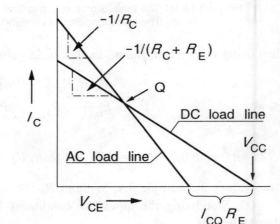

Figure 3.20

AC and DC load lines for a CE amplifier with bypassed emitter resistor

As shown in figure 3.20, the slope of the DC load line is now $-1/(R_C + R_E)$, while its intercept on the voltage axis is still V_{CC}. When R_E is bypassed the DC voltage dropped across it, $I_{EQ}R_E \approx I_{CQ}R_E$, is not available for the AC output. Then the slope of the AC load line becomes $-1/R_C$ and it passes through the same Q-point as the DC load line, as shown in figure 3.20. The maximum output voltage swing is now $V_{CC} - I_{EQ}R_E$ when the amplifier is not loaded. If the Q-point is optimally placed

$$v_o(\text{max}) = \pm \tfrac{1}{2}(V_{CC} - I_{EQ}R_E)$$

For this reason R_E should be small compared to R_C.

3.2.8 Q-points for maximum gain

Before R_E was bypassed the optimal Q-point for maximum output voltage with no load was at $V_{CEQ} = \frac{1}{2}V_{CC}$, whereas with bypass capacitor it is at $V_{CEQ} = \frac{1}{2}(V_{CC} - I_{EQ}R_E)$. However, in many cases the amplifier drives a load with a finite resistance, R_L, and this must be considered. For AC at the output of the amplifier, R_L is in parallel with R_C, as shown in figure 3.21a (we can ignore both the coupling capacitor and the DC supply), and this means that the output resistance is $R_o = R_C \, // \, R_L$. The slope of the AC load line is now $-1/R_o$ and it still passes through the DC Q-point, so that the AC load line is now as shown in figure 3.21b.

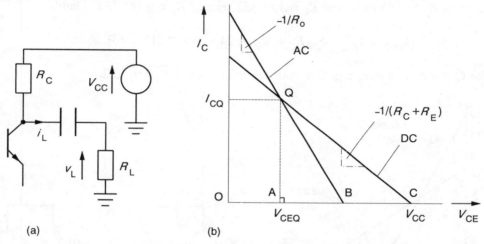

(a)   (b)

Figure 3.21 (a) R_C and R_L are in parallel for AC at the collector terminal (b) AC and DC load lines

The largest unclipped output voltage swing is then obtained when OA = AB, that is when $V_{CEQ} = I_{CQ}R_o$. But in triangle QAC of figure 3.21b, QA/AC = slope of DC load line = $1/(R_C + R_E)$, so that AC = QA$(R_C + R_E) = I_{CQ}(R_C + R_E)$. And AC is $V_{CC} - V_{CEQ}$, hence

$$V_{CEQ} = V_{CC} - I_{CQ}(R_C + R_E) = I_{CQ}R_o$$

$$\Rightarrow \quad I_{CQ} = \frac{V_{CC}}{R_C + R_E + R_o} \tag{3.20}$$

Example 3.6

A CE amplifier with a bypassed emitter resistor has $V_{CC} = 24$ V, a transistor with $h_{fe} = 160$ and $I_{CQ} = 8$ mA. If it is to drive a resistive load, $R_L = 1.5$ kΩ and if $R_C = 3R_E$, what must R_E and R_C be for maximum unclipped output voltage swing? Find the voltage, current and power gains of the amplifier if the source resistance is zero and the bias resistances are large compared to h_{ie}. What is the largest input voltage that can be amplified without clipping? (Take V_{th} to be 30 mV.)

Using equation 3.20

$$8 = \frac{24}{3R_E + R_E + R_o} \Rightarrow 4R_E + R_o = 3$$

But $R_o = R_C \; // \; R_L = 1.5R_C/(R_C + 1.5) = 4.5R_E/(3R_E + 1.5)$, and substitution into the equation above leads to

$$4R_E + \frac{4.5R_E}{3R_E + 1.5} = 3 \Rightarrow 12R_E^2 + 1.5R_E - 4.5 = 0$$

which gives $R_E = 0.553$ kΩ and $R_C = 1.66$ kΩ. Hence $R_o = 0.788$ kΩ. Then

$$V_{CEQ} = V_{CC} - I_{CQ}(R_C + R_E) = 24 - 8 \times 2.212 = 6.3 \text{ V}$$

and the Q-point is shown, together with the load lines in figure 3.22a.

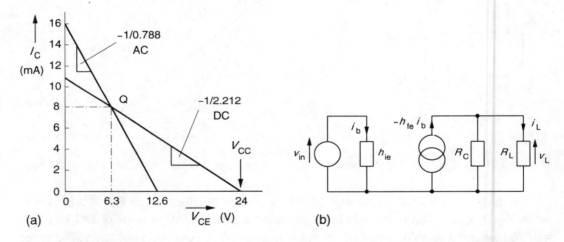

Figure 3.22 (a) The load lines and (b) The small-signal equivalent circuit for example 3.6

The equivalent circuit for calculating the gains is that of figure 3.22b, in which the base-biasing resistances have been omitted since we are told they are large. The input resistance of the amplifier is $h_{ie} = V_{th}/I_{BQ} = V_{th}\beta/I_{CQ} = 0.03 \times 160/8 = 0.6$ kΩ. The voltage gain is

$$A_{vL} = \frac{-h_{fe}R_o}{h_{ie}} = \frac{-160 \times 0.788}{0.6} = -210 = 46.4 \text{ dB}$$

The load current is $i_L = -h_{fe}i_b R_C/(R_C + R_L)$ and the current gain is

$$A_{iL} = \frac{i_L}{i_{in}} = \frac{-h_{fe}i_b R_C}{(R_C + R_L)i_b} = \frac{-160 \times 1.66}{1.66 + 1.5} = -84 = 38.5 \text{ dB}$$

as $i_{in} = i_b$. The power gain is therefore $A_{pL} = A_{vL} \times A_{iL} = 17\,640 = 42.5$ dB.

The maximum peak-to-peak output voltage is 12.6 V_{pp}, so the maximum peak-to-peak input voltage is $12.6/A_{vL} = 12.6/210 = 60$ mV$_{pp}$ or $60/2\sqrt{2} = 21$ mV$_{rms}$, for a sinusoidal input. If gross distortion is to be avoided the maximum input voltage should be 80% of this or 48 mV$_{pp}$ and 17 mV$_{rms}$.

3.2.9 Gain is not dependent on transistor parameters

The voltage-divider bias circuit is a means of making the CE amplifier's gain independent of the transistor parameters, though the formal equations for gain appear to include h_{ie} and h_{fe} — both parameters that are widely quoted in specification sheets and elsewhere. But if we look a little closer, a different picture emerges.

The voltage gain is nearly

$$A_v \approx \frac{-h_{fe}R_o}{h_{ie}}$$

where $R_o = R_C \mathbin{//} R_L$. Here are the two h-parameters, but we can replace h_{ie} by $V_{th}/I_{BQ} = V_{th}h_{fe}/I_{CQ}$ to give

$$A_v \approx \frac{-h_{fe}R_o}{V_{th}h_{fe}/I_{CQ}} = \frac{-I_{CQ}R_o}{V_{th}}$$

But $I_{CQ}R_o$ depends on V_{CC} — in an unloaded amplifier $I_{CQ}R_o \approx \frac{1}{2}V_{CC}$ and in a loaded amplifier somewhat less, ideally. Let us write $I_{CQ}R_o = \zeta V_{CC}$ where $\frac{1}{4} < \zeta < \frac{1}{2}$. Also $V_{th} = \eta kT/q = 0.025\eta$, where η depends on the temperature and the ideality of the base-collector p-n junction: normally $1 < \eta < 1.5$. The voltage gain is then

$$A_v \approx \frac{-\zeta V_{CC}}{0.025\eta} = -\gamma 40V_{CC}$$

where $\gamma = \zeta/\eta$. Usually γ lies between a sixth and a half; taking $\gamma \approx \frac{1}{4}$ gives $A_v \approx -10V_{CC}$.

It is assumed in this analysis that h_{fe} is large, consequently $I_{CQ} \gg I_{BQ}$ and that the transistor has been properly biased with a voltage divider; if this is so then the result is adequate for practical purposes. Note that in the preceding example the voltage gain with load was found to be −210; taking it to be $-10V_{CC}$ would have given $A_v = -240$.

3.3 The emitter follower or common-collector (CC) amplifier

Instead of taking the output of the amplifier from the transistor's collector, it can be taken from the emitter; rather surprisingly the amplifier has quite distinct characteristics differing markedly from the CE amplifier and these will be examined next. The configuration is sometimes called 'common collector' since for AC the collector is common to input and output and it then forms one of the amplifier trinity: CE, CC and CB. But as its behaviour is better described by 'emitter follower' that is what we shall call it.

Figure 3.23a shows an emitter follower which has a base-bias resistor, R_B, and an emitter resistor, R_E. What happens when a signal, v_{in}, is applied to the base? The base current, i_b ($= v_{be}/h_{ie}$), will vary[3] and produce a change in the emitter current, i_e, of almost βi_b which in turn produces an emitter voltage, $v_e = i_e R_E \approx \beta i_b R_E$. Thus

$$v_{in} = v_{be} + v_e = h_{ie} i_b + \beta R_E i_b = R_{in} i_b$$

We can see that the input resistance, R_{in}, is $h_{ie} + \beta R_E$ and if $R_E = 1$ kΩ, $\beta = 100$ and $h_{ie} = 1$ kΩ, then

$$R_{in} \approx \beta R_E = 100 \text{ k}\Omega$$

The input resistance of the emitter follower is fairly high.

To find the voltage gain of the amplifier the approximate small-signal equivalent circuit of figure 3.23b is used. The input voltage is $v_{in} = i_b(h_{ie} + \beta R_E)$, while the output voltage is $v_o = v_e = i_e R_E = (\beta + 1)i_b R_E \approx \beta i_b R_E$ and then the no-load voltage gain becomes

$$A_{v0} = \frac{v_o}{v_{in}} = \frac{\beta i_b R_E}{i_b(h_{ie} + \beta R_E)} \approx 1 \qquad (3.21)$$

since $h_{ie} \ll \beta R_E$, normally. Notice that v_o is in phase with v_{in} — there is no inversion as happens with the CE amplifier. It is because the emitter voltage, v_o, 'follows' v_{in} that the amplifier is given its popular descriptive name.

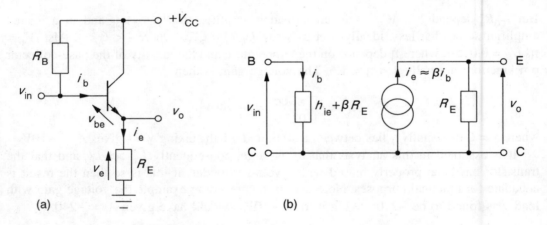

Figure 3.23 (a) An emitter follower (b) its small-signal equivalent circuit

When $R_E = h_{ie}/\beta = r_e$, equation 3.21 shows that $v_o = \frac{1}{2}v_{in}$ and thus that the output resistance[4] of the emitter follower is r_e, that is of the order of 5-10 Ω. The combination of high input resistance and low output resistance makes the emitter follower a good buffer amplifier for driving relatively low-impedance loads.

---

[3] Strictly speaking we should use the common-collector hybrid parameters h_{ic} and h_{fc}, but as these are almost identical to h_{ie} and h_{fe} ($\approx \beta$) we shall use the latter as is done in practice everywhere.

[4] For convenience the quantity h_{ie}/β is denoted r_e, the small-signal emitter resistance of the transistor.

Example 3.7

In the circuit of figure 3.23a, what must R_B and R_E be to make $V_{CEQ} = 6$ V when $V_{CC} = 12$ V, $V_{BEQ} = 0.7$ V and $\beta = 120$, and the output resistance of the amplifier is to be 5 Ω? (Take V_{th} to be 25 mV.)

On the input side by KVL we find that

$$V_{CC} = I_{BQ}R_B + V_{BEQ} + I_{EQ}R_E \approx I_{BQ}R_B + V_{BEQ} + \beta I_{BQ}R_E$$

where the usual approximation, $I_{EQ} \approx I_{CQ} = \beta I_{BQ}$, has been made. This equation leads to

$$I_{BQ} = \frac{V_{CC} - V_{BEQ}}{R_B + \beta R_E} = \frac{11.3}{R_B + 120R_E} \tag{3.22}$$

And as $V_{CEQ} = V_{CC} - I_{EQ}R_E \approx V_{CC} - \beta I_{BQ}R_E$, we find $I_{BQ} = (V_{CC} - V_{CEQ})/\beta R_E = 1/20R_E$. Comparing this to equation 3.22 gives

$$\frac{1}{20R_E} = \frac{11.3}{R_B + 120R_E} \quad \Rightarrow \quad R_B = 106R_E$$

Now the output resistance is $r_e = h_{ie}/\beta = V_{th}/\beta I_{BQ} = 0.025/120I_{BQ} = 5$ Ω, so that $I_{BQ} = 42$ μA. Then substitution for R_B and I_{BQ} into equation 3.21 yields

$$42 \times 10^{-6} = \frac{11.3}{106R_E + 120R_E} \quad \Rightarrow \quad R_E = 1190 \ \Omega$$

and so $R_B = 126$ kΩ.

3.3.1 The emitter follower with load

Consider now the circuit of figure 3.24, showing a signal source of internal resistance, R_s, connected to the base of an emitter follower via a coupling capacitor of negligible reactance. The base bias is now set by a voltage divider to guard against Q-point shifts due to variations in β and the load resistance, R_L, is also capacitively coupled.

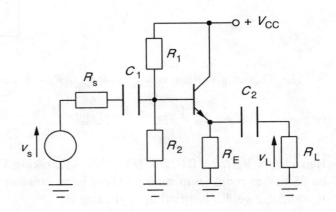

Figure 3.24

An emitter follower with capacitively coupled load and source and voltage divider bias

As far as the source is concerned, R_E and R_L are in parallel and can be combined to form R_o in the equivalent circuit of figure 3.25. Then the transistor's input resistance is approximately βR_o, not βR_E as in the unloaded amplifier of figure 3.23 (neglecting h_{ie}, that is). The base-bias resistors are in parallel for the input signal since R_1 connects to ground via the positive supply, V_{CC}. Thus R_B is R_1 // $R_2 = R_1R_2/(R_1 + R_2)$ and is in parallel with βR_o. In the next example the voltage, current and power gains are calculated.

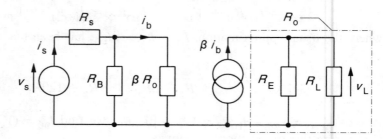

Figure 3.25

The small-signal equivalent circuit of the emitter follower circuit of figure 3.24

Example 3.8

Find the voltage current and power gains for the emitter follower of figure 3.24, where R_1 = 10 kΩ, R_2 = 12 kΩ, R_s = 2.2 kΩ, R_E = 1 kΩ, R_L = 500 Ω, V_{CC} = +15 V and β = 75.

As usual we work in V, mA and kΩ. First we find R_B = 10 kΩ // 12 kΩ = 5.45 kΩ, and R_o = 1 kΩ // 0.5 kΩ = 0.333 kΩ, making βR_o = 75 × 0.333 = 25 kΩ, which is probably large enough to let us neglect h_{ie}. We calculate the base current using the equivalent circuit of figure 3.26, which is the Thévenin equivalent of the voltage-divider bias resistors, just as for the common-emitter amplifier (figure 3.16).

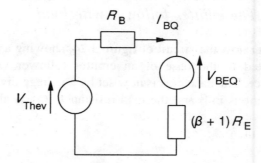

Figure 3.26

The circuit for calculating I_{BQ}; normally $(\beta + 1)R_E$ can be taken as βR_E

The Thévenin voltage is $V_{Thev} = R_2V_{CC}/(R_1 + R_2) = 8.18$ V, and so

$$I_{BQ} = \frac{V_{Thev} - V_{BEQ}}{R_B + \beta R_E} = \frac{8.18 - 0.7}{5.45 + 75 \times 1} = 0.093 \text{ mA}$$

Hence $h_{ie} = V_{th}/I_{BQ} = 0.025/0.093 = 0.27$ kΩ, taking V_{BEQ} to be 0.7 V and V_{th} (= kT/q) to be 25 mV at room temperature. (There is no point in using $\beta + 1$ instead of β here.) As $h_{ie} \ll \beta R_o$, we are justified in neglecting it.

The source current is

$$i_s = \frac{v_s}{R_s + R_B // \beta R_o} = \frac{v_s}{2.2 + 4.475} = 0.1498 v_s \qquad (3.23)$$

Thus the current in the base, i_b, can be found

$$i_b = \frac{R_B i_s}{R_B + R_{in}} = \frac{5.45 \times 0.1498 v_s}{5.45 + 25} = 0.0268 v_s$$

The load voltage is $v_L = \beta i_b R_o = 75 \times 0.0268 v_s \times 0.333 = 0.67 v_s$ and the voltage gain becomes

$$A_{vL} = v_L / v_s = 0.67 = -3.5 \text{ dB} \qquad (3.24)$$

The voltage gain is substantially less than unity because ($= R_B // \beta R_o$) is comparable to R_s, which is a result primarily of the low value of R_B.

The current gain is

$$A_{iL} = \frac{i_L}{i_s} = \frac{v_L / R_L}{0.2246 v_s} = \frac{A_{vL}}{0.1498 R_L} = \frac{0.67}{0.1498 \times 0.5} = 8.95 = 19 \text{ dB}$$

where i_s has been substituted for using equation 3.23 and A_{vL} using equation 3.24. The power gain is then $0.67 \times 8.95 = 6 = 7.8$ dB.

3.3.2 *The bootstrapped emitter follower*

One of the drawbacks of the voltage-divider bias circuit is that the current through the base-biasing resistors has to be much larger than I_{BQ} and this can only be so if the bias resistors are relatively low values. This consideration does not apply to the fixed-bias circuit of figure 3.23a, which, however, suffers from a migrating Q-point as β changes. One way of maintaining the high input resistance to AC signals but keeping the bias resistances low is to use the circuit of figure 3.27a, in which three resistors are used in biasing the base, one of which, R_1, is connected via a capacitor (C_2) to the output. Thus the AC voltage across R_1 is $v_{in} - v_o$ and the AC current drawn by it and the rest of the biasing network (R_1 and R_2) is

$$i_1 = (v_{in} - v_o)/R_1 = v_{in}(1 - A_v)/R_1$$

The effective resistance of the bias network to AC is then

$$R_b = v_{in}/i_1 = R_1/(1 - A_v)$$

Since A_v is only a little less than one for an emitter follower, $R_b \gg R_1$. The input resistance for AC signals is therefore

$$R_{in} = R_b \,//\, [h_{ie} + (\beta + 1)R_o] \approx \beta R_o$$

where $R_o = R_2 \,//\, R_3 \,//\, R_E \,//\, R_L$, as shown in the small-signal equivalent circuit of figure 3.27b, since the bias resistors R_2 and R_3 are capacitively coupled to the output via C_2. This assumes that $R_b \gg R_o$ and $h_{ie} \ll \beta R_o$.

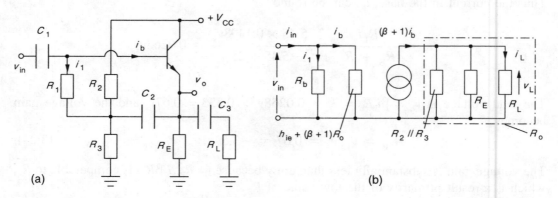

(a)   (b)

Figure 3.27 (a) A bootstrapped emitter follower (b) Its small-signal equivalent circuit

Example 3.9

If R_2 = 15 kΩ, R_3 = 22 kΩ, R_1 = 10 kΩ and $R_L = R_E$ = 2.2 kΩ in figure 3.27, and the transistor has h_{fe} = 175 and V_{CC} = +12 V, what will be the small-signal input resistance, the voltage gain, A_{vL}, and the current gain, A_{iL}, of the circuit? Assume the source resistance is zero and also the reactances of the capacitors. Take V_{th} = 26 mV and V_{BEQ} = 0.7 V.

Because $A_v \approx 1$ and we require $(1 - A_v)^{-1}$ it is necessary to calculate A_v rather accurately, which means finding h_{ie} among other things. For this we require I_{BQ} (as shown in figure 3.26). The equivalent base bias resistance comprises $R_2 \,//\, R_3$ in series with R_1, that is R_B = 15 // 22 + 10 = 18.92 kΩ, which is the Thévenin resistance in figure 3.26. The Thévenin voltage is $R_3 V_{CC}/(R_2 + R_3)$ = 7.135 V and I_{BQ} is given by

$$I_{BQ} = \frac{V_{Thev} - V_{BEQ}}{R_B + (\beta + 1)R_E} = \frac{7.135 - 0.7}{18.92 + 176 \times 2.2} = 0.01585 \text{ mA}$$

and so h_{ie} = 26/0.01585 = 1.64 kΩ. The output resistance is given by

$$R_o = (G_2 + G_3 + G_E + G_L)^{-1} = (1/15 + 1/22 + 1/2.2 + 1/2.2)^{-1} = 0.97923 \text{ k}\Omega$$

Then the input resistance at the transistor's base is

$$R'_{in} = h_{ie} + (\beta + 1)R_o = 1.64 + 176 \times 0.97923 = 173.98 \text{ k}\Omega$$

And the voltage gain is

$$A_{vL} = \frac{(\beta + 1)R_o i_b}{v_{in}} = \frac{(\beta + 1)R_o}{R'_{in}} = \frac{176 \times 0.97923}{173.98} = 0.9906$$

Thus the value of R_b is $10/(1 - 0.9906) = 1064$ kΩ and the overall input resistance of the amplifier is

$$R_{in} = R_b // R'_{in} = (1/1064 + 1/174)^{-1} = 149.5 \text{ k}\Omega$$

The current gain is

$$A_{iL} = \frac{i_L}{i_{in}} = \frac{G_L(\beta + 1)i_b}{G_o \times i_b R'_{in}/R_{in}} = \frac{0.4546 \times (175 + 1)}{1.021 \times 174/149.5} = 67$$

3.4 The common-base amplifier

The common-base amplifier was the first transistor amplifier examined (in section 3.1) to explain some of the physical workings of the BJT. It is the least used of the three transistor configurations because of its low input resistance and high output resistance. Its low input resistance is, however, an advantage in the *cascode* amplifier (see section 3.9.3). Figure 3.28a shows the configuration, which has both a positive collector supply, V_{CC}, and a negative emitter supply, V_{EE}, with a grounded base. The input is the emitter terminal and the output is taken at the collector terminal. Eliminating the negative supply requires a circuit such as that of figure 3.28b, in which the emitter is the DC ground and the base-emitter junction is forward biased, but grounded through C_B for AC signals.

Having the input at the emitter means that the input resistance is the same as the output resistance of the emitter follower, that is r_e — a few ohms; and taking the output from the collector means that the output resistance is the same as the CE amplifier's: R_C.

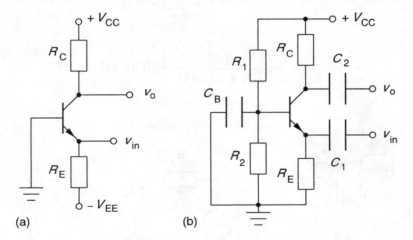

Figure 3.28 (a) A CB amplifier with two DC supplies (b) A CB amplifier with one DC supply

The small-signal equivalent circuit of the CB amplifier is shown in figure 3.29, where it can be seen that the input and output currents are nearly equal since $0.98 < \alpha < 0.998$. The no-load voltage gain is

$$A_{v0} = \frac{v_o}{v_{in}} = \frac{\alpha i_e R_C}{i_e r_e} \approx \frac{R_C}{r_e} = \frac{\beta R_C}{h_{ie}}$$

Usually R_C and h_{ie} are of comparable magnitude and the no-load voltage gain is of the same order as β. The current gain is $\alpha \approx 1$ and the power gain is roughly equal to β.

Figure 3.29

The small-signal equivalent circuit of the CB amplifier

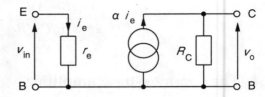

Example 3.10

Find the voltage, current and power gains for the CB amplifier of figure 3.28b, which is used to drive a load of 2 kΩ. The circuit values are $V_{CC} = 5$ V, $R_1 = 10$ kΩ, $R_2 = 4.7$ kΩ, $R_C = 560$ Ω, $R_E = 330$ Ω, $\beta = 200$ and the input signal is a voltage source of internal resistance is 10 Ω. (Take $V_{th} = 30$ mV, $V_{BEQ} = 0.6$ V and neglect the capacitive reactances.)

We must first find r_e, which means finding $I_{EQ} \approx I_{CQ} = \beta I_{BQ}$ using the Thévenin equivalent of the base bias, shown in figure 3.30a, just as we did before. In this circuit $R_B = R_1 /\!/ R_2 = 3.2$ kΩ and $V_{Thev} = V_{CC} R_2 / (R_1 + R_2) = 1.6$ V, so that

$$I_{BQ} = \frac{V_{Thev} - V_{BEQ}}{R_B + \beta R_E} = \frac{1.6 - 0.6}{3.2 + 200 \times 0.33} = 0.01445 \text{ mA}$$

Then

$$r_e = \frac{V_{th}}{\beta I_{BQ}} = \frac{0.03}{200 \times 0.01445} = 0.01 \text{ k}\Omega = 10 \text{ }\Omega$$

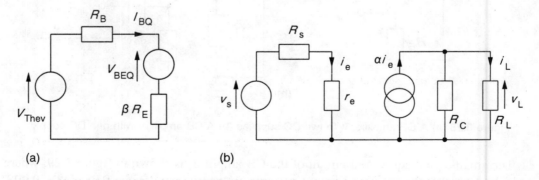

Figure 3.30 (a) The circuit for finding I_{BQ} (b) The equivalent circuit for calculating the gain

The equivalent circuit is as shown in figure 3.30b with $R_s = 10\ \Omega$, hence

$$i_e = \frac{v_s}{(R_s + r_e)} = \frac{v_s}{(0.01 + 0.01)} = 50v_s$$

working in V, mA and kΩ. And the load voltage is

$$v_L = \frac{\alpha i_e R_C R_L}{R_C + R_L} = \frac{\alpha i_e \times 1 \times 2}{3} = 0.667\alpha i_e = 33.3v_s$$

Hence $A_{vL} = 33.3$, taking $\alpha = 1$ instead of 0.995 ($= 1 - 1/\beta$), while the current gain is

$$A_{iL} = \frac{i_L}{i_e} = \frac{R_C \alpha i_e}{i_e(R_C + R_L)} = 0.333$$

The power gain is $A_{vL} \times A_{iL} = 33.3 \times 0.333 = 10$, which is low mainly because the source resistance, though rather low, is equal to r_e.

3.5 The complete h-parameter equivalent circuit

The complete hybrid equivalent circuit is shown in figure 3.31 and is seen to contain four h-parameters: h_i, h_r, h_f and h_o, which can be defined with reference to this circuit. First, on the input (left hand) side

$$V_{in} = h_i I_{in} + h_r V_o \qquad (3.25)$$

Keeping V_o constant, we can vary V_{in} by a small amount, δV_{in}, and the input current then varies by δI_{in} and the ratio is h_i, the input resistance:

$$h_i = \frac{\delta V_{in}}{\delta I_{in}}\bigg|_{V_o = \text{constant}}$$

Figure 3.31

The complete h-parameter equivalent circuit

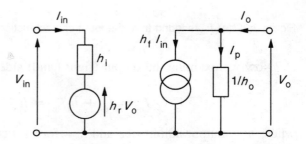

If the circuit were the CE amplifier's equivalent circuit, then the input terminals would be the base and the emitter (ground), so that $\delta V_{in} = \delta V_{BE} = v_{be}$ and $\delta I_{in} = \delta I_B = i_b$. Also the output terminals would be the collector and the emitter (ground), so that $V_o = V_{CE}$, making the input resistance

$$h_{ie} = \left.\frac{\delta V_{BE}}{\delta I_B}\right|_{V_{CE} = \text{constant}} = \left.\frac{v_{be}}{i_b}\right|_{V_{CE} = \text{constant}}$$

We have seen previously (in section 3.2) that the slope of the input characteristics of the CE amplifier can be used to find h_{ie}.

Returning to equation 3.25, we can now keep I_{in} constant and vary V_{in}, so that

$$h_r = \left.\frac{\delta V_{in}}{\delta V_o}\right|_{I_{in} = \text{constant}} \Rightarrow h_{re} = \left.\frac{v_{be}}{v_{ce}}\right|_{I_B}$$

This parameter is the reverse voltage transfer ratio. It can be found from the input characteristics, for example those of the CE amplifier in figure 3.32a, where for $I_B = 80\ \mu A$, $V_{BE} = 0.65$ when $V_{CE} = 6$ V and $V_{BE} = 0.68$ when $V_{CE} = 16$ V. Thus $\delta V_{BE} = 0.03$ V and $\delta V_{CE} = 10$ V giving $h_{re} = 0.03/10 = 3 \times 10^{-3}$ (the change in V_{BE} has been exaggerated for the purpose of illustration). Since $h_{re} \approx 0.001\text{-}0.002$, it is usually neglected.

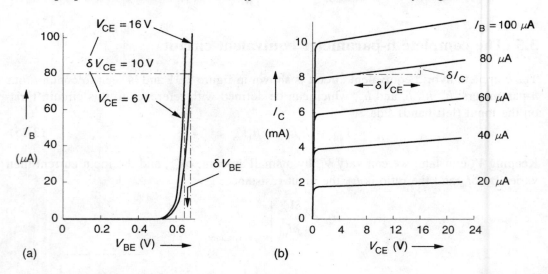

Figure 3.32 (a) h_{re} is found from the input characteristics and (b) h_{oe} from the output characteristics

Now we examine the output (right hand) side of figure 3.31 and see that

$$I_o = I_p + h_f I_{in} = h_o V_o + h_f I_{in} \tag{3.26}$$

where h_o = output admittance and h_f = forward current transfer ratio. In terms of the CE amplifier equation 3.26 leads to

$$h_o = \left.\frac{\delta I_o}{\delta V_o}\right|_{I_{in} = \text{constant}} \Rightarrow h_{oe} = \left.\frac{i_c}{v_{ce}}\right|_{I_B}$$

and

$$h_f = \frac{\delta I_o}{\delta I_{in}}\bigg|_{V_o = \text{constant}} \quad\Rightarrow\quad h_{fe} = \frac{i_c}{i_b}\bigg|_{V_{CE}}$$

Both h_{oe} and h_{fe} can be found from the CE output characteristics of figure 3.32b. The latter has already been calculated in section 3.2, but h_{oe} has not. It is the slope of one of the lines of constant I_B, such as that where $I_B = 80$ μA, where we have $I_C = 8$ mA when $V_{CE} = 6$ V and $I_C = 8.5$ mA when $V_{CE} = 16$ V, so making $\delta I_C = 0.5$ mA and $\delta V_{CE} = 10$ V. Thus $h_{oe} = 0.5 \times 10^{-3}/10 = 5 \times 10^{-5} = 50$ μS. The *Early effect* (discussed in section 3.10.2) is responsible for h_{re} and h_{oe} being non-zero, though they are usually small enough to neglect. The full set of h-parameters can be used to find the voltage, current and power gains of an amplifier as the next example illustrates.

Example 3.11

A CE amplifier has the following set of h-parameters:

$$h_{ie} = 1.1 \text{ k}\Omega \quad h_{re} = 1 \times 10^{-3} \quad h_{oe} = 20 \text{ μS} \quad h_{fe} = 120$$

Find the no-load voltage, current and power gains if $R_C = 2.2$ kΩ and the input signal is from a voltage source of zero internal resistance. Given $R_B = 4$ kΩ in parallel with h_{ie}, and $R_L = 1.5$ kΩ in parallel with R_C in the equivalent circuit and the signal voltage source has an internal resistance of 200 Ω, what are the voltage, current and power gains?

Figure 3.33

The equivalent circuit for a CE amplifier with no load and zero source resistance

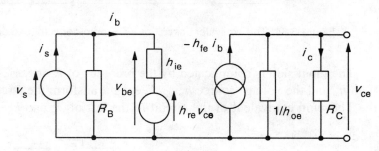

The equivalent circuit is that of figure 3.33 giving

$$G_o = h_{oe} \mathbin{/\!/} 1/R_C = 0.02 + 1/2.2 = 0.4745 \text{ mS or } R_o = 1/G_o = 2.107 \text{ k}\Omega$$

Then

$$v_{ce} = -h_{fe}i_bR_o = -120i_b \times 2.107 = -252.8i_b$$

On the input side we find

$$v_{be} = h_{re}v_{ce} + h_{ie}i_b = 10^{-3}v_{ce} + 1.1i_b$$

But $i_b = -v_{ce}/252.8$, so that

$$v_{be} = 10^{-3}v_{ce} - 1.1v_{ce}/252.8 = -3.35 \times 10^{-3}v_{ce}$$

and then

$$A_{v0} = v_{ce}/v_{be} = -1/3.35 \times 10^{-3} = -298$$

The current gain is

$$A_i = i_c/i_b \text{ and } i_c = v_{ce}/R_C = v_{ce}/2.2$$

while i_b is $-v_{ce}/252.8$ as before, yielding $A_i = -252.8/2.2 = -115$. The power gain is

$$A_p = A_{v0} \times A_i = -298 \times -115 = 34\,270 \text{ or } 45.3 \text{ dB}$$

When loaded and driven by a signal source with internal resistance, the equivalent circuit of the amplifier becomes that of figure 3.34, which shows that the output side is augmented by R_L in parallel with h_{oe} and R.

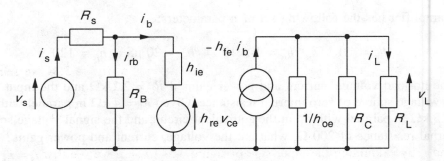

Figure 3.34 The equivalent circuit of a CE amplifier with load and source resistance

The input side is complicated by the presence of R_s in series with v_s and R_B in parallel with h_{ie} and the voltage source $h_{re}v_{ce}$. We can transform the circuit into that of figure 3.35, the Thévenin equivalent of the circuit to the left of h_{ie} and $h_{re}v_{ce}$.

Figure 3.35

The circuit for calculating i_b

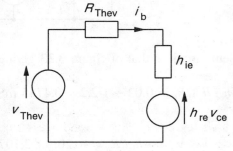

R_{Thev} is R_s // $R_B = 0.1905$ kΩ and v_{Thev} is $4v_s/4.2 = 0.9524v_s$ and then

$$i_b = \frac{v_{Thev} - h_{re}v_{ce}}{R_{Thev} + h_{ie}} \tag{3.27}$$

But $v_{ce} = -h_{fe}i_bR_o = -h_{fe}i_b/G_o$, where

$$G_o = \frac{1}{R_o} = \frac{1}{R_L} + \frac{1}{R_C} + h_{oe} = \frac{1}{1/5} + \frac{1}{2.2} + 0.02 = 1.141 \text{ mS}$$

giving

$$v_{ce} = \frac{-h_{fe}i_b}{G_o} = \frac{-120i_b}{1.141} = -105.2i_b \tag{3.28}$$

working in V, mA and mS. Substituting equation 3.28 into equation 3.27 produces

$$i_b = \frac{0.9524v_s - (10^{-3}) \times (-105.2i_b)}{0.1905 + 1.1} = 0.738v_s + 0.0815i_b$$

Whence $i_b = 0.803v_s$ and therefore

$$A_{vL} = \frac{-h_{fe}i_bR_o}{v_s} = \frac{-h_{fe}i_b}{v_sG_o} = \frac{-120 \times 0.803v_s}{1.141v_s} = -84.5$$

The current gain is i_L/i_s and looking at the output side of figure 3.34 we see that

$$i_L = \frac{-h_{fe}i_bG_L}{G_o} = \frac{-h_{fe}i_b}{R_LG_o} = \frac{-120i_b}{1.5 \times 1.141} = -70.1i_b$$

We find i_s by examining the input side of figure 3.34, where we see that $i_s = i_{rb} + i_b$ and $i_{rb} = v_{be}/R_B$. Now

$$v_{be} = h_{re}v_{ce} + i_bh_{ie} = 10^{-3} \times (-105.2i_b) + 1.1i_b = 0.995i_b$$

using equation 3.27 for v_{ce} and so $i_{rb} = 0.995i_b/4 = 0.249i_b$ to give $i_s = 1.249i_b$. Hence

$$A_{iL} = \frac{i_L}{i_s} = \frac{-70.1i_b}{1.249i_b} = -56.1$$

The overall power gain is

$$P_L = |A_{vL}| \times |A_{iL}| = 84.5 \times 56.1 = 4751 = 36.8 \text{ dB}$$

It is instructive to recalculate the gains using the approximate equivalent circuit of figure 3.18b, which ignores h_{re} and h_{oe} and uses only h_{ie} and h_{fe} (= β). The solutions are obtained more easily and are not significantly different.

Example 3.12

If, in example 3.11, h_{re} and h_{oe} are ignored, what are the voltage, current and power gains? The equivalent circuits are those of figures 3.36a and 3.36b.

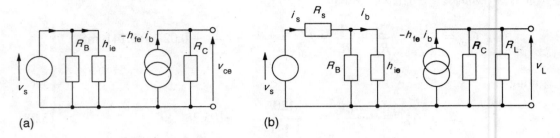

Figure 3.36 Equivalent circuits for example 3.12 (a) With no load and zero source resistance
(b) With load and source resistance

With a source of zero internal resistance $v_s = i_b h_{ie} = 1.1 i_b$ and on the output side we have $v_{ce} = -h_{fe} i_b R_C = -120 \times 2.2 i_b = -264 i_b$. The no-load voltage gain is

$$A_{v0} = v_{ce}/v_s = -264 i_b/1.1 i_b = -240$$

The current gain is $A_i = i_c/i_b = -h_{fe} i_b/i_b = -h_{fe} = -120$ and $A_p = 240 \times 120 = 28\,800 = 44.6$ dB, which is 0.7 dB less than the result obtained using the full h-parameter circuit.

When loaded, the resistance across the output is $R_o = R_C \,/\!/\, R_L = 2.2 \times 1.5/3.7 = 0.892$ kΩ, and then $v_L = -h_{fe} i_b R_o = -120 \times 0.892 i_b = -107 i_b$. The base current is found from equation 3.26 with $h_{re} = 0$

$$i_b = \frac{v_{Thev}}{R_{Thev} + h_{ie}} = \frac{0.9524 v_s}{0.1905 + 1.1} = 0.738 v_s \qquad (3.29)$$

since neither v_{Thev} nor R_{Thev} is altered. The voltage gain with load then becomes

$$A_{vL} = v_L/v_s = -107 i_b \times 0.738/i_b = -79.0$$

The current gain is $A_{iL} = i_L/i_s$ where

$$i_L = \frac{-h_{fe} i_b R_C}{R_C + R_L} = \frac{-120 \times 2.2 i_b}{2.2 + 1.5} = -71.4 i_b$$

The source current is

$$i_s = \frac{v_s}{R_s + R_B /\!/ h_{ie}} = \frac{v_s}{0.2 + (4 \times 1.1)/(4 + 1.1)} = 0.941 v_s$$

The ratio of these is

$$A_{iL} = \frac{i_L}{i_s} = \frac{-71.4 i_b}{0.941 v_s} = \frac{-71.4 i_s \times 0.738}{0.941 i_s} = -56.0$$

using equation 3.29 for v_s ($= i_s/0.738$). The power gain is $79 \times 56 = 4424 = 36.5$ dB, only 0.3 dB higher than the 'exact' solution. For practical purposes the approximate solutions are as good as the 'exact'.

3.6 Comparison of the CE, CC and CB amplifiers

The characteristics of the three amplifiers are compared in tabular form below. The common-emitter amplifier is best for general purposes, and the emitter follower (common-collector amplifier) is a reasonably good buffer, especially when bootstrapped. The common-base amplifier has few uses, apart from the cascode amplifier, which is discussed in section 3.9.4.

Table 3.2 *Characteristics of the CE, CC and CB amplifiers*

Parameter	CE	CC	CB
Voltage gain, A_v	High	≈ 1	High
Current gain, A_i	High	High	≈ 1
Power gain, A_p	High	Medium	Medium
Input resistance, R_{in}	Medium	High	Low
Output resistance, R_o	Medium	Low	Medium

3.7 The effect of temperature

Temperature affects the BJT chiefly in three ways: firstly, the leakage current through the reverse-biased collector-base junction, I_{CBO}, increases exponentially, secondly β increases (by about 1%/K) and thirdly, the base-emitter junction voltage, V_{BEQ}, falls by about 2 mV/K. The first of these effects has potentially the more serious consequence: thermal runaway. This used to be a much worse problem in germanium transistors where the leakage current was already significant at room temperature, but with silicon transistors it is less serious and should be eliminated with proper biasing.

We can calculate the effect of leakage current by inserting a current source of magnitude I_{CBO} between the collector and base terminals of the transistor as shown in figure 3.37a. The current into the base then becomes $I'_{BQ} = I_{BQ} + I_{CBO}$ and the current in the collector is $I'_{CQ} = I_{CQ} - I_{CBO}$. The transistor amplifies the true base current, I'_{BQ}, to form the true collector current, I'_{CQ}, that is

$$I'_{CQ} = \beta I'_{BQ} \;\Rightarrow\; I_{CQ} - I_{CBO} = \beta(I_{BQ} + I_{CBO})$$

and hence

$$I_{CQ} = \beta I_{BQ} + (\beta + 1)I_{CBO} \approx \beta(I_{BQ} + I_{CBO})$$

The collector current increases by βI_{CBO} because of the leakage current. At 25°C the leakage currents in silicon are almost always negligible, but as we saw in section 2.1.4, I_s — which is the same as I_{CBO} — doubles for each 7 K rise in temperature. Thus at 95°C, a temperature easily reached by a p-n junction in a transistor dissipating rated power, the leakage current can be 2^{10} or 1000 times greater than at room temperature. A negligible current of 10 nA, becomes a significant one of 10 µA, especially if the quiescent base current is only a few tens of µA.

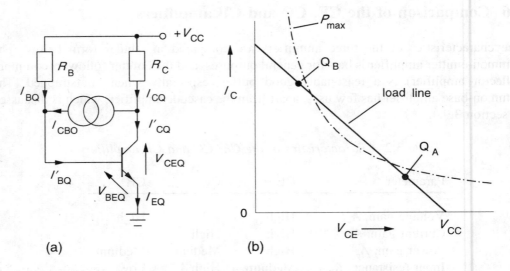

Figure 3.37 (a) A common-emitter amplifier with leakage current, I_{CBO}, passing from collector to base
(b) Leakage will move the Q-point higher, leading to *increased* power consumption at Q_A but *decreased* power consumption at Q_B

Since the effect of increased leakage current is to increase the collector current in the fixed-bias circuit of figure 3.37a, we can see that the power dissipation in the transistor is altered. This power is given by

$$P = V_{CEQ}I_{CQ} = (V_{CC} - I_{CQ}R_C)I_{CQ}$$

Differentiating with respect to I_{CQ} gives

$$dP/dI_{CQ} = V_{CC} - 2I_{CQ}R_C$$

which is positive if $I_{CQ} < \frac{1}{2}V_{CC}/R_C$ and negative if $I_{CQ} > \frac{1}{2}V_{CC}/R_C$. In the former case an increase in I_{CQ} will cause an increase in power dissipation, leading to further increases in I_{CBO} and hence I_{CQ}. The equilibrium point is reached when $I_{CQ} = \frac{1}{2}V_{CC}/R_C$ and $V_{CEQ} = \frac{1}{2}V_{CC}$. Provided the power dissipation at this point is within the transistor's rating (with suitable adjustments for the actual junction temperature) there will be no harm done.

The Q-point of the voltage-divider bias circuit, which fixes I_{EQ}, is not affected by changes in I_{CBO} and is preferable, especially as it overcomes difficulties due to variable β.

3.8 The frequency response

Hitherto the amplifier gain has been assumed to be independent of frequency (a 'flat response'), whereas in reality a plot of gain against frequency has the shape of that in figure 3.38. The central part — the mid-band gain — is indeed flat, but at both lower and higher frequencies the graph declines at a rate of 20 dB per decade, a first-order response that is typical of an RC circuit.

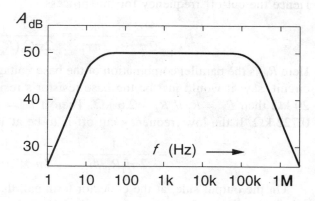

Figure 3.38

The frequency response of a CE amplifier

3.8.1 The low-frequency response

The low-frequency response of the BJT amplifier is determined by the external circuit RC components alone, whereas the high-frequency response is primarily a function of the device's intrinsic capacitances.

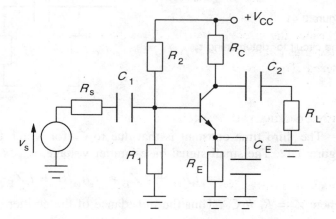

Figure 3.39

A fully AC-coupled, CE amplifier with emitter bypass and voltage-divider bias

We will determine the low-frequency response of the CE amplifier of figure 3.39, which is voltage-divider biased. There are three capacitors in the circuit external to the device: the input coupling capacitor, C_1, the output coupling capacitor, C_2, and the emitter bypass capacitor, C_E, with associated time constants τ_1, τ_2 and τ_E.

The input side of the amplifier is shown in figure 3.40, from which it is seen that the capacitor charging current flows from source to ground via R_s in series with $R_B \ // \ h_{ie}$.

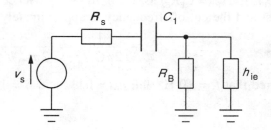

Figure 3.40

The circuit for determining τ_1

Hence the cut-off frequency for this process is

$$f_1 = \frac{\omega_1}{2\pi} = \frac{1}{2\pi\tau_1} = \frac{1}{2\pi C_1 (R_s + R_B /\!/ h_{ie})} \qquad (3.30)$$

Here R_B is the parallel combination of the base voltage-divider resistors, but in a fixed-bias circuit, say, it would just be the base resistor's resistance. Suppose $R_2 = 3$ kΩ and $R_1 = 20$ kΩ then $R_B = R_1 /\!/ R_2 = 2.6$ kΩ. Then if $h_{ie} = 1$ kΩ and $R_s = 0$, we find $R_B /\!/ h_{ie} = 0.722$ kΩ. If the low-frequency cut-off is to be at 10 Hz, equation 3.30 yields

$$C_1 = \frac{1}{2\pi f_1 R_B /\!/ h_{ie}} = \frac{1}{2\pi \times 10 \times 772} \approx 20 \ \mu\text{F}$$

On the output side, at the collector terminal, the circuit is that of figure 3.41 and the coupling capacitor is seen to charge via R_C and R_L in series. The cut-off frequency is thus

$$f_2 = 1/2\pi C_2 (R_C + R_L)$$

If $R_C + R_L = 2$ kΩ and $f_2 = 10$ Hz, then C_2 is 8 μF.

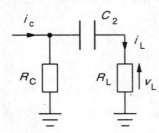

Figure 3.41

The circuit for determining τ_2

The third time constant is that due to C_E for which the relevant circuit is shown in figure 3.42. The small-signal base-emitter voltage, v_{be}, is given by

$$v_{be} = i_b h_{ie} + i_e Z_E \approx i_e h_{ie}/\beta + i_e Z_E = i_e(r_e + Z_E)$$

where $Z_E = R_E /\!/ C_E$. Thus the impedance of the emitter circuit is

$$Z_e = \frac{v_e}{i_e} = r_e + Z_E = r_e + \frac{R_E}{1 + j\omega C_E R_E}$$

$$= \frac{(r_e + R_E) + j\omega C_E R_E r_e}{1 + j\omega C_E R_E} \approx \frac{R_E(1 + j\omega \tau_e)}{1 + j\omega \tau_E}$$

where $\tau_e = C_E r_e \ll \tau_E \ (= C_E R_E)$ as $r_e = h_{ie}/\beta \ll R_E$. Hence with falling frequency τ_e comes into play first and the cut-off frequency is approximately

$$f_e = \frac{1}{2\pi C_E r_e} \approx \frac{\beta}{2\pi C_E h_{ie}}$$

If it is required that $f_e = 10$ Hz and $h_{ie} = 1$ kΩ and $\beta = 100$, then $C_E \approx 1600$ μF.

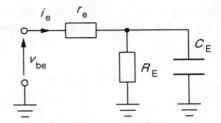

Figure 3.42

The circuit for determining τ_E

One of the three cut-off frequencies is usually substantially greater than the others and dominates the low-frequency response, to give its characteristic rise with increasing frequency of 20 dB/decade. If all three cut-offs were to coincide, then the resultant corner frequency — the point where the response is 3 dB down from the mid-band response — would be about five times higher and its rate of increase would be much greater.

3.8.2 The high-frequency response

The circuit model used most often for the high-frequency response of the BJT is the hybrid-π model, shown in figure 3.43b for the CE amplifier. The resistance, r_x, is that between the base terminal, B, and the transistor base, B$'$, and is about 10-50 Ω in a small transistor. r_π is the resistance of the base-emitter junction as seen by the base current and is identical to h_{ie} in the h-parameter model. The most important additional features of the hybrid-π model are the device capacitances shown in figure 3.43a. The base-emitter capacitance, C_{be}, becomes C_π in figure 3.43b. Then there is C_{cb}, the collector-base capacitance which is C_μ in figure 3.43b, while the collector-emitter capacitance ($= C_{ce}$) is left out of the hybrid-π model for reasons which will become clear later. The current source on the output side is given the value $g_m v_{b'e}$ (equivalent to $h_{fe} i_b$ in the h-parameter model), where g_m, the *transconductance*, is the conductance of the base-emitter junction as seen by i_e, and is of the order of 100 mS. If $g_m v_{b'e}$ is equated to $h_{fe} i_b$ in the h-parameter model, we find at room temperature

$$g_m = \frac{h_{fe} i_b}{v_{b'e}} = \frac{h_{fe}}{h_{ie}} \approx 40\beta I_{BQ}$$

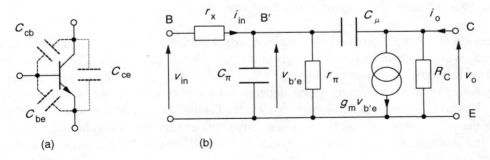

Figure 3.43 (a) The internal device capacitances (b) The hybrid-π model which incorporates them

The bridging capacitance, C_μ, complicates the analysis and is replaced by its equivalent in the circuit of figure 3.44, a much easier circuit to analyse.

Figure 3.44

The transformed hybrid-π equivalent circuit

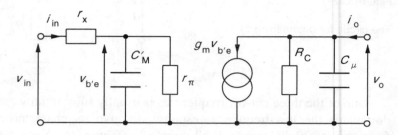

To find this equivalent circuit we look at the input side of the circuit of figure 3.43b and by nodal analysis find

$$i_{in} = v_{b'e}y_\pi + (v_{b'e} - v_o)y_\mu$$

Where y_π stands for r_π // C_π and y_μ for $j\omega C_\mu$. Dividing by $v_{b'e}$ gives

$$y_{in} = i_{in}/v_{b'e} = y_\pi + (1 - A_v)y_\mu$$

where A_v ($= v_o/v_{b'e}$) is the voltage gain of the amplifier. The term $(1 - A_v)y_\mu$ can be modelled as a capacitance of $(1 - A_v)C_\mu$ ($= C_M$) in parallel with y_π. Thus the capacitance across the input becomes $C'_M = C_M + C_\pi$. C_M is known as the Miller[5] capacitance. We can see too from figure 3.42b that $A_v = v_o/v_{b'e} = -g_m v_{b'e}R_C/v_{b'e} = -g_m R_C$. Thus A_v is negative, its magnitude $\gg 1$, so that $(1 - A_v)C_\mu = C_M \gg C_\pi$ and $C_M \approx A_v C_\mu \approx C'_M$. The impedance on the output side can be deduced similarly:

$$i_o = v_o y_c + g_m v_{b'e} + (v_o - v_{b'e})y_\mu$$

$$\Rightarrow \quad y_o = i_o/v_o = y_c + g_m K_R + (1 - K_R)y_\mu$$

where $y_c = 1/R_C$ and K_R ($= v_{b'e}/v_o \ll 1$) is the reverse voltage transfer ratio. Thus $(1 - K_R)y_\mu \approx y_\mu$ and the output side is modelled as a capacitance of C_μ in parallel with R_C. C_μ then includes the collector-emitter capacitance apparently omitted in figure 3.43b.

How to estimate the capacitances of the model

In section 2.2.2 we estimated the capacitances associated with p-n junctions and that analysis will suffice to estimate the hybrid-π capacitances, C_μ and C_M. The collector-base junction is normally reverse biased in the CE amplifier and its capacitance is the depletion capacitance of a p-n junction, which depends on the bias voltage but is of the order of 100 pF/mm^2 in a small, general-purpose device. Taking the junction area to be 0.04 mm^2 gives $C_{cb} = C_\mu = 4$ pF, a typical value. The base-emitter junction is conducting in forward bias and its capacitance is diffusion capacitance which depends on the emitter current. In

---

[5] The magnification of the bridging component by the amplifier's gain is known as the *Miller effect*.

section 2.2.2 this capacitance was given as $C_d = 40I_E q_s$, where $q_s = 20$ nC/A/mm², making C_d about $40 \times 20 \times 10^{-9} \times 5 \times 10^{-3} \times 0.01 = 40$ pF, if the junction area is 0.01 mm² and $I_E = 5$ mA. Again this is a typical value for $C_{be} (= C_\pi)$.

The amplifier's gain, A_v, is $-g_m R_C \approx -0.1 \times 1000 = -100$, giving $C_M = (1 - A_v)C_\mu = 101 \times 4 = 404$ pF and $C'_M = C_M + C_\pi = 404 + 40 = 444$ pF. In practice, one knows neither C_π nor C_μ accurately, nor is g_m usually given in data sheets, so the approximation $C'_M \approx C_M \approx A_v C_\mu \approx \beta C_\mu$ must suffice.

Example 3.13

A CE amplifier is driven by a voltage source of internal resistance 100 Ω, has an equivalent base resistance, R_B, of 2 kΩ, $R_C = 1$ kΩ and is capacitively coupled to a load of 1.5 kΩ. If the circuit model of figure 3.44 applies and $g_m = 110$ mS, $C_\mu = 4$ pF, $C_\pi = 90$ pF, $r_x = 30$ Ω and $r_\pi = 800$ Ω, find A_{vL} at high frequencies as a function of ω.

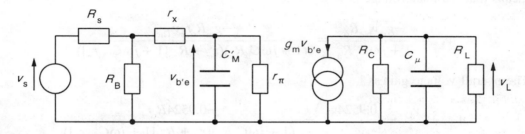

Figure 3.45 The circuit for example 3.13

Figure 3.45 shows the equivalent circuit, where the coupling capacitors have been omitted for they have negligible reactances at high frequencies. The Miller capacitance is

$$C_M = (1 + g_m R_C)C_\mu = (1 + 110 \times 1) \times 4 = 444 \text{ pF}$$

working in mS, kΩ and pF. And then $C'_M = 444 + 90 = 534$ pF.

The source can be transformed as in figure 3.46a into its Thévenin equivalent, with $v_{Thev} = v_s R_B/(R_B + R_s) = 2v_s/2.1 = 0.9524v_s$ and $R_{Thev} = R_B /\!/ R_s = 95.2$ Ω.

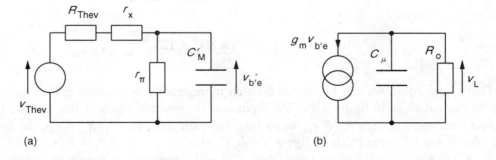

(a)

(b)

Figure 3.46 (a) The Thévenin equivalent of the source and R_B (b) The output side with $R_o = R_C /\!/ R_L$

Combining R_{Thev} and r_x gives $R_{\text{eq}} = R_{\text{Thev}} + r_x = 95.2 + 30 = 125.2 \ \Omega$. Then $v_{b'e}$ by voltage division is:

$$v_{b'e} = \frac{v_{\text{Thev}} \times r_\pi // C'_M}{R_{\text{eq}} + r_\pi // C'_M}$$

where

$$r_\pi // C'_M = \frac{r_\pi}{1 + j\omega C'_M r_\pi}$$

so that

$$v_{b'e} = \frac{v_{\text{Thev}} r_\pi}{r_\pi + j\omega C'_M r_\pi}$$

Looking at the output side, we can combine R_C and R_L in parallel to form $R_o = 1.5/2.5 = 0.6 \ \text{k}\Omega$ as in figure 3.46b. The output impedance is $R_o // C_\mu = R_o/(1 + j\omega C_\mu R_o)$ which means that the load voltage is

$$v_L = \frac{-g_m v_{b'e} R_o}{1 + j\omega C_\mu R_o} = \frac{-g_m R_o r_\pi v_{\text{Thev}}}{(1 + j\omega C_\mu R_o)(r_\pi + R_{\text{eq}}[1 + j\omega C'_M r_\pi])}$$

The overall voltage gain is

$$A_{vL} = \frac{v_L}{v_s} = \frac{0.9524 v_L}{v_{\text{Thev}}} = \frac{-0.9524 R_o r_\pi}{(1 + j\omega C_\mu R_o)(r_\pi + R_{\text{eq}}[1 + j\omega C'_M r_\pi])}$$

which can be written

$$A_{vL} = \frac{K}{(1 + j\omega\tau_1)(1 + j\omega\tau_2)}$$

where

$$K \equiv \frac{-0.9524 g_m R_o r_\pi}{r_\pi + R_{\text{eq}}} = \frac{-0.9524 \times 0.11 \times 600 \times 800}{800 + 125.2} = -54.4$$

and

$$\tau_1 \equiv C_\mu R_o = 4 \times 10^{-12} \times 600 = 2.4 \ \text{ns}$$

$$\tau_2 \equiv \frac{C'_M r_\pi R_{\text{eq}}}{r_\pi + R_{\text{eq}}} = \frac{534 \times 10^{-12} \times 800 \times 125.2}{925.2} = 57.8 \ \text{ns}$$

These time constants can be obtained much more readily by examining the relevant parts of the circuit as in figure 3.46. The input side is shown in figure 3.46a, and we can see that the capacitor sees $R_{\text{eq}} // r_\pi$, so its time constant must be $C'_M(R_{\text{eq}} // r_\pi)$, the same as τ_2 above. On the output side in figure 3.46b, C_μ charges via $R_C // R_L = R_o$ to give τ_1 as above. Since τ_2 is over ten times larger than τ_1 it will dominate the response, so that the upper cut-off will be at

$$f_2 = \frac{1}{2\pi\tau_2} = \frac{1}{2\pi \times 5.78 \times 10^{-8}} = 2.75 \text{ MHz}$$

From here on the response will fall at 20 dB/decade up to f_1 (= $1/2\pi\tau_1$ = 66 MHz). Beyond f_1 the response falls at 40 dB/decade. The mid-band gain (where $f \ll f_2$) is just $K = -54.4$ = 34.7 dB. At f_2 the gain is 31.7 dB and at f_1 it is

$$34.7 - 20\log(f_2/f_1) - 3 = 4.1 \text{ dB}.$$

The β-cut-off frequency

This is the corner frequency of the current response of the CE amplifier with the collector short-circuited to the emitter. The equivalent circuit is that of figure 3.47, from which we can deduce that

$$A_i = \frac{i_{sc}}{i_b} = \frac{g_m v_{b'e} - v_{b'e} y_\mu}{v_{b'e} y_\pi + y_\mu} \approx \frac{g_m}{y_\mu + y_\pi}$$

since y_μ (= $j\omega C_\mu$) $\ll y_\pi$ (= $1/r_\pi + j\omega C_\pi$). The expression above can be written

$$A_i \approx \frac{g_m r_\pi}{1 + j\omega r_\pi(C_\pi + C_\mu)} \approx \frac{g_m r_\pi}{1 + j\omega r_\pi C_\pi} = \frac{g_m r_\pi}{1 + j\omega\tau_\pi}$$

as $C_\pi \gg C_\mu$ and so the β cut-off is at $f_\beta \approx 1/2\pi\tau_\pi$. The values in example 3.12 give

$$f_\beta = \frac{1}{2\pi r_\pi C_\pi} = \frac{1}{2\pi \times 800 \times 80 \times 10^{-12}} = 2.5 \text{ MHz}$$

Figure 3.47

The equivalent circuit for determining the β cut-off

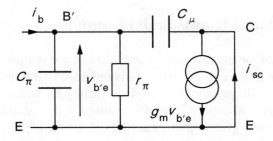

The emitter-follower's high-frequency response

The high-frequency response of the EF amplifier can be found with the aid of figure 3.48, which shows the hybrid-π equivalent circuit. Here the bridging component now goes between base and emitter, so must be the parallel combination of C_π and r_π or y_π, while that between base and collector is C_μ (= y_μ). By using nodal analysis at E we find

$$v_o g_E + (v_o - v_{b'c})y_\pi = g_m v_{b'e} = g_m(v_{b'c} - v_o)$$

since $v_{b'e} = v_{b'c} - v_o$, and so this equation becomes

$$v_o(g_E + g_m + y_\pi) = v_{b'c}(g_m + y_\pi)$$

Since $g_E = 1/R_E \ll g_m$, the equation above leads to $v_o \approx v_{b'c}$. Nodal analysis at B′ yields

$$v_{b'c}y_\mu + (v_{b'c} - v_o)y_\pi + (v_{b'c} - v_s)g_x = 0$$

$$\Rightarrow \quad v_o(g_x + y_\mu) = v_s g_x \quad \Rightarrow \quad \frac{v_o}{v_s} = \frac{g_x}{g_x + y_\mu} = \frac{1}{1 + j\omega r_x C_\mu}$$

Taking $v_{b'c} = v_o$ and $g_x = 1/r_x$. Thus the high-frequency break point is at

$$f_H = \frac{1}{2\pi r_x C_\mu}$$

If $r_x = 50\ \Omega$ and $C_\mu = 5$ pF, $f_H = 640$ MHz — much higher than the high-frequency break point of the CE amplifier.

Figure 3.48

The equivalent circuit for the EF amplifier at high frequencies

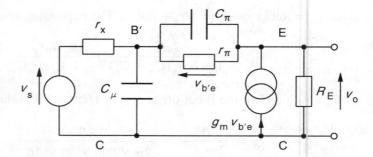

3.9 Multi-transistor BJT amplifiers

The gain obtainable from a single transistor is limited: in the case of the CE amplifier the voltage gain is about $10V_{CC}$, or about 35-45 dB with supply voltages of 5-18 V. Especially in low-voltage circuits therefore, it may be necessary to connect ('cascade' is the term often used, though it is easily confused with 'cascode' by the innocent). Two cascaded CE stages will give higher gain, though at the expense of frequency response. Two stages might also be needed to increase the input resistance of the CE amplifier — by using a CC stage on the input, for example. In addition to these multi-stage, cascaded amplifiers, there are other ways of using several transistors to achieve particular ends. Power transistors for instance tend to draw large base currents and have correspondingly small input resistances. The input base current can be reduced, and the input resistance simultaneously increased, by connected a pair of them together as in the Darlington configuration discussed next. High-power amplifiers are considered separately in chapter eight, which is particularly concerned with efficiency, a particularly weak point in CE amplifiers.

3.9.1 The Darlington connection

This pair of transistors, shown in figure 3.49a, is available in a single 3-terminal package with a variety of different specifications. Its advantages over a single transistor are:

- high input resistance
- low output resistance
- high beta

Darlingtons are often used in power amplifier output stages (see section 8.2) while its high input resistance is put to good effect in many op amps.

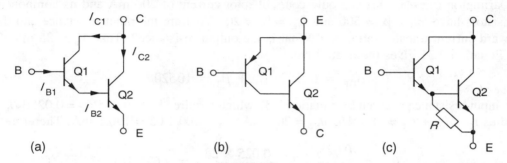

Figure 3.49   (a) The Darlington connection (b) The Sziklai connection: this acts like a discrete pnp transistor whose terminals are labelled B, C and E (c) R allows leakage current from Q1 to bypass Q2 and also reduces the switching time of Q2

The emitter of the input transistor is connected to the base of the output transistor so that $I_{B2} = I_{E1} = (\beta_1 + 1)I_{B1}$ and as $I_{C2} = \beta_2 I_{B2}$. Therefore the overall β is

$$\beta_{DP} = \frac{I_C}{I_B} = \frac{I_{C1} + I_{C2}}{I_{B1}} = \beta_1 + \frac{\beta_2 I_{B2}}{I_{B1}} = \beta_1 + \frac{\beta_2 I_{E1}}{I_{B1}}$$

$$= \beta_1 + \beta_2(\beta_1 + 1) = \beta_1\beta_2 + \beta_1 + \beta_2 \approx \beta_1\beta_2 \qquad (3.31)$$

The range of betas might be from 20 to 300 and so β_{DP} can be anywhere from 400 to 90 000. The manufacturer will give either a typical value, or a minimum value, or a range of values. For example, the ZTX605 is an E-line, plastic-packaged device with an $h_{fe}(\min)$, or minimum β, of 2 000 at $I_C = 50$ mA; it can operate at supply voltages up to 140 V and a maximum power dissipation of 1 W. The cost is only 40p in single quantities.

The input resistance of Q1 is h_{ie1}, and since the output current of Q1 goes into Q2's base, the overall input resistance is $h_{ie1} + (\beta_1 + 1)h_{ie2}$. However,

$$h_{ie2} \approx \frac{0.025}{I_{B2}} = \frac{0.025}{I_{E1}} = \frac{0.025}{(\beta_1 + 1)I_{B1}} = \frac{h_{ie1}}{\beta_1 + 1}$$

Thus

$$h_{ieDP} = h_{ie1} + (\beta_1 + 1)h_{ie2} = 2h_{ie1} = 2(\beta_1 + 1)h_{ie2} \qquad (3.32)$$

And so the input resistance is increased by a factor of about $2\beta_1$.

The output resistance of an emitter follower is h_{ie}/β, so the output resistance of a Darlington emitter follower will be

$$r_{oDP} = \frac{h_{ieDP}}{\beta_{DP}} = \frac{2(\beta_1 + 1)h_{ie2}}{\beta_1(\beta_2 + 1)} \approx \frac{2h_{ie2}}{\beta_2} \tag{3.33}$$

which is about double that of a normal emitter follower, but still low.

Example 3.14

A Darlington transistor takes a quiescent collector current of 200 mA and its component transistors have $h_{fe1} = \beta_1 = 500$ and $h_{fe2} = \beta_2 = 20$. What are the input resistance and the forward current transfer ratio, β_{DP}? What is the output resistance? (Take $V_{th} = 25$ mV.)

Equation 3.31 gives the overall beta:

$$\beta_{DP} = \beta_1\beta_2 + \beta_1 + \beta_2 = 10\,520$$

The input resistance is given by equation 3.32 which requires $h_{ie2} = 0.025/I_{B2} = 0.025\beta_2/I_{C2}$. And as $I_C = I_{C1} + I_{C2} = (1 + 1/\beta_2)I_{C2} = 200$ mA, $I_{C2} = 200/1.05 = 190.5$ mA. Therefore

$$h_{ie2} = \frac{0.025\beta_2}{I_{C2}} = \frac{0.025 \times 20}{190.5 \times 10^{-3}} = 2.625 \ \Omega$$

Substituting this into equation 3.31 gives

$$r_{in(DP)} = 2(\beta_1 + 1)h_{ie2} = 2 \times 501 \times 2.625 = 2.63 \ k\Omega$$

The output resistance of the Darlington pair is given by equation 3.33:

$$r_{o(DP)} = \frac{2h_{ie2}}{\beta_2 + 1} = \frac{2 \times 2.625}{21} = 0.125 \ \Omega$$

The *Sziklai connection* is shown in figure 3.49b, and acts exactly like a Darlington pair. The pnp input transistor means that overall it behaves like a pnp transistor. It is sometimes called a *pseudo-complementary Darlington*, since the Sziklai-connected complementary Darlington has a pnp transistor in place of Q2 and an npn in place of Q1, the pair then acting overall like an npn transistor.

The leakage current flowing in the input transistor of the Darlington pair is amplified by β_{DP} when it appears at the emitter terminal. Since β_{DP} can be up to $100\,000$ it may produce a large output current, especially at elevated temperatures, and Q2 is effectively turned on. Placing a resistor between the base and emitter terminals of Q2, as in figure 3.49c, prevents this turn on due to leakage current. The value of R required is about 1 kΩ; see problem 3.23. The switching speed of Q2 is also increased by R. Darlington packages usually contain R as part of a monolithic circuit.

3.9.2 The push-pull amplifier

The push-pull amplifier makes use of complementary BJTs to push and then pull current alternately into and out of a load, as shown in figure 3.50. When v_{in} is positive the npn transistor (Q1) is turned on, drawing current from the positive supply and sending it through the load, while Q2 (a pnp transistor) is turned off. When v_{in} is negative, Q2 turns on and draws current through the load from earth, returning it to the negative supply while Q1 is turned off. The advantage of the push-pull, or class B, amplifier over single-transistor amplifiers is their far higher efficiencies. The input resistance is quite low when single transistors are used, so these are usually complementary, matched Darlingtons. Push-pull amplifiers are discussed further in section 8.2.2.

Figure 3.50

A BJT push-pull (class-B) amplifier

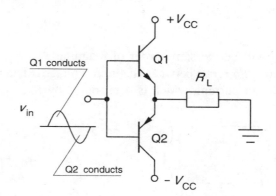

3.9.3 The BJT differential amplifier

When the difference between two voltages must be amplified, the configuration of figure 3.51a can be employed, which is the BJT differential amplifier (sometimes known as a 'long-tailed pair'). It comprises two transistors in CE configuration with a single, common, emitter resistor, R_E, connected to a negative supply, $-V_{EE}$. The negative supply means that the base needs no biasing resistances and the inputs are DC coupled. Because no coupling capacitors are required, and for other reasons[6], the differential amplifier is the usual form of amplifier in ICs. Capacitors, other than very small values, tend to occupy a large area of silicon in an IC and are therefore avoided. The circuit made as an IC has the great advantage also that the transistors are almost exactly matched in characteristics.

The small-signal equivalent circuit of figure 3.51a is shown in figure 3.51b, which we shall analyse next, assuming that the transistors are identical and in particular, $h_{ie1} = h_{ie2} = h_{ie}$ and $\beta_1 = \beta_2 = \beta$; we shall also make the customary assumption that h_{oe} and h_{re} are negligible.

On the lower half of figure 3.51b, we find by KVL that

$$v_1 = i_{b1}h_{ie} + v_{RE}$$

---

[6] Its output is a more linear function of its input than that of a single-transistor amplifier, and it can be configured to avoid the Miller effect, so operating at much higher frequencies than a CE amplifier.

and

$$v_2 = i_{b2}h_{ie} + v_{RE}$$

And hence

$$v_1 - v_2 = (i_{b1} - i_{b2})h_{ie} \tag{3.34}$$

By KVL on the upper half of figure 3.51b we can see that

$$v_o = \beta i_{b2}R_C - \beta i_{b1}R_C = (i_{b2} - i_{b1})\beta R_C$$

and substitution for $(i_{b2} - i_{b1})$ from equation 3.34 gives

$$v_o = (v_2 - v_1)\beta R_C/h_{ie} \tag{3.35}$$

Since R_C will be the same order of magnitude as h_{ie}, the output voltage will be roughly β times the *difference* between the input voltages. The gain achieved depends on the quiescent collector current, which is determined by R_E, as the next example illustrates.

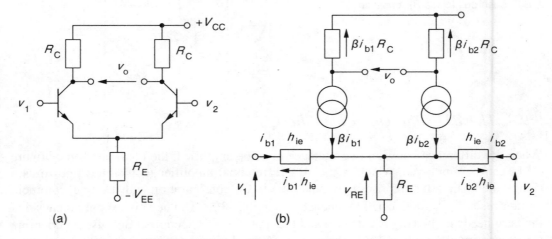

Figure 3.51 (a) The BJT differential amplifier, or long-tailed pair (b) Its small-signal equivalent circuit

Example 3.15

The BJT differential amplifier of figure 3.51a uses matched transistors with $\beta = 120$, $R_C = 2.2$ kΩ, $V_{CC} = +15$ V, $V_{EE} = -15$ V. What must R_E be to make the differential gain equal to 46 dB? (Take $kT/q = 30$ mV and $V_{BEQ} = 0.6$ V.)

The gain is given by equation 3.35 as $\beta R_C/h_{ie}$, that is

$$20\log_{10}(\beta R_C/h_{ie}) = 46 \implies \beta R_C/h_{ie} = 200$$

Therefore

$$h_{ie} = \beta R_C/100 = 120 \times 2.2/200 = 1.32 \text{ k}\Omega$$

Then I_{BQ} can be found from

$$h_{ie} = \frac{V_{th}}{I_{BQ}} = \frac{0.03}{I_{BQ}} \Rightarrow I_{BQ} = \frac{0.03}{h_{ie}} = \frac{0.03}{1.32} = 0.0227 \text{ mA}$$

Part of figure 3.51a is redrawn in figure 3.52, with the AC signal source, v_1, replaced by a short-circuit to ground. KVL on the left-hand side of this produces

$$V_{BEQ} + V_{REQ} + V_{EE} = 0$$

But the current through R_E is $2I_{EQ} = 2(\beta + 1)I_{BQ}$, making $V_{BEQ} = 2I_{EQ}R_E$; and $V_{EE} = -15$ V, while $V_{BEQ} = 0.6$ V. Substituting these into the equation above gives

$$2I_{EQ}R_E = 2(\beta + 1)I_{BQ}R_E = -0.6 - (-15) = 14.4 \text{ V}$$

$$\Rightarrow R_E = \frac{14.4}{2(\beta + 1)I_{BQ}} = \frac{14.4}{2 \times 121 \times 0.0227} = 2.62 \text{ k}\Omega$$

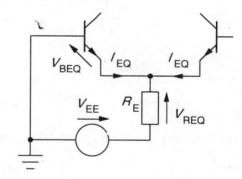

Figure 3.52

The circuit for determining R_E in example 3.15

3.9.4 The cascode amplifier

The cascode amplifier's name derives from a once-popular vacuum-tube circuit. It is used for high-frequency amplifiers, where the Miller capacitance of the CE amplifier would normally degrade performance, and in numerous ICs. Figure 3.53 shows the circuit, which consists of an input CE amplifier stage (Q2) with a CB stage (Q1) inserted into its collector circuit.

The input resistance of the CB transistor is low, and as this is the effective load for the CE amplifier, its gain and hence its Miller capacitance is small. Resistors R_1 and R_2 provide base bias for Q2, R_3 and R_4 do the same for Q1; and R_5 and R_6 are the collector and emitter resistors respectively. C_1 is a coupling capacitor for the input signal, C_2 is a bypass capacitor to ensure that Q1's base is grounded for AC signals, and C_3 is the bypass capacitor for R_6 that maintains the voltage gain. The circuit's performance is best illustrated by an example.

Example 3.16

In figure 3.53 the resistances are $R_1 = 22$ kΩ, $R_2 = R_3 = R_4 = 6.8$ kΩ, $R_5 = 2.2$ kΩ and R_6 = 1.2 kΩ. The transistors are identical with $\beta = 140$ and $V_{CC} = +15$ V. What are the quiescent collector voltages, V_{CE1} and V_{CE2}? What is the no-load voltage gain? What is the break frequency due to the shunt capacitances of the input transistor, Q2, if it has $C_{bc} = $ 3 pF and $C_{be} = 6$ pF and the signal source has an internal resistance of 500 Ω? (Ignore the capacitive reactances of C_1, C_2 and C_3, and assume the base currents are negligible compared to the base-biasing currents. Take $V_{BE} = 0.6$ V and $V_{th} = 28$ mV.)

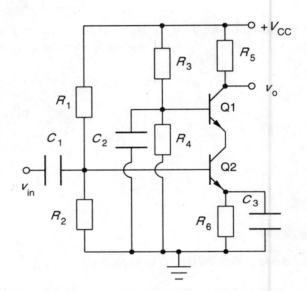

Figure 3.53

The cascode amplifier

The base-to-ground voltage of Q2 is found by the voltage-divider rule to be

$$V_{B2} \approx R_2 V_{CC}/(R_1 + R_2) = 6.8 \times 15/28.8 = 3.54 \text{ V}$$

Thus the voltage across R_6 is

$$V_{R6} = V_{B2} - V_{BE} = 3.54 - 0.6 = 2.94 \text{ V}$$

This voltage is $I_{E2}R_6 = 1.2 I_{E2}$, so that $I_{E2} = 2.94/1.2 = 2.45$ mA.

The base voltage of Q1 is also fixed by a voltage divider, R_3 and R_4, which are equal so that $V_{B1} = \frac{1}{2}V_{CC} = 7.5$ V and the voltage at the emitter of Q1 is 0.6 V less than this: 6.9 V. Thus the collector of Q2 (connected to the emitter of Q1) is at +6.9 V and so

$$V_{CE2} = 6.9 - V_{R6} = 6.9 - 2.94 = 3.96 \text{ V}$$

The collector of Q1 is at a potential of $V_{C1} = V_{CC} - I_{C1}R_5$ above ground. Taking $I_{C1} = I_{E1}$ = $I_{C2} = I_{E2}$ means that $I_{C1} = 2.45$ mA and then

$$V_{CE1} = V_{C1} - 6.9 = V_{CC} - I_{C1}R_5 - 6.9 = 15 - 2.45 \times 2.2 - 6.9 = 2.71 \text{ V}$$

Now the no-load voltage gain of a CE amplifier is $A_v \approx -\beta R_C/h_{ie}$, but Q2 is no longer connected to the collector resistor, R_s, but to the emitter of a CB amplifier, whose input resistance we have found previously (section 3.4) to be $r_e = h_{ie}/\beta$, hence the no-load gain of Q2 is

$$A_{v2} = \frac{-\beta r_e}{h_{ie}} = \frac{-\beta(h_{ie}/\beta)}{h_{ie}} = -1$$

Then the no-load gain of Q1, the CB amplifier is (see section 3.4)

$$A_{v1} = R_C/r_e = \beta R_C/h_{ie}$$

But

$$h_{ie1} = \frac{0.028}{I_{B1}} \approx \frac{0.028\beta}{I_{E2}} = \frac{0.028 \times 140}{2.45} = 1.6 \text{ k}\Omega$$

which makes the overall no-load voltage gain

$$A_{v0} = A_{v2} \times A_{v1} = \frac{-\beta R_C}{h_{ie}} = \frac{-140 \times 2.2}{1.6} = -308$$

Finally the input shunt capacitance is $C'_M = C_M + C_{be}$ (section 3.8.3) where

$$C_M = (1 - A_{v2})C_{cb} = 2C_{cb} = 6 \text{ pF}$$

and so $C'_M = 6 + 6 = 12$ pF. The resistance through which C'_M is charged is $R_s \text{ // } R_{in2}$, where

$$R_{in2} = R_1 \text{ // } R_2 \text{ // } h_{ie2} = \frac{1}{1/22 + 1/6.8 + 1/1.6} = 1.22 \text{ k}\Omega$$

making

$$R_{in2} \text{ // } R_s = \frac{1220 \times 500}{1220 + 500} = 354 \text{ }\Omega$$

Thus the high-frequency cut off due to Q2's shunt capacitance is

$$f_{H2} = \frac{1}{2\pi C'_M(R_s\text{ // }R_{in2})} = \frac{1}{2\pi \times 12 \times 10^{-12} \times 354} = 37.5 \text{ MHz}$$

3.10 BJT current mirrors

Current mirrors are frequently used in circuits where a current of a specific value that is required in one part can be controlled by the current in another part. They can be produced from either bipolar or field-effect transistors (for the latter see section 4.11.2). In its simplest form a current mirror requires just two transistors, but to improve performance three, four or more are often used. Along with the cascode and differential amplifiers, the current mirror is one of the most important building blocks in analogue integrated circuits.

3.10.1 A simple current mirror

Figure 3.54 shows a simple, two-transistor, current mirror. The collector of Q1 is short-circuited to its base so that the voltage across Q1 is V_{BE1} and therefore the current through R_{C1} is given by

$$V_{CC} = I_{in}R_{C1} + V_{BE1} \; \Rightarrow \; I_{in} = (V_{CC} - V_{BE1})/R_{C1}$$

And by KCL

$$I_{in} = I_{C1} + I_{B1} + I_{B2}$$

The current through Q2's collector is I_{C2} and as $V_{BE1} = V_{BE2}$, provided the transistors are absolutely identical, $I_{B1} = I_{B2} = I_{C1}/\beta = I_{C2}/\beta$ and hence

$$I_{in} = I_{C2} + 2I_{C2}/\beta = (1 + 2/\beta)I_{C2} = (V_{CC} - V_{BE1})/R_{C1}$$

which rearranges to give

$$I_{C2} = \frac{V_{CC} - V_{BE1}}{R_{C1}(1 + 2/\beta)}$$

Thus if $V_{CC} = 9$ V, $V_{BE1} = 0.7$ V, $R_{C1} = 1$ kΩ and $\beta = 100$, then $I_{C2} = 8.3/1.02 = 8.14$ mA. Doubling R_{C1} will halve I_{C2} because the product $I_{C2}R_{C1}$ is constant.

Figure 3.54

A two-transistor current mirror: I_{C2} is the 'mirror image' of I_{C1}, which is controlled by the value of R_{C1} and V_{CC}

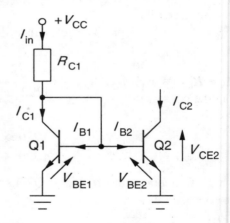

The current mirror is a constant-current source, so it is desirable that its output resistance is as high as possible. For a simple current mirror the output resistance is $1/h_{oe}$, the reciprocal of the slope of the CE output characteristic of the BJT, a fairly large value — typically 50-100 kΩ — which depends on the quiescent (DC) output current as the next section shows.

3.10.2 The Early effect

The above analysis has ignored the effect of changes in V_{CE2}, which will come into play because of the *Early effect*, even if the transistors are perfectly matched. Consider figure 3.55, which shows a BJT's CE output characteristics.

Figure 3.55

The Early voltage and its effect on the BJT's CE output characteristics. The vertical scale is exaggerated

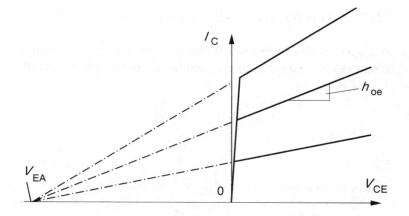

The collector current in the operating region is seen to increase with increasing V_{CE}, that is to say h_{oe}, the slope of the output characteristic, is not zero as we have usually assumed, but some small positive value. If the lines of I_C against V_{CE} in the operating region are extrapolated back into negative voltage, they are found to intersect the V_{CE} axis at a single point, known as the reverse Early voltage, $-V_{EA}$. If, in a small BJT, we were to find h_{oe} to be 50 μS when $I_C = 10$ mA and $V_{CE} = 1$ V, then we can see that $h_{oe} \approx I_C/V_{EA}$ or $V_{EA} = I_C/h_{oe} = 10\,000/50 = 200$ V, a typical value. The collector current in the operating region is then given by

$$I_C = I_s\left(1 + \frac{V_{CE}}{V_{EA}}\right)\exp\left(\frac{V_{BE}}{V_{th}}\right) \tag{3.36}$$

where $V_{th} \approx 25$ mV at 300 K.

Now in the current mirror, $V_{CE1} = V_{BE1}$, while V_{CE2} depends on what happens in the rest of the circuit, but is in general not equal to V_{CE1}. Using equation 3.36 we find

$$I_{C1} = I_s\left(1 + \frac{V_{CE1}}{V_{EA}}\right)\exp\left(\frac{V_{BE1}}{V_{th}}\right) \quad \text{and} \quad I_{C2} = I_s\left(1 + \frac{V_{CE2}}{V_{EA}}\right)\exp\left(\frac{V_{BE2}}{V_{th}}\right)$$

And therefore the current ratio is

$$\frac{I_{C2}}{I_{C1}} = \frac{1 + V_{CE2}/V_{EA}}{1 + V_{CE1}/V_{EA}} = \frac{V_{EA} + V_{CE2}}{V_{EA} + V_{BE1}}$$

If $V_{EA} = 200$ V and $V_{CE2} = 10$ V while $V_{BE1} = 0.7$ V, the current ratio is

$$\frac{I_{C2}}{I_{C1}} = \frac{210}{200.7} = 1.0463$$

That is 4.63% higher than expected; and this with *absolutely identical* BJTs. There are many ways of reducing the influence of the Early effect on the output current of a current mirror, some of which we shall examine in the next sections.

3.10.3 The BJT cascode current mirror

If cascoded BJTs are used in the current mirror, as shown in figure 3.56a, then V_{CE4} is constrained to be equal to V_{BE2} and the Early voltage drops out of the current ratio, making $I_{C2} = I_{C4}$.

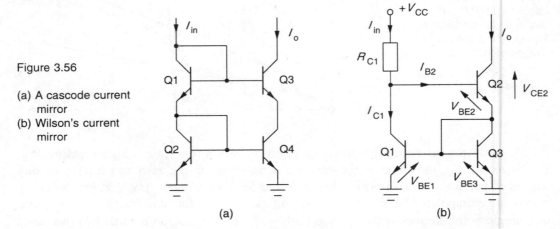

Figure 3.56

(a) A cascode current mirror

(b) Wilson's current mirror

(a)   (b)

It can be shown (problem 3.24) that, if the transistors are identical,

$$\frac{I_o}{I_{in}} = \frac{\beta^2}{\beta^2 + 4\beta + 2}$$

which is slightly less than one and does not vary with V_{CE3}. The small-signal output resistance can be shown to be β/h_{oe}, or $\beta V_{EA}/I_o$, which is in the order of 10 MΩ, a pretty high value so that the current ratio is almost independent of the current.

3.10.4 The Wilson current mirror

In this circuit, shown in figure 3.56b, negative current-shunt feedback (see section 5.2.2) is used to stabilise the ratio the input and output currents. Suppose V_{CE2} were to increase and cause an increase in I_{C2} by the Early effect, the change would cause increases in V_{BE2} and V_{BE3}. An increase in V_{BE3} would produce a corresponding increase in V_{BE1} and hence I_{C1}. But I_{in} is constant (determined by V_{CC} and R_{C1}) so an increase in I_{C1} would produce a decrease in I_{B2} and hence I_{C2} and V_{CE2}. Thus the output current will remain constant. It can be shown (see problem 3.24) that the ratio of output to input currents for perfectly matched transistors is

$$\frac{I_o}{I_{in}} = \frac{\beta^2 + \beta}{\beta^2 + \beta + 2} \approx 1$$

Since β is large (≈ 100), the approximation is a very good one (within 0.02%). The small-signal output resistance of the Wilson current mirror is only about half that of the cascode current mirror, but it is still large enough to keep the current ratio virtually independent of the current.

3.10.5 The Widlar current mirror

It may be that one does not require the output current of the 'mirror' to be the equal of the input, or nearly so, as in the examples above. One way of altering the current ratio is to use different transistors, which in ICs means differing active base areas. For example, suppose one transistor is ten times the other in active base area. The currents are

$$I_1 = I_{s1}\exp\left(\frac{V_{BE1}}{V_{th}}\right) \quad \text{and} \quad I_2 = I_{s2}\exp\left(\frac{V_{BE2}}{V_{th}}\right)$$

But if the active base area of Q1 is ten times that of Q2, $I_{s1} = 10I_{s2}$ and so $I_{C1}/I_{C2} = 10$ if $V_{BE1} = V_{BE2}$, as is the case in a current mirror. Unfortunately this is about as large a ratio as the technique will allow.

The Widlar current mirror, shown in figure 3.57a, enables large current ratios to be achieved. Q1 and Q2 are matched transistors and using KVL around the bottom emitter loop gives

$$V_{BE1} - V_{BE2} = I_{E2}R_{E2} \approx I_{C2}R_{E2}$$

Assuming β is large and the Early effect negligible means that the current ratio is

$$\frac{I_o}{I_{in}} \approx \frac{I_{C2}}{I_{C1}} = \exp\left(\frac{V_{BE2} - V_{BE1}}{V_{th}}\right) \implies \ln\left(\frac{I_o}{I_{in}}\right) \approx \frac{V_{BE2} - V_{BE1}}{V_{th}} \approx \frac{I_{C2}R_{E2}}{V_{th}}$$

hence

$$R_{E2} = \frac{V_{th}}{I_o}\ln\left(\frac{I_{in}}{I_o}\right) \tag{3.37}$$

It is clear that $I_o < I_{in}$ always. If $I_o = 0.01I_{in} = 100\ \mu\text{A}$ and $V_{th} = 30\ \text{mV}$, then $R_{E2} = 1.38\ \text{k}\Omega$. Unfortunately, I_o is very sensitive to changes in R_{E2}.

Figure 3.57

(a) The Widlar current mirror/amplifier
(b) A current reference based on the relative constancy of V_{BE}

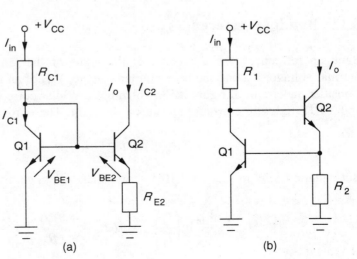

(a)   (b)

3.10.6 Supply-independent current reference

The current mirrors described above produce an output current which is directly dependent on the value of the supply voltage, V_{CC}, which may vary of course. A simple current reference which reduces the effect of variations in supply voltage is shown in figure 3.57b; however, this circuit is still fairly susceptible to changes in both V_{CC} and temperature. If we ignore the base currents then

$$V_{BE1} = I_o R_2$$

Thus if $I_o = 0.1$ mA and $V_{BE1} = 0\,7$ V, then $R_2 = 7$ kΩ. But V_{BE1} depends on I_{in} since

$$I_{in} = (V_{CC} - 2V_{BE})/R_1 \approx I_{C1} = I_{s1}\exp(V_{BE1}/V_{th})$$

$$\Rightarrow \quad V_{BE1} = V_{th}\ln(I_{C1}/I_{s1}) \approx V_{th}\ln(V_{CC}/I_{s1}R_1)$$

where we have taken $\ln(V_{CC} - 2V_{BE}) \approx \ln V_{CC}$. And so

$$I_o \approx (V_{th}/R_2)\ln(V_{CC}/I_{s1}R_1)$$

Differentiating I_o with respect to V_{CC} leads to

$$\frac{dI_o}{dV_{CC}} = \frac{V_{th}}{V_{CC}R_2} \quad \Rightarrow \quad \frac{\delta I_o}{I_o} = \frac{V_{th}}{I_o R_2}\frac{\delta V_{CC}}{V_{CC}} = \frac{V_{th}}{V_{BE1}}\frac{\delta V_{CC}}{V_{CC}}$$

With $V_{th} = 28$ mV and $V_{BE1} = 0.7$ V, we find that

$$\delta I_o/I_o = 0.04\,\delta V_{CC}/V_{CC}$$

This may be acceptable in some cases.

3.11 Bandgap references

Bandgap references use the fact that the signs of the temperature coefficients of the thermal voltage, V_{th}, and the base-emitter voltage, V_{BE}, of an npn BJT are of opposite sign. Consider the circuit of figure 3.58, which has a Widlar current source on the left hand side and a BJT fed with constant current on the right. The output voltage is

$$V_{REF} = V_{BE3} + I_2 R_2 \tag{3.38}$$

But equation 3.37 in section 3.10.5 shows that I_2 is given by

$$I_2 = \frac{V_{th}}{R_3}\ln\left(\frac{I_2}{I_1}\right) \quad \Rightarrow \quad I_2 R_2 = \frac{V_{th}R_2}{R_3}\ln\left(\frac{I_2}{I_1}\right) = GV_{th}$$

where $G = (R_2/R_3)\ln(I_2/I_1)$ and therefore equation 3.38 can be expressed as

$$V_{REF} = V_{BE3} + GV_{th} \tag{3.39}$$

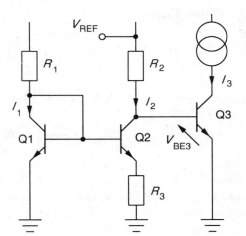

Figure 3.58

A bandgap reference based on a Widlar current source

Now V_{BE3} is given by

$$I_3 = I_s \exp(qV_{BE3}/kT)$$

where I_s, the reverse saturation current is given by

$$I_s = A \exp(-E_g/kT)$$

A being a constant. Taking natural logarithms of these two equations and rearranging gives

$$V_{BE3} = (kT/q)(\ln I_3 - \ln A) + E_g/q \tag{3.40}$$

Assuming that I_3 and E_g are constant, differentiating equation 3.40 with respect to T gives

$$\frac{dV_{BE3}}{dT} = (k/q)(\ln I_3 - \ln A)$$

But from equation 3.40 we can see that $\ln I_3 - \ln A = (qV_{BE3} - E_g)/kT$ so that the equation above is

$$\frac{dV_{BE3}}{dT} = \frac{V_{BE3} - E_g/q}{T} \tag{3.41}$$

Going back to equation 3.39, which is

$$V_{REF} = V_{BE3} + GV_{th} \tag{3.39}$$

and differentiating with respect to temperature gives

$$\frac{dV_{REF}}{dT} = \frac{dV_{BE3}}{dT} + G\frac{dV_{th}}{dT}$$

Now $V_{th} = kT/q$, so that $dV_{th}/dT = k/q$, and using this and equation 3.41 for dV_{BE3}/dT in the above yields

$$\frac{V_{BE3} - E_g/q}{T} + \frac{Gk}{q} = 0 \quad \Rightarrow \quad G = \frac{V_{BE3} - E_g/q}{kT/q}$$

on setting $dV_{REF}/dT = 0$ for zero temperature coefficient. This expression for G can be put into equation 3.38 to find V_{REF}

$$V_{REF} = V_{BE3} + \frac{(V_{BE3} - E_g/q)V_{th}}{kT/q} = E_g/q$$

as $V_{th} = kT/q$. Since $E_g = 1.1$ eV, we might expect this to be the bandgap reference voltage, but this turns out to be a little larger, ranging typically from 1.2 V to 1.26 V, because of the non-ideality of the p-n junction. By using a series of junctions this value can be multiplied several times over.

There has been a great proliferation of bandgap reference ICs of late: thirty or forty are listed in one popular stockist's catalogue. For example the Harris semiconductor's ICL8069DCZR bandgap reference has a tempco of 0.01%/°C and a reference voltage of 1.23 V; it costs £1.50 in single quantities. The REF25Z by GEC Plessey offers a 2.50 V ± 1% reference voltage with a tempco of 40 ppm/°C and costs £1.30. Others offer user-selectable reference voltages of up to 10 V, but the price is much greater.

3.12 BJT specifications

A typical BJT data sheet contains a mass of information and symbols which have not been seen before in this chapter, such as that listed below in table 3.3. This is followed by graphs indicating the changes in the various parameters with temperature, current or voltage. The definitions of the symbols in the table are as follows:

C_{obo}	C_{cb} with emitter open circuit, C_μ in hybrid-π terms
f_T	The transition frequency ('transistor cut-off' $= f_H$ for the EF amplifier). The product h_{fe} and f at a frequency high enough for h_{fe} to decrease at 20 dB/decade. Thus at f_T, $h_{fe} = 1$.
h_{FE}	The transistor β, the DC forward current transfer ratio: I_C/I_B
h_{fe}	The small-signal forward current transfer ratio: i_c/i_b; usually assumed equal to h_{FE} and β
h_{ie}	v_{be}/i_b with emitter grounded to AC
h_{oe}	i_c/v_{ce} with base open circuit to AC
h_{re}	v_{be}/v_{ce} with base open circuit to AC
I_{CBO}	The max current from collector into base when the junction is reverse biased and the emitter is open circuit. The 'leakage' current.
I_{Cmax}	The maximum permissible continuous DC collector current in saturation

P_{tot}	The total power dissipation in the transistor, in practice $\approx V_{CEQ}I_{CQ}$
V_{BE}	DC base-emitter voltage in the operating region
V_{CBO}	V_{CB} with the emitter open circuit. The value below is the *maximum* permitted.
V_{CEO}	V_{CE} with base open circuit. Do.
V_{EBO}	V_{EB} with collector open circuit. Do.
V_{CEsat}	DC collector-emitter voltage in the saturation region

Table 3.3 *Data for the BC108 general-purpose npn transistor at 25°C*

parameter	value
C_{obo}	6 pF max (V_{CB} = 5 V, f = 1 MHz)
f_T	150 MHz min., 300 MHz typical (V_{CE} = 10 V, I_C = 10 mA)
h_{FE}	110 min. 800 max. (I_C = 2 mA)
h_{fe}	125 min. 900 max. (I_C = 2 mA)
h_{ie}	1.3 kΩ min. 15.0 kΩ max. (V_{CE} = 5 V, I_C = 2 mA, f = 1 kHz)
h_{oe}	15 μS typical 75 μS max. (V_{CE} = 5 V, I_C = 2 mA, f = 1 kHz)
h_{re}	10^{-3} typical (V_{CE} = 5 V, I_C = 2 mA, f = 1 kHz)
I_{CBO}	15 nA max. (V_{CB} = 45 V)
	15 μA max. (V_{CB} = 45 V, T_j = 150°C)
I_{Cmax}	300 mA continuous
P_{tot}	360 mW max. continuous, derate linearly to zero at 200°C
V_{BE}	0.55 V min. 0.70 V max. (I_C = 2 mA, V_{CE} = 5 V)
	0.77 V max. (I_C = 10 mA, V_{CE} = 5 V)
V_{CBO}	30 V min. breakdown voltage
V_{CEO}	20 V min. breakdown voltage
V_{EBO}	6 V min. breakdown voltage
V_{CEsat}	250 mV (I_C = 10 mA, I_B = 0.5 mA)
	600 mV (I_C = 100 mA, I_B = 5 mA)

3.12.1 *Which parameters should I look for?*

It depends on what you are doing with the device. In practice, with a low-voltage DC supply, the voltage ratings are irrelevant, though occasionally the V_{EBO} rating can be exceeded by large input voltages as it is usually only a few volts. The power rating, P_{tot}, is important as it will restrict the operating range, especially in enclosed spaces in high ambient temperatures, as figure 3.59 shows. Even so, if I_{CQ} is kept to 10 mA or so and the DC supply is +15 V or less, the BC107 will be operating well within its rating.

When power amplifiers are required, the low efficiency of the amplifiers discussed here prevents their use: less wasteful amplification methods are required as discussed in chapter 8. Power transistors operating between saturation and cut-off dissipate an average power of about $\frac{1}{2}V_{CEsat}I_{Cmax}$ so these parameters and P_{tot} must be checked. A more important consideration here is the thermal impedance of the device so that its temperature rise can be estimated, but this depends on the package style. Device cooling is treated in section 8.1. High output voltages will require more expensive devices with adequate values for

V_{CEO}. If high frequencies are to be amplified, then f_{T} should be examined to see if the device can cope. The BC108 has $f_{\text{T}} = 300$ MHz, implying an upper frequency limit of 300 kHz with a 60 dB gain. The h-parameters are not important in transistor selection!

Figure 3.59

Maximum power dissipation curves plotted for various ambient temperatures on the CE output characteristics of a BC108 npn transistor

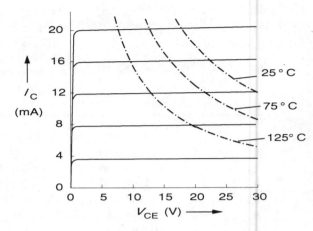

Notice that C_{π} and g_{m} are not given in the table: the only guides to high-frequency performance are f_{T} and C_{obo}. Table 3.4 shows an abbreviated list of data for various BJTs.

Table 3.4 *Selected data for some readily-available BJTs*

BJT No.	Type	Package	I_{Cmax}	V_{CEOmax}	P_{tot}	h_{fe}	f_{T}	Application
2SC3504D	npn	TO92	0.05	60	0.9	60-120	500	video amp.
BC108	npn	TO18	0.1	20	0.3	110-800	300	general purpose
BC178	pnp	TO18	−0.1	−25	0.3	125-500	200	BC108 complement
2N3906	pnp	TO92	−0.2	−40	0.3	100-300	250	switching
BCX70H	npn	SOT23	0.2	45	0.33	180-310	250	SMT, gen. purpose
ZTX302	npn	E-line	0.5	35	0.3	>100	>150	small-signal amp.
ZTX502	pnp	E-lone	−0.5	−35	0.3	>100	>150	ZTX302 complement
BC441	npn	TO39	2	60	1	40-250	50	general purpose
BUX84	npn	TO220	2	400	40	>15	20	motor controls
BU216	npn	TO220	4	800	70	12-40	10	power supplies
BD437	npn	TO126	4	45	36	>40	>7	audio amp.
BD438	pnp	TO126	−4	−45	36	>40	>7	BD437 complement
FZT853	npn	SOT223	6	100	3	100-300	130	SMT, gen. purpose
MJE18009	npn	TO220	10	450	150	14-34	12	ballast
BUV50	npn	TO3	50	125	250	>10	8	inductive switch

Notes: I_{Cmax} in A. V_{CEOmax} in V. P_{tot} in W at 25°C ambient. h_{fe} is for restricted range of I_{C} and V_{CE}. f_{T} in MHz, typical value. SMT = surface-mount device

3.12.2 Package styles

There are a number of popular transistor packaging styles, some of which are shown in figure 3.60. Low-power, general-purpose BJTs are sold very cheaply in plastic packages such as the E-line and TO92, or slightly more expensive metal cans such as the TO18 and TO39. The TO220 and TO3 packages, both of which require heat sinks of some kind, are used with high and medium power devices. Special-purpose transistors have other packages according to application and are not shown.

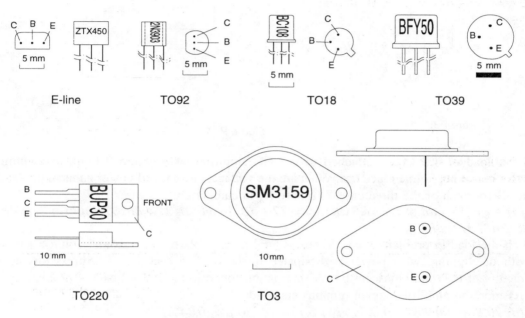

Figure 3.60 Six popular package styles for BJTs

Suggestions for further reading

See chapter four.

Problems

(In the following problems do not assume $I_E = I_C$ etc. unless told to.)

1 If in the circuit of figure P3.1, $R_C = 1.5$ kΩ, $R_B = 500$ kΩ, $V_{CC} = 6$ V and $V_{BEQ} = 0.6$ V, what must the transistor's β be to make V_{CEQ} (a) 5.5 V, (b) 3 V (c) 0.5 V. (Take I_{CQ} to be equal to I_{EQ}.) *[(a) 31 (b) 185 (c) 340]*

2 Using the simplified small-signal equivalent circuit of figure 3.12, estimate the no-load voltage gains of the amplifier of problem 3.1. (Take $h_{ie} = 0.03/I_{BQ}$, $\beta = h_{fe}$ and $R_s = 0$.) What are the largest r.m.s. sinusoidal input signals that can be amplified without clipping in each case if $V_{CEsat} = 0.2$ V?

[(a) −17 (b) −100 (c) −183 (a) 21 mV (b) 20 mV (c) 1.16 mV]

3 What power is dissipated in the quiescent transistor in each of the three cases of problem 3.1? And the total power dissipated by the quiescent circuit?

[(a) 1.83 mW (b) 6 mW (c) 1.83 mW (a) 2.2 mW (b) 12.2 mW (c) 22.2 mW]

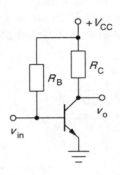

Figure P3.1

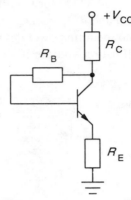

Figure P3.7

4 A load of 1.5 kΩ is attached to the CE amplifier of problem 3.1 via a coupling capacitor of negligible reactance. What are the voltage, current and power gains in the load in dB for each of the three cases? (Ignore R_B in calculating the gains.)

[(a) $A_{vL} = 18.4$ dB, $A_{iL} = 15.8$ dB, $A_{pL} = 17.1$ dB (b) 34 dB, 31.4 dB, 32.7 dB (c) 39.2 dB, 36.7 dB, 37.9 dB]

5 If the load in problem 3.4 is driven as hard as possible by a sinusoidal voltage source without clipping, what power is dissipated in the load in each case? Express this as a fraction of the total quiescent power dissipation found in the second part of problem 3.3. (Assume the amplified signal remains sinusoidal.)

[(a) 20.8 μW, 0.95% (b) 0.75 mW, 6.15% (c) 30 μW, 0.14%]

6 Where should the Q-point be in problem 3.5 if the load power is to be maximised without clipping? What must h_{fe} (= β) be for this? What is the power dissipation in the quiescent circuit then? What is the maximum load power without clipping as a percentage of the quiescent power dissipation?

[$V_{CEQ} = 2.133$ V, $I_{CQ} = 2.578$ mA; $h_{fe} = 95.5$; 15.6 mW; 8%]

7 In the shunt feedback bias circuit of figure P3.7, $V_{CC} = 12$ V and the Q-point is to be at $I_{CQ} = 12$ mA with $V_{CEQ} = 8$ V. If the transistor β is 230 and $R_C = 4R_E$, what should R_C, R_E and R_B be? Where is the Q-point when β is changed to 150?

(Take $V_{BEQ} = 0.7$ V, and assume $I_{EQ} = I_{CQ}$)

[$R_C = 267$ Ω, $R_E = 67$ Ω, $R_B = 140$ kΩ; $I_{CQ} = 8.9$ mA, $V_{CEQ} = 9$ V]

8 In the voltage-divider bias circuit of figure P3.8, $R_C = 2.2$ kΩ, $R_E = 620$ Ω, $R_1 = 33$ kΩ and $R_2 = 6.2$ kΩ. What must the supply voltage be if the no-load voltage gain is 50 dB? (Assume I_{BQ} is negligible compared to the current in R_1 and that $V_{BEQ} = 0.7$ V. Take h_{ie} to be $0.028/I_{BQ}$.) *[$V_{CC} = 20$ V]*

9 Repeat problem 3.8 taking β = 250 and not assuming I_{BQ} to be negligible compared to the current through R_1. What are the current and power gains if R_C is taken to be the load?

[$V_{CC} = 20.7$ V; $A_i = 45.5$ dB; $A_p = 47.7$ dB]

10 If the load resistance is 3.3 kΩ and the source resistance 1.5 kΩ in problem 3.9, what is the voltage gain, $A_{vL} = v_L/v_s$? What is the largest sinusoidal input signal that can be amplified without clipping? What is the ideal Q-point for this load and what is the largest sinusoidal input signal that will not be clipped on amplification with this Q-point?
[$A_{vL} = -88.3$. $v_s = 42.5$ mV. $I_{CQ} = 4.73$ mA, $V_{CEQ} = 6.245$ V, $v_s = 50$ mV]

11 The transistor in the CE circuit of figure P3.8 has $h_{fe} = 180$ and $h_{ie} = 0.027/I_{BQ}$. The transistor is to dissipate 100 mW and the voltage gain, $A_{vL} = v_L/v_s = -120$. Calculate the resistance values if $R_C = 3R_E$, $V_{CC} = +24$ V, $V_{BEQ} = 0.7$ V and $I_{R1} = 10I_{BQ}$.
[$R_C = 1.063$ kΩ, $R_E = 354$ Ω, $R_1 = 38.1$ kΩ, $R_2 = 8.57$ kΩ]

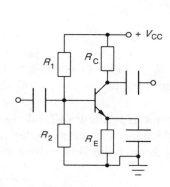

Figure P3.8

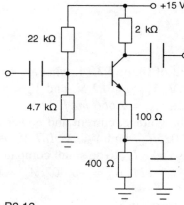

Figure P3.12

12 Find Q-point and the no-load voltage gain of the circuit of figure P3.12, if $h_{fe} = 175$, $h_{ie} = 0.026/I_{BQ}$ and $V_{BEQ} = 0.7$ V. What is the optimum load for maximum output voltage swing? What is the voltage gain with this load? If V_{BEQ} falls 2.2 mV/K, h_{fe} increases by 0.5%/K and h_{ie} increases by 0.35%/K, what effect will a temperature rise of 60°C have on the Q-point? What is the no-load voltage gain now? (Assume the capacitances have negligible reactances.) *[$V_{CEQ} = 5.7$ V, $I_{CQ} = 3.72$ mA; $A_{v0} = -18.7$; $R_L = 1.37$ kΩ; $A_{vL} = -7.6$; $\delta V_{CEQ} = -0.72$ V, $\delta I_{CQ} = +0.29$ mA; $A_{v0} = -18.5$]*

13 In the circuit of figure P3.13 $R_s = 1.4$ kΩ, $R_E = 500$ Ω, $R_B = 10$ kΩ, $R_L = 400$ Ω, β $(= h_{fe}) = 50$, $V_{CC} = +60$ V and $V_{BEQ} = 0.8$ V. Find the Q-point and (a) the voltage gain, $A_{vL} = v_L/v_s$ (b) the current gain, $A_{iL} = i_L/i_s$ and (c) the power gain, A_{pL}. If v_s is sinusoidal and as large as possible without clipping the output, what is the power dissipation in the load as a fraction of the total power consumed by the circuit? (Take $h_{ie} = 0.03/I_{BQ}$.)
[$V_{CEQ} = 17.5$ V, $I_{CQ} = 83$ mA; (a) 0.79 (b) 23.7 (c) 12.7 dB; 1.9%]

14 In the circuit of figure P3.14 at 25°C the zener has $V_Z = 4.7$ V when $I_Z = 2$ mA, with $r_z = 60$ Ω. Calculate the collector current if $V_{BEQ} = (0.6 + 500I_{BQ})$ V and β is (a) 50 (b) 100 (c) 400. What is the minimum that R_C can be for the transistor to be saturated when β = 100? (Take $V_{CEsat} = 0.2$ V.) *[(a) 3.945 mA (b) 4.006 mA (c) 4.052 mA; 1.94 kΩ]*

15 Repeat the problem 3.14 for a zener with a temperature coefficient of breakdown voltage, $\theta_{VZ} = +0.1\%/K$, and a transistor whose V_{BEQ} changes by −0.3%/K, when the temperature increases to 55°C. *[(a) 4.135 mA (b) 4.197 mA (c) 4.244 mA; 1.80 kΩ]*

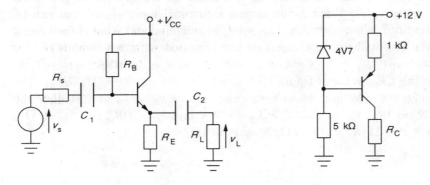

Figure P3.13   Figure P3.14

16 In the circuit of figure P3.14, what are I_{BQ} and I_{CQ} when $R_C = 3$ kΩ? (Take V_{BEQ} to be constant at 0.8 V, $V_{CEsat} = 0.2$ V, and $V_z = 4.7$ V at a current of 2 mA with $r_z = 60$ Ω.) $[I_{BQ} = 1.134$ mA, $I_{CQ} = 2.666$ mA$]$

17 Calculate the voltage, current and power gains of the circuit of figure P3.17, taking $\beta = 180$, $h_{ie} = 0.025/I_{BQ}$ and $V_{BEQ} = 0.7$ V. Assume the capacitances to have negligible reactances. (Also take I_{BQ} to be small compared to the current through the 5kΩ resistor and $I_{CQ} = I_{EQ}$.) $[A_{vL} = -111, A_{iL} = -0.607, A_p = 18.3$ dB$]$

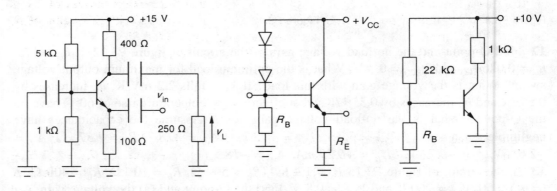

Figure P3.17   Figure P3.18   Figure P3.19

18 In the circuit of figure P3.18 there are n diodes in series, each having forward voltage drops of V_{AK} (= V_{BEQ}) independent of current. Show that I_{CQ} is given by

$$I_{CQ} = \frac{V_{CC} - (n+1)V_{AK}}{(1 + 1/\beta)R_E}$$

and I_D — the current through the diodes — by

$$I_D = \frac{V_{AK}}{R_B} + \frac{V_{CC} - (n+1)V_{AK}}{[(\beta + 1)R_E]//R_B}$$

If $V_{CC} = +5$ V, $V_{AK} = 0.6$ V, $V_{CEQ} \approx \frac{1}{2}V_{CC}$ and $I_{CQ} = 10$ mA, what must n and R_E be? Are the formulae for I_{CQ} and I_D valid if $n = 0$? *[n = 3; R_E = 260 Ω]*

19 At 20°C the circuit in figure P3.19 has $V_{BEQ} = 0.6$ V and $\beta = 200$. What should R_B be to make $V_{CEQ} = \frac{1}{2}V_{CC}$? With this value of R_B what temperature rise would drive the transistor into saturation, given that β increases by 0.9%/K and V_{BEQ} decreases by 2 mV/K? (Take $V_{CEsat} = 0.25$ V.) *[1.4915 kΩ; 13 K]*

20 In figure P3.20a only C_E need be considered at low frequencies. Show that the low-frequency equivalent circuit is that of figure 3.20b, in which $R_B = R_1 \parallel R_2$, $R'_E = (\beta + 1)R_E$ and $C'_E = C_E/(\beta + 1)$. Hence show that the voltage gain is

$$A_{vL} = \frac{v_L}{v_{in}} = \frac{-\beta(R_C \parallel R_L)(1 + j\omega\tau_E)}{[h_{ie} + (\beta + 1)R_E](1 + j\omega\tau_e)}$$

where $\tau_E = C_E R_E$ and $\tau_e = \tau_E[1 + (\beta + 1)R_E/h_{ie}]^{-1}$. If $C_E = 1000$ μF, $R_E = 100$ Ω, $R_C = R_L = 1.5$ kΩ, $\beta = 55$ and $h_{ie} = 600$ Ω, what is the lower cut-off frequency? What is A_{vL} at $f_E = 1/2\pi\tau_E$? What is the mid-band voltage gain? *[16.2 Hz; 19.5 dB; 36.75 dB]*

21 Find the upper cut-off frequency for the amplifier of figure P3.21, in which $g_m = 110$ mS, $R_L = R_C = 1.5$ kΩ, $r_\pi = 600$ Ω, $r_x = 45$ Ω, $C_\mu = 6$ pF, $C_\pi = 110$ pF, $R_s = 75$ Ω, $R_1 = 22$ kΩ and $R_2 = 5.6$ kΩ. What is the mid-band voltage gain? At what frequency is the voltage gain unity? *[2.65 MHz; 36.9 dB; 184.5 MHz]*

22 What value of R is required in a Darlington pair shown in figure 4.47c if it is operated at 100°C with an output emitter current of 10 A, given that the leakage current of Q1 is 1 nA at 25°C, doubling every 8°C, and that $\beta_1 = 200$ and $\beta_2 = 50$? With this value of R, at what input current will the Darlington turn on? (Take $V_{BE} = 0.8$ V.) *[4 Ω << R << 6 kΩ, so take R = √(4 × 6000) = 155 Ω; 26 μA]*

23 Show that, for the Wilson current mirror, the current ratio,

$$I_o/I_{in} = (\beta^2 + 2\beta)/(\beta^2 + 2\beta + 2)$$

and that for the cascode current mirror this ratio is $\beta^2/(\beta^2 + 4\beta + 2)$.

24 The circuit of figure P3.24 has matched transistors with $\beta = 100$ and $V_{BE} = 0.7$ V at 25°C. Given $V_{EE} = +15$ V (constant), $R_C = 5.6$ kΩ, $V_{AK} = 0.7$ V and $V_{Z0} = 6$ V, if the TCR of R_C is +100 ppm/K, $\theta_{VZ0} = 3.2 \times 10^{-4}$ V/V/K, $r_z = 100$ Ω, β increases by 0.5%/K and both V_{AK} and V_{BE} fall by 2 mV/K, by what fraction does V_{ref} change at 45°C? (The dynamic resistance of the ordinary diode is negligible.) *[−88 ppm]*

25 The circuit of figure P3.25 is a cascaded, DC coupled, CC-CE amplifier. What is the input resistance? Find the overall voltage, current and power gains if both transistors have $h_{re} = h_{oe} = 0$ and $\beta = 120$; $V_{CC} = +5$ V, $R_1 = R_2 = 22$ kΩ, $R_3 = 1$ kΩ, $R_4 = 470$ Ω and $R_5 = 220$ Ω. (Take $V_{BEQ1} = 0.6$ V, $V_{BE2} = 0.7$ V and $V_{th} = 28$ mV. Perform a DC analysis first to find I_{BQ1} and I_{BQ2} and hence h_{ie1} and h_{ie2}. Then draw the small-signal equivalent circuit for both stages together and find the gains. Take R_4 as the load for the current gain.) *[R_{in} = 10 kΩ; A_v = −32.4, A_i = −8421 and A_p = 54.4 dB]*

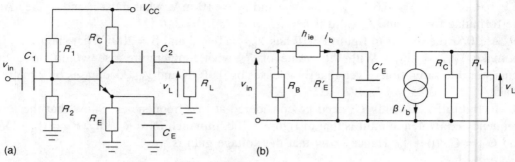

Figure P3.20

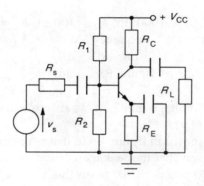

Figure P3.21

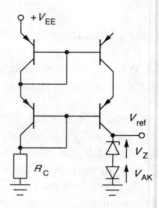

Figure P3.24

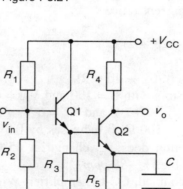

Figure P3.25

4 Field-effect transistors

T HE FIELD-EFFECT transistor, or FET, is a three-terminal device externally just
like the BJT but in fact it is quite different — if not inside the packaging then
underneath the silicon. Field-effect transistors use an electrode called the *gate* to
control the flow of current between two terminals known as the *source* and the *drain*.
Because only one type of charge carrier is involved in conduction the FET is a *unipolar*
device, though this is not apparent in its name. The structure is simpler than that of the
BJT and can be made with fewer steps in integrated circuit (IC) production, and with
rather more transistors to the square millimetre. For these reasons FETs are the most
common transistors to be found in large ICs, particularly in memory chips, where millions
are packed into tiny areas of silicon. Although FETs come in a variety of types with
various symbols to denote their differing modes of operation, the principles of operation
are very similar for all. We shall begin with a description of the type known as the
junction field-effect transistor or JFET (sometimes called a JUGFET — JUnction-Gate
FET), discussed in *Proc. IRE* **40**, 1365 (1952) by W Shockley.

4.1 The principles of JFET operation

Figure 4.1 shows a notional JFET in cross-section. It is seen to be a bar of semiconductor
material uniformly doped n-type or p-type throughout except for two areas of opposite
dopant type on the sides underlying the gate electrodes. The JFET shown is made of p-
type material with n-type regions under the gate electrodes and is called a *p-channel* JFET.

When a positive voltage is applied to the source (S) electrode with the drain grounded
(V_{DS} is negative) a positive current carried by holes, known as the *drain current*, I_D, passes
along the bar from source to drain. If a positive voltage is now applied to the gate
electrodes, it reverse biases the gate-source and gate-drain junctions. The depletion regions
are enlarged in the p-type channel beneath the heavily-doped n-type regions, and the
channel is thereby constricted. This causes the resistance between source and drain to
increase so that, for small values of V_{DS}, the drain current is proportional to V_{DS}. The
constant of proportionality, the channel conductance, depends on V_{GS}. However, as V_{DS}
becomes larger (increasingly negative) the depletion zones are drawn out towards the drain
electrode and the channel is pinched off. Increasing V_{DS} (in a more negative direction) does
not therefore produce an increase in drain current, I_D, which remains constant for a given
gate bias, V_{GS}.

As the gate bias is increased, the depletion zones get larger and larger until the channel
is completely cut-off and no drain current can pass. This cut-off voltage, denoted V_P, is
an important parameter which, like β in the case of BJTs, is critically dependent on slight

variations in manufacturing processes. Consequently V_P varies widely from one device to another, even if they bear the same type number.

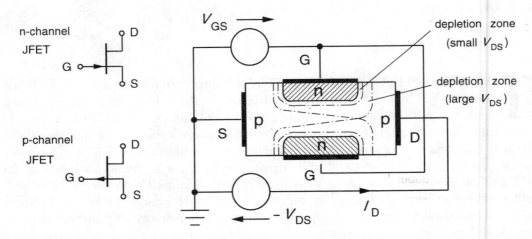

Figure 4.1 Symbols for both types of JFET and a cross-section of a p-channel JFET showing the depletion zones. Note that V_{DS} and V_{GS} are usually of opposite polarity in a JFET

Keeping the gate reverse biased means the current flowing through it is negligible (< 1 nA) and the current through the drain, I_D, is the same as the current through the source. However, if the gate-source junction is inadvertently forward biased (V_{GS} positive for n-channel JFETs) it will conduct, which is not desirable.

4.2 JFET characteristics

The characteristics of a JFET can be found using a circuit such as that of figure 4.2, in which V_{GS} is held constant while V_{DS} is varied and I_D is measured. Here an n-channel device has been chosen and we note that V_{GS} is now negative and so V_{DS} is positive — the opposite polarities to a p-channel device.

Figure 4.2

A circuit for determining the characteristics of an n-channel JFET

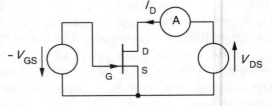

Each value of V_{GS} produces a different graph of I_D versus V_{DS} as shown in the right-hand graph in figure 4.3, but they all flatten out when V_{DS} exceeds ($V_{GS} - V_P$), to a value of I_D determined only by the gate bias, V_{GS}. These graphs are known as the *drain characteristics*; from them we can plot I_D in the flat part against V_{GS} as in the graph on the left-hand side of figure 4.3, which is called the *transfer characteristic*. It can be seen from this that $I_D = 0$ when $V_{GS} = -5$ V, that is $V_P = -5$ V. Also we note that when $V_{GS} = 0$, $I_D = 6$ mA, a value denoted by I_{DSS}: the drain-to-source saturation current. These two parameters, I_{DSS} and V_P, suffice to predict the circuit behaviour of any JFET. There is no need to measure the gate current, which is practically zero as it is the reverse saturation current of a p-n junction.

We can see that, unlike the BJT, the current through the JFET (I_D) is not a linear function of bias (V_{GS}), but actually varies in a quadratic manner in the operating region (the flat part of the drain characteristics). It can be shown that the drain current is given approximately[1] by

$$I_D = I_{DSS}\left(1 - \left|\frac{V_{GS}}{V_P}\right|\right)^2 \tag{4.1}$$

The vertical lines about V_{GS} and V_P mean that we treat them as positive quantities. In the characteristics given in figure 4.3 we can see that this equation is obeyed, for example when $V_{GS} = -2$ V equation 4.1 gives $I_D = 2.16$ mA, if $I_{DSS} = 6$ mA and $V_P = -5$ V.

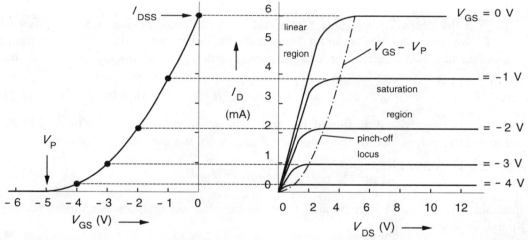

Figure 4.3 The drain characteristics (right) and the transfer characteristic (left) of an n-channel JFET

4.3 Biasing the JFET: the common-source (CS) amplifier

The JFET needs to be biased like the BJT if it is to operate in the flat part of the drain characteristics (the saturation region in figure 4.3). To establish a Q-point, that is DC values for I_{DQ} and V_{DSQ}, using only a single voltage source, V_{DD}, we must add some resistors to the JFET.

In the so-called *self-bias* circuit a resistor is attached in series with each terminal as in figure 4.4 and the source and gate resistors are grounded. C_1 and C_2 are the input and

---

[1] The actual relationship is more complicated than equation 4.1, which is close enough for most purposes; see problems 4.1 and 4.2

output decoupling capacitors and C_S is the source resistor's bypass capacitor, which serves the same purpose as C_E in the common-emitter amplifier, preventing gain degeneration. The source resistor, R_S, is the only real bias resistor: R_G merely provides a path to ground for the gate to ensure that it stays at 0 V, and R_D is a pull-up resistor for the drain, analogous to R_C in the CE amplifier.

Figure 4.4

The self-biased common-source amplifier

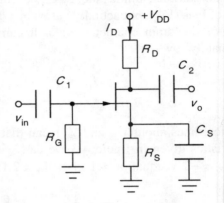

To find the necessary resistances required in the common-source (CS) amplifier of figure 4.4 we must analyse the circuit. Examining the lower part of figure 4.4 as in figure 4.5a, we see that as the current through the gate is zero, $V_{RG} = 0$, then $V_{GSQ} + V_{RS} = 0$, that is

$$V_{GSQ} = -V_{RS} = I_{DQ}R_S \tag{4.2}$$

Thus V_{GSQ} is negative, as it should be, and is determined by the value of I_D and R_S. Looking at the output side of the amplifier, figure 4.5b, we can see that by KVL

$$V_{DD} = I_{DQ}R_D + V_{DSQ} + I_{DQ}R_S \tag{4.3}$$

Equations 4.1, 4.2 and 4.3 are enough for us to calculate the resistances required for any Q-point we want.

Figure 4.5

(a) The lower part of the CS amplifier, showing that $V_{GSQ} + V_{Rs} = 0$
(b) The output side of the CS amplifier, establishing equation 4.3

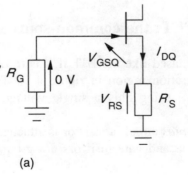

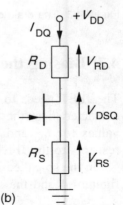

(a)   (b)

Example 4.1

A certain n-channel JFET has $I_{DSS} = 6$ mA and $V_P = -5$ V and is operated from a 12V supply in a self-biased common-source amplifier (figure 4.4). It is desired that $I_{DQ} = 3$ mA while $V_{DSQ} = 6$ V. What resistances are required?

From equation 4.1 we find

$$I_{DQ} = 3 = I_{DSS}(1 - |V_{GSQ}/V_P|)^2 = 6(1 - |V_{GSQ}/5|)^2$$

working in V and mA as usual. The above equation can be solved to give $V_{GSQ} = -1.465$ V. Then from equation 4.2

$$V_{GSQ} = -1.465 = -I_{DQ}R_S = -3R_S \;\Rightarrow\; R_S = 0.488 \text{ k}\Omega$$

Finally R_D comes from equation 4.3:

$$V_{DD} = 12 = I_{DQ}R_D + V_{DSQ} + I_{DQ}R_S$$
$$= 3R_D + 6 + 1.465 \;\Rightarrow\; R_D = 1.512 \text{ k}\Omega$$

The value of R_G does not affect the biasing, but as it is in effect the input resistance of the amplifier, it is made fairly large; 1 MΩ for example. If R_G is made too big then the input bias current, though small, might cause significant voltage drop across it.

4.3.1 The equivalent circuit of the common-source amplifier

The equivalent circuit of the FET by itself is an open circuit on the input side (as the gate draws virtually no current) which is connected to an AC signal source. The signal source acts on the gate bias and is written v_{gs}, since it is the alternating component of the gate-source voltage. On the output side the alternating voltage at the drain terminal is $-i_d(R_D + R_S)$, or $-i_dR_D$ if R_S is provided with a by-pass capacitor. The output voltage is negative, indicating a 180° phase change like that of the CE amplifier. The alternating part of the drain voltage, i_d, is related to the input signal, v_{gs}, by equation 4.1, which can be differentiated with respect to V_{GS} to give

$$\frac{dI_D}{dV_{GS}} = \frac{2I_{DSS}}{|V_P|}\left(1 - \left|\frac{V_{GS}}{V_P}\right|\right) = \frac{2\sqrt{I_{DSS}I_D}}{|V_P|} \tag{4.4}$$

because $(1 - |V_{GS}/V_P|) = \sqrt{(I_D/I_{DSS})}$, as can be found by rearranging equation 4.1. Replacing the differentials by small-signal values, we obtain from equation 4.4

$$\frac{\delta I_D}{\delta V_{GS}} = \frac{i_d}{v_{gs}} = \frac{2\sqrt{I_{DSS}I_{DQ}}}{|V_P|} = g_m \tag{4.5}$$

where g_m is the *transconductance* of the JFET at the Q-point, as we have replaced I_D by the quiescent value. Thus equation 4.5 becomes

$$i_d = g_m v_{gs} \tag{4.6}$$

From these considerations the equivalent circuit of the CS amplifier can be shown to be that of figure 4.6 in which the resistances are external components. The JFET by itself is reduced to an open-circuit input and a current source output.[2] Using this equivalent circuit the small-signal voltage gain can be found.

Figure 4.6

The small-signal equivalent circuit of the common-source amplifier

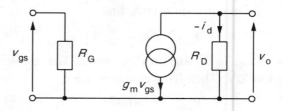

4.3.2 The small-signal gain of the CS amplifier

From figure 4.6 it can be seen that the output voltage without load of the CS amplifier is

$$v_o = -i_d R_D = -g_m v_{gs} R_D$$

And as the input voltage is just v_{gs}, we find the no-load voltage gain is

$$A_{v0} = \frac{v_o}{v_{in}} = \frac{-g_m v_{gs} R_D}{v_{gs}} = -g_m R_D$$

The current gain, taking R_D as the load, is

$$A_i = \frac{-i_d}{i_{in}} = \frac{-g_m v_{gs}}{v_{gs}/R_G} = -g_m R_G$$

Since R_G is usually very large the current gain is large too.

The power gain is the product of the voltage and current gains

$$A_p = A_{v0} \times A_i = g_m^2 R_D R_G$$

The following example provides an idea of the magnitudes.

Example 4.2

A CS amplifier as in figure 4.4 employs a JFET with $I_{DSS} = 10$ mA and $V_P = -3$ V. What values of R_D and R_S are required to give a quiescent drain current of 6 mA and a quiescent drain-source voltage of 8 V if $V_{DD} = +15$ V? Using the nearest preferred values, draw the AC and DC load lines and the Q-point on the drain characteristics. From this graph derive

---

[2] Strictly speaking we should include a large resistance, r_d, the reciprocal of the slope of the drain characteristic, the counterpart of $1/h_{oe}$ in the CE amplifier; but as $r_d \gg R_D$ and the pair are in parallel, there seems little point, especially as equations 4.1 and 4.5 are only approximations. Typically $r_d \approx 100$ kΩ and $R_D \approx 5$ kΩ.

the shape of the output signal for a sinusoidal input of 1 V_{pp}. Calculate the no-load current, voltage and power gains if $R_G = 500 \text{ k}\Omega$.

We can first find V_{GSQ} from equation 4.1 since it is the only unknown:

$$I_{DQ} = I_{DSS}(1 - |V_{GSQ}/V_P|)^2 \quad \Rightarrow \quad 6 = 10(1 - |V_{GSQ}|/3)^2$$

whence $V_{GSQ} = -0.676$ V. This can be substituted into equation 4.2 to find R_S:

$$V_{GSQ} = -I_{DQ}R_S \quad \Rightarrow \quad -0.676 = -6R_S$$

giving $R_S = 113 \text{ }\Omega$. Finally equation 4.3 is used to find R_D:

$$V_{DD} = V_{DSQ} + I_{DQ}(R_D + R_S) \quad \Rightarrow \quad 15 = 8 + 6(R_D + 0.113)$$

which yields $R_D = 1.054 \text{ k}\Omega$.

The nearest preferred values are $R_D = 1 \text{ k}\Omega$ and $R_S = 120 \text{ }\Omega$. We must recalculate the Q-point from equation 4.2: $V_{GSQ} = -I_D R_S = -0.12I_D$, which can then be put into equation 4.1 to give

$$I_{DQ} = I_{DSS}(1 - |V_{GSQ}/V_P|)^2 \quad \Rightarrow \quad I_{DQ} = 10(1 - 0.04I_{DQ})^2$$

Solving this quadratic equation yields $I_{DQ} = 5.86$ mA (the other solution is physically impossible), and equation 4.3 gives

$$V_{DD} = V_{DSQ} + I_{DQ}(R_D + R_S) \quad \Rightarrow \quad 15 = V_{DSQ} + 5.86 \times 1.12$$

whence $V_{DSQ} = 8.44$ V.

The DC load line's equation comes from rearranging equation 4.3:

$$I_{DQ} = \frac{-1}{R_D + R_S}V_{DSQ} + \frac{V_{DD}}{R_D + R_S}$$

which is a straight line relating I_{DQ} to V_{DS} whose slope is $-1/(R_D + R_S)$. The AC load line will have a slope of $-1/R_D$ due to the bypassing of R_S, and will pass through the same Q-point as the DC load line. Figure 4.7a shows the load lines and Q-point on the drain characteristics constructed from equation 4.1.

Now if the input signal is 1 V_{pp} it means that the peak is 0.5 V, so that the gate voltage is -0.7 ± 0.5 V, that is from -0.2 to -1.2 V. These voltages and the corresponding values of V_{DS} are plotted on figure 4.7 where we can see that V_{DS} will vary from about 5.6 V to about 10.7 V. Thus, with $V_{DSQ} = 8.44$ V, we see that the output voltage will have a maximum of $10.7 - 8.44 = +2.26$ V and a minimum of $5.6 - 8.44 = -2.84$ V, as shown in figure 4.7b.

Now it is possible calculate g_m at the Q-point, and hence the small-signal voltage gain, using equation 4.5

$$g_m = \frac{2\sqrt{I_{DQ}I_{DSS}}}{|V_P|} = \frac{2\sqrt{5.861 \times 10}}{3} = 5.1 \text{ mS}$$

Then the no-load voltage gain is $-g_mR_D = -5.1 \times 10^{-3} \times 1000 = -5.1$, which can be compared to the voltage gain found graphically of $-2.86/0.5 = -5.7$ for the positive half cycle of the input and $-2.24/0.5 = -4.48$ for the negative. The average of these is -5.09, essentially the same as the small-signal voltage gain.

The current gain is $-g_mR_G = -5.1 \times 10^{-3} \times 500 \times 10^3 = -2550$. The power gain is therefore $10\log(2550 \times 5.1) = 41.1$ dB, a respectable figure because of the high input resistance.

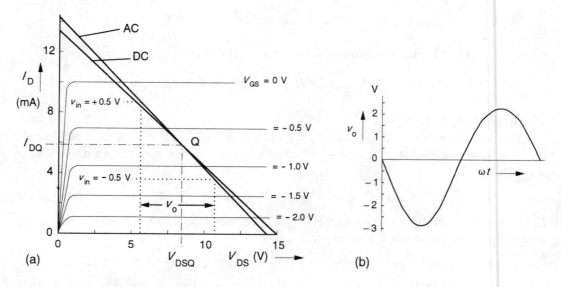

Figure 4.7 (a) Load lines and Q-point for example 4.1 (b) The output waveform for a 1V$_{pp}$ input

4.3.3 The voltage-divider bias circuit

The self-bias circuit of the CS amplifier suffers from a similar drawback to the fixed-bias circuit of the CE amplifier: it is highly sensitive to changes in the transistor's parameters. In the case of the JFET both the cut-off voltage, V_P, and the saturated drain current, I_{DSS}, can vary widely, as illustrated by the next example.

Example 4.3

The JFET in example 4.2 is replaced by one with $V_P = -6$ V and $I_{DSS} = 15$ mA, what is the Q-point now if R_S and R_D remain unchanged at 120 Ω and 1 kΩ respectively?
From equation 4.1 we have

$$I_{DQ} = I_{DSS}(1 - |I_{DQ}R_S/V_P|)^2 = 15(1 - |0.12I_{DQ}/6|)^2$$

where $|V_{GSQ}|$ has been replaced by $|I_{DQ}R_S|$ as usual in the self-bias circuit. The solution of this equation is $I_{DQ} = 9.73$ mA, so that equation 4.3 yields

$$V_{DD} = 15 = V_{DSQ} + I_{DQ}(R_D + R_S) = V_{DSQ} + 9.73 \times 1.12$$

whence $V_{DSQ} = 4.1$ V. The Q-point has shifted by $+3.87$ mA and -4.34 V.

If the voltage-divider bias circuit of figure 4.8 is used some amelioration of the problem can be obtained.

Figure 4.8

The voltage-divider bias circuit for the JFET

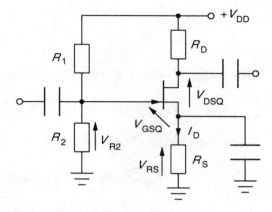

In this circuit the bias resistors hold the gate voltage at V_{R2} and since $I_G = 0$, R_1 and R_2 form a perfect voltage divider:

$$V_{R2} = \frac{R_2}{R_1 + R_2} \times V_{DD}$$

Then we can see by KVL that

$$V_{R2} = V_{GSQ} + V_{RS} = V_{GSQ} + I_{DQ}R_S$$

Hence $V_{GSQ} = V_{R2} - I_{DQ}R_S$. Since V_{GSQ} must be negative, it is clear that $I_{DQ}R_S > V_{R2}$. The value of R_S should be as large as possible to maximise the stability of the Q-point, but as it is by-passed, we do not want it to be larger than R_D or too much of the supply voltage will be 'lost' across it; as a rule of thumb we take $R_S = R_D$. The bias resistors must be as large as possible to keep up the input resistance, but again as a rule of thumb, it is not desirable to have any resistor of more than about 1 MΩ. With these rules of thumb we can easily find all the resistances to establish the desired Q-point as the next example shows.

Example 4.4

If the JFET in figure 4.8 has $V_P = -3$ V and $I_{DSS} = 2$ mA, what must R_1, R_2, R_D and R_S be to produce a Q-point at $I_{DQ} = 1$ mA and $V_{DSQ} = 8$ V given that $V_{DD} = 16$ V?

Equation 4.3 gives $R_D + R_S$:

$$V_{DD} = V_{DSQ} + I_{DQ}(R_D + R_S)$$

Thus

$$R_D + R_S = \frac{16 - 8}{1} = 8 \text{ k}\Omega$$

If we take R_D to be about the same magnitude as R_S or a little larger, then $R_D = 4.7$ kΩ and $R_S = 3.3$ kΩ would suit.

V_{GSQ} is found from equation 4.1

$$I_{DQ} = 1 = I_{DSS}\left(1 - \left|\frac{V_{GSQ}}{V_P}\right|\right)^2 = 2\left(1 - \left|\frac{V_{GSQ}}{3}\right|\right)^2$$

whence $V_{GSQ} = -0.877$ V. Now the voltage across R_2 by KVL is

$$V_{R2} = V_{GSQ} + V_{RS} = V_{GSQ} + I_{DS}R_S = -0.877 + 1 \times 3.3 = 2.423 \text{ V}$$

And by the voltage-divider rule it is

$$V_{R2} = 2.423 = V_{DD}R_2/(R_1 + R_2) = 16R_2/(R_1 + R_2)$$

$$\Rightarrow \qquad R_2/(R_1 + R_2) = 2.423/16 = 0.1514$$

If R_1 is chosen to be 1 MΩ, the rule-of-thumb maximum, then $R_2 = 178$ kΩ; the nearest preferred value being 180 kΩ. These resistor values almost give the Q-point required.

What now if the JFET is replaced by one of the same type with different parameters?

Example 4.5

The JFET in example 4.4 is replaced with one having $V_P = -5$ V and $I_{DSS} = 3$ mA. What is the Q-point now? What are the small-signal, no-load, voltage gains of the amplifiers?

The voltage divider will give $V_{R2} = 16 \times 0.18/1.18 = 2.44$ V, so that

$$V_{R2} = 2.44 = V_{GSQ} + I_{DQ}R_S \Rightarrow V_{GSQ} = 2.44 - 3.3I_{DQ}$$

This value of V_{GSQ} is substituted into equation 4.1 to yield

$$I_{DQ} = 3\left[1 - \left|\frac{2.44 - 3.3I_{DQ}}{5}\right|\right]^2$$

The solution to this is $I_{DQ} = 1.27$ mA (the other solution is physically impossible; it makes $|V_{GSQ}| > |V_P|$). Then $V_{DSQ} = V_{DD} - I_{DQ}(R_D + R_S) = 16 - 1.27 \times 8 = 5.84$ V. The Q-point has shifted up the load line as I_{DQ} has increased by 27%.

Now we can calculate the no-load voltage gains from $A_{v0} = -g_m R_D$ after first calculating g_m from equation 4.5

$$g_{m1} = \frac{2\sqrt{I_{DQ}I_{DSS}}}{|V_P|} = \frac{2\sqrt{1 \times 2}}{3} = 0.943 \text{ mS}$$

And so the gain of the the amplifier in example 4.4 is $-g_m R_D = -0.943 \times 4.7 = -4.43$, while that in example 4.5 is

$$g_{m2} = \frac{2\sqrt{I_{DQ}I_{DSS}}}{|V_P|} = \frac{2\sqrt{1.27 \times 3}}{5} = 0.781 \text{ mS}$$

Thus the gain is $A_{v0} = -0.781 \times 4.7 = -3.67$. Both voltage gains are reduced because of the voltage lost across R_S.

4.3.4 A graphical method for determining Q-points

We can readily grasp what is happening if we plot the Q-points on the transfer characteristics of the JFETs. Figure 4.9 shows the transfer characteristics of two JFETs, nominally the same device. The straight line from the origin corresponds to a self-bias circuit and that starting from the point V_{R2} on the V_{GS}-axis corresponds to a voltage-divider bias circuit. We can prove this by considering the self-bias equation for V_{GS}, namely

$$V_{GS} = -I_D R_S \quad \Rightarrow \quad I_D = -V_{GS}/R_S$$

which is the equation of a straight line from the origin of slope $-1/R_S$. And the equation for V_{GS} in the voltage-divider bias circuit of figure 4.8 is

$$V_{GS} = V_{R2} - I_D R'_S \quad \Rightarrow \quad I_D = -V_{GS}/R'_S + V_{R2}/R'_S$$

which is the equation of a straight line of slope $-1/R'_S$ starting from $V_{GS} = V_{R2}$, where R'_S is the source resistance in the voltage-divider bias circuit.

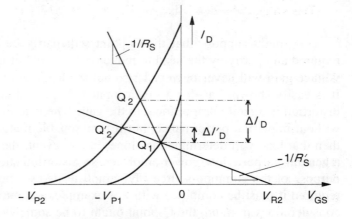

Figure 4.9

Q-points for the self-bias and voltage-divider bias circuits. The much steeper slope of the line for the self-bias circuit results in a much larger shift in I_{DQ}

The Q-points of both circuits are the same for the device with the smaller value of $|V_P|$, that is the point Q_1, but the self-bias circuit's Q-point becomes Q_2 when the device is changed, while that of the voltage-divider bias circuit becomes Q'_2. The self-bias circuit's Q-point is shifted by ΔI_D while that of the voltage-divider bias circuit is shifted by $\Delta I'_D$. It is clear that $\Delta I'_D < \Delta I_D$, mainly because $R'_S \gg R_S$. It is because we are free to make the source resistance much larger in the voltage-divider bias circuit that the Q-point is more stable, though at some expense in gain.

4.3.5 Where to place the Q-point?

Until now we have not considered where to put the Q-point for optimal performance. As we have seen, the voltage gain is modest, and distortion limits the output voltage long before clipping is encountered; nevertheless, there must be an ideal Q-point for any given circumstances. Figure 4.10 shows the operating region. It is constrained within the following limits: the maximum permissible drain-source voltage, V_{DSmax}, which effectively is the limit for V_{DD} only; the maximum value of I_D, which is I_{DSS}; the maximum transistor power limit, $P_{max} = V_{DSQ}I_{DQ}$, bearing in mind the temperature derating required; the locus of $V_{GS} - V_P$ on the left, which affects matters only slightly; and finally v_{GS} is limited by the cut-off voltage, V_P, which effectively means that $i_{Dmin} = 0$.

Figure 4.10

The operating region for a CS amplifier

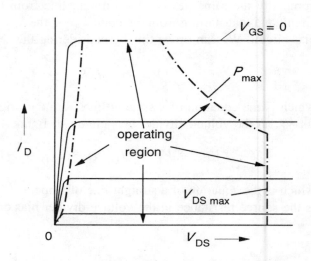

One might suppose that the Q-point will partly be determined by the voltage gain required and partly by the need to reduce distortion, but in practice these matter little. The voltage gain will never be very large and the larger it is the more the output is distorted. It is easily shown that, for a given input voltage and an unloaded amplifier, the output distortion is directly proportional to the gain. The load line can vary considerably in slope without impairing performance, but if a reasonable match to an external load is required, then that load will determine the slope. In any event, the input signal should be small for a general-purpose, low-power amplifier, so distortion should not be significant. The main purpose of the common-source FET amplifier is as a buffer input stage. If large gain is required it must be combined with, for example, a common-emitter amplifier. With these considerations in mind, the Q-point ought to be somewhere in the region of $I_{DQ} = 0.5I_{DSS}$ and $V_{GSQ} = 0.5V_P$, the latter being the same as $I_{DQ} = 0.25I_{DSS}$. Of course, if the JFET is damaged and must be changed, the Q-point may be moved considerably even when a voltage-divider bias method is used.

If the Q-point is chosen so that $I_{DQ}R_D = c_1V_{DD}$, where c_1 is a constant of proportionality such that $0 < c_1 < 1$, and $I_{DQ} = c_2I_{DSS}$, where c_2 is another constant of proportionality such that $0 < c_2 < 1$, then

$$R_D = \frac{c_1 V_{DD}}{I_{DQ}} = \frac{c_1 V_{DD}}{c_2 I_{DSS}}$$

and the transconductance is given by

$$g_m = \frac{2\sqrt{I_{DQ} I_{DSS}}}{|V_P|} = \frac{2\sqrt{c_2} I_{DSS}}{|V_P|}$$

Thus the no-load voltage gain becomes

$$A_{v0} = -g_m R_D = -\frac{2\sqrt{c_2} I_{DSS}}{|V_P|} \frac{c_1 V_{DD}}{c_2 I_{DSS}} = -\frac{2c_1 V_{DD}}{\sqrt{c_2}|V_P|}$$

With 'normal' values of c_1 (≈ 0.5) and c_2 (≈ 0.25) the voltage gain is $-2V_{DD}/|V_P|$, that is independent of all but the supply voltage and the JFET's cut-off voltage.

4.3.6 *The loaded CS amplifier*

When the CS amplifier is loaded, the load goes in parallel with R_D in the small-signal equivalent circuit. Figure 4.11 shows both this and the source, v_s, and its internal resistance, R_s. The gate resistance, R_G, is either the equivalent resistance of the voltage-divider, $R_1 \,//\, R_2$ in figure 4.8, or the value of R_G in figure 4.4. Our next example demonstrates the use of the equivalent circuit.

Figure 4.11

The small-signal equivalent circuit of a CS amplifier with load

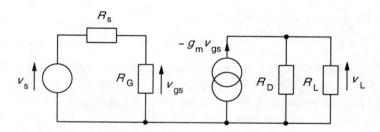

Example 4.6

A CS amplifier with voltage-divider bias as in figure 4.8 is fed by a sinusoidal voltage source of r.m.s. voltage 0.12 V and internal resistance 50 kΩ. It is to drive a load of 2 kΩ from a supply voltage, $V_{DD} = 24$ V. The JFET has $I_{DSS} = 7$ mA and $V_P = -2.5$ V and the resistor values are $R_D = 2.2$ kΩ, $R_S = 1.8$ kΩ, $R_1 = 1$ MΩ and $R_2 = 200$ kΩ. What is the voltage gain? How much distortion is there in the load voltage? What will happen if the JFET is replaced by one with $V_P = -1$ V and $I_{DSS} = 4$ mA?

The voltage divider produces a voltage across R_2 of

$$V_{R2} = \frac{V_{DD} R_2}{R_1 + R_2} = \frac{24 \times 0.2}{1.2} = 4 \text{ V}$$

This voltage is equal to $V_{GSQ} + I_{DQ}R_S$, so that

$$V_{GSQ} = V_{R2} - I_{DQ}R_S = 4 - 1.8I_{DQ}$$

Substituting this value for V_{GSQ} into equation 4.1 leads to

$$I_{DQ} = I_{DSS}\left(1 - \left|\frac{4 - 1.8I_{DQ}}{V_P}\right|\right)^2 = 7(1 - |1.6 - 0.72I_{DQ}|)^2$$

which gives $I_{DQ} = 2.742$ or 4.756 mA; the larger value makes $V_{GSQ} = -4.56$ V and is therefore unacceptable since with this gate voltage the device would not conduct.

The value of R_G in the small-signal equivalent circuit of figure 4.11 is $R_1 \mathbin{/\mkern-5mu/} R_2 = 1000$ k$\Omega \mathbin{/\mkern-5mu/} 200$ k$\Omega = 166.7$ kΩ and then the AC voltage at the gate is

$$v_{gs} = R_G v_s/(R_G + R_s) = 166.7 \times 0.12/216.7 = 0.0923 \text{ V} \tag{4.7}$$

The output resistance is $R_o = R_D \mathbin{/\mkern-5mu/} R_L = 2.2 \mathbin{/\mkern-5mu/} 2 = 1.048$ kΩ. The transconductance is

$$g_m = \frac{2\sqrt{I_{DQ}I_{DSS}}}{|V_P|} = \frac{2\sqrt{2.741 \times 7}}{2.5} = 3.504 \text{ mS} \tag{4.8}$$

from equation 4.5, and then the load voltage is

$$v_L = -g_m v_{gs} R_o = -3.504 \times 0.0923 \times 1.048 = -0.339 \text{ V}$$

The overall voltage gain becomes

$$A_{vL} = v_L/v_s = -0.339/0.12 = -2.825$$

The distortion can be calculated from equation 4.1 using the maximum and minimum values of v_{gs} derived from equation 4.7, which gives $v_{gs}(\text{peak}) = 0.0923 \times \sqrt{2} = 0.1306$ V. Thus the peak value of V_{GS} in equation 4.1 is $V_{GSQ} + 0.1306$ V. The quiescent gate voltage is

$$V_{GSQ} = V_{R2} - I_{DQ}R_D = 4 - 2.742 \times 1.8 = -0.9356 \text{ V}$$

making the peak gate voltage

$$V_{GSmax} = -0.9356 + 0.1306 = -0.805 \text{ V}$$

and the minimum gate voltage is

$$V_{GSmin} = -0.9356 - 0.1306 = -1.0662 \text{ V}$$

Substituting these into equation 4.1 gives $I_{Dmax} = 3.218$ mA and $I_{Dmin} = 2.303$ mA. Since $I_{DQ} = 2.742$ mA, the peak current is $I_{Dmax} - I_{DQ} = +0.483$ mA and the trough is $I_{Dmin} - I_{DQ} = -0.434$ mA. The load voltage is $-i_d R_o = -1.048i_d$, so that $v_{Lmax} = 1.048 \times 0.434 = 0.455$ V and $v_{Lmin} = -0.483 \times 1.048 = -0.506$ V. We can define a *distortion index* by

$$DI = \frac{|q_{max}| - |q_{min}|}{|q_{max}| + |q_{min}|} \tag{4.9}$$

From equation 4.9, $DI = (0.455 - 0.506)/(0.455 + 0.506) = -0.053$ or -5.3% (the minus sign indicates that the negative-going peak is larger than the positive-going peak).

Repeating the calculations for a JFET with $V_P = -1$ and $I_{DSS} = 4$ mA leads to $I_{DQ} = 2.352$ mA, $g_m = 6.135$ mS, $A_{vL} = -4.95$, $V_{GSQ} = -0.2336$ V, $I_{Dmax} = 3.218$ mA, $I_{Dmin} = 1.617$ mA, $v_{lmin} = -0.9076$ V, $v_{Lmax} = 0.7703$ V, $DI = -8.2\%$. Greater gain has resulted in more distortion, but the performances of the two, rather dissimilar, devices are comparable.

4.4 The source follower or common-drain amplifier

Just as the output of a BJT amplifier can be taken from the emitter to form an emitter follower, so the output of the JFET can be taken from the source rather than the drain to form a source follower or common-drain amplifier. Figure 4.12 shows the source follower, in which the drain resistor has been omitted, since it is no longer necessary to maintain the drain terminal of the JFET at a different potential to the supply. The output voltage will now be in phase with the input signal voltage and we shall see that, like the emitter follower, the voltage gain is slightly under unity.

Figure 4.12

The source follower or common-drain amplifier. Note that $v_{in} \neq v_{gs}$ but $v_{gs} + v_o$

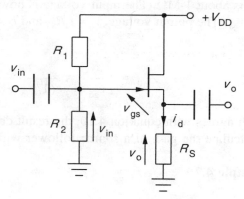

We can see from figure 4.12 that $v_{in} = v_{gs} + v_o$ since the input and output decoupling capacitors are assumed to have zero reactance. We also see that $v_o = i_d R_S = g_m v_{gs} R_S$ as I_D is still given by equation 4.1 and equation 4.6 therefore still holds. Hence

$$v_{in} = v_{gs} + v_o = v_{gs} + g_m v_{gs} R_S = v_{gs}(1 + g_m R_S)$$

Then

$$v_{gs} = v_{in}/(1 + g_m R_S)$$

And so

$$v_{in} = v_{gs} + v_o = \frac{v_{in}}{1 + g_m R_S} + v_o$$

giving

$$A_{v0} = \frac{v_o}{v_{in}} = \frac{g_m R_S}{1 + g_m R_S} = \frac{R_S}{1/g_m + R_S} \qquad (4.10)$$

When $R_S \gg 1/g_m$, the no-load voltage gain, $A_{v0} \approx 1$.

Equation 4.10 shows that $v_o = \frac{1}{2}v_{in}$ when $R_S = 1/g_m$, that is the output resistance of the common-drain amplifier is $1/g_m$, which the small-signal equivalent circuit of figure 4.13 shows.

Figure 4.13

The small-signal equivalent circuit of the common-drain amplifier

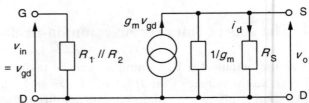

The value of g_m depends on the JFET parameters and I_{DQ} but is usually in the range 1-10 mS, so that the output resistance lies between 100 Ω and 1 kΩ — neither particularly low nor particularly high. The input resistance is $R_1 \parallel R_2$, that is of the order of 500 kΩ, if R_1 is about 1 MΩ. The input voltage is now v_{gd} and the current source has a magnitude of $g_m v_{gd}$. The output voltage, $v_o = i_d R_S$, and $i_d = v_{gd}/(1/g_m + R_S)$ from the equivalent circuit. Hence

$$A_{v0} = \frac{i_d R_S}{v_{gd}} = \frac{v_{gd}}{(1/g_m + R_S)} \frac{R_S}{v_{gd}} = \frac{R_S}{1/g_m + R_S}$$

which agrees with equation 4.10, the result derived earlier. The next example shows how to calculate the gain of a source follower with load.

Example 4.7

A source follower as in figure 4.12 has $R_1 = R_2 = 1$ MΩ, $R_S = 8.2$ kΩ, $R_L = 10$ kΩ and $V_{DD} = 16$ V. The JFET parameters are $I_{DSS} = 5$ mA and $V_P = -5$ V. What are the voltage, current and power gains if the signal source has negligible internal resistance? What would these become if the JFET was replaced with one having $I_{DSS} = 3$ mA and $V_P = -2$ V?

We shall solve graphically this time. The defining equation for the load line from the partial circuit diagram in figure 4.14 is

$$V_{R2} = V_{GS} + I_D R_S \;\Rightarrow\; I_D = \frac{-V_{GS}}{R_S} + \frac{V_{R2}}{R_S}$$

Now $V_{R2} = \frac{1}{2}V_{DD} = 8$ V, so the load line is

$$I_D = -0.122 V_{GS} + 0.9756$$

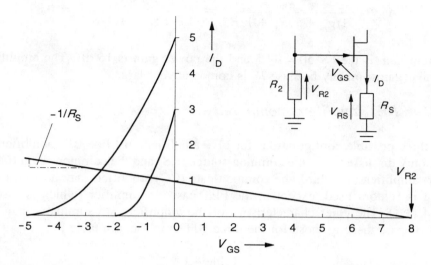

Figure 4.14 The load line for example 4.7 and the circuit from which it is derived

This line is plotted on the transfer characteristics of the JFETs in figure 4.14, from which it can be deduced that for the first JFET the Q-point is at $I_{DQ} = 1.28$ mA. Then the transconductance is

$$g_m = \frac{2}{|V_P|} \sqrt{I_{DQ}I_{DSS}} = \frac{2}{5}\sqrt{1.28 \times 5} = 1.012 \text{ mS}$$

From the equivalent circuit of figure 4.13, the voltage gain with load is

$$A_{vL} = \frac{g_m}{g_m + 1/R_S + 1/R_L} = \frac{1.012}{1.012 + 1/8.2 + 1/10} = 0.82$$

since R_L is in parallel with R_S and g_m.
The currents are

$$i_L = v_L/R_L \quad \text{and} \quad i_{in} = v_{in}/(R_1 /\!/ R_2)$$

making the current gain

$$A_{iL} = \frac{i_L}{i_{in}} = \frac{v_L/R_L}{v_{in}/(R_1 /\!/ R_2)} = \frac{A_{vL}(R_1 /\!/ R_2)}{R_L} = \frac{0.82 \times 500}{10} = 41$$

So that the power gain is $10\log(0.82 \times 41) = 15.3$ dB.
The replacement JFET can be seen from figure 4.14 to have $I_{DQ} = 1.08$ mA and thus

$$g_m = 2\sqrt{(1.08 \times 3)/2} = 1.8 \text{ mS}$$

and the voltage gain is

$$A_{vL} = \frac{g_m}{1/g_m + 1/R_L + 1/R_S} = \frac{1.8}{1.8 + 1/10 + 1/8.2} = 0.89$$

The current gain is $0.89 \times 50 = 44.5$ and the power gain is 16 dB. The amplifiers have very similar gains, mainly because R_S is comparatively large.

4.4.1 The FET amplifiers compared

The are three possible configurations for FET amplifiers just like BJT amplifiers, though we have only discussed two: the common-source (CS) and the common-drain (CD, source follower) amplifiers; the third, the common-gate (CG) amplifier, may be seen in problem 4.13, but is seldom used, except in the FET cascode amplifier (which is discussed in section 4.11.1). The chief characteristics of the amplifiers are summarised in table 4.1 below. Compare these to those for the three BJT amplifiers in table 3.2.

Table 4.1

Approximate magnitudes for some parameters of the CS, CD and CG amplifiers

Parameter	CS	CD	CG
Voltage gain	$g_m R_D$	1	$g_m R_D$
Current gain	$g_m R_G$	R_G / R_L	1
Power gain	high	medium	low
Input resistance	R_G	R_G	$1/g_m$
Output resistance	$1/g_m$	$1/g_m$	r_d

4.5 Other types of FET: MOSFETs and IGFETs

It had been realised by Shockley among others as long ago as 1950 that current flow near a semiconductor surface could be controlled by altering the field under a gate electrode, as in figure 4.15. The gate electrode is insulated from the semiconductor by a very thin layer of insulating material and so the layer sequence at the gate electrode are metal-insulator-semiconductor and the device is an IGFET (Insulated Gate FET). If the insulator is made from silicon by oxidation in situ then the gate layers are metal-oxide-semiconductor and the device is a MOSFET. Though this process was recognised as simpler in principle than bipolar operation, it was not capable of being put into commercial practice because the impurities — particularly sodium ions — in the insulating layer tended to migrate under the influence of the gate bias to form a layer of charge at the silicon surface thereby turning the device permanently on or off. By about 1969 this problem was solved and the manufacture, particularly in integrated circuits, of MOSFETs rapidly accelerated.

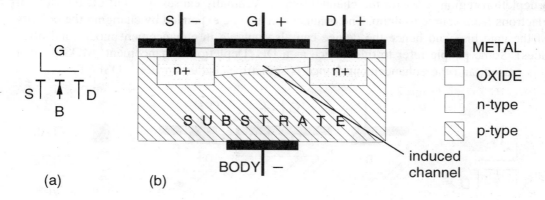

Figure 4.15 An n-channel, enhancement-mode MOSFET (a) Circuit symbol (b) Cross-section

4.5.1 *Principles of operation*

If a positive charge is placed on the gate electrode in figure 4.15, the p-type semiconducting material under the gate oxide will tend to become negatively charged as electrons are attracted towards the gate and holes repelled from it. The material under the gate is said to have become *inverted* and an inversion layer, one of opposite conductivity type to the original, is said to have formed. If the material under source and drain electrodes has been doped n-type, the inversion layer forms an *induced channel* between drain and source through which conduction can occur by the passage of electrons alone: the device is unipolar. The channel is of non-uniform thickness because the channel at the drain end is more positive than that at the source end. Thus the gate-channel voltage is higher at the source end and the induced channel thicker. The actual gate voltage at which the channel begins to conduct is rather variable and is known as the *threshold voltage*, V_T — usually denoted by $V_{GS(th)}$ in manufacturer's data books. The threshold voltage is almost the same as V_P in JFETS, but is of opposite sign.

The MOSFET in figure 4.15 is called an n-channel, enhancement-mode MOSFET, the symbol for which is shown on the left. A p-channel, enhancement-mode MOSFET has the same symbol, but the arrow points in the opposite direction. The substrate is also connected to a terminal (the bulk or B terminal in figure 4.15) which is usually connected to the source so that it is at a lower potential than the drain in the n-channel MOSFET. If this is done the transistor is effectively a three-terminal device like a BJT or JFET. The bulk terminal is not normally available externally.

Instead of inducing a channel beneath the gate where none existed before, it is possible to make a channel under the gate so connecting drain to source. Then the application of a voltage of the right polarity to the gate electrode can produce a depletion region in the channel which constricts it and reduces the current flow. This type of device is shown in figure 4.16 and is known as a depletion-mode MOSFET.

Only n-channel depletion devices are made and have the symbol depicted on the left. Making the gate negative will tend to attract holes to the surface of the semiconductor and a depletion region forms in the channel which eventually causes it to cut off the flow of electrons from source to drain. The channel can also be expanded by changing the polarity of the gate bias and hence the device can also operate in enhancement mode (and often does). Some people refer to this device as a DE (Depletion-Enhancement) MOSFET for this reason, and the enhancement device is sometimes called an EN MOSFET.

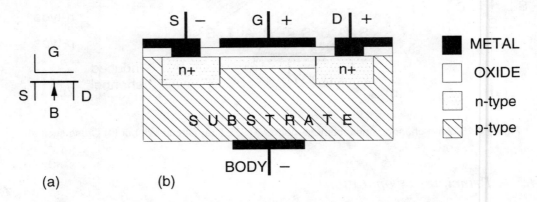

Figure 4.16 An n-channel, depletion-mode MOSFET (a) circuit symbol (b) Cross-section

4.5.2 Characteristics of MOSFETS

The characteristics described are for the substrate (or bulk, B) terminal connected to the source (S) terminal as is normal, but not invariably so. The p-channel devices work with their drains at a lower potential than their sources, but are otherwise identical in operation to the n-channel devices which we shall describe here. The transfer characteristics, shown in figure 4.17, of NMOS (n-channel MOS) transistors are the same as those of the JFET, except for the position of V_T for enhancement (EN) devices, which lies to the right of the I_D axis.

Figure 4.17

The transfer characteristics of n-channel enhancement (EN) and n-channel depletion (DE) MOSFETs

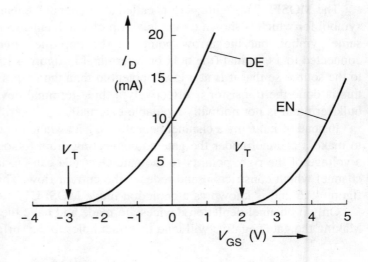

Because MOSFET gates can be forward biased without conducting, unlike JFET gates, there is no drain current limit, I_{DSS}, at $V_{GS} = 0$. In fact I_{DSS} for EN MOSFETs signifies the leakage current flowing when the device is turned off, a very small value, but to confuse matters, I_{DSS} for DE MOSFETs can mean the drain current for $V_{GS} = 0$; the magnitude of I_{DSS} should indicate which is meant. The gate voltage for any MOSFET can be positive or negative and the n-channel DE MOSFET is usually operated with $V_{GS} = 0$ so that the gate goes positive or negative with the input, v_{gs}.

The PMOS EN FET has a transfer characteristic which is the mirror image of the NMOS device's about the I_D axis, hence a negative V_T. The drain characteristics are like those of the JFET with a linear part at the beginning up to a knee at $V_{DS} = V_{GS} - V_T$. The saturation drain current after the knee is given by

$$I_D = k(V_{GS} - V_T)^2 \qquad (4.11)$$

where k is a constant and $V_{GS} > V_T$. This equation, unlike the corresponding equation 4.1 for a JFET, is more nearly exact. Usually one can calculate the constant, k, from the drain current given at some stated value of V_{GS}. To take a random example: the IRFPG30 power HEXFET[3], which is an n-channel enhancement device, has $V_T = 4$ V and $I_D = 4$ A when $V_{GS} = 5.5$ V at 25°C. Substitution of these values in equation 4.11 leads to $k = 1.8$ A/V^2, so enabling us to find I_D at any value of V_{GS}. From equation 4.11 we can also find the transconductance as we did for the JFET:

$$g_m = dI_D/dV_{GS} = 2k(V_{GS} - V_T) \qquad (4.12)$$

Thus at $I_D = 2$ A, $V_{GS} = 5.05$ V from equation 4.11 and $g_m = 7.6$ S from equation 4.12.

Example 4.8

The MOSFET in figure 4.18 has $V_T = -2.4$ V and its drain current is 100 mA when $V_{GS} = 5$ V. Where is the Q-point? What is the voltage gain v_L/v_s ? What power is delivered to the load? What is the distortion in v_L ? (The capacitors have negligible reactances.)

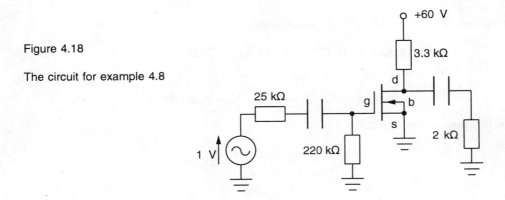

Figure 4.18

The circuit for example 4.8

---

[3] HEXFETs are so called because of the hexagonal structure of the devices, like a honeycomb; see figure 4.32.

The device is an n-channel DE MOSFET with its gate grounded and the substrate (b) connected to the source (s). We must find I_{DQ} from equation 4.11, which can be done by finding k given $V_T = -2.4$ V and $V_{GS} = 5$ V when $I_D = 100$ mA. Substitution into equation 4.11 yields

$$0.1 = k[5 - (-2.4)]^2 \quad \Rightarrow \quad k = 1.826 \text{ mA/V}^2$$

Then setting $V_{GS} = 0$ gives $I_{DQ} = 10.5$ mA and the quiescent drain-source voltage is

$$V_{DSQ} = V_{DD} - I_{DQ}R_D = 60 - 10.5 \times 3.3 = 25.3 \text{ V}$$

The transconductance from equation 4.12 is

$$g_m = 2k(V_{GSQ} - V_T) = 2 \times 1.826 \times 10^{-3} \times 2.4 = 8.765 \text{ mS}$$

The voltage gain is found from the equivalent circuit of figure 4.11 in the same way as in example 4.6, that is

$$A_{vL} = \frac{-g_m(R_D /\!/ R_L)R_G}{R_G + R_s} = \frac{-8.765 \times 1.245 \times 220}{220 + 25} = -9.8$$

The r.m.s. input voltage is 1 V so that the peak input voltage is $\pm\sqrt{2}$ V, that is $V_{GS} = \pm\sqrt{2}$ V. When $V_{GS} = +\sqrt{2}$ V,

$$I_D = 1.826 \times 10^{-3} \times (2.4 - \sqrt{2})^2 = 1.775 \text{ mA}$$

and

$$i_d(\text{max}) = I_{DQ} - I_D = 10.5 - 1.775 = 8.725 \text{ mA}$$

And when $V_{GS} = -\sqrt{2}$ V,

$$I_D = 1.826 \times 10^{-3} \times (2.4 + \sqrt{2})^2 = 26.57 \text{ mA}$$

making

$$i_d(\text{min}) = 10.5 - 26.57 = -16.07 \text{ mA}$$

The load voltage will be $-i_d (R_L /\!/ R_D)$, which means

$$v_L(\text{max}) = 16.07 \times 1.245 = +20.00 \text{ V}$$

and

$$v_L(\text{min}) = -8.725 \times 1.245 = -10.86 \text{ V}$$

The distortion index from equation 4.9 is

$$DI = \frac{100 \times (20 - 10.86)}{20 + 10.86} = 29.6\%$$

There is a substantial amount of distortion.

4.6 The frequency response of the FET amplifier

The high and low frequency responses of the FET amplifiers can be found in the same way as we found those of the BJT amplifiers in section 3.8.

4.6.1 The low-frequency response

At low frequencies this is determined by the coupling and by-pass capacitors exactly as for the BJT. Examining figure 4.19 we can see that the time constant for the input decoupling capacitor is

$$\tau_1 = C_1 R_G$$

When the voltage divider bias circuit is used we can replace R_G by R'_G where $R'_G = R_1 \,//\, R_2$. Since R_G is usually large, C_1 can be relatively small. Suppose we want $f_1 \, (= 1/2\pi C_1 R_G) = 10$ Hz, then if $R_G = 1$ MΩ, $C_1 = 16$ nF.

On the output side we see that the time constant for C_2 is given by

$$\tau_2 = C_2(R_D + R_L)$$

Then if $R_D = R_L = 2.2$ kΩ and we wish $f_2 = 10$ Hz, then $C_2 = 3.6$ μF.

The third and final time constant to determine is that for C_S, which is charged via R_S and $1/g_m$ in parallel, since it 'looks back' into the FET and 'sees' the transconductance, g_m, as another path to ground. Thus the time constant is

$$\tau_S = C_S(R_s \,//\, 1/g_m)$$

Now g_m is typically a few mS, say 5 mS, and R_S might be about 330 Ω, making $R_S \,//\, 1/g_m$ about 125 Ω. To obtain a cut-off frequency at 10 Hz requires $C_S = 127$ μF, which is much the biggest of the three.

Figure 4.19

The circuit for determining the low frequency response of the FET CS amplifier. The input source and the DC supply have been replaced by connections to ground; the internal resistances of these should be negligible compared to R_G and R_D respectively

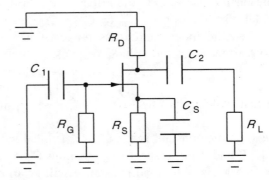

4.6.2 The high-frequency response

The high-frequency response of FET amplifiers is determined by the parasitic shunt capacitances shown in figure 4.20a for a JFET: those between drain and gate, C_{gd}, between gate and source, C_{gs}, and between drain and source, C_{ds}. In MOSFETs there are three further parasitic capacitances between the substrate (B) and the other terminals, which we shall ignore. The high-frequency equivalent circuit of the common-source amplifier that results is shown in figure 4.20b and its transformation is in figure 4.20c.

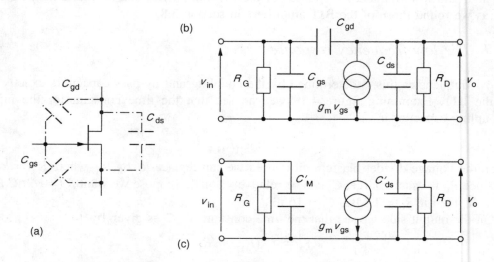

Figure 4.20 (a) JFET parasitics (b) the high-frequency model (c) transformed high-frequency model

The bridging capacitance, C_{gd}, is amplified (like C_{bc} in the CE amplifier) by the Miller effect, that is it becomes

$$C_M = (1 - A_v)C_{gd}$$

on the input side, where A_v is the voltage gain of the amplifier, in this case $-g_m R_D$. Then C_M is in parallel with C_{gs} and the pair can be replaced by $C'_M (= C_M + C_{gs})$ in figure 4.20c, while on the output side the effect of the bridging capacitance is merely to place C_{gd} in parallel with C_{ds} to give $C'_{ds} (= C_{gd} + C_{ds})$, also shown in figure 4.20c.

When the input source has no resistance associated with it the Miller capacitance cannot affect v_{in} or the voltage response, which is

$$A_v = -g_m Z_o = -g_m \left(R_D \, // \, \frac{1}{j\omega C'_{ds}} \right) = \frac{-g_m R_D}{1 + j\omega C'_{ds} R_D}$$

Thus the time constant is $\tau = C'_{ds} R_D$. Typical values for a small-signal FET of C_{ds} and C_{gd} are 1 pF and 4 pF, respectively, making C'_{ds} = 5 pF; thus if R_D = 2.2 kΩ, τ = 11 ns and f_H = 14.5 MHz. Any load resistance will go in parallel with R_D which will increase f_H.

If the input's (or the FET's) internal resistance is not negligible, then the effective input capacitance, C'_M, charges via R_s in parallel with R_G. Then the input side's time constant is

$$\tau = C'_M (R_s \, // \, R_G) \quad \Rightarrow \quad f_H = \frac{1}{2\pi C'_M (R_s \, // \, R_G)}$$

If the source resistance is low the time constant is $\tau \approx C'_M R_s \approx A_v C_{gd} R_s$.

For example, suppose $A_v = 8$, $C_{gd} = 4$ pF and $R_s = 50$ Ω, then $f_H \approx 100$ MHz. But if a voltage-divider bias is used with $R_1 = 100$ kΩ and $R_2 = 20$ kΩ (unusually low), and the source resistance is 2.5 kΩ (unusually high), we find $R'_G = R_1 // R_2 = 17$ kΩ and $R'_G // R_s = 2.2$ kΩ. With the same Miller capacitance and again ignoring C_{gs}, $f_H \approx 2.3$ MHz. In the equivalent circuit of figure 4.20c, it is therefore the output side's time constant that will normally determine the high-frequency response of the FET CS amplifier; the Miller capacitance is of little consequence.

4.7   FET switches and inverters

Most FETs are used as switches, whether in high-power circuits or in extremely low-power logic ICs. Their extremely high gate resistance means that the power drawn from the signal source is always negligible. The power consumption of FET switches is then entirely due to the product of the drain current and the drain-source voltage when conducting in the non-saturation region. This latter is not usually given, but instead the parameter $R_{DS}(\text{on})$ is used, and the average power consumption becomes $\frac{1}{2}I_D^2 R_{DS}(\text{on})$. In power devices great efforts have been made during manufacture to reduce $R_{DS}(\text{on})$ as far as possible.

4.7.1   The FET analogue switch

FETs are often used for switching a signal on or off by means of a control voltage applied at the gate, and often large matrices of FETs are used in devices such as multiplexers. FET switches are now used instead of resistors in new ICs such as switched-capacitor filters (see section 6.3). Usually FETs are used as the JFET gate biasing is complicated by the need to avoid forward bias. Figure 4.21 shows an n-channel enhancement FET analogue switch.

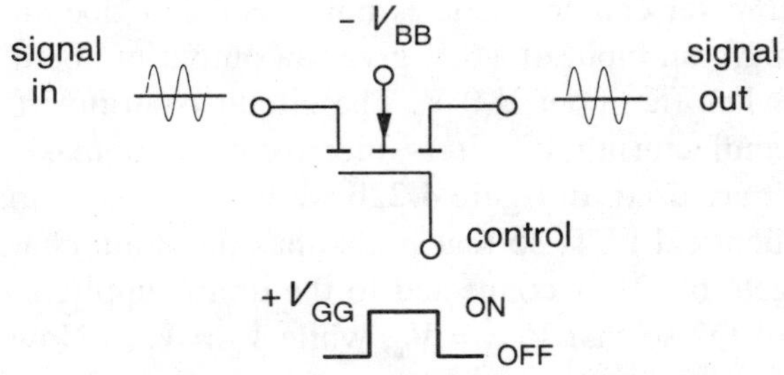

Figure 4.21

A MOSFET analogue switch

The bulk terminal is connected to the most negative supply voltage available (which can be ground), while the ON control voltage is the most positive supply voltage available (and this must be substantially greater than V_T). When ON the FET has low V_{DS} and high V_{GS} so that the drain current lies in the linear and not the saturation region of the drain characteristics. The input signal need not be biased positive, but must lie between $\pm |V_{GG} - V_T|$ and as it approaches these limits there will be distortion as the FET begins to turn off. The ON resistance of a small FET can be 100 Ω or more, which may have to be taken into account, especially as there must be a path to ground on the output side to prevent capacitively coupled feedthrough of the signal when the FET is OFF. The OFF resistance is very large: perhaps as much as 1 TΩ ($= 10^{12}$ Ω), but in practice the leakage current in the OFF state, though small, is high enough to dominate the OFF resistance. The effective OFF resistance may be 1-10 GΩ because of this.

4.7.2 The FET inverter

Let us consider the circuit of figure 4.22a and find what v_o is as a function of v_{in} ($= V_{GS}$), given the FET has drain characteristics given in figure 4.23a. The output voltage is

$$v_o = V_{DS} = V_{DD} - I_D R_D$$

which means we can plot a load line of slope $-1/R_D$ on the drain characteristics as shown in figure 4.23a. We then read off the values of V_{DS} corresponding to V_{GS} on the drain current intercepts with the load line. This process gives the voltage transfer characteristic, that is v_o against v_{in}, of figure 4.23b; the line marked (a).

Figure 4.22

(a) An n-channel enhancement
   FET switch with a drain resistor
(b) The FET switch with another
   FET, Q1, instead of the drain
   resistor

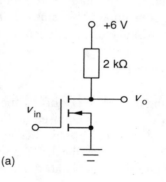

(a)

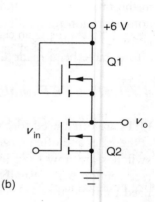

(b)

The transfer characteristic is not linear, nor does it go down to zero volts when the input is high: an input of +6 V gives an output of about 2.9 V, though the output does go to +6 V when the input is 0 V. The circuit is an inverter that will not turn fully off.

In manufacturing ICs it is inconvenient to make resistors and wherever possible transistors are used. In figure 4.22b we have replaced the drain resistor of figure 4.22a by another identical FET, so that it also has the drain characteristics of figure 4.23a.

The gate of Q1 is connected to the drain supply, so that $V_{GS1} = V_{DS1}$. The input is to the gate of Q2 so that $V_{GS2} = V_{in}$, while $V_o = V_{DS2}$. Now $V_{DS1} + V_{DS2} = V_{DD} = 6$ V, so that $V_{DS2} = V_o = 6 - V_{DS1} = V_{GS1}$. The parabolic loci of V_{DS1} ($= V_{DGS1}$) and V_{DS2} are drawn on the drain characteristics in figure 4.23a. From the intercept of V_{DS2} locus with the drain current, which must be the same for both FETs, one can read off V_{DS2} ($= V_o$) for any value of V_{GS2} ($= V_{in}$). For example, when $V_{in} = V_{GS2} = 5$ V, the intercept on the V_{DS} axis is $V_o = 1.6$ V and when $V_{in} = 7$ V, $V_o = 1.4$ V. The output falls very slowly as V_{in} increases beyond 2 V. The voltage transfer characteristic is plotted as curve (b) in figure 4.23b, and is also seen to be an inverter that is linear over a restricted range (+2 V < V_{in} < +4 V), will turn fully on when $V_{in} < 2$ V but does not turn fully off: when $V_{in} = +6$ V, $V_o = +1.5$ V. It performs, however, better than the first circuit.

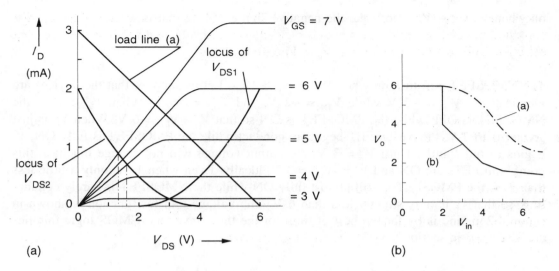

Figure 4.23 (a) The drain characteristics of all the FETs in figure 4.22, together with the load line for the circuit of figure 4.22a and the loci of the drain voltages of Q1 and Q2 in figure 4.22b
(b) The resultant voltage transfer characteristics

4.7.3 The complementary MOS (CMOS) inverter

CMOS circuits, as their name indicates, use pairs of p-channel (PMOS) and n-channel (NMOS) devices. The circuit of figure 4.24a is a CMOS inverter having two complementary enhancement MOSFETs in series.

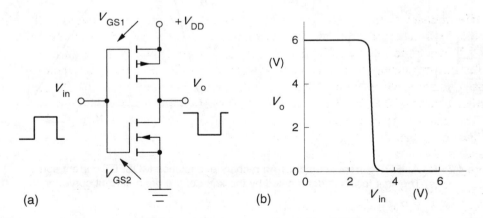

Figure 4.24 (a) A CMOS inverter (b) Its transfer function when V_{DD} =+6 V, V_{TP} = –2 V and V_{TN} = + 2 V

The PMOSFET has its source and substrate connected to the positive supply, $+V_{DD}$, while the NMOSFET's source and substrate are grounded. Thus the gate voltages are V_{GS1} = $V_{in} - V_{DD}$ for the PMOSFET and $V_{GS2} = V_{in}$ for the NMOSFET. The PMOSFET will turn on when $V_{GS1} < -|V_{TP}|$ (to make the negative sign explicit), that is when

$$V_{in} < V_{DD} - |V_{TP}|$$

The NMOSFET will turn on when $V_{GS2} = V_{in} > V_{TN}$. Let us suppose that the devices are identical and $V_{TP} = -2$ V while $V_{TN} = +2$ V, and $V_{DD} = +6$ V. When $V_{in} < 2$ V, the NMOSFET is OFF while the PMOSFET is ON so that $V_o = V_{DD} = 6$ V. When V_{in} is just greater than 2 V, the NMOSFET begins to conduct while the PMOSFET is fully ON; V_o begins to fall a little. Until $V_{in} = 3$ V, the output voltage will not change much; at this point both FETs are ON and $V_o = \frac{1}{2}V_{DD} = 3$ V, ideally. Even when V_{in} is only a little less than 3 V, the PMOSFET is still almost fully ON while the NMOSFET is barely conducting, so that V_o is only slightly less than 6 V. The voltage transfer function is shown in figure 4.24b, and is by far the best of those of the three inverters. CMOS logic families are discussed in section 9.11.3.

4.8 FET current sources and voltage references

FET constant-current sources are easily made by short-circuiting the gate to the source in a JFET or an EN MOSFET, as in figure 4.25a and 4.25b.

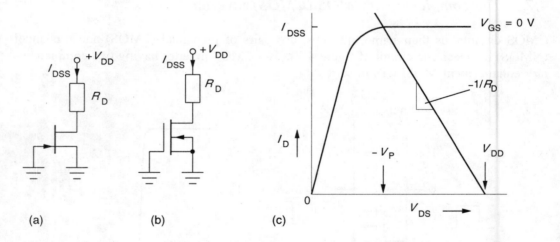

Figure 4.25 (a) A JFET current source (b) An n-channel depletion MOSFET current source
(c) The limit for R_D is determined by the knee of the drain current curve for $V_{GS} = 0$.

The current in the JFET will be I_{DSS} but, if R_D becomes too large and causes the load line to stray from saturation into the linear region, the drain current will fall. Figure 4.25c shows that the limiting resistance is

$$R_D < \frac{V_{DD} - |V_P|}{I_{DSS}}$$

With a constant current, a voltage reference is easily obtained from the drain terminal of the FET. Alternatively the gate bias can be derived from V_{DS} of an n-channel enhancement MOSFET as shown in figure 4.26, so that

$$V_{GS} = \frac{R_2 V_{REF}}{R_1 + R_2} \quad \Rightarrow \quad V_{REF} = \left(1 + \frac{R_1}{R_2}\right) V_{GS}$$

Example 4.9

In figure 4.26, the n-channel EN MOSFET has $V_T = 1$ V and $I_D = 16$ mA when $V_{GS} = 5$ V. If $V_{DD} = 15$ V, $R_D = 22$ kΩ, $R_1 = 330$ kΩ and $R_2 = 100$ kΩ, what is V_{REF} and by how much will it change if V_{DD} changes by 0.1 V? How does this compare with a simple voltage-divider reference?

Figure 4.26

The circuit for example 4.9

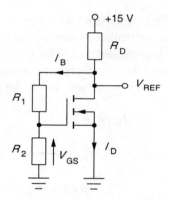

By the voltage-divider rule

$$V_{GS} = \frac{R_2 V_{REF}}{(R_1 + R_2)} \quad \Rightarrow \quad V_{REF} = \left(1 + \frac{R_1}{R_2}\right) V_{GS} = 4.3 V_{GS}$$

The gate bias current, I_B, is $V_{GS}/R_2 = V_{GS}/100$, and might not be negligible compared to I_D. The current through R_D is $I_B + I_D$ and is given by

$$I_B + I_D = \frac{(V_{DD} - V_{REF})}{R_D} = \frac{(15 - 4.3 V_{GS})}{22} \tag{4.13}$$

Finally I_D is

$$I_D = k(V_{GS} - V_T)^2 = (V_{GS} - 1)^2$$

Since $I_D = 16$ mA when $V_{GS} = 5$ V, we find $k = 1$ mA/V². Then substitution for I_B and I_D into equation 4.13 gives

$$\frac{V_{GS}}{100} + (V_{GS} - 1)^2 = \frac{(15 - 4.3 V_{GS})}{22}$$

which rearranges to

$$V_{GS}^2 - 1.7945V_{GS} + 0.3182 = 0$$

Solving this equation leads to $V_{GS} = 1.595$ V (the other root being less than V_T and therefore unacceptable) and $V_{REF} = 6.8585$ V.

When V_{DD} changes by 0.1 V it is easily shown, by solving the equation above with 14.9 or 15.1 instead of 15, that V_{REF} changes by 14 mV. A simple voltage-divider reference would have changed by $0.1 \times 6.86/15 = 46$ mV — slightly over three times more. (See also problem 4.10.)

4.9 The FET as a voltage-controlled resistor (VCR)

In ICs especially, the FET is used as an active resistor, that is the FET acts as a resistor whose resistance can be controlled to some extent. This property of the FET is also useful in circuits made from discrete devices. The initial part of the drain characteristics, when V_{DS} is small, is the linear or ohmic region when the resistance of the FET channel is almost proportional to the gate voltage. It can be shown that for a JFET in this state

$$I_D = g_0(1 - \sqrt{|V_{GS}/V_P|})V_{DS}$$

where g_0 is the initial slope of I_D-V_{DS} curve for $V_{GS} = 0$, and is given by

$$g_0 = 3I_{DSS}/|V_P|$$

Thus the conductance of the JFET for $V_{DS} < \frac{1}{2}|V_P - V_{GS}|$ — well before the knee of the drain characteristic — is

$$g_{ds} = \frac{1}{r_{ds}} = \frac{\delta I_D}{\delta V_{DS}}\bigg|_{V_{DS} \to 0} = g_0\left(1 - \sqrt{\left|\frac{V_{GS}}{V_P}\right|}\right)$$

Figure 4.27 shows a plot of this equation for a JFET with $|V_P| = 3$ V and $I_{DSS} = 4$ mA; it is reasonably linear for $\frac{1}{2}|V_P| < |V_{GS}| < |V_P|$.

MOSFETs follow a slightly different equation

$$g_{ds} = 2k(V_{GS} - V_T - |V_{DS}|) \approx 2k(V_{GS} - V_T)$$

when $0 < |V_{DS}| < (V_{GS} - V_T)$. A plot of g_{ds} against V_{GS} for an n-channel DE MOSFET with $V_T = -3$ V, $V_{DS} = 0$ and $k = 0.4$ mA/V^2 is also shown in figure 4.27.

The linear property of a MOSFET's ON conductance is a very useful feature for control and other purposes. It can be used for example in automatic gain control (AGC), in voltage-controlled RC oscillators and in many analogue ICs, where MOSFETs are used instead of resistors for almost every purpose, because they are much smaller and easier to make to specification than resistors.

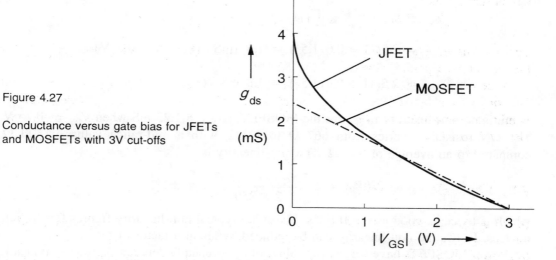

Figure 4.27

Conductance versus gate bias for JFETs and MOSFETs with 3V cut-offs

Example 4.10

A JFET used as a VCR has $I_{DSS} = 2$ mA and $V_P = -1.5$ V and operates with -0.6 V $< V_{GS}$ < -0.9 V, how does r_{ds} vary with V_{GS}? What is the non-linearity? If it is replaced by a DE MOSFET with $V_T = -1.5$ V and $g_0 = 2.5$ mS, how will r_{ds} vary now as V_{GS} is varied over the same range, assuming V_{DS} is negligible?

For the JFET g_0 is given by

$$g_0 = \frac{3I_{DSS}}{|V_P|} = \frac{3 \times 2}{1.5} = 4 \text{ mS}$$

When $V_{GS} = -0.6$ V,

$$g_{ds} = g_0\left(1 - \sqrt{\frac{V_{GS}}{|V_P|}}\right) = 4\left(1 - \sqrt{\frac{0.6}{1.5}}\right) = 1.47 \text{ mS}$$

thus $r_{ds} = 1/g_{ds} = 680$ Ω. And when $V_{GS} = -0.9$ V, $g_{ds} = 0.9$ mS and $r_{ds} = 1.11$ kΩ. Halfway between these two values, $V_{GS} = -0.75$ V, $g_{ds} = 1.172$ mS and $r_{ds} = 853$ Ω, while the average of the previous two values is 895 Ω. Thus the non-linearity in the middle of the range is

$$NL\% = \pm\frac{100(853 - 895)}{2 \times 895} = \pm2.35\%$$

(The factor 2 in the denominator arises because we can choose a straight line between the two values 853 Ω and 895 Ω.)

When the MOSFET is used and V_{DS} is negligible we find

$$g_{ds} = g_0(|1 - V_{GS}/V_T|)$$

$$= 2.5(|1 - 0.6/1.5|) = 1.5 \quad mS \quad (V_{GS} = -0.6 \ V)$$

$$= 2.5(|1 - 0.9/1.5|) = 1.0 \ mS \quad (V_{GS} = -0.9 \ V)$$

There is no non-linearity in g_{ds} for the MOSFET, so $g_{ds} = 1.25$ mS when $V_{GS} = -0.75$ V. The ON resistance varies from 667 Ω to 1 kΩ, with a mid-range value of 800 Ω, compared to an average of 833 Ω. The non-linearity is

$$NL\% = \pm\frac{100(850 - 833)}{2 \times 850} = \pm1\%$$

which is twice as good as the JFET's. These are typical non-linearity figures for discrete devices. In ICs the non-linearity can be reduced by about a factor of ten.

Power MOSFETs have very small values of r_{ds} — usually denoted $R_{DS}(on)$ — typically $R_{DS}(on) \approx \frac{1}{2}P_{max}/I_{Dmax}^2$, so that for a 100W, 10A device, $R_{DS}(on) = 0.5 \ \Omega$.

4.10 Temperature effects

In any semiconductor device it is important to consider the effects of temperature on its operation and the FET is no exception. The parameters that are temperature-dependent are V_T (or V_p) and the constant, k, in equation 4.11, which is the MOSFET I_D–V_{GS} relation:

$$I_D = k(V_{GS} - V_T)^2$$

Taking natural logarithms and differentiating with respect to T produces

$$\ln I_D = \ln k + 2\ln(|V_T - V_{GS}|)$$

$$\frac{1}{I_D}\frac{dI_D}{dT} = \frac{1}{k}\frac{dk}{dT} + \frac{2}{|V_T - V_{GS}|}\frac{d|V_T - V_{GS}|}{dT} \tag{4.14}$$

For silicon FETs the temperature coefficient (tempco) of V_T and V_p is about −5 mV/K, though somewhat variable, so that the cut-off voltages move to the left of the V_{GS} axis. However, our interest is only in the magnitude of $(V_T - V_{GS})$, which *increases* with T:

$$\frac{d|V_T - V_{GS}|}{dT} = +5 \ mV/K$$

The constant, k, in equation 4.14, depends on the mobility, which goes as $T^{-1.5}$, so that

$$k = AT^{-1.5} \ \Rightarrow \ \ln k = \ln A - 1.5\ln T$$

where A is a constant. Differentiating $\ln k$ with respect to T gives

$$\frac{1}{k}\frac{dk}{dT} = \frac{-1.5}{T} \approx -0.5\%/K \quad (at \ T = 300 \ K)$$

And so the overall temperature dependence of I_D is found from equation 4.14 to be

$$\frac{1}{I_D}\frac{dI_D}{dT} = -0.005 + \frac{2 \times 0.005}{|V_T - V_{GS}|} = -0.005\left(1 - \frac{2}{|V_T - V_{GS}|}\right)$$

This means that at a certain value of V_{GS} the temperature coefficient of the drain current is zero. In any event the drain current will not be subject to thermal runaway; FETs are quite different from BJTs in this particular. Figure 4.28 shows the way in which the $\sqrt{I_D}$ v. V_{GS} graph varies with temperature for a typical FET.

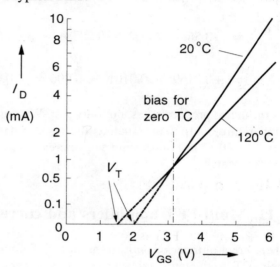

Figure 4.28

Variation of I_D (in the saturation region) with gate bias and temperature for a typical small-signal MOSFET. The plot for a JFET would look similar. The vertical scale is linear in $\sqrt{I_D}$

Example 4.11

A current source is used to provide a reference voltage as in figure 4.25a where the JFET has $V_P = -3$ V and $I_{DSS} = 5$ mA. If $V_{DD} = 12$ V and $R_D = 1.2$ kΩ, what is V_{REF} and by how much will it change if the temperature increases by 20°C? Take the TCR of R_D to be +0.1%/K and that of V_P to be −3 mV/K.

The reference voltage is given by

$$V_{REF} = V_{DD} - I_D R_D$$

And $I_D = I_{DSS} = 5$ mA, hence $V_{REF} = 12 - 5 \times 1.2 = 6$ V. For a JFET, I_D is given by equation 4.1, which can be written in the same form as for MOSFETs (equation 4.11):

$$I_D = I_{DSS}\left(1 - \left|\frac{V_{GS}}{V_P}\right|\right)^2 = k(V_P - V_{GS})^2$$

where $k = I_{DSS}/V_P^2 = 5/3^2 = 0.56$ mA/V². Hence

$$V_{REF} = V_{DD} - k(V_P - V_{GS})^2 R_D$$

Differentiating with respect to T yields

$$\frac{dV_{REF}}{dT} = \frac{dV_{DD}}{dT} - (V_P - V_{GS})^2 R_D \frac{dk}{dT} - 2k(V_P - V_{GS})R_D \frac{dV_P}{dT} - k(V_P - V_{GS})^2 \frac{dR_D}{dT}$$

We must assume $dV_{DD}/dT = 0$ (though voltage regulators for example will have tempcos of about $-0.01\%/K$). Taking the tempco of k to be $-0.005/K$ as before, then $dk/dT = -0.005k = -0.0028$ mA/V^2/K, $dV_P/dT = -0.003$ V/K and $dR_D/dT = +0.001R_D = 0.0012$ kΩ/K, which can be substituted into the equation above to give

$$\frac{dV_{REF}}{dT} = -0.56(-3)^2(1.2)(-0.0028) - 2(0.56)(-3)(-0.003) - (0.56)(-3)^2(0.0012)$$

$$= +0.0169 - 0.0101 - 0.006 = +0.0008 \text{ V/K} = +0.8 \text{ mV/K}$$

k's influence predominates, but only just. We now see that components with compensating temperature coefficients, such as the resistor has here, can be beneficial. In this example a rise of 20°C in temperature will increase V_{REF} by $20 \times 0.8 = 16$ mV to 6.016 V, only 0.27% greater.

4.11 Multi-FET amplifiers and current mirrors

There are a number of multi-transistor FET circuits, some of which are similar in form to their BJT equivalents. BJT current mirrors, for example, all have FET counterparts which look and act in much the same way. FET current mirrors and cascode circuits are extensively used in analogue integrated circuits.

4.11.1 The FET cascode amplifier

Figure 4.29 shows a self-biased cascode amplifier made from two identical n-channel JFETs. Since the two gates are at the same potential and the drain current is the same in both, then the gate-source voltages must be the same, that is $V_{GS1} = V_{GS2}$. But by KVL we see that

$$V_{GS2} + V_{RS} = V_{GS2} + I_{DQ}R_S = 0$$

Then equation 4.1 can be used to find I_{DQ} and all the voltages, as in the next example.

Example 4.12

The JFETs in the cascode circuit of figure 4.29a are identical with $I_{DSS} = 6$ mA and $V_P = -2$ V. If $R_D = 3.3$ kΩ and $R_S = 1.2$ kΩ, find I_{DQ} and all the voltages, given that $V_{DD} = 9$ V.
We find $V_{GS1} = V_{GS2} = -I_{DQ}R_S = -1.2I_{DQ}$ and so

$$I_{DQ} = I_{DSS}\left(1 - \left|\frac{V_{GS}}{V_P}\right|\right)^2 = 6\left(1 - \left|\frac{1.2I_{DQ}}{2}\right|\right)^2$$

which gives $I_{DQ} = 1$ mA. (The other solution, 2.8 mA, leads to V_{GS2} being less than the cutoff voltage.) Hence $V_{GS1} = V_{GS2} = -1.2$ V and

$$V_{DS1} = V_{DD} - I_{DQ}(R_S + R_D) = 9 - 4.5 = 4.5 \text{ V}$$

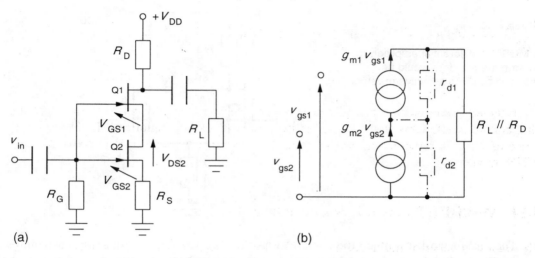

Figure 4.29 (a) A JFET cascode amplifier (b) Its small-signal equivalent circuit; r_d is usually neglected

The small-signal voltage gain can be calculated from the equivalent circuit of figure 4.29b quite readily, since R_D // R_L is normally much greater than r_{d1} or r_{d2}. (r_d, the reciprocal of the slope of the drain characteristics, is about 100 kΩ for small-signal JFETs such as those used here.) In that case, and given $v_{gs1} = v_{gs2}$, the current through R_D // R_L is $-g_{m1}v_{gs1} = -g_{m2}v_{gs2}$ and the voltage gain is

$$A_{vL} = \frac{v_L}{v_{gs}} = -g_m(R_D // R_L)$$

which is exactly the same as the CS amplifier.

As with the BJT cascode amplifier, the advantage lies not with increased gain, but with the elimination of the Miller capacitance and therefore improved high-frequency performance and also increased output resistance, which is made use of in the cascoded current mirror described below (section 4.11.3).

4.11.2 The FET differential amplifier

The circuit, shown in figure 4.30, and implemented with n-channel DE MOSFETs, looks like its BJT counterpart with FETs instead of BJTs. The advantage of the MOSFET differential amplifier is the ultra-high input impedance (effectively determined by parasitics as $R_{in} \approx \infty$) and hence ultra-low current drawn from the source. This is very important when sources, such as thermocouples, are being used where virtually no current can be taken from the device because this makes it inaccurate.

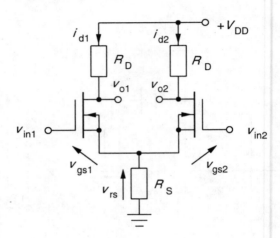

Figure 4.30

A MOSFET differential amplifier. In IC form the resistors would be replaced by FETs as active resistors

The circuit works in much the same way as the BJT version: the small-signal voltage across R_S is $v_{rs} = (i_{d1} + i_{d2})R_S$ and the differential input voltage is

$$\Delta v_{in} = v_{in1} - v_{in2} = v_{gs1} + v_{rs} - v_{gs2} - v_{rs} = v_{gs1} - v_{gs2}$$

The differential output voltage is

$$\Delta v_o = v_{o1} - v_{o2} = -i_{d1}R_D - (-i_{d2}R_D) = -(g_{m1}v_{gs1} - g_{m2}v_{gs2})R_D$$

With identical transistors, $g_{m1} = g_{m2} = g_m$, and the differential voltage gain is

$$A_{\Delta v} = \Delta v_o / \Delta v_{in} = -g_m R_D$$

4.11.3 FET current mirrors

Current mirrors made from FETs suffer from the disadvantage of requiring a relatively large voltage across the FET (somewhat larger than V_T) in order to operate in the constant-current (saturation) region of the FET drain characteristic, whereas the voltage across a BJT need only be a little greater than V_{CEsat}. This becomes quite a serious problem in ICs operated from low voltage supplies such as ±5 V, when V_T may be 2-3 V; for this reason most FET analogue ICs operate from 12-18 V supply rails. Figure 4.31a shows a simple two-FET current mirror operated from a bipolar supply (usually $|V_{DD}| = |V_{SS}|$). For Q1 to be in saturation $V_{DS1} > (V_{GS1} - V_T)$, and since $V_{DS1} = V_{GS1}$ this is always so for EN MOSFETs which must have positive V_T.

The simple 2-FET current mirror suffers from an effect analogous to the Early effect in BJTs, which means that I_o is not exactly equal to I_{in} but varies according to the voltages across Q1 and Q2. The current ratio is given by

$$\frac{I_o}{I_{in}} = \frac{1 + \lambda_2 V_{DS2}}{1 + \lambda_1 V_{DS1}}$$

where λ is a parameter (known as the channel-length modulation parameter), analogous to $1/V_{EA}$ for BJTs, which expresses the dependence of the drain current in the saturation region on the drain-source voltage, and is typically 0.005 V^{-1}. In other terms, if the slope of the drain characteristic in the saturation region is $1/r_d$, then $\lambda = 1/I_D r_d$. Note that for a given device it is λ that is constant, not r_d; the latter varies with I_D.

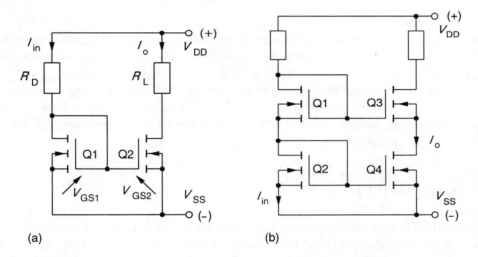

Figure 4.31 (a) A 2-FET current mirror. The current, I_{in}, is set by the left-hand side and is mirrored in the right-hand side, where $I_o = I_{in}$ ideally. (b) A 4-FET cascode current mirror

Example 4.13

In a current mirror as in figure 4.31a, R_D is chosen to make $I_{in} = 2$ mA, when the transistors are matched with $V_T = +3$ V, $\lambda = 0.005$ V^{-1} and I_D is 10 mA when $V_{GS} = 6$ V. Find the output current when $R_L = 330$ Ω and $V_{DD} = |V_{SS}| = 5$ V.

If $I_D = 10$ mA when $V_{GS} = 6$ V, then equation 4.11 gives

$$10 = k(6 - 3)^2 \quad \Rightarrow \quad k = 1.111 \text{ mA/V}^2$$

and thus for I_{in} to be 2 mA,

$$V_{GS1} - V_{T1} = \sqrt{\frac{I_D}{k}} = \sqrt{\frac{2}{1.111}} = 1.342 \text{ V}$$

using equation 4.11 again. This gives

$$V_{GS1} = V_{T1} + 1.342 = 4.342 \text{ V}$$

On the input side KVL leads to

$$V_{DD} = 5 = I_{in}R_D + V_{GS1} + V_{SS} = 2R_D + 4.342 + (-5)$$

whence $R_D = 2.83$ kΩ. On the output side, with $I_o = 2$ mA, the voltage across Q2 is

$$V_{DS2} = V_{DD} - V_{SS} - I_oR_L = 10 - 2 \times 0.33 = 9.34 \text{ V}$$

Thus the true current ratio will be

$$\frac{I_o}{I_{in}} = \frac{1 + \lambda V_{DS2}}{1 + \lambda V_{DS1}} = \frac{1 + 0.005 \times 9.34}{1 + 0.005 \times 4.342} = 1.0245$$

and so $I_o = 2.049$ mA.

There are a number of FET analogues to the BJT current mirrors which have better fidelity than the two-FET mirror; one of them is the 4-FET cascode current mirror shown in figure 4.31b. The effect of the extra transistors, Q1 and Q3, is to make $V_{DS2} = V_{DS4}$ and hence remove the influence of λ on the output current. Whereas the output resistance of the simple 2-FET current mirror is $r_d \approx 100$ kΩ, that of the cascode mirror is $g_m r_d^2 \approx 10$ MΩ, with output currents of a few mA. The current ratio is improved by the same degree — about a hundredfold.

4.12 Power MOSFETs

Power MOSFETs are designed to carry large currents and so must have low ON resistance; $R_{DS}(on)$ ranges from about 5 Ω for 10W devices down to 10 mΩ for 300W ones. To keep the on-state resistance low and to help dissipate the heat produced in the MOSFET, a so-called vertical structure has been developed. There are two principal types of vertical structure: the VMOSFET and the HEXFET, the latter being shown in figure 4.32.

The gate signal is carried by a hexagonal network of polycrystalline silicon completely enclosed in silicon dioxide for electrical insulation. The source metallisation covers the whole of the top surface and the drain metallisation the whole of the bottom, so reducing the current density and enabling effective heat transfer to a large external drain tab and its heat sink (neither of which is shown in figure 4.32). The channel which the poly-Si gate controls is the thin, vertical, p-type region of silicon indicated. Though current densities can be high in the channel, its hexagonal distribution throughout the whole of the device results in rapid heat transfer to the drain. If the supply polarity is reversed the drain-source junction is forward biased and acts as a diode of very low dynamic resistance, shown in the symbol for the device.

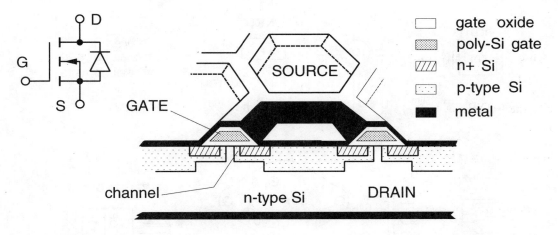

Figure 4.32 A small part of an n-channel, enhancement-mode, power MOSFET in cross-section

4.13 FET specifications

Table 4.2 gives data for some readily-obtainable FETs of various types, arranged in order of power, which is roughly the same as the order for price. All are low-voltage devices. Note that there are no standard pin designations for FETs and each device's data sheet must be examined to find the source, gate and drain pins.

Table 4.2 *Specifications for some popular FETs*

Device	Type[A]	Pack[B]	P(max)	V_{GS}(th)[C]	V_{DS}[D]	I_{DSS}[E]	I_{GSS}[F]	g_{fs}[G]	I_D(max)	Price[H]
BF981	N/DE	SOT103	225 mW	−2.5 V	20 V	4 mA	50 nA	10 mS	20 mA	£0.57
2N5486	N/JFET	TO92	310 mW	−6 V	25 V	8 mA	1 nA	4 mS	30 mA	£0.80
2N4861	N/JFET	TO18	360 mW	−4 V	30 V	8 mA	250 pA	10 mS	—	£1.40
BF245A	N/JFET	TO92	360 mW	—	30 V	2 mA	5 nA	3 mS	100 mA	£0.35
VN10LP	N/EN	E-line	625 mW	2.5 V	60 V	10 µA	100 nA	100 mS	300 mA	£0.27
VPO610L	P/EN	TO92	800 mW	−3.5 V	−60 V	−1 µA	10 nA	80 mS	180 mA	£0.49
IRFD113	N/EN	DIP	1.2 W	4 V	60 V	250 µA	500 nA	560 mS	800 mA	£0.73
VN88AFD	N/EN	TO220	15 W	2.5 V	80 V	10 µA	100 nA	170 mS	1.3 A	£1.58
BUZ71	N/EN	TO220	40 W	4 V	50 V	1 µA	100 nA	5 S	14 A	£1
BUZ271	P/EN	TO220	125 W	−4 V	−50 V	−1 µA	100 nA	1.5 S	22 A	£2
BUZ900	N/EN	TO3	125 W	1.5 V	160 V	—	—	700 mS	8 A	£4.75
IXTH75N10	N/EN	TO247	300 W	3.4 V	70 V	100 µA	100 nA	30 S	75 A	£15

Notes: [A] Type: n-channel (N) or p-channel (P), EN = enhancement MOSFET, DE = depletion MOSFET.
[B] Pack = package style (figure 4.33).   [C] V_{GS}(th) = max of $|V_T|$ or $|V_P|$
[D] In effect the maximum for V_{DD}.
[E] I_{DSS} is the maximum value and for EN MOSFETs is the drain-source leakage current when V_{GS} = 0 and the device is cut-off.
[F] I_{GSS} is the maximum gate leakage current.   [G] The minimum value; it is the same as g_m.
[H] Prices are for single quantities. R_{DS}(on) is the higher of $\frac{1}{2}P_{max}/I_{Dmax}^2$ or $1/g_{fs}$(min). All parameters at 25°C.

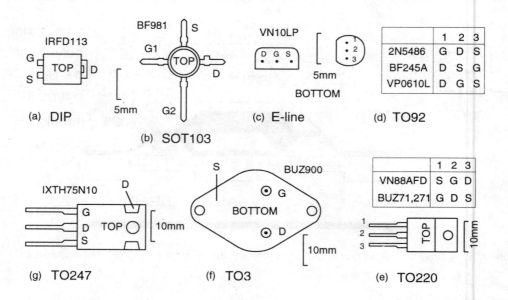

Figure 4.33 Some popular package styles for FETs shown either looking at the top or the bottom. The pin designations are NOT standard and apply only to the devices named. (a) to (d) are low-power devices. (b) is a UHF package. (e) to (f) are for power MOSFETs ($P_{max} > 10$ W)

Suggestions for further reading

Principles of transistor circuits by S W Amos (Butterworth/Heinemann, 7th ed., 1990)
A practical introduction to electronic circuits by M Hartley-Jones (Cambridge University Press, 3rd ed., 1995)
Electronic devices and circuits by T F Bogart Jr (Merrill, 3rd ed., 1993)
Designing with field-effect transistors by Siliconix Inc., revised by E Oxner (McGraw-Hill, 2nd ed., 1990)
Spice, practical device modeling by R Kielkowski (McGraw-Hill, 1995)

Problems

1 If the current in the saturation region ($V_{DS} > |V_P - V_{GS}|$) for a JFET is given by

$$I_D = I_{DSS}[1 - 3|V_{GS}/V_P| + 2(|V_{GS}/V_P|)^{1.5}]$$

show that the maximum error in using equation 4.1 instead occurs when $V_{GS} = \frac{1}{4}V_P$, and that the error in I_D is then $0.0625I_{DSS}$.
2 Show that the transconductance of the JFET, whose I_D(sat) vs. V_{GS} relation is that given in problem 4.1 above, is given by

$$g_m = g_0(1 - \sqrt{|V_{GS}/V_P|})$$

where $g_0 \equiv 3I_{DSS}/|V_P|$. Using these relationships, what is the transconductance of an n-channel JFET at $V_{GS} = -2$ V, given it has $I_{DSS} = 5$ mA and $I_D = 2$ mA when $V_{GS} = -1.4$ V? What would the transconductance be if the JFET obeyed equation 4.1 instead, with the same values of V_P, I_{DSS} and V_{GS}? [*1.11 mS; 1.24 mS*]

3 In the circuit of figure P4.3, $V_{GG} = 0$ V, $V_{DD} = +30$ V, $R_D = 8.2$ kΩ and $R_S = 2.2$ kΩ. If the JFET follows equation 4.1 and has $I_{DSS} = 4$ mA and $V_P = -3.5$ V, what are I_D, V_{GS} and V_{DS}? [*$I_D = 0.855$ mA, $V_{GS} = -1.88$ V, $V_{DS} = 21.1$ V*]

4 In the circuit of figure P4.3, $V_{DD} = +24$ V, $R_D = 5.6$ kΩ, $R_S = 3.3$ kΩ and the JFET obeys equation 4.1 with $I_{DSS} = 8$ mA and $V_P = -5$ V. What must V_{GG} be to make $I_D = 1.5$ mA and what are V_{GS} and V_{DS} then? What will I_D be if the JFET is changed to one with $I_{DSS} = 4$ mA and $V_P = -3$ V and V_{GG} is unchanged?
[*$V_{GG} = +2.115$ V, $V_{GS} = -2.835$ V, $V_{DS} = 10.65$ V. $I_D = 1.078$ mA*]

5 For the circuit of figure P4.3, $I_D = 2$ mA, $V_{GG} = +2$ V, $V_{DD} = +16$ V, $R_D = 3.3$ kΩ and the JFET has the transfer characteristic of figure P4.5. Use a graphical method to find R_S. What are V_{GS} and V_{DS}? With this value for R_S what should V_{GG} be to make $V_{DS} = 10$ V?
[*$R_S = 1.62$ kΩ, $V_{GS} = -1.24$ V, $V_{DS} = 6.16$ V. $V_{GG} = +0.39$ V*]

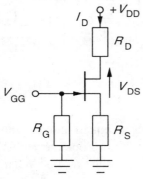

Figure P4.3

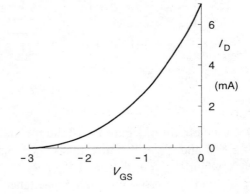

Figure P4.5

6 For the circuit of figure P4.6, $V_{DD} = +18$ V, $R_1 = 500$ kΩ, $R_D = R_S = 3.3$ kΩ and $V_{DSQ} = 8$ V. The JFET obeys equation 4.1 with $V_P = -3$ V and $I_{DSS} = 7$ mA. What are I_{DQ} and R_2? What would R_2 have to be if the JFET had the transfer characteristic of figure P4.5?
[*$I_{DQ} = 1.515$ mA, $R_2 = 116$ kΩ; 122 kΩ*]

7 In the CS amplifier of figure P4.6, the JFET obeys equation 4.1 with $I_{DSS} = 5$ mA and $V_P = -1.8$V and is to operate with $V_{DSQ} = 9$ V. What should R_1 and R_2 be to achieve this if $V_{DD} = +27$ V, $R_D = 5.6$ kΩ, $R_S = 3.3$ kΩ and the amplifier's input resistance is to be 100 kΩ. (Ignore the reactances of the capacitors.) [*$R_1 = 449$ kΩ, $R_2 = 129$ kΩ*]

8 A CS amplifier using the circuit of figure P4.6 uses a JFET which obeys equation 4.1 with $I_{DSS} = 4$ mA and $V_P = -1.5$ V. It also has $V_{DD} = +30$ V, $R_D = R_S = 4.7$ kΩ, $R_1 = 500$ kΩ and $R_2 = 100$ kΩ. What is I_{DQ}? What is the transconductance of the amplifier? What is the small-signal, no-load voltage-gain of the amplifier? What is the voltage gain with a load of 6.8 kΩ? If the input voltage source has an internal resistance of 20 kΩ what is the voltage gain, v_L/v_s, with a 6.8 kΩ load? What is this last voltage gain if r_{ds} (the slope of the drain characteristic at the Q-point) is 60 kΩ and goes in parallel with R_D? (The capacitances all have negligible reactances.) [*$I_{DQ} = 1.208$ mA, $g_m = 2.93$ mS, $A_{v0} = -13.8$, $A_{vL} = -8.14$, $v_L/v_s = -6.56$; $A_{vL} = -6.27$*]

9 The JFETs to be used in the CS amplifier of figure P4.6 are known to obey equation 4.1 and have $V_P = -0.9$ V with $I_{DSS} = 3$ mA, and $V_P = -6$ V with $I_{DSS} = 12$ mA. The resistance values are $R_D = 2.2$ kΩ, $R_1 = 500$ kΩ and $R_2 = 120$ kΩ, and the supply voltage is +20 V. Use a graphical method to find the value of R_S which will make the change in I_{DQ}, when one JFET is replaced by the other, equal to 1 mA. Where are the Q-points then? What are the no-load, small-signal, voltage-gains? *[$R_S = 2.4$ kΩ; $I_{DQ} = 1.7$ mA, $V_{DSQ} = 12.2$ V; $I_{DQ} = 2.7$ mA, $V_{DSQ} = 7.85$ V; $A_{v0} = -11.0$; $A_{v0} = -5.0$]*

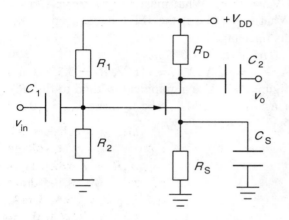

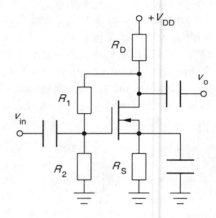

Figure P4.6   Figure P4.10

10 In the circuit of figure P4.10 the n-channel DE MOSFET obeys equation 4.11 with $V_T = -2.3$ V and $I_{DSS} = 3.5$ mA, that is $I_D = 3.5$ mA when $V_{GS} = 0$. It is to operate with $V_{DSQ} = 8$ V when $V_{DD} = +16$ V, and $R_D = R_S$. If $R_1 = 2R_2 = 1$ MΩ, what must R_D be? What are I_{DQ}, g_m and A_{v0}? (Assume each capacitance has no reactance.)
[$R_D = 1.143$ kΩ, $I_{DQ} = 3.5$ mA, $g_m = 3.044$ mS, $A_{v0} = -3.48$]

11 The MOSFET of figure P4.11 has $I_D = 1.6$ mA when $V_{GS} = 5$ V and $V_T = +4$ V. Write down V_{GE} as a function of V_{GS} and I_D and hence plot the gate-bias load-line on the transfer characteristic, given $V_{DD} = +24$ V, $R_D = 5.6$ kΩ, $R_S = 1$ kΩ, $R_1 = 330$ kΩ and $R_2 = 120$ kΩ, assuming the capacitances have negligible reactances. Use this diagram to find A_{v0} for an input signal of ±1 V. What is the small-signal, no-load voltage-gain obtained by calculation using g_m? *[$A_{v0} = -4.0$ and -4.22]*

12 The components in figure P4.11 are the same as in problem 4.11 and V_{DD} is too. The cut-off voltage of the MOSFET has a temperature coefficient of −3 mV/K and the tempco of k in equation 4.11 is −0.5%/K. If V_{DD} has a tempco of +0.04%/K and the resistors all have the same tempco, what must this be to produce a Q-point which does not change with a temperature increase of 10 K? What happens to I_{DQ} if, with this resistance tempco, the temperature falls by 10 K? *[+0.213%/K; $\delta I_{DQ} = -1.41$ μA]*

13 The common-drain amplifier of figure P4.13 employs a JFET which obeys equation 4.1 with $V_P = -2.7$ V and $I_{DSS} = 10$ mA. If $V_{DD} = 12$ V, the input resistance is to be 100 kΩ, $R_S = 2.2$ kΩ and $V_{DSQ} = 6$ V, what should R_1 and R_2 be? What is the no-load voltage (v_o/v_s) gain if the input is a voltage source, v_s, of internal resistance 10 kΩ? What is the current gain when a resistive load of 1 kΩ is attached?

$[R_1 = 255\ k\Omega,\ R_2 = 165\ k\Omega;\ A_{v0} = 0.814;\ A_{iL} = 72.7]$

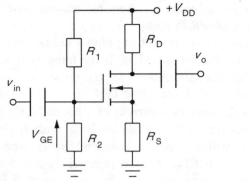

Figure P4.11

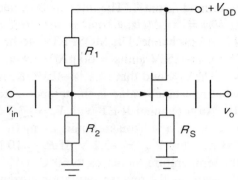

Figure P4.13

14 The circuit shown in figure P4.14a is a common-gate amplifier, which has the equivalent circuit of figure P4.14b, where $R_o = R_D\ /\!/\ R_L$. Find expressions for the voltage and current gains and show that $A_{vL} \approx g_m R_o$ and $A_{iL} \approx -g_m R_S R_o / R_L (g_m R_S + 1)$ when r_{ds} is large. If $R_1 = 2R_2$, $R_D = 3.3\ k\Omega$, $R_S = 4.7\ k\Omega$, $R_L = 5\ k\Omega$, r_{ds}, the slope of the drain characteristic, is large, $V_{DD} = +18$ V and the JFET has $V_P = -1.8$ V and $I_{DSS} = 4.5$ mA, what are the voltage, current and power gains? $[A_{vL} = (g_m r_{ds} + 1)R_o/(r_{ds} + R_o);\ A_{iL} = -\alpha R_S R_o / R_L(\alpha R_s + R_D + r_{ds})$, where $\alpha \equiv g_m r_{ds} + 1$; $A_{vL} = 5.6$, $A_{iL} = -0.37$, $A_p = 3.16\ dB]$

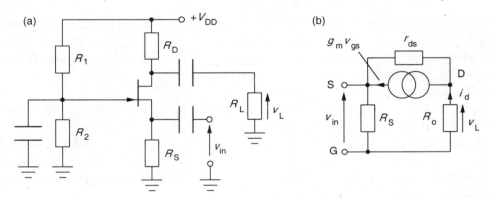

Figure P4.14

15 In a CS amplifier as in figure P4.6, the $R_D = 3.3\ k\Omega$ and $R_L = 4.7\ k\Omega$. What should the output coupling capacitor, C_2, be to make the lower cut-off frequency 5 Hz, if the reactances of the other capacitors can be ignored? If the source by-pass is to have its cut-off at 5 Hz, $g_m = 2.5$ mS and $R_S = 220\ \Omega$, what must C_S be? If $R_1 = 330\ k\Omega$ and $R_2 = 100\ k\Omega$ and the input cut-off frequency is to be 5 Hz, what must C_1 be? What then is the mid-band voltage gain? What will it be approximately at 10 Hz, 5 Hz and 2.5 Hz?
$[C_2 = 4\ \mu F;\ C_S = 224\ \mu F;\ C_1 = 0.4\ \mu F;\ 13.7\ dB,\ 10.7\ dB,\ 4.7\ dB,\ -7.3\ dB]$
16 A CS amplifier uses a device with $C_{gs} = 3$ pF, $C_{ds} = 1.2$ pF and $C_{dg} = 3.8$ pF. The effective gate resistance is 8 kΩ, $R_D = 4.7\ k\Omega$ and $R_L = 10\ k\Omega$. If $g_m = 2.3$ mS and the

input is a voltage source of internal resistance 800 Ω, what are the high-frequency cut-offs for input and output. What is the mid-band voltage gain? What are the voltage gains at $f_H(\text{in})$ and $f_H(\text{out})$? (The internal resistance of the source must be taken into account here.) *[$f_H(\text{in}) = 6.3$ MHz; $f_H(\text{out}) = 10$ MHz; 16.5 dB; 12.0 dB, 8.0 dB]*

17 An n-channel DE MOSFET is to be used as an active resistor of value 4 kΩ in an IC. If $V_T = -0.8$ V and $k = 60$ µA/V^2, what should V_{GS} be if $V_{DS} = 0$ V? If the tempco of V_T is -4 mV/K and that of k is -0.5%/K, what is the resultant tempco for the resistance? At what resistance value is the tempco zero? *[1.283 V; -0.3%/K; 10.4 kΩ]*

18 An n-channel JFET with $V_P = -3.3$ V draws an input bias current of 1 nA at 20°C. It is used in a self-bias circuit, as in figure 4.4, designed to operate at 20°C with $I_{DQ} = 1.3$ mA and $V_{GSQ} = -2.2$ V. If $R_G = 10$ MΩ, how will the quiescent current change when the temperature increases to 60°C if V_P and I_{DSS} are unaffected? (The input bias current is the same as the reverse saturation current of a p-n junction, which doubles with every 8 K that the temperature increases.) If V_P changes by -3 mV/K and I_{DSS} by $+0.5\%$/K in addition to the change of input bias current, what will the quiescent current be at 60°C? *[$\delta I_{DQ} = +0.152$ mA; $I_{DQ} = 1.54$ mA]*

5 Feedback and operational amplifiers

EEDBACK IS a means of improving a number of different aspects of circuit performance, especially that of amplifiers. For example it can be used to make an amplifier's gain independent of the parameters of its components, or to increase bandwidth, or to produce oscillations at a desired frequency, or to alter input or output resistance. Most operational amplifier circuits use feedback as a means of achieving some of these goals either singly or in combination. Feedback does not confer all these benefits without any drawbacks, as we shall see, but they are usually quantifiable and can therefore be allowed for. One advantage of using feedback is that it makes circuit performance more predictable as well as better in some chosen aspect.

5.1 The feedback loop and gain

Consider the block diagram of figure 5.1 which shows a circuit of some kind with an input, x_{in}, and output, x_o. The circuit has forward gain, A, so that *without feedback*

$$x_o = Ax_{in}$$

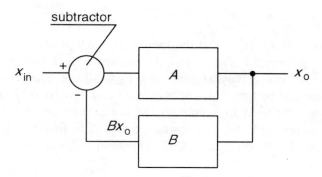

Figure 5.1

Block diagram of a circuit with feedback. A, the forward gain, and B, the feedback gain, are the path gains

However, some of the output is fed back into the input via a feedback path with a gain, B, and this feedback, Bx_o, is then *subtracted* from the input. The input to the circuit with feedback is then $x_{in} - Bx_o$, which the circuit multiplies by A to give x_o:

$$A(x_{in} - Bx_o) = x_o$$

which can be rearranged in the form

$$\frac{x_o}{x_{in}} = \frac{A}{1 + BA} = K \qquad (5.1)$$

K is the overall gain with feedback or the *closed-loop gain*, BA is the *loop gain*, since it is the gain that results from going round the loop comprising amplifier and feedback path, and A is the *open-loop gain*, that is the gain with no feedback.

There are four cases to consider according to the value of the loop gain:

(1) $AB = 0$. No feedback, $x_o/x_{in} = A$
(2) $-1 < AB < 0$. Positive feedback, $x_o/x_{in} > A$
(3) $AB > 0$. Negative feedback, $x_o/x_{in} < A$
(4) $AB = -1$. Positive feedback, $x_o/x_{in} = \infty$, an output with no input: an oscillator

Oscillators (case four) are discussed separately in section 7.1 and case one, no feedback, is of no interest. Case two, positive feedback has only limited application in electrical circuits, and so here we consider case three: negative feedback. It is termed negative because, though the loop gain, AB, is positive, the fraction that is fed back, Bx_o, is subtracted from the original input.

Example 5.1

What must the feedback gain, B, be to make the overall gain $K = 100$, if the circuit's forward gain, $A = 10000$?
We substitute these values in equation 5.1 to obtain

$$K = 100 = \frac{10000}{1 + 10000B}$$

whence $B = 0.0099$ and the loop gain is $AB = 99$.

Equation 5.1 shows how the influence of device parameters (which are not easy to control closely) may be moderated by feedback. The gain without feedback, A, is strongly influenced by the actual devices used and may be subject to wide variations even though the devices used are nominally the same. Suppose for instance that a similar circuit to the one in example 5.1 is constructed and A is found to be 12000, what will K be if the feedback gain stays at 0.0099? Substituting for A and B in equation 5.1 gives

$$K = \frac{12000}{1 + 0.0099 \times 12000} = 100.17$$

The difference in gains is only 0.17%. Looking at equation 5.1 we can see that, if $AB \gg 1$, then the closed-loop gain, $K \approx 1/B$, and is independent of A.

5.2 Types of feedback

Feedback can be one of four basic types

(1) Voltage feedback, Voltage subtractor (series input)
(2) Voltage feedback, Current subtractor (shunt input)
(3) Current feedback, Voltage subtractor (series input)
(4) Current feedback, Current subtractor (shunt input)

Voltage feedback tends to maintain output voltage, and therefore *reduces* the output resistance of the circuit, while current feedback tends to maintain output current, thereby *increasing* the output resistance of the circuit. Series input, in which the feedback voltage is subtracted from the source voltage, tends to make the circuit a better voltage amplifier, and therefore *increases* the input resistance. Shunt input, in which feedback current is subtracted from the input current, tends to make the circuit a better current amplifier and thus *decreases* the input resistance. As we shall see the factor by which the improvement is made is $(1 + AB)$.

5.2.1 Voltage-derived feedback, series applied

This is case (1) of those listed above. The circuit of figure 5.2a shows an amplifier with load, R_L, across which the voltage is v_o. This voltage is sampled and fed back in series with the voltage source, v_s. By KVL on the input side

$$v_{in} = v_s - Bv_o \tag{5.2}$$

With the output open circuit, $v_o = Av_{in}$, so that equation 5.2 becomes

$$v_{in} = v_s - ABv_{in} \;\Rightarrow\; v_s = (1 + AB)v_{in} = (1 + AB)i_{in}R_{in}$$

Thus the effective input resistance with feedback is

$$R_{inf} = v_s/i_{in} = (1 + AB)R_{in}$$

Series-applied feedback increases the input resistance by a factor of $(1 + AB)$.

On the output side by KVL we find

$$Av_{in} = i_oR_o + v_o \tag{5.3}$$

Substituting v_{in} from equation 5.2 into equation 5.3 and rearranging, we obtain

$$v_o = \frac{Av_s}{1 + AB} - \frac{i_oR_o}{1 + AB}$$

which leads to the equivalent circuit with feedback of figure 5.2b. The effect of voltage-derived feedback has been to *reduce* the output resistance from R_o to $R_o/(1 + AB)$.

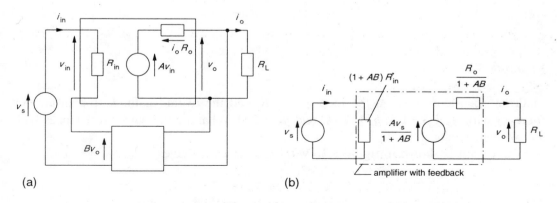

(a) (b)

Figure 5.2 (a) Voltage-derived, series-applied feedback (b) The resultant equivalent circuit

Examples of voltage feedback

The source-follower JFET amplifier of figure 5.3a is an example of voltage-derived, series-applied feedback with unity feedback gain ($B = 1$) because all the output voltage is fed back. Noting that v_{gs} is the same v_{in} in figure 5.2a, the analysis follows the same path as above; by KVL

$$v_s = v_{gs} + i_d R_S = i_d/g_m + i_d R_S$$

and $v_o = i_d R_S$ so that

$$K = \frac{v_o}{v_s} = \frac{i_d R_S}{i_d/g_m + i_d R_S} = \frac{g_m R_S}{1 + g_m R_S}$$

The feedback is *voltage derived*, since if $v_o = 0$, there would be no feedback signal, and it is series applied, since the voltage fed back is subtracted from the source voltage. The forward path gain, $A = g_m R_S$ and the loop gain, $AB = g_m R_S$. The output resistance with feedback is $1/g_m$, since when R_S ($\equiv R_L$ in figure 5.2) is equal to $1/g_m$ the output voltage is halved. The output resistance without feedback would be R_S and if we divide this by $(1 + AB) = (1 + g_m R_S)$, we find the resistance with feedback, by the more general analysis above, should be

$$R_{of} = R_S/(1 + g_m R_S) \approx 1/g_m$$

provided $R_S \gg 1/g_m$. The emitter follower is another example of voltage-derived, series-applied feedback.

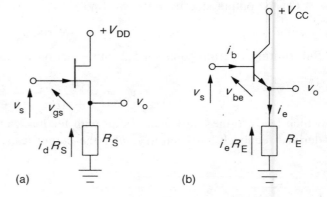

Figure 5.3

(a) The common-source amplifier: v_{gs} is v_{in} for the purposes of feedback
(b) The emitter-follower: v_{be} is v_{in} for feedback analysis

(a)　　　　(b)

Example 5.2

Find the open-loop gain, A, the loop gain, AB, and the overall gain, K, for an emitter follower with $R_E = 3.3$ kΩ, $h_{ie} = 2$ kΩ and $h_{fe} = 200$. What are the output resistances with and without feedback?

The circuit is shown in figure 5.3b with $v_{be} = v_{in}$. KVL gives

$$v_s = v_{be} + i_e R_E = v_{be} + v_o$$

Thus

$$K = \frac{v_o}{v_s} = \frac{v_o}{v_{be} + v_o} = \frac{(h_{fe} + 1)i_b R_E}{i_b h_{ie} + (h_{fe} + 1)i_b R_E} = \frac{(h_{fe} + 1)R_E/h_{ie}}{1 + (h_{fe} + 1)R_E/h_{ie}}$$

since $v_{be} = i_b h_{ie}$ and $v_o = i_e R_E = (h_{fe} + 1)i_b R_E$. Comparing the above expression for K with equation 5.1 we find $A = (h_{fe} + 1)R_E/h_{ie} = 201 \times 3.3/2 = 332$, and $AB = A = 332$, making $K = 332/333 = 0.997$.

With feedback the output resistance is $h_{ie}/(h_{fe} + 1) = 2000/201 = 9.95\ \Omega$. Therefore the output resistance without feedback must be

$$R_o = R_{of}(1 + AB) = \frac{h_{ie}}{h_{fe} + 1} \times \left(1 + \frac{(h_{fe} + 1)R_E}{h_{ie}}\right) = \frac{h_{ie}}{h_{fe} + 1} + R_E$$

which is approximately $R_E = 3.3$ kΩ.

We have thus been using feedback implicitly in all the BJT and FET amplifiers discussed prior to this chapter, but the feedback path has usually been obscured by the transistor's presence in the circuit.

5.2.2 *Current-derived, shunt-applied feedback*

Figure 5.4a shows the diagram for this case. On the input side KCL gives

$$i_s = i_{in} + Bi_o \qquad (5.4)$$

With the output open circuit, $i_o = Ai_{in}$ and the above equation becomes

$$i_s = i_{in} + ABi_{in} = (1 + AB)i_{in} = (1 + AB)v_{in}/R_{in}$$

The effective input resistance with feedback is then

$$R_{inf} = v_{in}/i_s = R_{in}/(1 + AB)$$

Shunt-applied feedback reduces the input resistance by a factor of $(1 + AB)$.

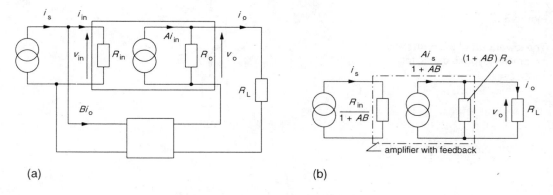

(a)   (b)

Figure 5.4 (a) Current-derived, shunt-applied feedback (b) The equivalent circuit

On the output side, the current through R_o is v_o/R_o and therefore by KCL

$$Ai_{in} = i_o + v_o/R_o$$

But from equation 5.4, $i_{in} = i_s - Bi_o$, and then the equation above becomes

$$A(i_s - Bi_o) = i_o + v_o/R_o \quad \Rightarrow \quad i_o(1 + AB) = Ai_s - v_o/R_o$$

Hence

$$i_o = \frac{Ai_s}{1 + AB} - \frac{v_o}{R_o(1 + AB)}$$

which leads to the equivalent circuit of figure 5.4b. The effective output resistance with feedback is therefore

$$R_{of} = (1 + AB)R_o$$

Current feedback has led to an increase in output resistance by a factor of $(1 + AB)$.

An example: the CE amplifier

Consider the CE amplifier of figure 5.5, which has no emitter bypass capacitor. The absence of this capacitor results in some of the output signal being subtracted from the input, since by KVL

$$v_{in} = v_{be} + i_e R_E \approx i_b h_{ie} + h_{fe} i_b R_E$$

taking $i_e \approx h_{fe} i_b$. And as $v_o = -h_{fe} i_b R_C$, the magnitude of the overall gain, K, is

$$K = \frac{v_o}{v_{in}} = \frac{h_{fe} R_C}{h_{ie} + h_{fe} R_E}$$

ignoring the minus sign.

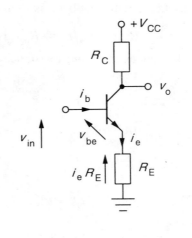

Figure 5.5

The lower part of a CE amplifier with no emitter bypass capacitor

This can be expressed as

$$K = \frac{h_{fe} R_C / h_{ie}}{1 + h_{fe} R_E / h_{ie}} = \frac{A}{1 + AB}$$

so that the open-loop gain $= A = h_{fe}R_C/h_{ie}$, the feedback gain, $B = R_E/R_C$ and the loop gain, $AB = h_{fe}R_E/h_{ie}$. The feedback in this case is current derived, since even if the collector were connected directly to V_{CC}, which would make v_o zero, there would still be current flowing through R_E and therefore a feedback signal. But it is series applied because the feedback signal, i_eR_E, is in series with v_{in}. The input resistance has clearly been increased from h_{ie} without feedback to $h_{ie} + h_{fe}R_E$ with it, a factor of $1 + h_{fe}R_E/h_{ie} = 1 + AB$.

The output resistance of the amplifier with R_E bypassed would be $1/h_{oe}$ (considering R_C to be part of the external circuit or load) and with no bypass it should be $(1 + h_{fe}R_E/h_{ie})/h_{oe}$. The calculation of the output resistance from the equivalent circuit shows that this is so, provided $R_E \ll h_{ie}$.

Further examples of feedback will be considered after we have discussed the properties of ideal operational amplifiers.

5.3 Ideal operational amplifiers

Most low-power analogue circuits and virtually all the digital circuits that are designed and built contain few discrete transistors; instead they use integrated circuits in one form or another. These of course contain transistors — FETs and BJTs — in abundance, but they are unseen and often overlooked, that is perhaps until the circuit fails to perform as expected. One of the commonest ICs used in analogue circuits of all kinds is the operational amplifier, frequently abbreviated to op amp. Until ICs were invented op amps were made up of discrete transistors and could be very expensive, even for modest performance. There are many types of op amp — one popular catalogue lists nearly 400 — but they all have characteristics in common. We shall begin by considering the ideal operational amplifier and then we shall consider limitations on performance brought about by the non-ideal behaviour of practical devices.

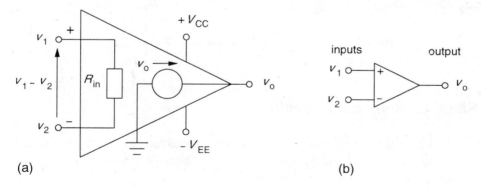

Figure 5.6 (a) An equivalent circuit for an op amp (b) The circuit symbol

Figure 5.6a show an equivalent circuit for an operational amplifier, which is represented in circuit diagrams by an equilateral triangle with two input terminals marked $+$ and $-$ on one side and an output terminal coming from the opposite vertex. Two DC supply terminals, $+V_{CC}$ and $-V_{EE}$ in bipolar circuits, are also usually present, and various other terminals for special purposes. The supply voltages are normally equal in magnitude and opposite in polarity. In the ideal op amp we shall omit all but the three input and output terminals that are shown in figure 5.6b.

The output voltage is given by

$$v_o = A(v_1 - v_2) = Av_{in}$$

provided $-V_{EE} < v_o < V_{CC}$, since the output voltage can never exceed the supply voltage. The voltage transfer function of an ideal op amp is shown in figure 5.7, where $v_{in} = v_1 - v_2$, and the device operates linearly when $-V_{EE}/A < v_{in} < V_{CC}/A$. However, the open-loop gain, A, of most op amps is very large — ranging from about 10 000 to perhaps 10 million — and as a result whenever the op amp is used without feedback the output voltage is $+V_{CC}$ or $-V_{EE}$, unless v_{in} is very small, that is unless $v_1 \approx v_2$. In addition, R_{in} is very large, at least 1 MΩ and possibly 1 TΩ (= 1 million MΩ), so that the current into the op amp is effectively nil.

When used with feedback, therefore, the ideal op amp obeys two golden rules:

1. The current into the op amp is zero.
2. The output voltage is whatever makes the input voltages equal.

Figure 5.7

The voltage transfer function of an ideal op amp without feedback: because *A*, the open-loop gain, is very large, v_{in}(min) and v_{in}(max) are both almost zero

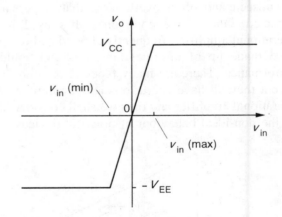

5.4 Some basic op amp circuits

The golden rules make the analysis of the basic op amp circuits very straightforward, since the device characteristics scarcely enter into the matter. We shall examine here a few circuits that contain only a single ideal op amp.

5.4.1 The inverting amplifier

The circuit of figure 5.8 shows an ideal op amp with a feedback resistance between the output terminal and the inverting input (−), while the non-inverting (+) input is connected directly to ground.

Figure 5.8

An inverting amplifier using an ideal op amp and negative feedback. The feedback is shunt applied via the current subtractor, node A

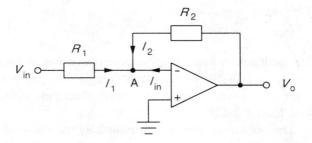

Applying KCL at node A, the inverting input, gives

$$I_{in} + I_1 + I_2 = 0 \qquad (5.5)$$

which is why node A is often called the *summing point*. But $I_{in} = 0$ by the first golden rule, and by the second

$$V_A = V_- = V_+ = 0$$

which is why node A is also known as a *virtual ground*. Then the current through R_2 by Ohm's law is

$$I_2 = \frac{V_o - V_A}{R_2} = \frac{V_o}{R_2}$$

And the current through R_1 is

$$I_1 = \frac{V_{in} - V_A}{R_1} = \frac{V_{in}}{R_1}$$

Substituting for I_f and I_1 in equation 5.5 leads to

$$\frac{V_o}{R_2} + \frac{V_{in}}{R_1} = 0 \;\Rightarrow\; K = \frac{V_o}{V_{in}} = \frac{-R_2}{R_1}$$

The output is opposite in sign to the input, hence the name inverting amplifier.

The input resistance is $V_{in}/I_1 = R_1$ and not the very high input resistance of the op amp, R_{in}, because the feedback is shunt applied via a current subtractor, which is node A. The feedback gain, $B = R_1/R_2 = -1/K$.

5.4.2 *The non-inverting amplifier*

This is shown in figure 5.9; here the signal voltage is applied to the non-inverting input, though the feedback is still to the inverting terminal. At node A, the second golden rule gives $V_A = V_{in}$, but since the current into the inverting input is zero by the first golden rule, R_1 and R_2 form a perfect voltage divider so that

$$V_A = V_{in} = \frac{V_o R_1}{R_1 + R_2} = BV_o$$

which gives

$$K = \frac{1}{B} = \frac{V_o}{V_{in}} = 1 + \frac{R_2}{R_1}$$

The feedback is now series applied since the voltage across the amplifier's terminals is $V_{in} - V_A$ and V_A is the feedback voltage. The input resistance is therefore increased by a factor of $1 + AB$, thereby becoming effectively infinite, since without feedback it would be at least 1 MΩ.

The output current is determined by the value of R_1 and V_{in} because it flows through both resistances and has the value $V_A/R_1 = V_{in}/R_1$. If R_2 is considered to be the load, then the inverting amplifier acts like an ideal constant-current source, that is its output resistance is infinite.

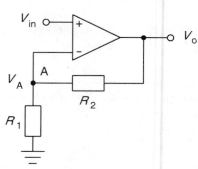

Figure 5.9

The non-inverting amplifier. Its major advantage over the inverting amplifier is its very high input resistance

5.4.3 The buffer

Figure 5.10 shows the circuit of a simple buffer or voltage follower. All of the output is fed back just as in the source and emitter followers and indeed its performance is analogous, though superior, to those circuits. By the second golden rule, $V_o = V_{in}$. With series-applied feedback, the input resistance is increased by a factor of $1 + AB$ to $(1 + AB)R_{in} = (1 + A)R_{in} \approx \infty$, since $B = 1$. The output resistance is low as the feedback is voltage-derived and therefore the output voltage is kept constant.

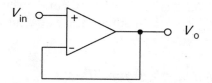

Figure 5.10

A buffer or voltage follower

Like the emitter and source followers, the circuit is used with high-impedance sources and low-impedance loads. In addition, op amps which have to drive large capacitive loads often go into unwanted oscillation which can be stopped by putting a buffer between the load and the op amp. The special characteristics required by some buffering tasks, such as a low output impedance, fast response and a high output current, are not obtainable in ordinary op amps and purpose-made ICs have been designed, but they are relatively expensive.

5.4.4 The summing amplifier or summer

The summing amplifier is exactly the same as the inverting amplifier, except that it has multiple, simultaneous inputs. The one shown in figure 5.11 has unity gain as the input and feedback resistances are equal. The summing node, A, is a virtual ground at which the application of KCL gives

$$I_1 + I_2 + I_3 + I_f = 0$$

and by Ohm's law $I_1 = V_1/R = V_1 G$ etc. Therefore

$$V_1 G + V_2 G + V_3 G + V_o G = 0$$

$$\Rightarrow V_1 + V_2 + V_3 = -V_o$$

The output voltage is equal to the sum of the input voltages, with a change of sign. If the voltages are all the same and the resistances are varied then the currents can be varied at will before summation: this principle is used in D/A converters (see section 11.1.2).

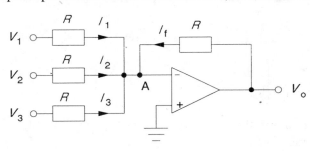

Figure 5.11

A summing amplifier

5.4.5 The integrator

The integrator, shown in figure 5.12, has a capacitor in the feedback path and a resistor at the input to the inverting terminal. The zeroing switch can be a FET for example. The non-inverting terminal is grounded making node A a virtual ground. The current through R is $I_1 = V_{in}/R$, and as $I_2 = -I_1$ the output voltage is given by

$$V_o = \frac{1}{C} \int I_2 \, dt = \frac{1}{C} \int (-I_1) \, dt = \frac{-1}{RC} \int V_{in} \, dt$$

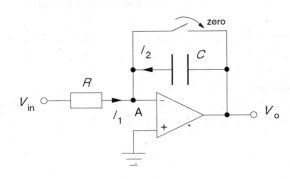

Figure 5.12

An integrator

The output is opposite in sign to the input. For the integrator to work well, the capacitor must have low leakage resistance and the RC product should not be too small, which is the same as saying the voltage gain should not be too big. Normally the gain on an integrator (= $1/RC$) is kept below about 100 to avoid amplifying noise. The input resistance is R because the feedback is shunt applied as in the inverting amplifier. If an initial condition is required, that is an initial output voltage, then a voltage source must be placed in parallel with the capacitor.

Example 5.3

If an integrator is built as in figure 5.13a with $R = 1$ kΩ, $C = 10$ nF and the capacitor has a leakage resistance, $R_p = 50$ MΩ, what is the output voltage as a function of time with an input of 100 μV (DC) and a capacitor that is initially uncharged?

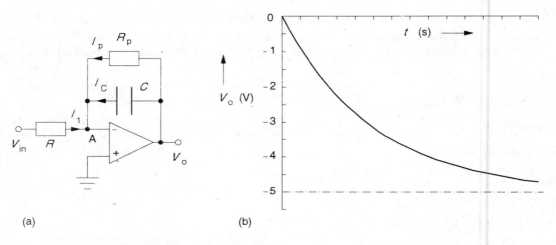

(a)   (b)

Figure 5.13 (a) An integrator with a leaky capacitor (b) The output voltage for example 5.3

Node A is the summing node for the currents, which are

$$I_1 = \frac{V_{in}}{R} = \frac{100}{1000} = 0.1, \quad I_p = \frac{V_o}{R_p} = 0.02V_o, \quad \text{and} \quad I_C = C\frac{dV_o}{dt} = 0.01\frac{dV_o}{dt}$$

working in μA. Thus as the sum of these is zero we have

$$0.1 + 0.02V_o + 0.01\frac{dV_o}{dt} = 0$$

Rearranging this gives

$$\int \frac{dV_o}{V_o + 5} = -2\int dt$$

Carrying out the integration yields

$$\ln(V_o + 5) = -2t + \alpha \quad \Rightarrow \quad V_o + 5 = \beta \exp(-2t) \text{ V}$$

where α is a constant of integration and $\beta = \exp(\alpha)$. Since $V_o = 0$ when $t = 0$, then $\beta = 5$ and the output voltage is

$$V_o = -5[1 - \exp(-2t)] \text{ V}$$

as shown in figure 5.13b. Instead of steadily ramping down to the negative rail, the output saturates at -5 V, because of the leakage in the capacitor. The output is linear for only about 0.2 s because the time constant of the process is $CR_p = 0.5$ s; integrating DC or 'slow' waveforms requires a much bigger time constant. Another point to note is that the input voltage is amplified eventually by a factor of $R_p/R = 50\,000$, so that the input has to be very small if the output is to stay within the limits of the supply voltages.

5.4.6 The differentiator

If the capacitor and the resistor of the integrator in figure 5.12 are exchanged, the circuit becomes a differentiator. This circuit is seldom used because it is very susceptible to noise: any rapidly changing signal — such as a noise spike — will send it into saturation, that is the output will be one of the supply voltages. The analysis is the same as for the integrator: node A is a virtual ground, so that $I_1 = -I_2$. And $I_1 = C dV_{in}/dt$, while $I_2 = V_o/R$ and then $V_o = I_2 R = -I_1 R = CR dV_{in}/dt$.

5.5 Practical operational amplifiers

In practice operational amplifiers have certain idiosyncrasies that must be understood before they can be properly employed. There are a great many different types of op amp to be found in suppliers' catalogues — several hundred — each of which has some special feature or features that makes its use in some applications preferable to others. Table 5.1 lists a few readily-available op amps and some manufacturers' data for them. The price depends largely on the performance required.

Many op amps are sold in DIL (dual-in-line) plastic packages with 8 pins, with many variants on the packaging for each device: ceramic, metal can, bare chip, mini-dip — all having a letter code in the device number. If no external nulling or frequency compensation is provided, then only 3 terminals per op amp are needed, plus two for the supply. Hence some of the cheaper op amps are in quad (four op amps on a chip) 14-pin or dual (two op amps) 8-pin packages. Figure 5.14a shows the μA741CP single-device package with its circuit diagram in figure 5.14b, indicating the input-offset nulling method, which requires an external 10kΩ potentiometer.

We shall now deal with each feature of op amp performance in turn, starting with the CMRR, common-mode rejection ratio.

Table 5.1 *A selection of operational amplifiers*

Device	Type[A]	GB[B]	SR[C]	V_{OS}[D]	θ_{VO}[E]	Drift[F]	I_{OS}[G]	I_B[H]	e_n[I]	R_{in}[J]	R_o[K]	A_v[L]	CM[M]	V_S[N]	I_S[P]	Price[Q]	Note
OP290F	BD	0.02	0.01	75	0.6	—	0.1	4	—	30	100	112	120	1.6/36	10	£6	2
MAX430	CS	0.5	0.5	1	0.02	0.1	0.01	0.01	—	10^6	5000	150	140	±2½/±15	1.3	£9	3
OP07DP	BS	0.6	0.3	60	0.3	0.5	0.8	2	0.4	31	60	112	110	±3/±18	5000	£1.40	4
AD648JN	FD	1	1.8	750	7	15	5 pA	5 pA	2	10^6	250	120	90	±4½/±18	170	£1.92	5
LM358N	BD	1	0.3	2000	7	—	2	20	—	2	100	100	80	3/30	250	£0.34	6
µA741CN	BS	1	0.5	1000	7	—	20	80	20	2	75	106	90	±2/±18	1700	£0.80	7
TL071CP	BF	3	13	3000	10	—	0.01	1	4	$>10^3$	—	110	90	±5/±15	1000	£0.80	8
TL084CN	BFQ	3	13	3000	10	—	0.01	0.2	4	$>10^3$	—	110	95	±3½/±18	700	£1.00	9
AD711JN	FS	4	20	300	7	15	0.01	0.01	2	3×10^6	250	112	90	±4½/±18	2500	£1.12	10
OPA404	FQ	6	35	260	3	—	0.5 pA	1 pA	1	10^7	—	100	100	±5/±18	2250	£12	11
OP275GP	BFD	9	22	300	5	—	2	100	—	$>10^3$	100	106	106	±5/±22	2000	£1.90	12
AD845JN	BFS	16	100	700	20	—	0.03	0.75	4	10^5	5	114	113	±5/±18	10^4	£4.70	13
MAX412CPA	BD	28	4.5	120	—	—	—	100	—	—	—	120	—	±2½/±5¼	2000	£5.50	14
AD844AN	BS	60‡	2000	50	1	—	—	150	—	10¶	15	—	—	±4½/±18	6500	£4.70	15
MAX408ACPA	BS	100	90	3000	—	—	—	—	—	—	—	80	—	±4/±6	—	£6.25	16
AD840JN	BS	400	400	100	3	—	100	3500	10†	0.03	15	102	110	±5/±18	10^4	£7.30	17

[A] B = bipolar, C = CMOS, F = FET, BF = Bi-FET, JFET input, remainder of circuitry bipolar, B = bipolar, S = single device package, D = dual package, Q = quad package [B] Gain-bandwidth product in MHz. [C] Slew rate in V/µs [D] Input offset voltage in µV [E] Offset voltage tempco in µV/K [F] Offset voltage drift with time in µV/√month [G] Input offset current in nA [H] Input bias current in nA [I] p-p noise in nA at 0.1 to 10 Hz [J] Input resistance in MΩ, so an entry of 10^6 means 10^{12} Ω or 1 TΩ [K] Output resistance in Ω [L] open-loop voltage gain in dB [M] CMRR in dB for DC [N] Supply voltage range: V_{min}/V_{max} single supply, bipolar supply will be ±½V$_{min}$/±½V$_{max}$ if used [P] Standby current from supply in µA/device [Q] Prices are for single quantities ¶ At non-inverting input, 50 Ω at inverting input ‡ At unity voltage gain; depends on feedback resistance and voltage gain. † broadband noise from 10 Hz to 10 MHz

Notes:

1. All values are typical at 25°C and with rated supply voltage. 2. For long-life operation from batteries; can be used with bipolar supply. 3. Chopper stabilised instrumentation amplifier. 4. 'Precision' op amp; thermocouple amplifier, low drift. 5. Low-cost photodiode preamp, instrumentation amp. 6. Cheapest medium-frequency op amp. 7. The most popular of op amps; many versions, the original dating back at least to 1970. 8. Cheapest general-purpose op amp. 9. Improved TL071. 10. For active filters, high speed integrators, D/A output I-V convertor, A/D output buffer. 11. Precision, instrumentation op amp. Ultra-low bias current, ultra-high input resistance. 12. Audio amplifier. 13. As AD711JN, but faster; good log amp. 14. Low-noise instrumentation amplifier for infra-red detectors, magnetic search-coil integrators, accelerometers. 15. Current-feedback amp; video buffer, pulse amplifier. 16. Video amplifier. 17. Fast video and pulse amplifier.

Figure 5.14

(a) Single-device, DIL, plastic-packaged, 741 op amp (b) The pin numbers on its circuit diagram and the offset-null-adjusting potentiometer

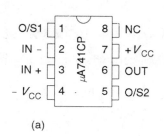

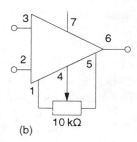

(a)

(b)

10 kΩ

5.5.1 The common-mode rejection ratio

A *common-mode* signal is one that is applied to both inputs and is usually an unwanted component of the input. Ideal op amps have an infinite CMRR so that the output signal is entirely free from common-mode signal. As a consequence of the slightly differing amplifications of the two input channels, common-mode signals are only partially rejected. Any two voltages at the inputs, v_1 and v_2, can be expressed as a common-mode signal, v_c, and a differential-mode signal, v_d:

and

$$v_1 = \tfrac{1}{2}v_d + v_c$$

$$v_2 = -\tfrac{1}{2}v_d + v_c$$

from which we see that

and

$$v_d = v_1 - v_2$$

$$v_c = \tfrac{1}{2}(v_1 + v_2)$$

If the differential signal is amplified by A_d and the common-mode signal by A_c then the CMRR in dB is given by

$$\text{CMRR} = 20\log_{10}(A_d/A_c)$$

Example 5.4

In a certain op amp the input to the inverting terminal is 30 μV and that to the non-inverting terminal is 46 μV. If the differential gain, A_d, is 10^5 and the CMRR is 60 dB what is the output voltage? What would the output voltage be if the input voltages were both reduced by 30 μV? What are the channel gains, A_1 and A_2?

When the CMRR is 60 dB then $A_d/A_c = 1000$ and so $A_c = 10^5/1000 = 100$. The differential voltage is $v_d = v_1 - v_2 = 46 - 30 = 16$ μV, and the common-mode input is $A_c = \tfrac{1}{2}(46 + 30) = 38$ μV.

The output voltage is the sum of the differently amplified differential and common-mode signals

$$v_o = A_d v_d + A_c v_c = 10^5 \times 16 \times 10^{-6} + 100 \times 38 \times 10^{-6} = 1.6038 \text{ V}$$

Reducing both inputs by 30 µV leaves v_d unchanged while v_c is 30 µV less at 8 µV. Then the output voltage will be

$$v_o = 10^5 \times 16 \times 10^{-6} + 100 \times 8 \times 10^{-6} = 1.6008 \text{ V}$$

The common-mode voltage will be zero when $v_1 = 8$ µV and $v_2 = -8$ µV. In most op amps the CMRR is 60 dB or greater.

The output voltage will also be given by

$$v_o = A_1 v_1 - A_2 v_2 = A_d v_d + A_c v_c = A_d(v_1 - v_2) + \tfrac{1}{2} A_c(v_1 + v_2)$$

whence $A_1 = A_d + \tfrac{1}{2} A_c = 10^5 + 50 = 100\,050$ and $A_2 = A_d - \tfrac{1}{2} A_c = 99\,950$.

In the differential amplifiers of sections 3.9.3 and 4.11.2 the two halves of the differential signal cancel out in the emitter or source resistor, but the common-mode signal does not. As a result the gain for differential signals is $h_{fe}R_C/h_{ie}$ (BJT) or $g_m R_D$ (FET), while the gain for common-mode signals is R_C/R_E (BJT) or R_D/R_S (FET). The ratio of these is the CMRR, which for the BJT two-transistor differential amplifier is $h_{fe}R_E/h_{ie}$ and for the FET version is $g_m R_S$. Taking $h_{fe} = 100$ and $R_E = h_{ie}$ leads to a CMRR of 100 or 40 dB for the BJT differential amplifier, while taking $g_m = 5$ mS and $R_S = 2$ kΩ leads to a CMRR of only 20 dB for the FET differential amplifier. Neither of these can be regarded as acceptable and so the IC op amp versions use current-source loads and current mirrors to improve the CMRR, though the basic configuration is still that of figures 3.51a and 4.30.

5.5.2 *Frequency response*

Many op amp circuits have been built that look fine on paper, but turn out to have much less gain than expected. The reason often lies in the misinterpretation of the op amp specification on the manufacturer's data sheet. The frequency response is normally given as a gain-bandwidth[1] product, or unity-gain bandwidth (*GB*), which can vary from 1 kHz to 1 GHz, but is normally about 1-10 MHz.

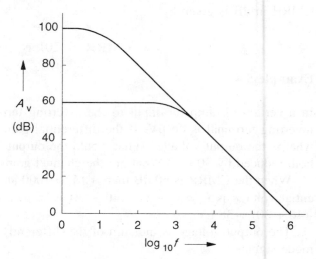

Figure 5.15

Gain as a function of frequency for an op amp with a gain-bandwidth product of 100 dB. With feedback and gain reduced to 60 dB, the frequency response is flat to 1 kHz

---

[1] Sometimes this is denoted f_T, the transition frequency, as for a transistor.

Though the high-frequency response may contain a number of break frequencies, the manufacturers are careful to ensure that one is much lower than the rest and causes the open-loop gain to fall off at 20 dB/decade until the gain is 0 dB. Then if $GB = 1$ MHz and the open loop gain is 10^5 the frequency response will be that shown in figure 5.15, staying constant at 100 dB to 10 Hz, then rolling off at 20 dB/decade until unity gain is reached at 1 MHz. If the gain is reduced by feedback to 1000, then the response will stay flat at 60 dB up to 1 kHz before falling, the gain-frequency product remaining constant (and equal to 1 MHz) above 1 kHz. To obtain constant gain up to 50 kHz requires it to be reduced to only 20 or just 26 dB: to achieve a gain of 60 dB at 50 kHz requires a device with $GB = 50$ MHz or greater.

5.5.3 Offset voltage

Most op amps have a specified offset voltage at the input terminals, which means that if the device is used as an integrator, even with zero input, it will ramp up (or down) to the supply voltage rail in time. If the input offset voltage is V_{os}, then

$$V_o = \frac{-1}{RC} \int V_{os} \, \mathrm{d}t = \frac{-V_{os}t}{RC}$$

If $V_{os} = 1$ mV and $RC = 1$ ms, then the output voltage will ramp to -15 V in 15 seconds! This aspect of op amp performance has been improved greatly in recent years with advances in IC manufacturing processes. Some ICs are chopper-stabilised at a moderately low frequency, typically 400 Hz, so that the input offset is zeroed automatically. However, the chopping frequency is usually carried through as noise on the output, which might not cause difficulties if the signals of interest are far from the chopping frequency. In a few op amps, such as the 741 of figure 5.14b, the input can be zeroed using an external potentiometer connected between two of the op amp pins and one of the supply rails.

Unfortunately, the offset voltage drifts with both temperature and time, so nulling it at one temperature will not eliminate or even reduce it much if the temperature changes. The temperature drift is about 2% of V_{os} per °C, implying that a 20°C change of temperature after nulling (easily possible) will still produce an offset of $0.4V_{os}$. The drift with time (if given, which is rare) is usually stated as so many µV/√month, which means if the drift is 100 µV/√month, then in 1 h = 0.0014 month, the drift will be $100 \times \sqrt{0.0014}$ = 3.7 µV. External nulling therefore must be repeated at frequent intervals if it is to be effective, and as devices have improved it has become virtually redundant.

5.5.4 Input bias and input offset currents

In FET-input op amps the input resistance is so large that these are negligible, but they can be significant in bipolar circuits, where the input resistance is smaller. The input bias current affects the output voltage by an amount depending on the external components used. The op amp configured as an amplifier with resistive feedback, shown in figure 5.16, has an input bias current, I_b, which sees a resistance of $R_1 \mathbin{/\!/} R_2$ at the inverting input, so that $V_1 = I_b(R_1 \mathbin{/\!/} R_2)$, while at the non-inverting input, $V_2 = I_b R_3$. Making $R_3 = R_1 \mathbin{/\!/} R_2$

means $V_1 = V_2$, which eliminates the effect of the input bias current. If $R_1 = 0$, the output voltage will be $I_b(R_1 /\!/ R_2) \times R_2/R_1$, even when $V_{os} = 0$ and no input is applied.

Figure 5.16

Showing the effect of input bias current on the output of an op amp with resistive feedback

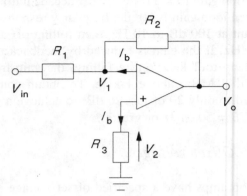

The input offset current, I_{os}, is caused by the differing input resistances of the two inputs giving a difference in the input bias currents. Thus the technique used to eliminate the effect of input bias current will not completely remove the problem; however, as $I_{os} \approx 0.1I_b$, it will reduce its impact by about 90%. With an integrator the input bias and offset currents will produce an output voltage of $C^{-1}\int(I_b \pm I_{os})\mathrm{d}t$, depending on the sign of I_{os}.

Example 5.5

A certain op amp is to be used as in integrator with $R = 15$ kΩ and $C = 1.8$ µF. The input bias current is 0.07 nA, and increases by 50% for every 8°C increase in temperature. The input offset current is negligible, and V_{os} is nulled externally to zero at 20°C, but drifts with temperature at a rate of 1 µV/K. If the temperature increases by 1°C per minute, what will the integrator output be after 1 minute and 5 minutes?

The offset current is given by

$$I_b = I_{b0}(1.5)^{\Delta T/8} = 0.07 \times (1.5)^{\Delta T/8} \text{ nA}$$

And $\Delta T = t/60$ K, working in seconds, so that the equation for I_b as a function of time is

$$I_b = 0.07 \times 1.5^{t/480} = 0.07\exp([t\ln 1.5]/480) = 0.07\exp(\alpha t) \text{ nA}$$

since $1.5^n = \exp(n\ln 1.5)$ and $\alpha = (\ln 1.5)/480 = 8.45 \times 10^{-4}$ s^{-1}. Therefore the output voltage change due to the input bias current is

$$V_o(I_b) = \frac{1}{C}\int_0^t 0.07\exp(\alpha t)\mathrm{d}t = \left[\frac{0.07\exp(\alpha t)}{\alpha C}\right]_0^t \text{ nV}$$

Substituting for α, C and $t = 60$ s and $t = 300$ s leads to $V_o(I_b) = 2.4$ and 13.3 mV.

The offset voltage-time equation is $V_{os} = t/60$ µV and thus the output voltage is

$$V_{\mathrm{o}}(V_{\mathrm{os}}) \; = \; \frac{1}{RC}\int_0^t \left(\frac{t}{60}\right)\mathrm{d}t \; = \; \frac{1}{RC}\left[\frac{0.5t^2}{60}\right]_0^t \; \mu V$$

Substituting for RC (= 0.027 s) and $t = 60$ s and $t = 300$ s produces V_{o} (V_{os}) = 1.1 mV and 27.8 mV. The total drifts are therefore 3.5 mV and 41.1 mV. Note that the quadratic nature of the effect of the offset voltage drift means that it predominates with long integration times.

5.5.5 Slew rate

The slew rate is the rate of change of output voltage and effectively limits the maximum output amplitude of a sinusoidal voltage at high frequencies, since

$$V = V_{\mathrm{m}}\sin \omega t \;\Rightarrow\; \frac{\mathrm{d}V}{\mathrm{d}t} = \omega V_{\mathrm{m}}\cos \omega t$$

Hence

$$\left(\frac{\mathrm{d}V}{\mathrm{d}t}\right)_{\max} = S = \omega V_{\mathrm{m}} = 2\pi f V_{\mathrm{m}} = 2\pi f A \sqrt{2} V_{\mathrm{in}} \tag{5.6}$$

because $V_{\mathrm{o}} = AV_{\mathrm{in}}$ and $V_{\mathrm{m}} = V_{\mathrm{o}}(\mathrm{peak}) = \sqrt{2}V_{\mathrm{o}}$.

Example 5.6

An LM358N op amp is used to amplify a sinusoidal signal using resistance feedback and a gain of 42 dB. If the supply voltage is ±15 V and the dropout is 1 V (that is the output can be ±14 V), use table 5.1 to find the maximum frequencies at which sinusoidal inputs of 1 mV, 10 mV and 7100 mV can be fully amplified. (These are r.m.s. voltages.)

Table 5.1 show that $S = 0.3$ V/µs = 300 kV/s. 'Full amplification' implies a voltage gain of $10^{2.1} = 126$. The maximum amplitude of the output voltage is 14 V = $14/\sqrt{2}$ = 9.9 V_{rms}, which means that the maximum input, whatever else happens, can only be 9.9/126 = 78.6 mV, or the output will go into saturation. Thus there is no frequency at which a 100mV input signal will be fully amplified. Using equation 5.6 leads to

$$f_{\max} = \frac{S}{2\pi V_{\mathrm{m}}} = \frac{S}{2\pi \times A\sqrt{2}V_{\mathrm{in}}} = \frac{3 \times 10^5}{2\pi \times 126\sqrt{2} \times 0.1} = 2.68 \text{ kHz}$$

for a 100mV input and 26.8 kHz for the 10mV input. However, table 5.1 shows that the gain-bandwidth product, $GB = 1$ MHz, so that at 42 dB (= $10^{2.1} = 126$) gain the amplifier's response will roll off below $10^6/126 = 7.9$ kHz, and this is the limiting frequency for the smallest input. Both slewrate, S, and gain-bandwidth product, GB, must be taken into account when assessing frequency response.

5.6 Stability

Operational amplifier (and many transistor amplifier) circuits sometimes unexpectedly go into unwanted oscillation at a frequency which seems to have no relation to any that should be in the equipment. The reason is that the loop gain, **AB**, is in general complex, making it possible for the phase shift round the loop to be 180° so that AB is negative. Therefore in the expression for the gain with feedback

$$\mathbf{G} = \frac{A}{1 + \mathbf{AB}} \quad \Rightarrow \quad G = \frac{A}{|1 + \mathbf{AB}|}$$

the magnitude of the denominator ($|1 + \mathbf{AB}|$) can become less than unity. The type of feedback has changed from negative to positive, and if $AB = -1$ the system is unstable, since the gain is infinite. Then it is possible to have an output with no input, that is an oscillator. When there are several amplifier stages in a system, it is more likely that large phase shifts leading to such instability will be produced at one frequency or another.

However, the system will still remain stable if AB falls below unity before the phase shift becomes 180°, and thus arise the terms *gain margin* and *phase margin*. The phase margin is the amount by which the phase is less than 180° when the loop gain (AB) is unity (0 dB), and the gain margin is the amount by which the loop gain has fallen below 0 dB when the phase is 180°. These margins can be calculated to give some idea of how close to instability the system is.

5.6.1 Nyquist diagrams

A Nyquist diagram is type of *locus diagram*, being a plot of the imaginary part of **AB** against the real part. The distance of a point on the locus from the origin is the magnitude of **AB**. Nyquist diagrams have characteristic shapes according to the number of high and low-frequency cut-offs in the system. Consider figure 5.17, a Nyquist plot for an amplifier with one low-frequency and one high-frequency cut-off.

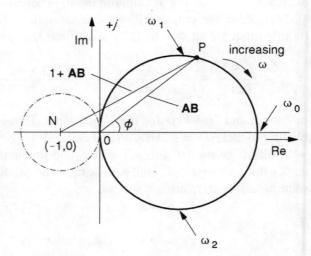

Figure 5.17

A Nyquist plot

A point on the locus such as P has a magnitude OP = AB while its phase angle is ϕ. The locus shown is a circle which is traced out as the frequency increases from zero (a point at the origin) to infinity (also at the origin). When $\omega = 0$ the phase angle, ϕ, is $+90°$, the most it can be with one low-frequency cut-off; then the locus passes successively through the lower cut-off at ω_1, where $\phi = 45°$; the mid-band frequency, ω_0, where $\phi = 0°$ and **AB** is wholly real; the upper cut-off at ω_2, with $\phi = -45°$; and finally back to the origin again when $\omega = \infty$ and $\phi = -90°$, the most negative it can be with one high-frequency cut-off.

Stability depends on the magnitude of $1 + $ **AB**, so that if a line, NP, is drawn from the point, $N = (-1,0)$ to the point, P, its length is $1 + AB$, which will be unity if it crosses a circle of unity radius centred on N. For points on the locus within this circle the feedback will be positive. Clearly the locus plotted in figure 5.17 never intersects the unit circle; thus the system always exhibits negative feedback and is always stable.

Consider the general case

$$\mathbf{AB} = \frac{K(j\omega)^n}{(1 + j\omega\tau_1)(1 + j\omega\tau_2)...}$$

The terms leading to low-frequency cut-offs are those like

$$H_1(j\omega) = \frac{j\omega}{1 + j\omega\tau_1}$$

which at low frequencies ($\omega\tau_1 \ll 1$) will give a phase shift of $+90°$. Thus two of these are required to produce a phase shift of $180°$ near the origin and three will produce a phase shift of $+270°$ at most. A phase shift of $180°$ therefore requires at least three terms like that above. Similarly the terms leading to high-frequency cut-offs are those like

$$H_2(j\omega) = \frac{1}{1 + j\omega\tau_2}$$

which will produce a phase shift of $-90°$ when $\omega\tau_2 \gg 1$. Thus the phase shift from the high-frequency terms will only be $180°$ (other than when $\omega = \infty$) if there are at least three of them. A Nyquist plot for an amplifier with one low-frequency and three high-frequency cut-offs is shown in figure 5.18.

Each low-frequency term moves the beginning of the Nyquist plot round one quadrant anticlockwise and each high-frequency term moves the end of the Nyquist plot round by one quadrant clockwise.

In summary, there are three chief types of Nyquist diagram:

(1)   The locus is always on the right-hand side of the j-axis and never enters the unit circle: the system is stable and always has negative feedback.
(2)   The locus enters the unit circle but does not enclose the point $(-1,0)$: the system is stable but exhibits positive feedback when in the unit circle.
(3)   The locus encloses the point $(-1,0)$: the system will oscillate.

The plot shown in figure 5.18 is clearly of the third type.

Figure 5.18

A Nyquist plot for a system
with one low-frequency cut-
off and three high-frequency
cut-offs

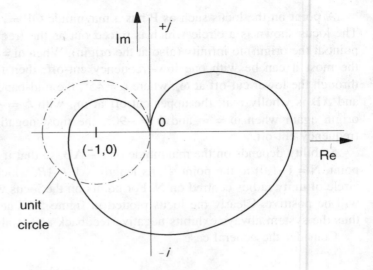

5.7 Comparators and Schmitt triggers

Comparators are a special type of op amp used for deciding when an input voltage exceeds some preset or reference level. Comparators are designed to change output state very rapidly as the input exceeds the reference level, that is their slew rates are very high. However, their specifications do not include slew rate, but give instead their response times for maximum output as a function of the input 'overdrive' — the amount by which the reference voltage is exceeded. For example, in table 5.2 the response time for the LM319N dual comparator is stated as 0.08 μs for a 5 mV overdrive.

Table 5.2 *A selection of comparators*

Device	Type[A]	Max i/p[B]	Resp time[C]	Supply[D]	I_{sc}[E]	Pins[F]	Price[G]
CA139	QBB	±36	1.3	36D	20	14	£1.40
LM311P	SBB	±30	0.2	5-36D	40	8	£0.44
LM339N	DBB	5	0.08	5-36	16	14	£2.35
TLC339CN	QFF	±18	2.5	3-16	20	14	£1.54
LT1016CN[H]	SBB	±5	0.01‡	±7¶	30	8	£4.44
CA3290E	DFB	±36	1.2	36D	30	8	£0.92

Notes: [A] No. of devices in package, input type, output type. S = Single, D = Dual, Q = Quad, B = Bipolar, F = FET. [B] Maximum differential input voltage [C] For 5 mV overdrive [D] min-max voltage on positive supply terminal, grounded negative supply; D = dual supply, 5-36D means from 5-36 V unipolar or $\pm 2\frac{1}{2}$ V to ±18 V bipolar. [E] Short-circuit current to ground in mA [F] No. of pins in plastic DIL package [G] For single quantities [H] Fast latch ‡ with 500 mV overdrive ¶ max, bipolar only

The output stage of nearly all comparators is a BJT with an open collector which must be supplied with a pull-up resistor as in figure 5.19 — about 1 kΩ is used when TTL logic is being driven, but the collector resistance can be some other load that one desires to drive. The response times vary not only with level of overdrive, but also with the polarity of the input signal.

Figure 5.19

A comparator with an open-collector output stage, which must be connected via a suitable resistor to the positive supply

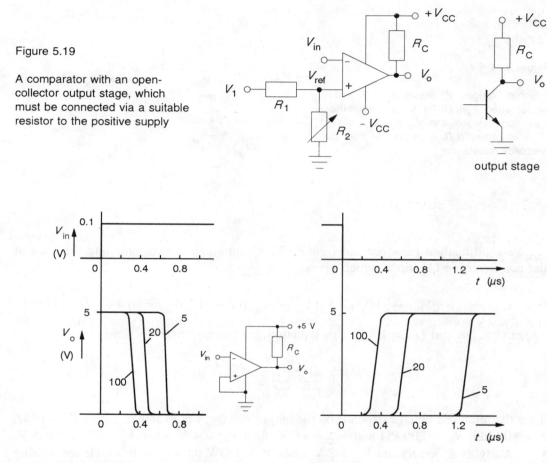

Figure 5.20 A comparator's response to various input overdrive levels (marked in mV on the graphs) for output transitions to HIGH and LOW. The circuit is shown inset

For example, figure 5.20 shows the response of a typical comparator such as the CA139 for overdrives of 5 mV, 20 mV and 100 mV and with the output going from high to low (faster) and from low to high (slower). In table 5.2 this comparator's response time would be given as 1.3 μs, though in most applications it would be appreciably faster.

5.7.1 Schmitt triggers

A Schmitt trigger is a comparator with hysteresis, so that noisy input signals do not cause multiple transitions on the output. Also the comparator is a fast-acting device often used to switch relatively heavy currents that can cause rapid variations to the input signal. These signal variations produce multiple output transitions in the comparator too. By using a feedback resistor as in figure 5.21 the comparator output shows hysteresis: once it has changed state it will only revert to its original state if the input signal is substantially less (or more) than the reference voltage.

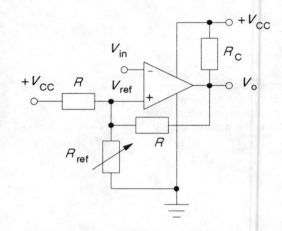

Figure 5.21

A Schmitt trigger. V_{ref} depends on V_o and is higher when V_o is HIGH. In this circuit the maximum hysteresis in V_{ref} is $\frac{1}{2}V_{CC}$ when $R_{ref} = \infty$, provided R_C is much less than the feedback resistance

We shall analyse the circuit of figure 5.21 by summing the currents into the node at the non-inverting (+) input, which gives

$$(V_{CC} - V_{ref})G + (V_o - V_{ref})G - V_{ref}G_{ref} = 0$$

where $G = 1/R$ and $G_{ref} = 1/R_{ref}$. This equation can be rearranged to yield

$$V_{ref} = \frac{(V_o + V_{CC})G}{2G + G_{ref}}$$

Thus the reference voltage depends on the output voltage. For example, taking $R_C = 1$ kΩ, $R = 100$ kΩ, $R_{ref} = 200$ kΩ and $V_{CC} = +5$ V, then $V_{ref} = 4$ V when V_o is HIGH (= 5 V, very nearly as $R_C \ll R$) and $V_{ref} = 2$ V when V_o is LOW (= $V_{CEsat} \approx 0$ V). Hence a rising edge on the input must exceed 4 V to cause an output transition from HIGH to LOW and a falling edge on the input must drop below 2 V to cause a transition from LOW to HIGH.

5.7.2 The comparator IC

The actual circuit depends on what performance is required, but the basic circuit is like that of figure 5.22 in all comparators. The input is a standard differential amplifier with a current sink. The current sources could be any of the standard circuits dealt with in sections 3.10 and 4.8. In the example shown the input devices are configured to increase the input resistance and voltage gain, but if a very high input resistamce is required JFETs or MOSFETs could be used instead. The output transistor is nearly always an open-collector npn BJT as shown.

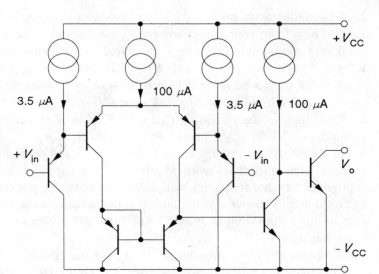

Figure 5.22

A simple, bipolar-input comparator with open-collector output. The negative supply is usually grounded and only a single positive supply is used. The output then swings from 0 to +V_{CC} or vice versa

Suggestions for futher reading

Introduction to operational amplifier theory and applications by J V Wait, L P Huelsman and G A Korn (McGraw-Hill, 2nd ed., 1992)
Introductory operational amplifiers and linear ICs: theory and experimentation by R F Coughlin and R S Villanucci (Prentice-Hall, 1990)
Analog electronics with op amps: a source book of practical circuits by A J Peyton and V Walsh (Cambridge University Press, 1993)

Problems

1 Show that the circuit of figure P5.1 is an integrator. If $L = 200$ mH and $R = 100 \ \Omega$, what is the output if the input is (a) a square voltage pulse of 4 mV for 2.5 s (b) a negative ramp from 0 to −100 mV in 0.1 s which returns to zero in 0.05 s?
[(a) −5 V (b) +7.5 V]

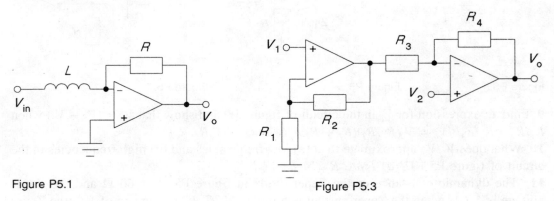

Figure P5.1

Figure P5.3

2 If the inductor in problem 5.1 has resistance r, what is V_o if V_{in} is the step voltage $Vu(t)$? Show from your expression for V_o that the circuit acts as a perfect integrator when $r = 0$, and its output is given by $V_o = -VRt/L$. What is the saturation value, V_{osat}, of V_o? If V_{in} is a 10 mV step, $L = 80$ mH, $R = 1$ kΩ and $r = 4$ Ω, what is V_{osat} and when is $V_o = \frac{1}{2}V_{osat}$? What must r be if the device is to work as an integrator with 5% accuracy for 1 s? $[V_o = -(VR/r)(1 - exp[-rt/L]u(t), V_{osat} = -VR/r, V_{osat} = -2.5$ V, $t = 13.7$ ms; $r = 8$ m$\Omega]$

3 Show that, in the circuit of figure P5.3, $V_o = V_2[1 + R_4/R_3] - V_1[R_4(R_1 + R_2)/R_1R_3]$. When does $V_o = V_2 - V_1$? *[When $R_1 = R_4$ and $R_2 = R_3$]*

4 The AD845JN op amp is chosen from table 5.1 and configured as an inverting amplifier using resistance feedback with 55 dB gain. The supply voltage is ±6 V and the dropout negligible. At what frequency will a 2mV sinusoidal input cease to be fully amplified? If it is used at 50 kHz, what is the largest signal which can be fully amplified? What will the gain be for a 20mV input at 200 kHz? How will all these answers change if the supply voltage is ±18 V.

5 Show that the input impedance, V_{in}/I_{in}, of the circuit of figure P5.5 is $-Z_1Z_3/Z_2$ and that it is therefore a negative impedance converter (NIC). If $Z_1 = Z_2 = R$ and $Z_3 = -j/\omega C$ then show that the equivalent inductance of the circuit is $L_{eq} = R^2C$. What is L_{eq} if $R = 22$ kΩ and $C = 47$ nF? *[22.75 H]*

6 Find the input impedance of the circuit of figure P5.6. By making this circuit into Z_3 of figure P5.5 and $Z_1 = Z_2$ what will the input impedance of figure P5.5 become? $[Z_{in} = -R^2/Z. Z_{in} = R^2/Z]$

7 What is the input impedance at 50 Hz of the circuit of figure P5.7 if $R_1 = 1$ kΩ, $R_2 = 1$ MΩ and $C = 0.1$ μF? What is the equivalent inductance of the reactive part? *[31.42∠86.4° kΩ, 99.8 H]*

8 What is the gain in dB of the amplifier in figure P5.8 if $R_1 = R_2 = R_3 = 56$ kΩ, $R_4 = 470$ Ω and $R_5 = 27$ kΩ? What is R_5 for? *[41.67 dB]*

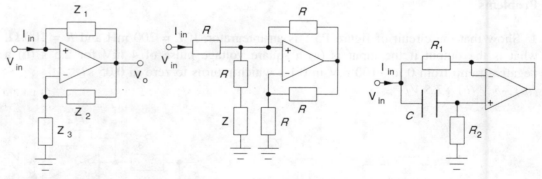

Figure P5.5   Figure P5.6   Figure P5.7

9 Find an expression for V_o in the circuit of figure 5.9 and show that $V_o \propto (V_1 - V_2)$ when $R_1/R_3 = R_2/R_4$. $[V_o = V_2(R_4/R_1)(R_1 + R_3)/(R_2 + R_4) - V_1R_3/R_1]$

10 What does V_o/V_{in} approximate to at (a) low frequencies and (b) high frequencies in the circuit of figure P5.10? $[(a) 1/\omega C_2R_1$ (b) $\omega C_1R_2]$

11 The dynamic resistance of the Zener diode in figure P5.11 is 60 Ω and the Zener voltage is 4.7 V when the Zener current is 5 mA at 20°C. If the tempco of V_{Z0} (the Zener

voltage at an extrapolated Zener current of 0 A) is +0.1%/°C and the resistors have negligible tempcos, what is the output voltage variation if the temperature varies from 20°C to 40°C and the supply voltage varies independently from 8 to 9 V?
[−10.84 V to −11.34 V]

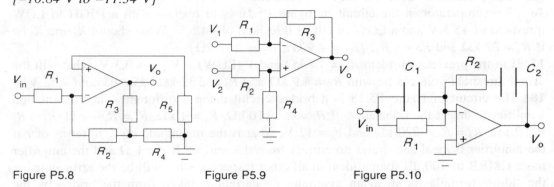

Figure P5.8   Figure P5.9   Figure P5.10

12 Draw a graph of V_o against $\log_{10}V_{in}$ for the circuit of figure P5.12 for values of V_{in} lying between 10 mV and 10 V, given that the current through the diode is $I_s\exp(qV_{AK}/kT)$ and $kT/q = V_{th} = 30$ mV, while $I_s = 1$ pA. If the tempco of I_s is +14%/K, what effect does a temperature rise of 5 K have on V_o? (In practice the graph of V_{AK} against $\ln I_D$ has a slope which varies with the current and is linear only over a restricted range.)

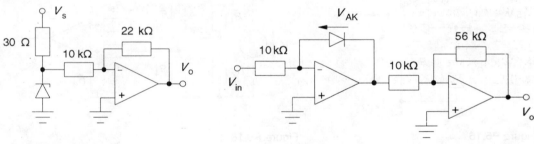

Figure P5.11   Figure P5.12

13 Draw a graph of V_o against V_{in} (the transfer function) for the circuit of figure P5.13 with values of V_{in} lying between −0.5 V and +0.5 V, given that the diodes have a constant voltage drop of 0.55 V across them when conducting. Draw the transfer function when D1 is replaced by a short-circuit.

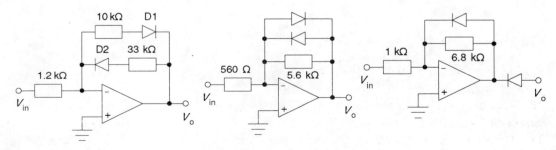

Figure P5.13   Figure P5.14   Figure P5.15

14 Graph the transfer function (V_o against V_{in}) for the circuit of figure P5.14, given that the diodes' *I-V* relation is $I = I_s \exp(35V)$ with $I_s = 5$ pA.

15 Graph the transfer function for the circuit of figure P5.15 given that the diodes have the same *V-I* relation as in the previous problem.

16 The comparator in the circuit of figure P5.16 is to operate with a HIGH to LOW threshold of +5.5 V and a LOW to HIGH threshold of +4.5 V. What should R_1 and R_2 be if $R_3 = 22$ kΩ and $R_C \ll R_3$? *[$R_1 = 4.9$ kΩ, $R_2 = 2.32$ kΩ]*

17 If in the previous problem $R_C = 4.7$ kΩ and $V_o(\text{LOW}) = V_{\text{CEsat}} = 0.3$ V, what will the actual threshold voltages be with $R_1 = 4.9$ kΩ and $R_2 = 2.32$ kΩ? *[5.39 V and 4.52 V]*

18 The circuit of figure P5.18 is a bridge circuit using a differential instrumentation amplifier to detect the imbalance. If $R_1 = R_3 = 10$ kΩ, $R_4 = 1$ kΩ, $R_2 = R_4 + r$ Ω, $R_5 = R_6 = 2.2$ kΩ, $R_7 = R_8 = 220$ kΩ and $V_s = 12$ V, what is the magnitude of V_o in terms of r if the amplifier is ideal and draws no current from the bridge? If $r = 1$ Ω and the amplifier has a CMRR of 100 dB (being ideal in all other respects), what will be the error in using the 'ideal' formula for it, again assuming no current is taken from the bridge by the amplifier? What will this error be if the CMRR is 120 dB? If R_1 is swapped with R_2 and R_3 with R_4 how will these figures change? *[+1.1% , +0.11%; −11%, −1.1%]*

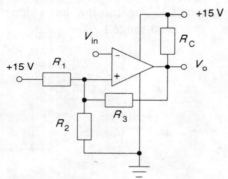

Figure P5.16

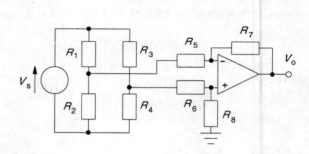

Figure P5.18

6 Filters

ANY CIRCUIT containing inductance or capacitance can act as a filter, that is to say it will not have a flat frequency response, but will attenuate or amplify at some frequencies more than at others. In many cases we desire to remove unwanted noise at certain frequencies or bands of frequencies so that the signals we do want are as clean as possible; this is the purpose of a filter, but in achieving its purpose it may introduce problems — such as phase shift, delay and distortion — of its own.

Ideal filters have frequency responses of four basic types: lowpass, highpass, bandpass and bandstop, which are illustrated in figure 6.1; these are the so-called 'brickwall' responses, with infinite gain-frequency slopes at the *corner frequency*, f_c (or corner frequencies, f_1 and f_2, in the case of the bandpass and bandstop filters). In practical filters the rate of increase or decrease in gain is finite and can only approximate a brickwall response at the expense of considerable complexity. As we shall see, simple filters are very limited in their rates of change of gain with frequency, known as roll-off; and in addition the gains in the passband and stopband may have some ripple which can, however, be controlled by judicious choice of components.

Practical filters may be divided into two main classes: passive filters that are inserted into a line or circuit and require no separate power supply, and active filters that do require their own power supply, being dependent usually on operational amplifiers for their functioning.

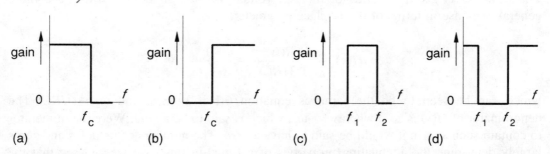

Figure 6.1 Ideal, brickwall, filter responses (a) lowpass (b) highpass (c) bandpass (d) bandstop

6.1 Passive filters

Passive filters use capacitors and/or inductors to provide the requisite frequency response since resistors ideally have no frequency-dependent property. In passive filters the passband gain must always be less than unity since these reactive elements will always be less than ideal. The simplest possible filter is one containing only one reactive element, which is known as a *first order* filter, though normally a passive filter will contain several reactive elements and its order will be the same as the number of these. As we shall see, to approximate a brickwall response the filter must be of a high order and must therefore contain many reactive elements, so becoming more costly to manufacture.

6.1.1 The first-order, highpass, RC filter

Consider the circuit of figure 6.2a containing a single capacitance, which acts as a short-circuit at high frequencies and which must therefore be a highpass filter whose approximate gain response is shown in figure 6.2b.

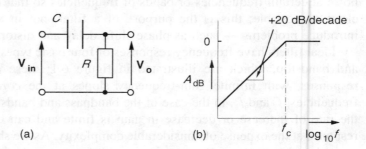

Figure 6.2

(a) A first-order, highpass, RC filter

(b) Its gain response on a log-log plot

This response is readily derived from figure 6.2a, since the output voltage for AC is

$$\mathbf{V_o} = \frac{R\mathbf{V_{in}}}{R + 1/j\omega C} \;\Rightarrow\; \mathbf{V_o}/\mathbf{V_{in}} = \mathbf{H}(j\omega) = \frac{j\omega RC}{1 + j\omega RC} \qquad (6.1)$$

The response is given in terms of the AC voltage transfer function, $\mathbf{H}(j\omega)$. The more general response in terms of the Laplace parameter, s, is

$$H(s) = \frac{RCs}{1 + RCs} = \frac{1}{s + \omega_c}$$

which can be seen to be the same as equation 6.1 if $s \equiv j\omega$ and $\omega_c \equiv 1/RC$. The denominator of $H(s)$ is zero when $s = -\omega_c$, which is said to be a *pole*. Were the numerator to contain such a term it would be said to have a *zero*. The numbers of its poles and zeros largely determine the attenuating properties of a filter. In this case we can see that the magnitude of $\mathbf{H}(j\omega)$ is given by

$$|\mathbf{H}(j\omega)| = H(j\omega) = \frac{\omega RC}{\sqrt{1 + (\omega RC)^2}} = \frac{\Omega}{\sqrt{1 + \Omega^2}} \qquad (6.2)$$

where $\Omega \equiv \omega RC = \omega\tau = \omega/\omega_c = f/f_c$ and $\tau = RC = 1/\omega_c$. The *corner frequency*, *break point* or *cut-off frequency* is f_c or ω_c; it is the most important characteristic of the filter and lies at $\Omega = 1$. Using the reduced frequency, Ω, is a convenient way of simplifying the mathematical expressions for frequency responses.

6.1.2 Bode diagrams

Frequency responses are usually plotted as gain in dB and phase in degrees against $\log_{10}\omega$ or $\log_{10}f$, and are known as Bode plots after their inventor[1]. To construct the Bode plot for the first-order, highpass filter of figure 6.2a, we take the magnitude as given by equation 6.2 and convert this into decibels:

$$A_{dB} = 20\log_{10}|\mathbf{H}(j\Omega)| = 20\log_{10}\Omega - 10\log_{10}(1 + \Omega^2)$$

The term $20\log\Omega$ (dropping the subscripted 10) will be zero at $\Omega = 1$ and increase linearly with frequency at a rate of 20 dB/decade[2]. This means it will increase by 20 dB when the frequency is increased ten-fold, for example to 20 dB at $\Omega = 10$ or fall by 20 dB when the frequency is decreased by a factor of 10, for example to -20 dB at $\Omega = 0.1$. The second term of A_{dB} is $-10\log(1 + \Omega^2)$, and we can see that if $\Omega << 1$ it becomes approximately $-10\log 1 = 0$; and if $\Omega >> 1$ it becomes $\approx -10\log\Omega^2 = -20\log\Omega$. We can approximate the second term's contribution to A_{dB} to zero when $\Omega < 1$, and to -20 dB/decade when $\Omega > 1$. Then up to the corner frequency only the first term operates, and after it the two terms cancel out to give the approximate plot which is the unbroken line in figure 6.3a.

The approximate gain response can be corrected to give the exact gain response by noting that at $\Omega = 1$ the gain response is

$$A_{dB}(1) = 20\log 1 - 10\log(1 + 1^2) = -3 \text{ dB}$$

For this reason the corner frequency is sometimes called the 3dB point. The exact gain response is also shown in figure 6.3a as the broken line.

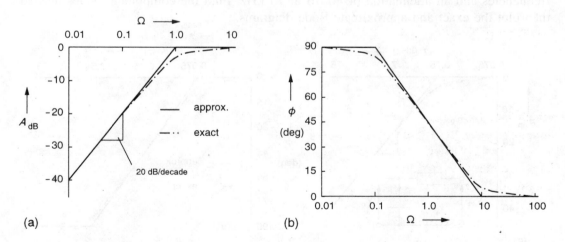

Figure 6.3 Bode diagram for a first-order, highpass, RC filter (a) gain in dB v. logΩ (b) phase v. logΩ

---

[1] H W Bode, of Bell Telephone Laboratories, published his filter theory in *J. Math. Phys.*, **13**, 275-362 (1934)

[2] Some texts use octaves — a doubling of frequency is an increase of an octave — and these will give the gain increase/decrease as ±6 dB/octave.

The phase response as a function of $\log_{10} f$ forms the other half of the Bode plot, and it is found from equation 6.1, whose numerator is wholly imaginary, so contributing a phase shift of $+90°$, and whose denominator has a phase angle of

$$\phi_d = \tan^{-1}(\omega RC) = \tan^{-1}\Omega$$

Thus the overall phase response is

$$\phi = 90° - \phi_d = 90° - \tan^{-1}\Omega$$

When $\Omega \ll 1$, $\phi \approx 90°$ and when $\Omega \gg 1$, $\phi \approx 0°$. In between, when $\Omega = 1$, $\phi = 45°$. The phase response of the first-order, highpass filter is shown in figure 6.3b. The corrected or exact phase plot is derived from the straight-line approximation by noting that when $\Omega = 0.1$, $\phi = 84.3°$ and when $\Omega = 10$, $\phi = 5.7°$.

Most Bode plots are more complicated than this, but the principles are exactly the same: use straight lines for the approximate plots and if necessary apply corrections at the corner frequencies. We shall be using Bode plots extensively in this chapter; the finer points will be dealt with where and when necessary.

6.1.3 The first-order, lowpass, RC filter

If we exchange resistor and capacitor in figure 6.2a we will obtain a lowpass filter with gain and phase responses like those of figure 6.4. This is shown in the following example.

Example 6.1

A lowpass, first-order, RC filter is required with an output impedance of 3.3 kΩ at low frequencies and an attenuation of 40 dB at 75 kHz. Find the component values needed, then plot the exact and approximate Bode diagrams.

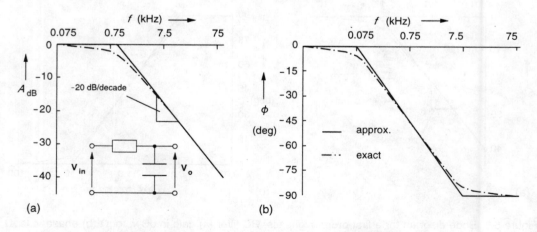

Figure 6.4 A first-order, lowpass, RC filter (a) gain response (b) phase response

The circuit is shown in figure 6.4a where we see that the output impedance is the Thévenin impedance looking into the output terminals, that is $R \;//\; 1/j\omega C$. At low frequencies the capacitance acts as an open circuit so the impedance is just $R = 3.3$ kΩ. The response will be 40 dB down at $\Omega = 100$ ($f = 75$ kHz) and thus $\Omega = 1$ when $f = f_c = 1/RC = 750$ Hz. Hence

$$RC = 750^{-1} \;\Rightarrow\; C = (750R)^{-1} = (750 \times 3300)^{-1} = 0.404 \;\mu\text{F}$$

The response function is

$$\mathbf{H}(j\omega) = \mathbf{V_o}/\mathbf{V_{in}} = \frac{1/j\omega C}{R + 1/j\omega C} = \frac{1}{1 + j\omega RC} = \frac{1}{1 + j\Omega}$$

The gain and phase responses are

$$A_v = H(j\Omega) = \frac{1}{\sqrt{1 + \Omega^2}} \quad \text{and} \quad \phi = -\tan^{-1}\Omega$$

The gain response in decibels is

$$A_{dB} = -10\log(1 + \Omega^2)$$

We see when $\Omega < 1$ that $A_{dB} \approx 0$ dB and $\phi \approx 0°$; and when $\Omega > 1$ that $A_{dB} \approx -20\log\Omega$ and $\phi \approx -90°$. The 3dB point is at 750 Hz so that below this A_{dB} is zero and above 750 Hz it falls at 20 dB/decade, giving the gain plot of figure 6.4a. The phase plot must be approximately zero below $\Omega = 0.1$ or 75 Hz, $-45°$ at 750 Hz and $-90°$ above $\Omega = 10$ or $f = 7.5$ kHz. The phase corrections are $-5.7°$ at 75 Hz and $+5.7°$ at 7.5 kHz, giving the phase plot of figure 6.4b.

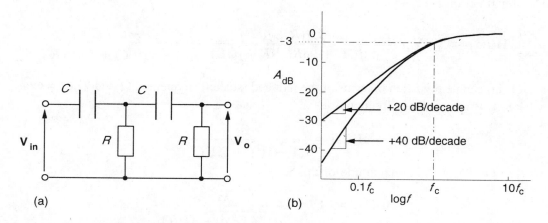

(a)   (b)

Figure 6.5 (a) Cascaded RC stages (b) Frequency-adjusted gain response compared to 1st-order circuit

First order filters are not often used because of the gentle roll-off of 20 dB/decade, though many circuits act like them in practice. First-order RL filters are used even less frequently because inductors are usually more expensive, more bulky and less ideal than capacitors. It is readily shown that replacing R by L and C by R in an RC filter will produce an RL filter of the same type (see problem 6.1). It is possible to improve the roll-off rate with two cascaded RC sections as in the highpass filter circuit of figure 6.5a, whose gain response is shown in figure 6.5b compared to the response of the first-order RC circuit. The cascaded, or ladder, filter has its 3dB point at $0.38f_c$, where f_c is the corner frequency for the first-order filter. When the responses are made to coincide at the 3dB points $f_c = 1/2\pi RC$ for the first-order circuit and $0.38/2\pi RC$ for the second-order circuit. In consequence, the responses at $0.1f_c$ are -20 dB and -25 dB for the first and second-order filters respectively, instead of the expected values of -20 dB and -40 dB. The cascaded filter takes longer to settle down to its 40 dB/decade roll-off than the single-stage circuit. In this aspect its performance is inferior to that of the equivalent, second-order, RLC circuit, which we shall examine next.

6.2 Second-order RLC filters

The resistive part of RLC filters is sometimes a consequence of the non-ideal nature of the circuit elements — inductors particularly at low frequencies and capacitors at high — used in their construction, but it can be used deliberately as a further means of controlling the Q-factor and hence the shape of the amplitude response. The four types of filter will be examined in turn, starting with highpass and lowpass filters, which have almost identical characteristics.

6.2.1 Second-order highpass and lowpass RLC filters

Consider the RLC circuit in figure 6.6a, which is configured as a highpass filter. The transfer function is

$$\mathbf{H}(j\omega) = \mathbf{V_o}/\mathbf{V_{in}} = \frac{j\omega RL/(R + j\omega L)}{1/j\omega C + j\omega RL/(R + j\omega L)} = \frac{-\omega^2 LC}{1 - \omega^2 LC + j\omega L/R}$$

$\mathbf{H}(j\omega)$ can be rewritten in terms of the reduced (scaled) frequency, $\Omega \equiv \omega\sqrt{(LC)} = \omega/\omega_0$, where ω_0 is the resonant frequency:

$$\mathbf{H}(j\Omega) = \frac{-\Omega^2}{(1 - \Omega^2) + j\Omega/Q}$$

where $Q \equiv R/\omega_0 L$. The magnitude of this is

$$|\mathbf{H}(j\Omega)| = A_v = \frac{\Omega^2}{\sqrt{(1 - \Omega^2)^2 + \Omega^2/Q^2}}$$

Now if R is large, Q is large and A is nearly $\Omega^2/(1 - \Omega^2)$, which tends to infinity at $\Omega = 1$, that is when $\omega = \omega_0$, since the circuit is then resonant. However, if Q is made equal to $1/\sqrt{2}$, the magnitude becomes

$$A_v = \frac{\Omega^2}{\sqrt{1 + \Omega^4}}$$

The circuit resonance is critically damped and the response is said to be *optimally flat*, as shown in figure 6.6b, with the response of the 2-section RC filter shown for comparison. The 3dB points are at $f = f_c = \omega_0/2\pi$ for the RLC circuit and $f_c = 0.38/2\pi RC$ for the 2-stage RC circuit. The RLC circuit's gain is exactly -40 dB at $\Omega = 0.1$, while the cascaded RC circuit's response is -25 dB. The knee of the gain response is much sharper for the RLC circuit, even though both eventually roll off at 40 dB/decade.

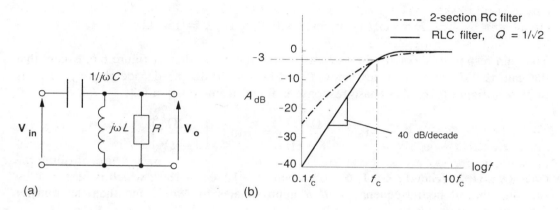

Figure 6.6 (a) Second-order, highpass, RLC filter (b) Its gain response compared to a 2-stage RC filter

In general the gain roll-off is 40 dB/decade for each LC high or lowpass section, so that a 3-section RLC filter will have a roll-off of 120 dB/ decade. These roll-off rates are halved in bandpass RLC filters. A second-order, lowpass, RLC filter is readily designed by exchanging capacitor and inductor in figure 6.6a, as shown in the following example.

Example 6.2

Design a second-order, lowpass, optimally-flat, RLC filter with an attenuation of 10 dB at 1 kHz using an inductor which has an inductance of 60 mH and negligible resistance. Draw the Bode diagram for the circuit.

The circuit is shown in figure 6.7a, from which the voltage-transfer function, or response, can be found:

$$\mathbf{H}(j\Omega) = \frac{1}{(1 - \Omega^2) + j\Omega/Q} \quad \checkmark = \frac{1}{S^2 + \sqrt{2}S - 1} \quad (6.3)$$

For an optimally-flat gain response we require $Q = R/\omega_0 L = 1/\sqrt{2}$. The attenuation, α, is then given by

$$A_v = \frac{1}{\sqrt{1 + \Omega^4}} \quad \Rightarrow \quad \alpha = -A_{dB} = 10\log(1 + \Omega^4)$$

This will be 10 dB when $\log(1 + \Omega^4) = 1$, that is when $1 + \Omega^4 = 10$ or $\Omega = 9^{0.25} = \sqrt{3}$. From this requirement we see that

$$\Omega = \omega/\omega_0 = \sqrt{3} \implies \omega_0 = \frac{\omega}{\sqrt{3}} = \frac{2\pi f}{\sqrt{3}} = \frac{2000\pi}{\sqrt{3}} = 3.628 \text{ krad/s}$$

and therefore

$$C = 1/\omega_0^2 L = 1/(3628^2 \times 0.06) = 1.266 \ \mu F$$

The resistance required is found from

$$R = \omega_0 L Q = 3628 \times 0.06 \times 1/\sqrt{2} = 154 \ \Omega$$

The gain response is exactly the same as for the highpass filter of figure 6.6, except that the gain is 0 dB at frequencies below f_0 and falls at 40 dB per decade above f_0. This is plotted in figure 6.7b. The phase response is found from equation 6.3:

$$\phi = -\tan^{-1}\left(\frac{\Omega/Q}{1 - \Omega^2}\right) = -\tan^{-1}\left(\frac{\sqrt{2}\Omega}{1 - \Omega^2}\right)$$

since $Q = 1/\sqrt{2}$. When $\Omega < 0.1$, $\phi \approx 0°$; when $\Omega > 10$, $\phi \approx -180°$, which is best seen by realising that at high frequency $C \ /\!/ \ R$ approximates to $1/j\omega C$, and then the transfer function approximates to

$$\mathbf{H}(j\omega) \approx \frac{1/j\omega C}{j\omega L} = \frac{-1}{\omega^2 LC} = \frac{-1}{\Omega^2}$$

The phase shift when $\Omega = 1$ must be $-90°$. The straight line approximation in figure 6.7b follows. The error in this can be estimated from the phase at $\Omega = \sqrt{0.1}$, which is

$$-\tan^{-1}\left(\frac{\sqrt{2} \times \sqrt{0.1}}{1 - 0.1}\right) = -\tan^{-1}0.497 = -26.4°$$

This is considerably different from the straight-line approximation of $-45°$; in fact when Q is larger than unity, a better straight-line approximation for the phase is that shown by the dotted line in figure 6.7b, labelled high-Q ϕ.

6.2.2 Second-order notch RLC filter = bandstop

The notch, or band-reject, filter is shown in figure 6.8a, with an ideal inductor of negligible resistance. The transfer function is

$$\mathbf{H}(j\omega) = \frac{\mathbf{V}_o}{\mathbf{V}_{in}} = \frac{R}{R + j\omega L/(1 - \omega^2 LC)} = \frac{1 - \omega^2 LC}{(1 - \omega^2 LC) + j\omega L/R}$$

Taking $\Omega = \omega/\omega_0$ and $Q = R/\omega_0 L$ gives the simpler form

$$\mathbf{H}(j\Omega) = \frac{1 - \Omega^2}{(1 - \Omega^2) + j\Omega/Q} \tag{6.4}$$

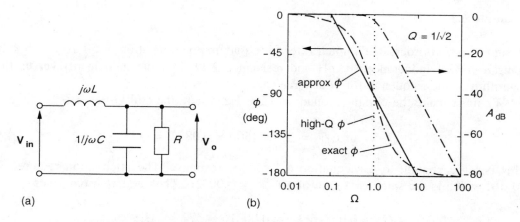

Figure 6.7 (a) A lowpass, second-order, RLC filter (b) Its Bode plot when $Q = 1/\sqrt{2}$

Regardless of the value of Q this filter will completely attenuate at the resonant frequency, f_0, since the inductor's Q-factor is infinite (it has no resistance). If we take the inductor's Q-factor, Q_c, into account, it can be shown (problem 6.3) that the response is

$$\mathbf{H}(j\Omega) = \frac{(1 - \Omega^2) + j\Omega/Q_c}{(1 + r/R - \Omega^2) + j\Omega(1/Q_c + 1/Q)} \tag{6.5}$$

where $Q_c = \omega_0 L/r$ and r ($\ll R$) is the coil resistance in series with its inductance. This reduces to equation 6.4 when $Q_c \gg Q$. When $\Omega = 1$ and $Q_c \gg Q$, equation 6.5 becomes

$$\mathbf{H}(1) \approx \frac{j/Q_c}{r/R + j/Q} \approx \frac{Q}{Q_c} \tag{6.6}$$

With a given inductor, the attenuation at $\Omega = 1$ is made large by making Q small. Regardless of the size of Q the bandwidth can be shown to be f_0/Q (see problem 6.4).

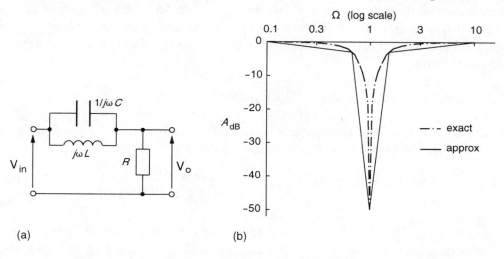

Figure 6.8 (a) 2nd-order, RLC, notch filter (b) Its gain response when $Q_c/Q = 316$

Example 6.3

Design a second-order, RLC, notch filter for a maximum attenuation of 50 dB at 1.5 kHz using a coil of inductance 80 mH and resistance 2 Ω. Plot the gain in dB versus the logarithm of the frequency for your circuit.

We must make the circuit resonant at 1500 Hz, where the coil Q-factor is

$$Q_c = \omega_0 L / r = 2\pi \times 1500 \times 0.08/2 = 377$$

This is large enough to use equation 6.4 for the response. The attenuation required is 50 dB, which is the same as a gain of $10^{-50/20} = 0.00316$. Then equation 6.6 gives

$$Q = 0.00316 Q_c = 0.00316 \times 377 = 1.19$$

We can now find the required resistance from

$$R = Q\omega_0 L = 1.19 \times 2\pi \times 1500 \times 0.08 = 897 \ \Omega$$

And the capacitance is found from

$$C = (\omega_0^2 L)^{-1} = [(2\pi \times 1500)^2 \times 0.08]^{-1} = 0.141 \ \mu F$$

The gain response is symmetrical about $\Omega = 1$ on a log plot and, except when $\Omega \approx 1$, is given by equation 6.4, from which we can find the 3dB points:

$$\frac{1 - \Omega_{3dB}^2}{\sqrt{(1 - \Omega_{3dB}^2)^2 + (\Omega_{3dB}/1.19)^2}} = \frac{1}{\sqrt{2}}$$

The solutions are $\Omega_{3dB} = 0.6645$ and $\Omega_{3dB} = 1/0.6645 = 1.505$, giving the straight-line approximation to the gain response shown in figure 6.8b, in which the gain at $\Omega = 0.1$ (and $\Omega = 10$) is taken as 0 dB. The error in this can be seen to be large when compared to the exact gain response, but a simple approximation such as this could not be expected to do much better.

6.2.3 Second-order, bandpass, RLC filter

The second-order bandpass filter suffers from a poor roll-off rate and a passband which is not very flat. The circuit is shown in figure 6.9a, and its response is

$$\mathbf{H}(j\Omega) = 1 - \mathbf{H}(j\Omega)_{notch} = \frac{j(\Omega/Q)}{(1 - \Omega^2) + j(\Omega/Q)} \tag{6.7}$$

provided $Q_c \gg Q$. This response is 0 dB at $\Omega = 1$ and rolls off at 20 dB/decade above and below this frequency. If $Q \ll 1$ the 3dB points are at $\Omega_{3dB} \approx Q$ and $1/Q$, as shown in the gain plot of figure 6.9b; but if $Q \gg 1$, then the 3dB points are at $\Omega_{3dB} \approx 1 \pm 1/2Q$.

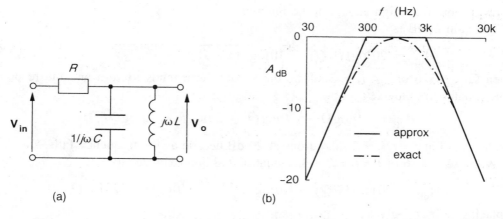

Figure 6.9 (a) A 2nd order, bandpass, RLC filter (b) Its gain response with $f_0 = 949$ Hz and $Q = 0.351$

Example 6.4

Design a second-order RLC bandpass filter with a passband between 300 Hz (f_1) and 3 kHz (f_2) using an inductor with $r = 3\ \Omega$ and $L = 120$ mH. Draw the gain response as a function of frequency from 30 Hz to 30 kHz. What is the attenuation at 50 Hz?

The passband is the part between the points (f_1 and f_2) where the response is 3 dB down from its maximum response. The maximum gain is at the resonant frequency, f_0, given by

$$f_0 = \sqrt{f_1 f_2} = \sqrt{300 \times 3000} = 949 \text{ Hz}$$

The capacitance is

$$C = 1/\omega_0^2 L = [(2\pi \times 949)^2 \times 0.12]^{-1} = 0.234 \ \mu F$$

The 3dB points are given by

$$\frac{\Omega_{3dB}/Q}{\sqrt{(1 - \Omega_{3dB}^2)^2 + (\Omega_{3dB}/Q)^2}} = \frac{1}{\sqrt{2}}$$

where $\Omega_{3dB} = \omega_{c1}/\omega_0 = f_1/f_0 = 300/949 = 0.316$ or $\Omega_{3dB} = f_2/f_0 = 3000/949 = 3.16$. Letting $1/Q = a$ and choosing $\Omega_{3dB} = 0.316$ leads to

$$\frac{0.316a}{\sqrt{(1 - 0.316^2)^2 + (0.316a)^2}} = \frac{1}{\sqrt{2}} \quad \Rightarrow \quad 0.2a^2 = 0.9^2 + 0.1a^2$$

which gives $a = 2.846$ and $Q = 1/a = 0.351$, not much less than unity, though the error in taking it to be so is only about 1 dB, as we shall see. Q determines the value of R since

$$R = \omega_0 L Q = 2\pi \times 949 \times 0.15 \times 0.351 = 314 \ \Omega$$

The coil's Q-factor is

$$Q_c = \omega_0 L/r = 2\pi \times 949 \times 0.15/3 = 298$$

which is more than high enough to be ignored.

The gain in dB is

$$A_{dB} = 20\log(\Omega/Q) - 10\log\sqrt{(1 - \Omega^2)^2 + (\Omega/Q)^2} \qquad (6.8)$$

When $\Omega < 0.316$ or $f < f_1 = 300$ Hz, the second term tends to zero and the response approximates to $20\log(\Omega/Q)$ or

$$A_{dB} = 20\log\Omega - 20\log Q = 20\log f - 20\log Qf_0$$

which is zero at $f = Qf_0 \approx f_1$ then drops at 20 dB/decade as the frequency falls.

When $\Omega > 3.16$ or $f = f_2 = 3$ kHz, equation 6.8 approximates to

$$A_{dB} \approx 20\log(\Omega/Q) - 10\log\sqrt{\Omega^4} = -20\log\Omega - 20\log Q$$

Replacing Ω by f/f_0 gives

$$A_{dB} \approx -20\log f + 20\log f_0 - 20\log Q = -20\log f + 20\log(f_0/Q)$$

which is zero when $f = f_0/Q \approx f_2$ and then falls at 20 dB/decade as the frequency goes up. The gain Bode plot is shown in figure 6.9b.

At 50 Hz the gain will be approximately $20\log(f/f_1) = 20\log(50/300) = -15.6$ dB making the attenuation 15.6 dB. A more accurate approximation would be to take the 0dB point to be f_0Q and then the gain is $20\log(f/f_0Q) = 20\log(50/949 \times 0.351) = -16.5$ dB, which is identical to that given by equation 6.8.

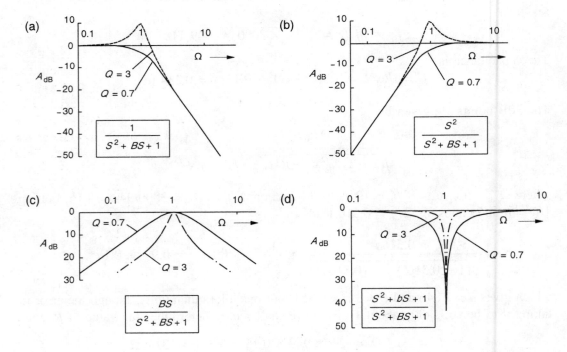

Figure 6.10 The gain responses of second-order RLC filters. They are functions of $S = j\Omega = j\omega/\omega_c$ and $B = 1/Q$. (a) Lowpass (b) highpass (c) bandpass (d) notch or band-reject, where $b = 1/Q_c = 0.01$

The gain responses of second-order RLC filters are summarised in figure 6.10, for two different values of Q. In the case of the lowpass and highpass filters the higher circuit Q-factor leads to pronounced excursions above the 0dB line, and with the bandpass filter the response becomes markedly more peaked. The notch filter responses are both for a coil Q-factor of 100 but the notch width is proportional to $1/Q$.

6.3 Types of filter

Filters can be classified according to their response functions which may be among others: Butterworth, Bessel (or Thomson, or time-delay), Chebyshev (or Tchebycheff), Gaussian and elliptic (or Cauer). Each of these types has characteristics which make it more desirable than the others for a particular application. The principal advantages and disadvantages of each type are given in Table 6.1 below.

Table 6.1 *Characteristics of some types of filter*

Type	Advantages	Disadvantages
Butterworth	Flattest pass band	Medium transition rate, medium distortion
Chebyshev	Fast transition rate	High distortion, high passband ripple
Elliptic	Fastest transition rate	High distortion, high ripple
Bessel	Least distortion	Slow transition rate
Gaussian	Low distortion	Medium transition rate

The table does not tell the whole truth: for example, the Butterworth filter is said to be flattest in the pass band, but it will still have a transition from the 3dB point (where the passband is customarily said to start) into the passband. Distortion arises from the variation in time delay with frequency and for this reason the Bessel filter, which is designed to have constant time delay in the passband, is sometimes called a time-delay filter. The speed of transition from passband to stopband is usually set by the requirements for attenuation at two frequencies, one in the passband and one in the stopband. That requirement may lead to a lower order for a Chebyshev filter or for an elliptic filter, which has stopband as well as passband ripple, and therefore to a less complex design. But in practice Butterworth filters are often chosen because they are less sensitive to component tolerances that inevitably degrade the performance of supposedly superior designs. Speed of transition in the Chebyshev and elliptic filters can be traded for ripple, and it is necessary to specify the ripple when giving performance curves for these.

The RC, Bessel, Butterworth and Chebyshev lowpass filters are compared in figure 6.11, which shows gain responses for fourth-order filters. The Chebyshev filter has 1 dB ripple in the passband, and the transition can be made more (or less) rapid if more (or less) ripple is allowable. The Chebyshev filter's response actually falls faster at first than the asymptotic rate of 80 dB/decade, the Butterworth settles down at about $\Omega = 1.2$ and the Bessel at about $\Omega = 2$, while the RC filter only settles to the asymptotic rate around $\Omega = 30$.

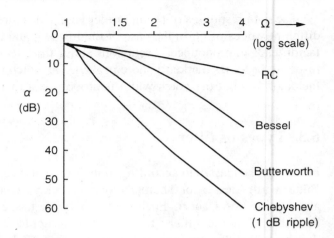

Figure 6.11

Attenuation (α) in the transition region as a function of frequency for various fourth-order, lowpass, filters

6.3.1 *Butterworth filters*

Butterworth filters are among the most popular — especially with students and novices — because they are easiest to design, have flat passbands, and simple response functions. The gain responses for lowpass and highpass Butterworth filters can be written

$$H_{LP} = \frac{1}{\sqrt{1 + \Omega^{2n}}} \quad \text{and} \quad H_{HP} = \frac{\Omega^n}{\sqrt{1 + \Omega^{2n}}} \tag{6.9}$$

where n is the order of the filter, normally an integer from two and upwards. The attenuation goes as $20n$ dB/decade in the stopband, and the 3dB cut-off is at $\Omega = 1$.

As previously stated, an RLC filter of a given order is superior to an RC filter of equal order because the 'knee' at the cut-off is much sharper. This point is illustrated in figure 6.12 where second, third and fourth-order, lowpass, Butterworth filters are compared to two, three and four-section, lowpass, RC filters. It can be seen that though the slopes for equal order are eventually all the same, the Butterworth filters achieve the limiting slope far more quickly — not far past the 3dB cut-off in fact. Thus for a given cut-off and attenuation requirement, more sections are required in an RC filter.

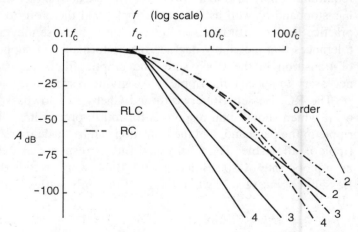

Figure 6.12

Gain responses of lowpass, RLC, Butterworth filters compared to those of RC ladders, all normalised to the 3dB cut-off frequency, f_c

Filters in practice are designed by looking at the attenuations required in the passband and stopband, how much ripple is tolerable in these and the delay caused to the signal in passing through the filter. The ripple requirement generally determines the type of filter required, being greater in Chebyshev and Bessel filters than in Butterworth filters. Consider the Butterworth lowpass filter response of equation 6.9 and suppose that the attenuation required at a frequency, f_p, in the passband is at most α_p dB and that at a frequency, f_s, in the stopband is at least α_s dB. The cut-off frequency is f_c and the attenuation is 1/gain, or $-A_{dB}$, so that

$$-A_{dB} = 10\log(1 + \Omega^{2n}) = 10\log[1 + (f/f_c)^{2n}] = \alpha$$

The stopband requirement means that

$$\alpha_s = 10\log[1 + (f_s/f_c)^{2n}] \quad \Rightarrow \quad (f_s/f_c)^{2n} = 10^{0.1\alpha_s} - 1 \tag{6.10}$$

and the passband requirement leads to

$$(f_p/f_c)^{2n} = 10^{0.1\alpha_p} - 1 \tag{6.11}$$

Dividing equation 6.10 by 6.11 produces

$$(f_s/f_p)^{2n} = \frac{10^{0.1\alpha_s} - 1}{10^{0.1\alpha_p} - 1} \quad \Rightarrow \quad n = \frac{\log\left(\dfrac{10^{0.1\alpha_s} - 1}{10^{0.1\alpha_p} - 1}\right)}{2\log(f_s/f_p)} \tag{6.12}$$

Hence the order required can be found, bearing in mind that n must be an integer. For a highpass filter the same equation can be used for n with a minus sign, since f_s is then lower than f_p.

Example 6.5

What is the order of the Butterworth filter that is to give an attenuation of at most 0.1 dB at 10 kHz and at least 60 dB at 1 kHz? What is the cut-off frequency of this filter? What are the actual attenuations at 1 kHz and 10 kHz? Draw the gain Bode plot.

The order is given by equation 6.12 with $\alpha_s = 60$ dB, $\alpha_p = 0.1$ dB, $f_s = 1$ kHz and $f_p = 10$ kHz. Substitution leads to $n = -3.82$, that is a highpass filter, and since n must be a positive integer $n = 4$ will do. The highpass analogue of equation 6.11 is

$$(f_c/f_p)^{2n} = 10^{0.1\alpha_p} - 1 = 0.02329$$

$$\Rightarrow \quad f_c = (0.02329)^{0.125} f_p = 0.625 \times 10000 = 6250 \text{ Hz}$$

The gain response of a highpass Butterworth filter is given in equation 6.9, and is

$$A_v = H(\Omega) = \frac{\Omega^n}{\sqrt{1 + \Omega^{2n}}}$$

which in dB becomes, for a fourth-order filter,

$$A_{dB} = 20\log\Omega^4 - 20\log\sqrt{1 + \Omega^8} = 80\log\Omega - 10\log(1 + \Omega^8)$$

That is it rises at 80 dB/decade when $\Omega < 1$, or $f < f_c = 6.25$ kHz. Now 1 kHz is $\log 6.25$ = 0.796 of a decade below 6.25 kHz, so the attenuation is $80 \times 0.796 = 63.7$ dB, 3.7 dB better than required. At 10 kHz, $\Omega = 10/6.25 = 1.6$ and the attenuation is

$$\alpha_p = -A_{dB} = -80\log 1.6 + 10\log[1 + (1.6)^8] = 0.1 \text{ dB}$$

as specified. The Bode plot of gain is shown in figure 6.13, in which we see that the transition from zero slope to the limiting slope of 80 dB/decade at low-frequency — the 'knee' — lies between about 5 kHz and 10 kHz or between $\Omega = 0.8$ and $\Omega = 1.6$.

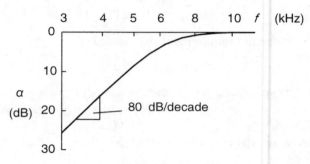

Figure 6.13

The gain-frequency plot for example 6.5

Of course, finding the order of the filter does not specify the components, which have to be calculated from those tabulated in normalised form according to the type of filter. The LC elements of lowpass filters are arranged in ladder networks such as shown in figure 6.14, and the components for a given filter type and a cut-off angular frequency of $\omega_c = 1$ rad/s are found from tables: table 6.2 gives some of these values for lossless[3] Butterworth filters. These values must be scaled according to the actual cut-off frequency and source resistance. If there is finite termination resistance or zero source resistance, then the component values must be found from appropriate tables.

Table 6.2

Normalised component values for lossless, lowpass, Butterworth filters. (Cut-off at 1 rad/s, source resistance = 1 Ω, open circuit output, values in H and F)

order, n	L_1	C_2	L_3	C_4	L_5	C_6	L_7	C_8	
2	0.7071	1.414	—	—	—	—	—	—	
4	0.3827	1.802	1.577	1.531	—	—	—	—	
6	0.2588	0.7579	1.202	1.553	1.759	1.533	—	—	
8	0.1951	0.5776	0.9371	1.259	1.528	1.729	1.824	1.561	

order, n	C_1	L_2	C_3	L_4	C_5	L_6	C_7	L_8	C_9
3	0.500	1.333	1.500	—	—	—	—	—	
5	0.309	0.8944	1.382	1.694	1.545	—	—	—	—
7	0.2225	0.656	1.054	1.397	1.659	1.799	1.588	—	—
9	0.1736	0.5155	0.8414	1.141	1.404	1.62	1.777	1.842	1.563

[3] 'Lossless' means that the reactive components are ideal.

The even-order filters start with a T-section and an inductor, while the odd-order filters start with a π-section and a capacitor as shown in figure 6.14.

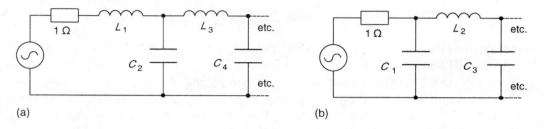

(a)   (b)

Figure 6.14 (a) Even-order filter configuration (b) Odd-order filter configuration

For example, suppose we desire a lowpass filter with a 3dB cut-off at 1.6 kHz and are working with a 50Ω source resistance, then the resistances and reactances have to be scaled up by a factor of 50, which means $R' = kR$, $L' = kL$ and $C' = C/k$ where the primed values are the scaled up values and k (= 50) is the scaling factor. The frequency is 1.6 kHz = 10.053 krad/s, so the reactive elements are scaled *down* by this amount[4]. As a result the conversion factors are

$$R' = kR = 50 \times 1 = 50 \ \Omega$$

$$L' = kL/\omega = 50L/10053 = 4.974 \times 10^{-3}L$$

$$C' = C/k\omega = C/(50 \times 10053) = 1.9895 \times 10^{-6}C$$

Example 6.6

Design a lowpass Butterworth filter which has an attenuation of at most 0.1 dB at 1 kHz and at least 55 dB at 5 kHz and which is connected to a source of internal resistance 50 Ω and has an open circuit termination. The components are lossless. What is the actual attenuation at 5 kHz?

Equation 6.12 gives the order as 5.1, so a sixth-order filter is required.

The attenuation at 5 kHz can be found from equation 6.9, if we know f_c, which must be calculated from equation 6.11 with $n = 6$:

$$f_p/f_c = (0.02329)^{1/12} = 0.731 \quad \Rightarrow \quad f_c = 1000/0.731 = 1368 \ \text{Hz}$$

as $f_p = 1000$ Hz. Then 5 kHz is at $\Omega = 5000/1368 = 3.655$, which is 0.563 of a decade above f_c. A 6-pole filter will attenuate at 120 dB/decade above f_c, giving an attenuation of $0.563 \times 120 = 67.6$ dB. The specification is greatly exceeded because a fifth-order filter would almost suffice.

Table 6.2 gives $L_1 = 0.2588$ H, $C_2 = 0.7579$ F, $L_3 = 1.202$ H, $C_4 = 1.553$ F, $L_5 = 1.759$ H and $C_6 = 1.533$ F. The scale factors for inductances and capacitances are

---

[4] One can see this by considering a circuit with $C = 1$ F and $L = 1$ H which has a resonant frequency of 1 rad/s. If the circuit is to have a resonant frequency of 100 rad/s it is clear that if $L' = kL$ and $C' = kC$ where k is the frequency scaling factor, then $L'C' = k^2LC = k^2 = 1/\omega'^2 = 1/100^2$ and $k = 1/100 = 0.01$.

$$k_L = R/\omega_c = 50/(2\pi \times 1368) = 5.817 \times 10^{-3}$$

$$k_C = 1/R\omega_c = 1/(50 \times 2\pi \times 1368) = 2.327 \times 10^{-6}$$

so these values become $L_1' = 1.505$ mH, $C_2' = 1.764$ μF, $L_3' = 6.992$ mH, $C_4 = 3.614$ μF, $L_5 = 10.23$ mH and $C_6 = 3.567$ μF.

If we require a Chebyshev filter with a specified passband ripple — or any other type of filter — we must look up the component values in appropriate tables, or use other design aids such as nomograms and graphs. There are books full of these. The design of bandpass and band-reject filters is more complicated than the example given, but relies on similar tables of normalised components. High-order bandpass filters often require large inductance values which are hard to make with sufficient Q. The design must then be modified by using transformers to scale down the required inductance.

For non-ideal, lossy components the filter dissipates power and the attenuation in the passband is non-zero. Similar tables to table 6.2 have been compiled for lossy filters, of which there are two main varieties: those with ideal capacitors but lossy inductors of known and constant Q-factor, and those in which all the components are lossy and of equal Q-factor. The latter are known as uniform-distribution networks. The effect of losses in components is not necessarily all bad as it can lead to smoothing of the passband ripple for Chebyshev and elliptic filters. If the Q-factor of the components is reasonably high, say fifty or more, then low-order passive filters can be considered nearly ideal. High-order filters are more sensitive to Q-factor and component tolerances; for that reason — and because inductors are large, relatively expensive, and time-consuming to produce — high-order filters tend to be active. Active filters are based on operational amplifiers which require a separate power source. Line filters are therefore made from passive components.

6.3.2 Chebyshev filters

A Chebyshev lowpass filter of order, n, is characterised by a gain response of the type

$$A_n = \frac{K}{\sqrt{1 + \varepsilon^2 T_n(\Omega)^2}} \tag{6.13}$$

where the parameter, ε, controls the ripple in the passband and T_n is a Chebyshev polynomial of order n. The Chebyshev polynomials satisfy the recursion relation

$$T_{n+1} = 2\Omega T_n - T_{n-1}$$

with $T_0 = 1$ and $T_1 = \Omega$. It is then soon shown from these and the recursion relation that $T_n(1) = 1$ and $T_2 = 2\Omega^2 - 1$, $T_3 = 4\Omega^3 - 3\Omega$ etc. The gain responses of highpass and bandpass Chebyshev filters follow analogous equations to the lowpass response of equation 6.13 and tables of normalised component values exist for passive filters with various values for passband ripple. The nomogram shown in figure 6.15 can be used to find the order of a Chebyshev lowpass (or highpass) filter when the attenuation in the stopband, $\alpha(\Omega)$, is known at the normalised frequency, Ω, and the ripple is known in the passband.

For instance if the passband ripple is to be 2 dB and the attenuation 40 dB at $\Omega = 2.5$, then one draws a line between $A_{PB} = 2$ and $\alpha(\Omega) = 40$ and extrapolates it to the nomogram. Next a horizontal line is drawn from that point on the nomogram, and a vertical line from the horizontal Ω-axis at $\Omega = 2.5$. The point of intersection lies below the line marked 4, so the filter order is 4.

Appropriate tables must then be consulted to determine the component parameters. If we look at the filter specification in example 6.6, we find from figure 6.15 that a fourth-order Chebyshev filter will fill the specification, instead of a sixth-order Butterworth.

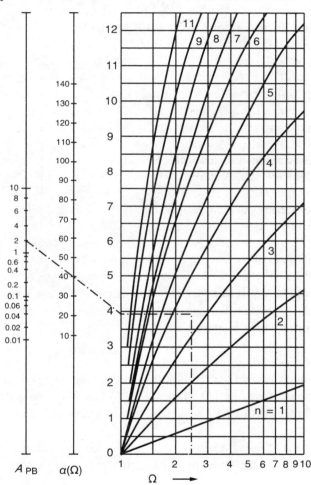

Figure 6.15

Nomogram for determining the order of a Chebyshev lowpass (or highpass) filter. The passband ripple is denoted by A_{PB} and the attenuation at the normalised frequency, Ω, is denoted $\alpha(\Omega)$. The lines labelled $n = 1, 2 \ldots$ are the boundaries between the various filter orders

Example 6.7

Find the value of ε required for a fourth-order, lowpass, Chebyshev filter having a passband between 0 and 1 kHz and a passband ripple of 1 dB. Compare its response to a fourth-order Butterworth filter at 3 kHz and at 10 kHz.

The gain response given by equation 6.13 with $K = 1$ and the relevant polynomial, T_4:

$$A = \frac{1}{\sqrt{1 + \varepsilon^2(8\Omega^4 - 8\Omega^3 + 1)^2}}$$

The passband lies by definition between $\Omega = 0$ and $\Omega = 1$, though this Ω is not quite the same as the one previously used, since the 3dB point is not at $\Omega = 1$. The maximum of A is seen to be 1 when $T_4 = 0$ (that is when $\Omega = 0.383$ and 0.924) and the minimum is $1/\sqrt{(1 + \varepsilon^2)}$ when $\Omega = 0$ and 1. Thus the peak-to-peak ripple in the passband is $10\log(1 + \varepsilon^2)$, which is in fact the passband ripple for Chebyshev filters of any order, since $T_n(1) = 1$.

The ripple in dB is given by

$$10\log(1 + \varepsilon^2) = 1 \text{ dB} \quad \Rightarrow \quad \varepsilon^2 = 0.259$$

and for this filter $\Omega = 1$ when $f = 1$ kHz.

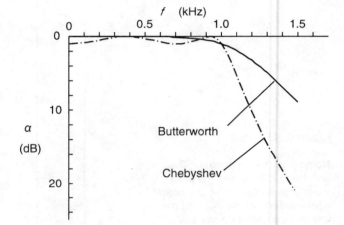

Figure 6.16

Fourth-order amplitude responses of the filters in example 6.7

The equivalent Butterworth filter will have a flat passband, falling by 3 dB at its cut-off, f_c, which we can find by requiring its response to be down by 1 dB at 1 kHz. The Butterworth response is

$$A_{dB} = -10\log(1 + \Omega^8) = -1$$

which gives $\Omega = 0.8446$, that is the response is 1 dB down at $f = 0.8446f_c = 1$ kHz. This means that $f_c = 1/0.8446 = 1.184$ kHz. The Chebyshev Ω is normalised to 1 kHz and the Butterworth to 1.184 kHz. Thus in terms of frequency in Hz the two responses are

$$A_{CH} = \frac{1}{\sqrt{1 + 0.259[8(f/1000)^4 - 8(f/1000)^2 + 1]^2}}; \quad A_{BU} = \frac{1}{\sqrt{1 + (f/1184)^8}}$$

Substituting $f = 3$ kHz into these gives $A_{CH} = 0.00341 \equiv -49.4$ dB, and $A_{BU} = 0.0243 \equiv -32.3$ dB; and $f = 10$ kHz gives $A_{CH} = -92.1$ dB and $A_{BU} = -74.1$ dB. Figure 6.16 shows these responses in and near the passband. The difference in responses is 18 dB in the Chebyshev's favour in the stopband; it can be shown (see problem 6.7) that the difference between Butterworth and Chebyshev filters in the stopband is nearly $6(n - 1)$ dB, where n is the order of the filter, if the filters have the same responses at the limit of the Chebyshev's passband (or ripple channel, as it is sometimes called).

Example 6.8

Show that the circuit of figure 6.17 has a Chebyshev response and find the values of ε, the passband ripple and the passband cut-off frequency.

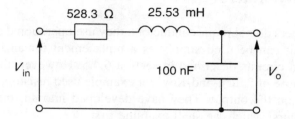

Figure 6.17

The circuit for example 6.8

It is best to work in terms of R, C and L and substitute the actual values later. The response is

$$\mathbf{H}(j\omega) = \frac{1/j\omega C}{j\omega L + R + 1/j\omega C} = \frac{1}{1 - \omega^2 LC + j\omega RC}$$

The magnitude of this is

$$A_v = \frac{1}{\sqrt{(1 - \omega^2 LC)^2 + (\omega RC)^2}} \tag{6.14}$$

But a second-order Chebyshev response is

$$A_v = \frac{K}{\sqrt{1 + [\varepsilon T_2(\Omega)]^2}} = \frac{K}{\sqrt{1 + [\varepsilon(2\Omega^2 - 1)]^2}} \tag{6.15}$$

since $T_2 = 2\Omega^2 - 1$. Comparing responses at $\omega = 0$ we find that they are unity according to equation 6.14 and $K/\sqrt{(1 + \varepsilon^2)}$ according to equation 6.15, so that K must be $\sqrt{(1 + \varepsilon^2)}$ and equation 6.15 can then be rewritten in the form

$$A = \frac{1}{\sqrt{1 + \beta(\Omega^4 - \Omega^2)}} = \frac{1}{\sqrt{1 + \beta[(\omega/\omega_c)^4 - (\omega/\omega_c)^2]}}$$

where $\beta = 4\varepsilon^2/(1 + \varepsilon^2)$ and ω_c is the Chebyshev (ripple band) cut-off, not the 3dB cut-off, nor ω_0. Comparing this to equation 6.14 we find that

$$(LC)^2 = \beta/\omega_c^4 \quad \text{and} \quad 2LC - (RC)^2 = \beta/\omega_c^2$$

But $LC = 2.553 \times 10^{-9}$ and $RC = 5.283 \times 10^{-5}$ so that

$$\beta/\omega_c^4 = 6.518 \times 10^{-18} \quad \text{and} \quad \beta/\omega_c^2 = 2.315 \times 10^{-9}$$

Hence $\omega_c^2 = 2.315 \times 10^{-9} \div 6.518 \times 10^{-18} = 3.552 \times 10^8$, making $\omega_c = 18.85$ krad/s and $f_c = 3$ kHz. We can then find β to be 0.8223 and $\varepsilon^2 = \beta/(4 - \beta) = 0.2588$. The passband ripple is $10\log(1 + \varepsilon^2) = 1$ dB. Provided it is underdamped, that is $Q > 1/\sqrt{2}$, any series RLC circuit like figure 6.17 could be described as a 2nd-order Chebyshev filter with varying degrees of passband ripple.

6.4 Active filters

Active filters do not require inductors: they use operational amplifiers with an RC network. The circuit can be used simply as a replacement for an inductor, as in the inductance-simulation circuits described in section 6.4.5. However, the earliest active filters did not do this; those of Sallen-and-Key[5] for example used resistors and capacitors with a feedback amplifier on the output. They have developed into an important family often known as VCVS filters, which we shall examine first.

6.4.1 Sallen-and-Key or VCVS filters

Voltage-controlled voltage-source filters use op amps, capacitors and resistors to obtain the desired transfer characteristics. The circuit of figure 6.18 is a VCVS filter, containing one op amp with two capacitors in one feedback path. It is therefore likely to be a 2-pole or second-order filter. If we replace the capacitances by short circuits we can see that at high frequencies the output will be zero. At low frequencies we can see by replacing the capacitances by open circuits, that the output voltage will be $(1 + R_4/R_3)V_{\text{in}} = KV_{\text{in}}$, that is the device is a lowpass filter with a gain of $K\ (= 1 + R_4/R_3)$. We shall now analyse the circuit in more detail.

The voltage at the inverting input of the op amp is $V_oR_3/(R_3 + R_4) = V_o/K$ and this is also the voltage at the non-inverting input. At the latter, the current through C_2 is simply $j\omega C_2V_o/K$, and is equal to the current through R_2 which is

$$(V_A - V_+)G_2 = (V_A - V_o/K)G_2 = j\omega C_2V_o/K$$

$$\Rightarrow \qquad V_A = (V_o/K)(1 + j\omega C_2R_2)$$

where $G_1 = 1/R_1$ etc.

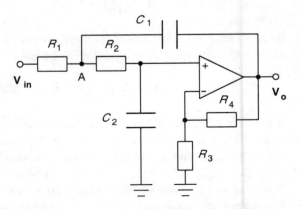

Figure 6.18

A Sallen-and-Key or VCVS, second-order, lowpass filter

---

[5] The historic paper written by Sallen and Key appeared in *IRE Trans. Circuit Theory* **CT2** 74-85 (1955)

Summing currents out of A yields

$$(V_A - V_{in})G_1 + (V - V_o/K)G_2 + (V_A - V_o)j\omega C_1 = 0$$

$$\Rightarrow \qquad V_A(G_1 + G_2 + j\omega C_1) - V_o(G_2/K + j\omega C_1) = V_{in}G_1$$

$$\Rightarrow \qquad (V_o/K)[(1 + j\omega C_2 R_2)(G_1 + G_2 + j\omega C_1) - G_2 - j\omega C_1 K] = V_{in}G_1$$

Hence the transfer function is

$$\mathbf{H}(j\omega) = \frac{V_o}{V_{in}} = \frac{K}{1 - \omega^2 C_1 C_2 R_1 R_2 + j\omega[C_1 R_1(1 - K) + C_2 R_2 + C_2 R_1]} \qquad (6.16)$$

In practice some flexibility in design is usually sacrificed by making $R_1 = R_2 = R$ and $C_1 = C_2 = C$ so that the transfer function above becomes

$$\mathbf{H}(j\omega) = \frac{K}{1 - \omega^2 C^2 R^2 + j\omega(3 - K)CR}$$

which can be written as

$$\mathbf{H}(j\Omega) = \frac{K}{(1 - \Omega^2) + j\Omega/Q} \qquad (6.17)$$

where $\Omega = \omega CR = \omega/\omega_c$ and $Q = 1/(3 - K)$. The filter's properties depend mainly on Q, which must be less than about five in practice. Once the Q-factor is fixed, then so is the gain. It is possible to vary both K and Q only by having $R_1 \neq R_2$ as shown in the next example.

Example 6.9

Design a VCVS, lowpass filter with a Butterworth response, a gain of 10 dB in the passband, and a cut-off frequency of 3 kHz.

The gain is $10^{10/20} = 3.162 = K = 1 + R_4/R_3$. We could choose $R_4 = 43$ kΩ and $R_3 = 20$ kΩ as the nearest convenient preferred resistance values, which gives $K = 3.15$ and a gain of 9.97 dB. A Butterworth response means that Q must be $1/\sqrt{2}$. The Q can be found from the full response, equation 6.16, to be given by

$$1/Q = (2 - K)\sqrt{R_1/R_2} + \sqrt{R_2/R_1} \qquad (6.18)$$

It is preferable to find which values are available of the type of capacitor required, since capacitance is less easy to adjust, the values available are more widely spaced and the tolerance is much wider than for resistance. Suppose we choose low-voltage, polyester capacitors with 5% tolerance, costing around 10p apiece, with a working DC voltage of 100 V and a capacitance nominally of 10 nF. We now find the resistances R_1 and R_2. Equation 6.18 gives

$$-1.15\sqrt{R_1/R_2} + \sqrt{R_2/R_1} = \sqrt{2}$$

Replacing $\sqrt{(R_1/R_2)}$ by x and rearranging leads to

$$1.15x^2 + x\sqrt{2} - 1 = 0$$

whence $x = 0.5021$ and $R_1 = 0.252R_2$.

The cut-off is at $\Omega = 1$, so that $\omega_c = 2\pi \times 3 = 18.85$ krad/s and then the resistances are found from

$$R_1R_2 = 0.252R_2^2 = 1/\omega_c^2 C^2 = \frac{1}{(18850 \times 10^{-8})^2}$$

which gives $R_2 = 10.6$ kΩ and $R_1 = 2.66$ kΩ. If the actual capacitance is 5% out the cut-off frequency will be 5% out too.

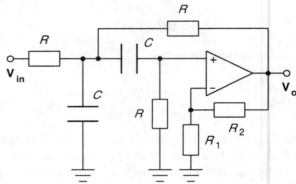

Figure 6.19

A VCVS, second-order, band-pass filter

Higher-order filters can be achieved by cascading more stages, the order being double the number of stages, and adjusting the values of R to give the response that is wanted. Highpass filters can be designed by exchanging resistances and capacitances in the lowpass filter. A second-order VCVS bandpass filter is shown in figure 6.19 in which only the gain resistances, R_1 and R_2, are different from the other three. It can be shown (see problem 6.8), that the transfer function for this circuit is

$$\mathbf{H}(j\Omega) = \frac{jK\Omega}{(1 - \Omega^2) + j\Omega/Q}$$

where $\Omega = \omega/\omega_0$, $\omega_0 = \sqrt{2}/CR$ is the centre frequency, $K = (1 + K_1)/\sqrt{2}$, $K_1 = 1 + R_2/R_1$, and $Q = \sqrt{2}/(3 - K_1)$. The bandwidth of the filter is $B = \omega_0/Q$ and becomes very sensitive to small changes in R_1 and R_2 when $Q > 5$ when it is better to use a so-called biquad filter.

Cascaded, higher-order, VCVS filters are soon designed from standard tables such as table 6.3, which is for lowpass filters. Each stage of these filters has equal values for its capacitances and resistances, apart from those controlling K. The cut-off frequency is f_c and the values for R and C for each stage are found from

$$R_m C_m = 1/2\pi f_c F_m$$

where m is the number of the stage. For Chebyshev filters f_c is the cut-off for the ripple band (that is the α-dB point for an α-dB Chebyshev) and not the 3 dB point which it is for the Butterworth. The table is easy to use (see problem 6.9). Highpass filters can be constructed from the table using the circuit of figure 6.18 with C and R interchanged. For Chebyshev filters the highpass F_n values are the reciprocal of the lowpass ones in the table.

Table 6.3 *Design parameters for Butterworth and Chebyshev VCVS filters*

		Butterworth	$\tfrac{1}{2}$-dB Chebyshev		1-dB Chebyshev		2-dB Chebyshev	
n	m	K_m	K_m	F_m	K_m	F_m	K_m	F_m
2	1	1.586	1.842	1.231	1.955	1.050	2.114	0.907
4	1	1.152	1.582	0.597	1.725	0.529	1.924	0.471
	2	2.235	2.660	1.031	2.719	0.993	2.782	0.964
6	1	1.068	1.537	0.396	1.686	0.353	1.891	0.316
	2	1.586	2.448	0.768	2.545	0.747	2.648	0.730
	3	2.483	2.847	1.011	2.875	0.995	2.904	0.983
8	1	1.038	1.522	0.297	1.672	0.265	0.238	1.879
	2	1.337	2.379	0.599	2.489	0.584	0.572	2.605
	3	1.889	2.712	0.861	2.766	0.851	0.842	2.821
	4	2.610	2.913	1.006	2.930	0.997	0.990	2.946

n is the order of the filter, m is the stage number.

6.4.2 Tow-Thomas filters

Tow-Thomas filters require more op amps than the VCVS filters described above, but they are much less sensitive to component tolerances and are frequently used for high-order filters for this reason. Figure 6.20 shows a Tow-Thomas filter. It employs capacitive feedback with its first two stages followed by an inverting third stage, and we assume it therefore to be a second-order, lowpass filter. In the version shown in figure 6.20, the capacitances are of equal value, as are three of the resistors. Analysis is straightforward when it is recognised that the uppermost op amp (OA3) feeds back $-V_o$ into OA1, where it is amplified by a factor $-Z_2 G_3$ and added to V_{in} multiplied by $-Z_2 G_1$, where $G_1 = 1/R_1$ etc. and $Z_2 = 1/(j\omega C + G_2)$. The output of OA1 is therefore

$$\mathbf{V}_1 = -\mathbf{Z}_2 \mathbf{V}_{in} G_1 + -\mathbf{Z}_2(-\mathbf{V}_o G_3)$$

which is fed into OA2 and multiplied by $-1/j\omega CR$ to give $\mathbf{V}_o$:

$$\mathbf{V_o} = \frac{-\mathbf{V_1}}{j\omega CR} = \frac{\mathbf{Z_2}\mathbf{V_{in}}G_1}{j\omega CR} - \frac{\mathbf{Z_2}\mathbf{V_o}G_3}{j\omega CR}$$

$$\Rightarrow \quad \mathbf{V_o}/\mathbf{V_{in}} = \frac{\mathbf{Z_2}G_1}{j\omega CR + \mathbf{Z_2}G_3} = \frac{G_1}{j\omega CR(j\omega C + G_2) + G_3}$$

This can be rearranged into the form

$$\mathbf{V_o}/\mathbf{V_{in}} = \frac{R_3/R_1}{(1 - \omega^2 C^2 RR_3) + j\omega RR_3/R_2} = \frac{R_3/R_1}{(1 - \Omega^2) + j\Omega/Q}$$

in which $\Omega = \omega/\omega_c$, $\omega_c = 1/C\sqrt{(RR_3)}$ and $Q = R_2/\sqrt{(RR_3)}$. The use of these equations can be illustrated with another example.

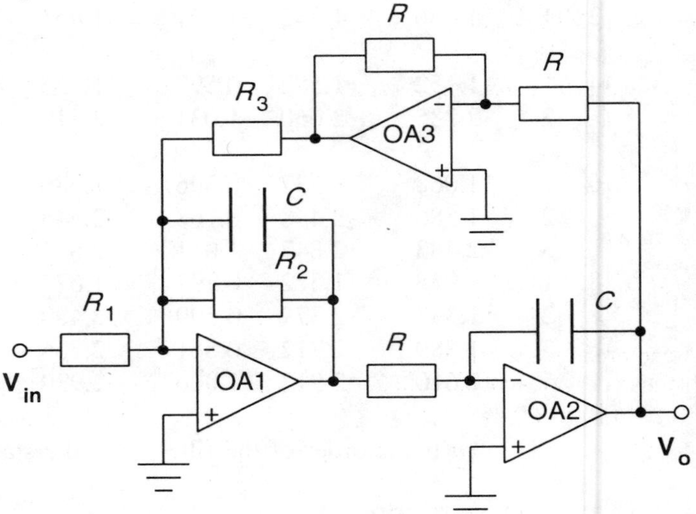

Figure 6.20

A second-order Tow-Thomas filter, with lowpass output from OA2

Example 6.10

Design a second-order, lowpass, Butterworth, Tow-Thomas filter with a cut-off at 5 kHz and a passband gain of 20 dB using 1nF capacitors.

A Butterworth filter requires $Q = 1/\sqrt{2}$ and the cut-off is at $\Omega = 1$, or $\omega = \omega_c$. Now Q is given by

$$Q = R_2/\sqrt{RR_3} = \omega_c CR_2 = 1/\sqrt{2}$$

$$\Rightarrow \quad R_2 = (\omega_c C\sqrt{2})^{-1} = (2\pi \times 5000 \times 10^{-9} \times \sqrt{2})^{-1} = 22.5 \text{ k}\Omega$$

And then we require that

$$\sqrt{RR_3} = (\omega_c C)^{-1} = (10000\pi \times 10^{-9})^{-1} = 31.83 \text{ k}\Omega$$

The passband gain is 20 dB, so that $R_3/R_1 = 10$. These conditions mean that we have the freedom to choose any value for R, so we can conveniently make it 10 kΩ and then $R_3 = 31.83^2/10 = 101$ kΩ, while $R_1 = R_3/10 = 10.1$ kΩ. The Tow-Thomas filter is easier to design, has more flexibility, is less susceptible to component tolerances, and allows more choice than a Sallen-and-Key filter. It has also led to the next concept, the state-variable filter.

6.4.3 *The state-variable or KHN filter*

The name derives from the use of state-variables in solving the differential equations that are used to produce the filtering functions. The alternative name KHN comes from the its originators[6]. If we examine the biquad circuit in figure 6.20 we could deduce that the output from OA1 is the same as a second-order bandpass filter, and with slight modification the circuit can be made to give outputs from its three op amps that are lowpass, highpass and bandpass: the circuit is a universal filter in that sense. The circuit of figure 6.21 shows a KHN circuit with three op amps that gives three types of filtering by means of feedback from two integrators: it is virtually an analogue computer.

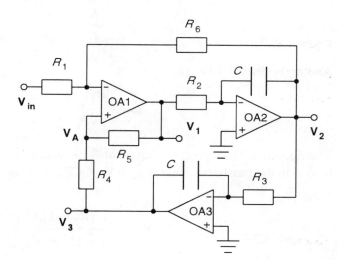

Figure 6.21

A 2nd-order, state-variable filter with highpass (V_1), bandpass (V_2) and lowpass (V_3) outputs

The voltage, V_A, at the input terminals of OA1 is found in terms of V_1 and V_3 by nodal analysis at the non-inverting input:

$$V_A = V_3(1 - 1/K_1) + V_1/K_1$$

where $K_1 = 1 + R_5/R_4$. And V_A is found in terms of V_2 by similar means at the inverting input

$$V_A = V_{in}(1 - 1/K_2) + V_2/K_2$$

---

[6] Kerwin, Huelsman and Newcomb in *IEEE J. Solid-state Circuits* **SC2** 87-92 (1967).

where $K_2 = 1 + R_6/R_1$. Hence

$$\mathbf{V_{in}} = \frac{\mathbf{V_1}K_2}{K_1(K_2 - 1)} - \frac{\mathbf{V_2}}{K_2 - 1} - \frac{\mathbf{V_3}K_2(K_1 - 1)}{K_1(K_2 - 1)}$$

But $\mathbf{V_2} = -\mathbf{V_1}/j\omega CR_2$ and $\mathbf{V_3} = -\mathbf{V_2}/j\omega CR_3 = -\mathbf{V_1}/\omega^2C^2R_2R_3$, so that

$$\mathbf{V_{in}} = \mathbf{V_1}\left[\frac{K_2}{K_1(K_2 - 1)} + \frac{1}{j\omega CR_2(K_2 - 1)} - \frac{K_2(K_1 - 1)}{\omega^2C^2R_2R_3K_1(K_2 - 1)}\right]$$

which can be written as

$$\mathbf{V_1}/\mathbf{V_{in}} = \frac{K\Omega^2}{1 - \Omega^2 + j\Omega/Q} \tag{6.19}$$

where $K = -K_1(1 - 1/K_2)$, $\Omega = \omega/\omega_c$ and Q is given by

$$Q = \frac{K_2}{K_1}\sqrt{\frac{R_2R_5}{R_3R_4}} \tag{6.20}$$

And ω_c is given by

$$\omega_c = \frac{1}{C}\sqrt{\frac{K_1 - 1}{R_2R_3}} \tag{6.21}$$

The output of OA1 given in equation 6.19 is that of a second-order highpass filter with gain K in the passband. Since $\mathbf{V_2} = -\mathbf{V_1}/j\omega CR_2$, we can soon find that the output from OA2 is

$$\mathbf{V_2}/\mathbf{V_{in}} = \frac{-(K_2 - 1)j\Omega/Q}{1 - \Omega^2 + j\Omega/Q} \tag{6.22}$$

which is the transfer function of a second-order bandpass filter of gain $(1 - K_2)$ in the centre of the passband. The output from OA3 is $\mathbf{V_3} = -\mathbf{V_2}/j\omega CR_2$ and thus

$$\mathbf{V_3}/\mathbf{V_{in}} = \frac{K_3}{1 - \Omega^2 + j\Omega/Q} \tag{6.23}$$

where $K_3 = (1 - 1/K_2)/(1 - 1/K_1)$. This is the transfer function of a second-order lowpass filter with a gain of K_3 in the passband.

State-variable filters can be obtained as ICs such as the Burr Brown UAF42AP, which is a second-order filter in 16-pin DIL form costing about £12 in single quantities. The 20-pin DIL MAX275ACPP made by Maxim, which can be configured as a fourth-order filter, costs about £7 and the 8th order MAX274ACNG costs £9. By suitably choosing external resistance values Butterworth, Chebyshev and Bessel filters can be made. The cut-off frequency range is typically from 100 Hz to 150 kHz and can be achieved within about ±1%. These are often called universal filters because the outputs can be obtained in the non-inverted forms and with variable gains, but they are only slightly modified versions of the state-variable filter of figure 6.21.

Example 6.11

Design a state-variable filter with a Butterworth response and a gain of 16 dB in both bandpass and lowpass passbands. The cut-off frequency is to be 2 kHz. Plot the responses at each of the outputs.

Equations 6.20 and 6.21 are used for Q and ω_c, the latter being 4π krad/s and the former must be $1/\sqrt{2}$ for a Butterworth response. Considering ω_c first, then equation 6.21 gives

$$\omega_c = \frac{1}{C}\sqrt{\frac{K_1 - 1}{R_2 R_3}} = \frac{1}{C}\sqrt{\frac{R_5}{R_2 R_3 R_4}} = 4\pi \times 10^3 \text{ rad/s}$$

Now we look for values of resistance in the order of 10 kΩ, so that to satisfy the above equation we need

$$C \approx \frac{\sqrt{10^4/(10^4)^3}}{4\pi \times 10^3} \approx 8 \text{ nF}$$

We choose $C = 10$ nF and use this in our subsequent calculations. The value of K_2 will determine the mid-band gain of the bandpass filter, and that depends only on the ratio of R_6 to R_1; we have therefore a degree of freedom. The mid-band gain is 16 dB, or a factor of 6.31, so that

$$-(K_2 - 1) = -6.31 \quad \Rightarrow \quad K_2 = 7.31$$

Thus $R_6/R_1 = K_2 - 1 = 6.31$, and it is best now to find K_1 from equation 6.23 which gives the lowpass gain

$$K_3 = \frac{1 - 1/K_2}{1 - 1/K_1} = \frac{1 - 1/7.31}{1 - 1/K_1} = 6.31$$

whence $K_1 = 1.1585$. In order to achieve the right gain, K_1 has to be close to this value, which means that R_5 and R_4 must be carefully adjusted to give $R_5/R_4 = 0.1585$.

From the equation for ω_c and taking $C = 10$ nF we find that

$$\sqrt{\frac{R_5}{R_2 R_3 R_4}} = \sqrt{\frac{0.1585}{R_2 R_3}} = \omega_c C = 4\pi \times 10^3 \times 10 \times 10^{-9}$$

whence $R_2 R_3 = 1 \times 10^7 \ \Omega^2$.

The equation for Q is

$$Q = \frac{1}{\sqrt{2}} = \frac{K_2}{K_1}\sqrt{\frac{R_2 R_5}{R_3 R_4}} = \frac{7.31}{1.1585}\sqrt{\frac{0.1585 R_2}{R_3}}$$

hence $R_2/R_3 = 0.07923$ and therefore $R_2 = 890\ \Omega$ and $R_3 = 11.23\ \text{k}\Omega$. If these resistances are considered to be too low, we can increase them by reducing C to 2.2 nF, which leads to $R_2 = 4.05\ \text{k}\Omega$ and $R_3 = 51\ \text{k}\Omega$. We can arbitrarily choose $R_1 = 10\ \text{k}\Omega$ and $R_4 = 33\ \text{k}\Omega$, and then $R_6 = 63.1\ \text{k}\Omega$ and $R_5 = 5.23\ \text{k}\Omega$ from the required values for K_1 and K_2.

The passband gain for the highpass filter is

$$K = -K_1(1 - 1/K_2) = -1.1585(1 - 1/7.31) = -1$$

The output is inverted and the passband gain is 0 dB. Note that we can only specify the gains of two of the outputs and the third is then fixed by that choice; for this reason, and because the highpass and lowpass outputs are inverted, the filter is not strictly a universal filter. Figure 6.22 shows the three responses on a log-log plot.

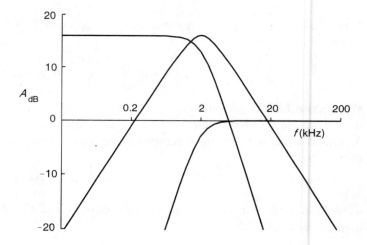

Figure 6.22

Responses of the filter of example 6.11

6.4.4 The biquad filter

The name arises because of the biquadratic form of the transfer function which for a second-order section can be written

$$H(s) = \frac{s^2 + (\omega_z/Q_z)s + \omega_z^2}{s^2 + (\omega_p/Q_p)s + \omega_p^2} \tag{6.24}$$

This type of response is necessary for elliptic filters, but it can be used for other types as well. One form of the biquad is shown in figure 6.23, which is a modified Tow–Thomas filter. The circuit analysis is straightforward but tedious and leads to the transfer function

$$\mathbf{V_o}/\mathbf{V_{in}} = H(s) = -\frac{s^2 + s(1/CR_6 - R_4/CR_3R_5) + R_4/C^2R_1R_2R_3}{s^2 + s/CR_7 + R_4/C^2R_2R_3R_8} \qquad (6.25)$$

using s instead of $j\omega$ and taking all the capacitances to be the same, which is usual practice. Comparing the equations 6.24 and 6.25 we see that $\omega_z^2 = R_4/C^2R_1R_2R_3$, $\omega_z/Q_z = (1/CR_6 - R_4/CR_3R_5)$, $\omega_p^2 = R_4/C^2R_2R_3R_8$ and $\omega_p/Q_p = 1/CR_7$. The biquadratic nature of the output means that any type of second-order filter can be obtained by suitable choice of components.

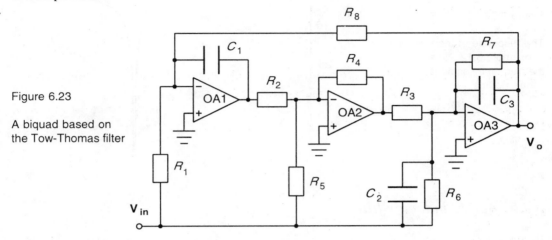

Figure 6.23

A biquad based on the Tow-Thomas filter

Example 6.12

A biquad filter as in figure 6.23 is required to realise the transfer function

$$H(s) = -\frac{s^2 + 1.21 \times 10^8}{s^2 + 3 \times 10^3 s + 10^8} \qquad (6.26)$$

Choose suitable component values using equation 6.25, then draw a graph of the response as a function of frequency.

Comparing equations 6.26 and 6.25, we see that the term in s is missing in the numerator, that is the coefficient of s is zero, and so $1/R_6 = R_4/R_3R_5$. It is convenient to make $R_1 = R_2 = R_3 = R_4 = R$, and $R_5 = R_6 = \infty$ (these last two are omitted). The $\omega_z^2 = R_4/C^2 R_1R_2R_3 = 1/(CR)^2 = 1.21 \times 10^8$. Choosing $C = 10$ nF leads to $R = 9.09$ kΩ, a reasonable enough value. Only R_7 and R_8 remain to be found and these are determined by the requirements that $\omega_p^2 = 1/C^2 RR_8 = 10^8$, making $R_8 = 11$ kΩ and $\omega_p/Q_p = 3 \times 10^3 = 1/CR_7$, which leads to $R_7 = 33.3$ kΩ. This biquad design allows very considerable freedom of choice and is therefore a popular choice.

In terms of $\Omega = \omega/\omega_p$ equation 6.26 becomes

$$H(j\Omega) = -\frac{-\Omega^2 + 1.21}{(1 - \Omega^2) + j0.3\Omega} \quad \Rightarrow \quad |H(j\Omega)| = \frac{\Omega^2 - 1.21}{\sqrt{\Omega^4 - 1.91\Omega^2 + 1}}$$

The graph of this function is shown in figure 6.24 which shows that it is a notch-filter response with centre frequency, $\Omega = 1.1$ or $\omega_z = 11$ krad/s, that is $f_z = \omega_z/2\pi = 11\,000/2\pi = 1.75$ kHz and the -3dB points are at about $\Omega = 1$ and $\Omega = 1.4$, so that $f_1 = 1.592$ kHz and $f_2 = 2.228$ kHz. The filter is an elliptic type.

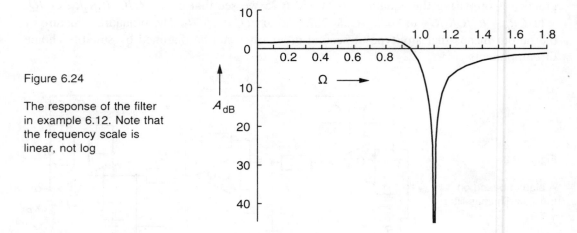

Figure 6.24

The response of the filter in example 6.12. Note that the frequency scale is linear, not log

6.4.5 Active-inductor filters

A different approach to active filter design is to use an inductance simulator, or active inductor, instead of the real article. Then the normalised inductance values from tables can be used to calculate the value of the inductance required and hence the component values in the simulation circuit. There are several different ways of simulating inductance all of which use op amps, capacitors and resistors (or simulated resistances — see the following section). Some of these circuits are shown in chapter five, for example in figures P5.5 and P5.6, which are respectively a negative impedance converter and a gyrator. A third example, shown in figure 6.25, is the Antonionou inductance simulator.

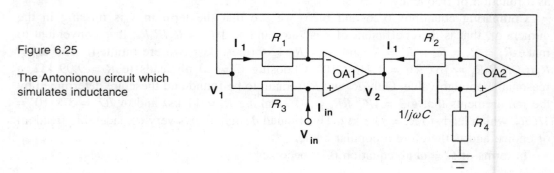

Figure 6.25

The Antonionou circuit which simulates inductance

The Antonionou circuit is analysed by first noting that all four inputs to the two op amps are at the same potential, $\mathbf{V_{in}}$, and that the current through R_1 and R_2 is the same, $\mathbf{I_1}$, and is given by

$$\mathbf{I}_1 = (\mathbf{V}_1 - \mathbf{V}_{in})G_1 = (\mathbf{V}_{in} - \mathbf{V}_2)G_2$$

leading to

$$\mathbf{V}_1 = \mathbf{V}_{in}(1 + G_2/G_1) - \mathbf{V}_2 G_2/G_1 \tag{6.27}$$

where $G_1 = 1/R_1$ etc. At the non-inverting input to OA2 we see that

$$(\mathbf{V}_2 - \mathbf{V}_{in})j\omega C = \mathbf{V}_{in}G_4 \quad \Rightarrow \quad \mathbf{V}_2 = \mathbf{V}_{in}(1 + G_4/j\omega C)$$

which is substituted into equation 6.27 to give

$$\mathbf{V}_1 = \mathbf{V}_{in}(1 - G_2 G_4/j\omega C G_1) \tag{6.28}$$

And at the non-inverting input to OA1 we find

$$\mathbf{I}_{in} = (\mathbf{V}_{in} - \mathbf{V}_1)G_3$$

Substituting $\mathbf{V}_1$ into this from equation 6.28 gives the final result

$$\mathbf{I}_{in} = \mathbf{V}_{in}G_2 G_3 G_4/j\omega C G_1 \quad \Rightarrow \quad \mathbf{Z}_{in} = \mathbf{V}_{in}/\mathbf{I}_{in} = j\omega C R_2 R_3 R_4/R_1$$

The equivalent inductance of the circuit is $L = C R_2 R_3 R_4/R_1 = C R^2$ if all the resistances are made equal, so if $R = 100$ kΩ and $C = 10$ nF, $L = 100$ H. Example 6.13 illustrates an application of this formula to switched-capacitor filters.

6.4.6 *Switched-capacitor and g_m-C filters*

There is no need to use actual resistances in making a filter: instead the resistance can be effectively simulated by a number of techniques among which are switched capacitors, MOSFETs operating in the VCR mode (see chapter four), and transconductor-capacitor (g_m-C) filters. Simulated-resistance filters are a comparatively new development in fabricating complete filters as integrated circuits.

The oldest method uses a charging capacitor as a pseudo-resistance (a technique invented by James Clerk Maxwell in the mid-19th century). Since capacitors are much more easily made and accommodated on IC chips, the advantages are clear: only transistors and capacitors need fabricating and the saving in chip area is considerable. Also close control of capacitance values is not required as it is the capacitance ratios that matter. Now the capacitance ratios are determined by geometry, and that can be adjusted within fine limits, so the IC process is well suited to producing these devices. The major problem lies with clocking speeds and the limited range of capacitance ratios (which must be less than about 100 to 1). Clocking speeds must be much higher than the highest frequency of the output waveform, which limits the latter to about 1 MHz. In contrast, g_m-C filters have been made which operate at 100 MHz and above and they are less affected by constraints on capacitance ratios.

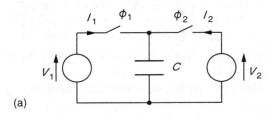

Figure 6.26

(a) A switched-capacitor circuit
(b) Its equivalent circuit
(c) Clock wave-forms for (a)

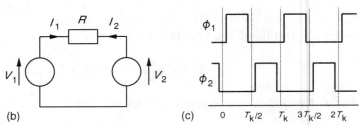

We can see how the switched-capacitor filters work by considering the circuits in figure 6.26, which we shall show to be equivalent. The voltage across the capacitor in figure 6.26a is controlled by two switches, ϕ_1 and ϕ_2, which are closed (ON) and opened (OFF) alternately as shown in figure 6.26c with a clock frequency of $f_k = 1/T_k$. The ON times are arranged not to overlap. Let us assume that the capacitor is initially charged to potential V_2, and then when ϕ_1 is closed the potential will change to V_1 instantaneously if the switch is ideal and has no resistance. The average current flowing through ϕ_1 during one clock cycle is given by

$$I_1(\text{av}) = \frac{1}{T_k} \int_{\frac{1}{2}T_k}^{1\frac{1}{2}T_k} I_1 \, dt = \frac{\Delta Q}{T_k} = \frac{Q_1 - Q_2}{T_k} = \frac{C(V_1 - V_2)}{T_k}$$

And the average current flowing through ϕ_2 is $I_2(\text{av}) = -I_1(\text{av})$. Looking at figure 6.26b we see that the average currents are given by

$$I_1(\text{av}) = -I_2(\text{av}) = \frac{V_1 - V_2}{R}$$

though these are actually constant. If the currents are to be the same in figures 6.26a and 6.26b, then we require that

$$\frac{C(V_1 - V_2)}{T_k} = \frac{V_1 - V_2}{R}$$

that is

$$R = \frac{T_k}{C} = \frac{1}{Cf_k}$$

This is the fundamental equation for switched-capacitor filters. However, the voltages, V_1 and V_2, will normally not be constant but time-varying. If we consider them to be sinusoidal of frequency, f, we require that $f \ll f_k$ for them to approximate to constant voltages, which is an important limitation for switched-capacitor filters. The closeness of the switched-capacitor approximation to resistive behaviour depends on the circuit configuration. Taking a simple RC circuit, as in figure 6.27b, and its switched-capacitor equivalent as in figure 6.27a, with a time constant

$$\tau = R_1 C_2 = C_2/C_1 f_k = c/f_k$$

then it can be shown[7] that the magnitude of the frequency response is given by

$$H(j\omega) = \frac{1}{\sqrt{1 + 2c(1 + c)[1 - \cos(2\pi\omega/\omega_k)]}}$$

where $c = C_2/C_1$. The response of the conventional RC circuit of figure 6.27b is

$$H(j\omega) = \frac{1}{\sqrt{1 + (\omega\tau)^2}} = \frac{1}{\sqrt{1 + (\omega c/f_k)^2}} = \frac{1}{\sqrt{1 + (2\pi c\omega/\omega_k)^2}}$$

since $\tau = R_1 C_2 = c/f_k$. These responses, normalised to the clock frequency, are plotted in the lower part of figure 6.27 for $c = 1$ and we can see that the deviation of the switched-capacitor circuit from the desired response is significant when $\omega/\omega_k > 0.03$.

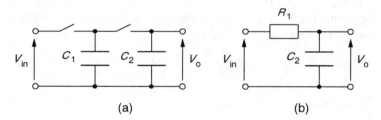

Figure 6.27

(a) A switched-capacitor simulation of (b) an RC circuit. C_1 is the equivalent of R_1 in (b). The circuits' amplitude responses are shown below as a function of $\log(\omega/\omega_k)$ with $c = 1$, which is the same as making $R_1 C_2 = 2\pi/\omega_k$

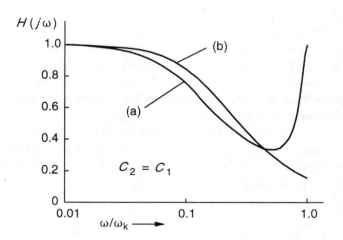

[7] See, for example, *VLSI design techniques for analog and digital circuits* by Geiger, Allen and Strader (McGraw-Hill) 1990

In addition to the requirement of high clock rates for accurate simulation, switched-capacitor filters also suffer from clock-frequency breakthrough on the output, especially at high frequencies when the effect of parasitic capacitances becomes more appreciable.

Example 6.13

A switched capacitor IC requires an Antonionou circuit to simulate an inductances using a clock frequency, f_k, of 3 MHz. If the largest capacitor that can be accommodated is 20 pF and the largest capacitance ratio is 25:1, what range of inductance values can be simulated?

The smallest capacitance is 20/25 = 0.8 pF, since the largest is 20 pF and the greatest capacitance ratio is 25. The inductance is given by

$$L = \frac{CR_2R_3R_4}{R_1} = \frac{CC_1f_k}{C_2C_3C_4f_k^3} = \frac{CC_1}{C_2C_3C_4f_k^2}$$

where we have used the switched-capacitor resistance formula, $R_1 = 1/C_1f_k$, to replace the resistances. Making CC_1 as large as possible and $C_2C_3C_4$ as small as possible makes L a maximum:

$$L_{max} = \frac{(20 \times 10^{-12})^2}{(0.8 \times 10^{-12})^3(3 \times 10^6)^2} = 86.8 \text{ H}$$

Then making CC_1 as small as possible and $C_2C_3C_4$ as large as possible gives the smallest value of L

$$L_{min} = \frac{(0.8 \times 10^{-12})^2}{(20 \times 10^{-12})^3(3 \times 10^6)^2} = 8.9 \text{ } \mu\text{H}$$

In terms of capacitance ratio, $c = C/C_2 = C_1/C_3$ etc., the inductance is

$$L = \frac{c^2}{f_k^2 C_4}$$

The important thing to notice is that the capacitance ratio can be inverted and C_4 can also have a range of c, giving a range of inductance values of c^5:1, in this case $25^5 = 9.8 \times 10^6$. The Antonionou circuit enables designers to make any filter required and to do so with considerable accuracy in achieving the desired parameters, since the capacitance ratio, c, can be very precise.

Gm-C filters have fewer limitations than switched-capacitor filter because there is no switching, just the replacement of a resistor by a CMOS or bipolar circuit with transconductance, g_m, as in figure 6.28a. The output current of the transconductor circuit is directly proportional to its input voltage:

$$i_o = g_m v_{in}$$

and then the voltage across the capacitor on the output side is

$$v_o = \frac{1}{C} \int i_o \, dt = \frac{g_m}{C} \int v_{in} \, dt$$

and so the circuit in figure 6.28a is equivalent to the integrator of figure 6.28b (apart from a change of sign) with $g_m = 1/R$. In this way all of the active filter circuits can be replaced by gm-C equivalents. Gm-C filters have been made to work at up to 100 MHz.

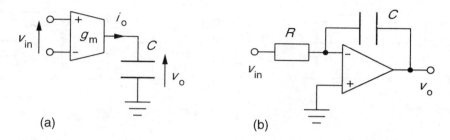

Figure 6.28 (a) Transconductor-capacitor integrator (b) its op amp equivalent circuit

Monolithic switched-capacitor filters include the LTC1063CN8 by Linear Technology, which offers a fifth-order, Butterworth, lowpass response with a cut-off determined by the external clock frequency. It costs about £4 and has a maximum cut-off frequency at about 50 kHz. National Semiconductor's LMF100CIN is a 20-pin, fourth-order, state-variable filter which has a cut-off from 0.1 Hz to 100 kHz and can be configured as a Butterworth, Chebyshev, Bessel or elliptic filter. It has external clock connections and costs £6.

Suggestions for further reading

Active and passive analog filter design by L P Huelsman (McGraw-Hill, 1993)
Electronic filter design handbook: LC, active and digital filters by A B Williams (McGraw-Hill, 2nd ed., 1988)
Designing and building electronic filters by D T Horn (Tab Books, 1992)
Filter design by S Winder (Newnes, 1997)

Problems

1 Show that the normalised voltage transfer function of the circuit of figure P6.1 is

$$\mathbf{H}(j\Omega) = j\Omega/(1 - \Omega^2 + j2.01\Omega)$$

where $\Omega = \omega/\omega_c$ and $\omega_c = 10$ krad/s. At what frequencies is the response 3 dB down from the maximum? *[657 Hz and 3.856 kHz]*

2 Show that the voltage transfer function of the circuit of figure P6.2 is given by equation 6.3, with $\Omega = \omega/\omega_c$, $\omega_c = \sqrt{(R_1 R_2 C_1 C_2)}$ and $Q = \omega_c/(C_1 R_1 + C_2 R_2)$. Show that the maximum value of Q is 0.5 and that the response's magnitude is then

$$|\mathbf{H}(j\Omega)| = 1/(1 + \Omega^2)$$

Compare this response to that of the second-order Butterworth filter which has the same 3dB frequency, f_{3dB}. What is the additional attenuation in dB of the Butterworth at $f = 10 f_{3dB}$? *[7.45 dB]*

3 Show that the voltage response of the circuit of figure 6.8a is given by equation 6.5 if the inductor has series resistance, r, and a Q-factor, $Q_c = \omega_0 L/r$ where $\omega_0 = 1/\sqrt{(LC)}$.

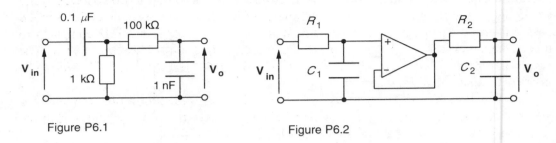

Figure P6.1   Figure P6.2

4 Show that if a notch filter's response is given by equation 6.4, then the 3dB bandwidth is f_0/Q for any value of Q.
5 The circuit of figure P6.5 is that of a twin-T notch filter. Show that the attenuation is infinite at a frequency of $f = 1/2\pi RC$. Find the circuit's Q-factor. *[Hint: use the star-delta transform. $Q = 0.25$]*

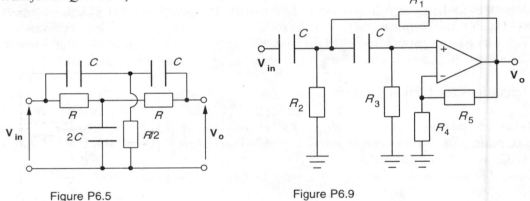

Figure P6.5   Figure P6.9

6 A passive Butterworth filter with open-circuit termination must have an attenuation of at least 50 dB at 30 kHz and at most 0.5 dB at 3 kHz. What order of filter is needed? What is the 3dB cut-off frequency of this filter? Use table 6.2 to determine the normalised components. If the source resistance is 600 Ω what are the actual component values required? What are the actual attenuations of this filter at 3 kHz and 30 kHz? Would a Chebyshev filter with a passband ripple of 0.5 dB require fewer components? What would the attenuation of the Chebyshev equivalent be at 30 kHz? *[3rd order; $f_c = 4.26$ kHz; $C_1 = 31.1$ nF, $L_2 = 29.9$ mH, $C_3 = 93.4$ nF; 0.1 dB, 51 dB; no; 63 dB]*
7 Show that the response in the stopband of a Chebyshev lowpass filter is $6(n-1)$ dB below that of the equivalent Butterworth filter of the same order, n. *('Equivalent' means its response is the same as the Chebyshev's at the end of the passband or ripple channel.)*
8 Show that the voltage transfer function of the circuit of figure 6.19 is that of a second-order, bandpass filter with $Q = \sqrt{2}/(3-K_1)$ and $K_1 = 1 + R_2/R_1$. By what percentage does the bandwidth change when $Q = 10$ and R_2/R_1 increases by 1%? *[−7%]*
9 Find the response, $\mathbf{H}(j\omega)$, and the normalised response, $\mathbf{H}(j\Omega)$, of the circuit of figure

P6.9 when $C = 220$ pF, $R_1 = R_2 = R_5 = 10$ kΩ and $R_3 = R_4 = 4.7$ kΩ. *(Hint: scale R and C, making $C' = 1$ F and $R' = 1$ Ω, then find the normalised response.)*
$[-3.13/(1 + (j\Omega)^2 + j1.516\Omega); -3.56 \times 10^{-12}(j\omega)^2/(1 + 1.137 \times 10^{-12}(j\omega)^2 + 1.617 \times 10^{-6}j\omega)]$

10 Design a fourth-order, Chebyshev, highpass, VCVS filter with a ripple of 2 dB in the passband and a passband from 15 kHz up. What is the attenuation at 5 kHz compared to that of the passband? *[$C_1 = C_2 = 1$ nF, $R_1 = 5$ kΩ, $R_2 = 10$ kΩ would be appropriate. The gain resistors could be $R_{11} = 13$ kΩ, $R_{21} = 12$ kΩ, $R_{12} = 5.1$ kΩ, $R_{22} = 9.1$ kΩ, sticking to preferred values; $\alpha = 52.9$ dB]*

11 Show that the K-values for the second and fourth-order Butterworth filters in table 6.3 are correct. (Use equation 6.16 for the mth section's response with $Q_m = 1/[3 - K_m]$; the response is the product of the section responses.)

12 The circuit of figure P6.12 is an Åkerberg-Mossberg filter. Show that its responses are given by

$$\frac{V_1}{V_{in}} = \frac{-jK\Omega/Q}{(1 - \Omega^2) + j\Omega/Q} \quad \text{and} \quad \frac{V_2}{V_{in}} = \frac{1}{(1 - \Omega^2) + j\Omega/Q}$$

where $\Omega = \omega/\omega_c$, $\omega_c = 1/\sqrt{(R_3 R_4 C_1 C_2)}$, $K = R_2/R_1$ and $Q^2 = C_1 R_2^2/C_2 R_3 R_4$. Show also that this reduces to the same response as the Tow-Thomas biquad of figure 6.20 if $R_3 = R$ and $C_1 = C_2 = C$.

Figure P6.12

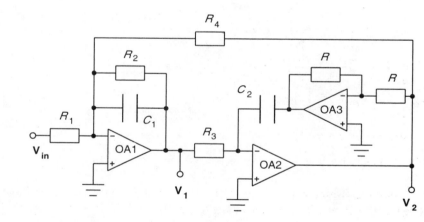

13 An Åkerberg-Mossberg filter with the responses given in problem 6.12 is required with a centre frequency, $f_0 = 4$ kHz, a Q of 16, a midband gain of 20 dB and has $C_1 = C_2 = 20$ nF and $R_3 = R$. What are the resistance values? What are the lowpass and bandpass responses at 400 Hz?

14 Design a second-order, highpass filter using the biquad of figure 6.24 which will have a cut-off at 1.6 kHz and a Butterworth response.

15 The circuit of figure P6.15 is a Fliege filter. Show that its response is

$$H(j\Omega) = \frac{jK\Omega/Q}{1 - \Omega^2 + j\Omega/Q}$$

where $\Omega = \omega/\omega_c$, $\omega_c^2 = (K - 1)/(RC)^2$ and $Q^2 = (K - 1)(R_1/R)^2$. Calculate component values for this filter so that its centre frequency is 2.5 kHz, the midband gain is 20 dB and its lower 3dB point is at 2 kHz.

Figure P6.15

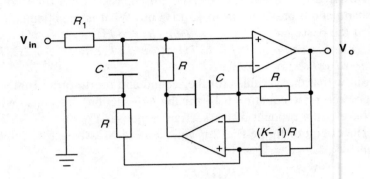

7 Waveform generators

I N MANY circuits we have no need to generate a particular waveform, but merely operate with what we are given in the way of signals. The natural waveform resulting from the generation of power by rotating machines is sinusoidal and this can be converted to any desired waveform and frequency with the appropriate electronics. But sometimes we wish to use a DC supply and generate from it the desired waveform, of which there are two main types: sinusoidal and non-sinusoidal. Electronic sinusoidal waveform generators are usually oscillators employing feedback.

7.1 Oscillators

Oscillators are based on the generalised feedback circuit of figure 7.1, which was discussed in chapter five. There we were mostly concerned with negative feedback in which the loop gain was positive and was subtracted from the input, x_{in}. In an oscillator the feedback is positive; the loop gain, AB, is made equal to -1, and then the closed-loop gain is

$$G = \frac{x_o}{x_{in}} = \frac{A}{1 + AB} = \frac{1}{1 - 1} = \infty$$

that is the circuit has an output with no input. The circuit is unstable and will go into oscillations of a frequency and type determined by its components. Very often circuits are built with unwitting feedback, usually capacitively coupled between output and input, and spontaneous oscillations arise to confound the designer. And in contrary fashion circuits purposely built as oscillators stubbornly refuse to start oscillating as required.

Figure 7.1

A circuit with feedback. A is the open-loop gain, the gain with no feedback. B is the fraction fed back, and AB is the loop gain

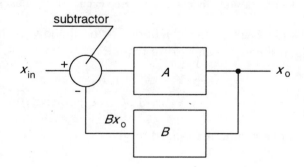

7.1.1 RC oscillators

RC oscillators are very convenient for varying the frequency of the output although the stability of this frequency is not particularly good. They are often based upon operational amplifiers with various RC networks providing feedback. Consider the general op amp circuit with feedback to both inputs as shown in figure 7.2a. The voltage transfer function of this circuit can be shown to be

$$\frac{V_o}{V_{in}} = \frac{Y_1(Y_3 + Y_4)}{Y_4(Y_1 + Y_2) - Y_2(Y_3 + Y_4)}$$

where $Y_1 = 1/Z_1$ etc. For this to be infinite we require that

$$Y_1 Y_4 = Y_2 Y_3 \quad \Rightarrow \quad Y_1/Y_2 = Y_3/Y_4$$

This condition arises from the requirement that the input terminals to the op amp should be at the same potential and is much the same as balancing an AC bridge. Looking at Wien's bridge, which is conveniently an RC bridge, we find that Y_1 is a capacitance in parallel with a resistance, and Y_2 a capacitance in series with a resistance, while Y_3 and Y_4 are just resistances. The resulting oscillator is shown in figure 7.2b with the input grounded, so that $V_{in} = 0$.

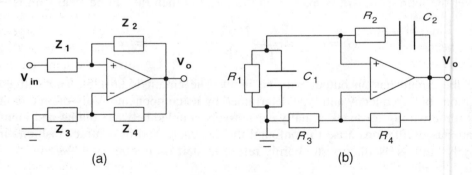

Figure 7.2 (a) Generalised op amp circuit with feedback (b) The Wien-bridge oscillator

Going back to the condition for oscillation we have $Y_1 = 1/R_1 + j\omega C_1$ and $Y_2 = j\omega C_2/(1 + j\omega C_2 R_2)$ and so we require that

$$\frac{Y_1}{Y_2} = \frac{1/R_1 + j\omega C_1}{j\omega C_2/(1 + j\omega C_2 R_2)} = \frac{Y_3}{Y_4} = \frac{R_3}{R_4}$$

The above can be rearranged to give

$$1 - \omega^2 C_1 C_2 R_1 R_2 + j(C_1 R_1 + C_2 R_2) = j\frac{C_1 R_2 R_4}{R_3}$$

Thus the oscillations can only occur at an angular frequency of

$$\omega = \frac{1}{\sqrt{C_1 C_2 R_1 R_2}}$$

which comes from the real part of the equation. There is a further condition from the imaginary part that

$$C_1 R_1 + C_2 R_2 = \frac{C_1 R_2 R_4}{R_3}$$

Normally we make $C_1 = C_2 = C$ and $R_1 = R_2 = R$ to facilitate the design and use of the oscillator and then the output frequency is

$$f = \frac{1}{2\pi RC}$$

and $R_4 = 2R_3$. It is then convenient to vary R by using ganged variable resistors over about a decade for each frequency range, and to change ranges by switching the ganged capacitors over the required number of decades. In this way frequencies from a few Hz to 1 MHz can be covered with a suitably chosen op amp, a pair of fixed resistors, a pair of variable resistors and six pairs of capacitors.

Wien's bridge is not particularly stable in the form shown in figure 7.2b and several adaptations have been proposed to improve it. The difficulty lies with the additional requirement that $R_4 = 2R_3$: when $R_4 < 2R_3$ the gain of the amplifier is insufficient to make $AB = -1$ and the circuit will not oscillate. And when $R_4 > 2R_3$ the gain is too large and the output becomes distorted. The simplest method to stabilise the gain is to make R_3 a lamp with a positive temperature coefficient of resistance. Then initially the lamp resistance is low and the output voltage is large, so sending a large current through R_3, which heats up until the output is reduced. If the output is reduced too much the oscillations cease and the lamp cools down till they start again. Alternatively a thermistor of negative temperature coefficient (NTC) can be placed in series with R_4 (see problem 7.1). Another method of stabilising the circuit is to restrict the voltage developed across R_4 by means of two diodes in parallel and antiparallel with it; then if R_3 is replaced by a potentiometer of total resistance, R_5, the output voltage is limited to $\pm(R_4 + R_5)V_{AK}$ where V_{AK} is the forward voltage drop of the diodes. (See problem 7.2.)

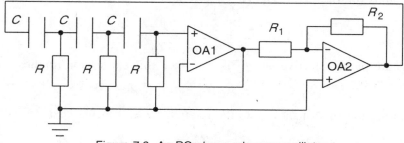

Figure 7.3 An RC phase-advance oscillator

A number of RC oscillator designs have been produced apart from Wien's bridge — the most popular — such as the phase advance circuits of figure 7.3 and problem 7.6, which are most useful at very low frequencies (< 10 Hz). The buffer OA1 stiffens the output and OA2 feeds back part of the output to the normal input of the RC network. It is left as an exercise (see problem 7.3) to show that for oscillations to occur the gain of OA2 must be −29. The angular frequency of oscillation can also be shown to be $\omega = 1/CR\sqrt{6}$.

The chief drawback of the Wien-bridge oscillator is its poor frequency stability (about 0.1%) caused partly by the temperature variations in R and C, but mainly by variations in the op amp's parameters which cause phase shift and this results in relatively large changes in frequency. Other AC bridges are less suitable for making oscillators because they use inductors as well as resistors and capacitors. However, there are a number of LC oscillators in use — especially at radio frequencies — including the Hartley and Colpitts oscillators described next. These have much more stable output frequencies (within about 0.01%), though it is rather less convenient to change these at will, especially over large ranges.

7.1.2 LC oscillators

LC oscillators employ transistors for amplification as these will work at higher frequencies than op amps. Figure 7.4a shows a generalised BJT amplifier with feedback from collector to base via $\mathbf{Z}_2$. The equivalent circuit of this is shown in figure 7.4b, which is just the small-signal CE amplifier equivalent circuit. We see from this that h_{ie} is in parallel with $\mathbf{Z}_2$ and so the small-signal base current, i_b, is

$$i_b = \frac{\mathbf{IZ}_2}{\mathbf{Z}_2 + h_{ie}}$$

Then by KCL we see that the current through $\mathbf{Z}_1$ is $\mathbf{I} + h_{fe}i_b$ and that through $\mathbf{Z}_2$ is $\mathbf{I} - i_b$. Using KVL round the loop containing $\mathbf{Z}_1$, $\mathbf{Z}_2$ and $\mathbf{Z}_3$ gives

$$\mathbf{IZ}_3 + (\mathbf{I} - i_b)\mathbf{Z}_2 + (\mathbf{I} + h_{fe}i_b)\mathbf{Z}_1 = 0$$

Substituting for i_b in this yields

$$\mathbf{I\Sigma Z} + \frac{h_{fe}\mathbf{IZ}_2\mathbf{Z}_1}{\mathbf{Z}_2 + h_{ie}} - \frac{(\mathbf{IZ}_2)^2}{\mathbf{Z}_2 + h_{ie}}$$

where $\mathbf{\Sigma Z}$ stands for $\mathbf{Z}_1 + \mathbf{Z}_2 + \mathbf{Z}_3$. This expression can be rearranged in the form

$$(1 + h_{fe})\mathbf{Z}_1 + \mathbf{Z}_1 + \frac{h_{ie}\mathbf{\Sigma Z}}{\mathbf{Z}_2} = 0$$

If all the $\mathbf{Z}$s are purely reactive and h_{ie} is real, then this expression implies

$$\mathbf{Z}_3 = -(1 + h_{fe})\mathbf{Z}_1 \quad \text{and} \quad \mathbf{\Sigma Z} = 0$$

And substituting the first of these into the second, $\mathbf{\Sigma Z} = 0$ means that

$$-(1 + h_{fe})\mathbf{Z}_1 + \mathbf{Z}_2 + \mathbf{Z}_1 = 0 \quad \Rightarrow \quad \mathbf{Z}_2 = h_{fe}\mathbf{Z}_1$$

Since h_{fe} is real, these equations can only be true if $\mathbf{Z}_1$ and $\mathbf{Z}_2$ are the same sort of reactances and opposite in sign to $\mathbf{Z}_3$. That is if $\mathbf{Z}_1$ and $\mathbf{Z}_2$ are inductors, $\mathbf{Z}_3$ must be a capacitor and vice versa.

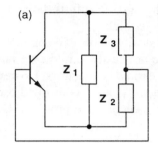
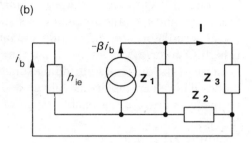

Figure 7.4

(a) General form of BJT oscillator
(b) Equivalent circuit of (a)

Two of the best known LC oscillators are the Hartley and the Colpitts oscillators. In the Hartley oscillator $\mathbf{Z}_3$ is a capacitor and in the Colpitts oscillator it is an inductor. The Hartley oscillator without any biasing or coupling components is shown in figure 7.5a, and a fully biased and coupled version in 7.5b. C_B and C_E are bypass capacitors and C_1 a DC decoupling capacitor, all of negligible reactance at the operating frequency, while L_{RF} is an AC decoupling inductor of large reactance. The resistors in figure 7.5b are the normal biasing and stabilising resistors which play no part in the AC circuit. Here the output is taken from the collector terminal, but it could be transformer-coupled to L_1.

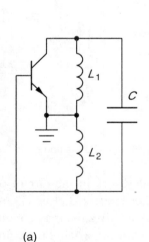
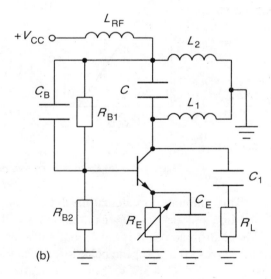

Figure 7.5

(a) The equivalent circuit of the Hartley oscillator with feedback via L_2
(b) The full circuit of the CE amplifier version

From figure 7.5a we see that the feedback ratio is

$$B = L_2/L_1$$

and so for AB to be -1 we require the amplifier's gain, $A = -L_1/L_2$. The condition that $\Sigma\mathbf{Z} = 0$ means that

$$j\omega L_1 + j\omega L_2 + 1/j\omega C = 0$$

from which we derive the angular frequency of oscillation as

$$\omega = \frac{1}{\sqrt{(L_1 + L_2)C}}$$

The circuit's Q-factor is large, so that the phase shift is large for small changes in frequency and thus the frequency's stability is high.

LC oscillators are self starting and intrinsically stable. The output is always undistorted if high-Q components are used, but they are not easily varied over large frequency ranges because of the difficulty of finding suitable variable capacitors. However, by using varicap diodes whose capacitance is voltage dependent, voltage-controlled oscillators (VCOs) can be made to cover a reasonable range of frequencies (see problem 7.4).

Example 7.1

Design a Hartley oscillator to work at 40 kHz with maximum output voltage using a transistor of $h_{fe} = 100$, a supply voltage of +12 V and a 1kΩ load. Assume $h_{oe} = h_{re} = 0$. (For the analysis of CE amplifiers see section 3.2.)

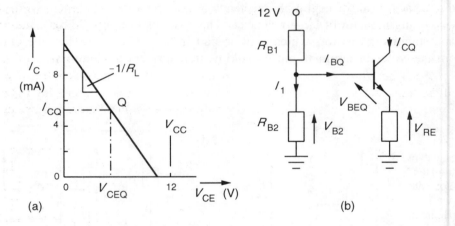

Figure 7.6 (a) Load line for example 7.1 (b) DC biasing conditions

This amounts to calculating the component values in the circuit of figure 7.5b. The load line's slope is $-1/R_L$ as there is no collector resistor. Let us choose R_E such that the voltage drop across it is small, but sufficient to stabilise the bias conditions: about 1.5 V. This leaves 10.5 V for the peak-to-peak output voltage swing at the collector terminal, that is $v_L = 10.5/2\sqrt{2} = 3.7$ V (r.m.s.). Thus the load line will start at $V_{CE} = 10.5$ V and finish at $I_C = 10.5$ mA, the mid-point being $I_{CQ} = 5.25$ mA, as shown in figure 7.6a. Then in figure 7.6b we have $V_{RE} \approx I_{CQ}R_E = 1.5$ V and $R_E = 1.5/5.25 = 0.29$ kΩ. Taking $V_{BEQ} = 0.65$ V, we require the voltage across R_{B2} to be $V_{B2} = 2.15$ V. The quiescent base current is $I_{BQ} = I_{CQ}/h_{fe} = 52.5$ µA, so we need a bias current, I_1, through R_{B1} and R_{B2} of at least 500 µA, which leads to $R_{B1} = 18$ kΩ and $R_{B2} = 3.9$ kΩ using preferred values and neglecting I_{BQ} compared to I_1.

For C_E we must estimate the transistor's input resistance, $h_{ie} = V_{th}/I_{BQ} = 0.026/52.5 \times 10^{-6} = 500\ \Omega$ taking $V_{th} = kT/q$ as 26 mV at 300 K. This means that on the output side $r_e = h_{ie}/h_{fe} = 5\ \Omega$ and the reactance of C_E needs to be much smaller than this at 40 kHz. That is

$$\frac{1}{\omega C_E} \ll 5\ \Omega \quad \Rightarrow \quad C_E \gg \frac{1}{5 \times 2\pi \times 40 \times 10^3} = 0.8\ \mu F$$

A value of 100 µF will more than suffice. The other capacitances are more readily found, since we need $X_{CB} \ll R_{B1}$ and hence $C_B \gg 220$ pF. A value of 22 nF will do. And for C_1 we need $X_{C1} \ll R_L$ or $C_1 \gg 4$ nF, so we choose $C_1 = 0.47\ \mu F$.

The voltage gain of the CE amplifier is given by

$$A_{vL} \approx -\beta R_L/h_{ie} = -100 \times 1000/500 = -200$$

Hence $L_1/L_2 = 200$, and if $C = 220$ pF we should make

$$L_1 + L_2 = \frac{1}{\omega^2 C} = \frac{1}{(2\pi \times 40 \times 1000)^2 \times 220 \times 10^{-12}} = 72\ \text{mH}$$

Then $L_1 = 71.6$ mH and $L_2 = 360$ µH. Finally we require the RF choke's reactance to be much greater than R_L, which implies $L_{RF} \gg 4$ mH. Any convenient value about 40 mH will do.

No matter if we miscalculate the bias conditions, the oscillator will still work at 40 kHz. R_E can be adjusted to whatever is needed to give a feedback ratio, $B = 1/A_{vL}$, provided the transistor's h_{fe} is high enough in the first place. The circuit's high Q ensures that the output is undistorted, even if the transistor is driven into saturation or cut-off.

Figure 7.7

(a) Equivalent circuit for a Colpitts oscillator
(b) Full circuit for a Colpitts oscillator. C_C is a DC blocking capacitor of negligible reactance at the operating frequency

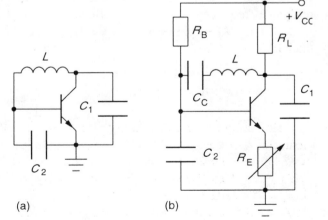

(a)   (b)

Figure 7.7a shows the equivalent circuit for AC of a Colpitts oscillator, whose complete circuit is shown in figure 7.7b. It is sometimes preferred to the Hartley oscillator because it uses only one inductor. The analysis is identical to that of the Hartley oscillator. The feedback ratio is

$$B = \frac{X_{C2}}{X_{C1}} = \frac{C_1}{C_2}$$

and thus $A_v = -C_2/C_1$. The condition that $\Sigma Z = 0$ this time gives

$$\frac{1}{j\omega C_1} + \frac{1}{j\omega C_2} + j\omega L = 0$$

Hence

$$f = \frac{1}{2\pi}\sqrt{\frac{1}{LC_1} + \frac{1}{LC_2}} = \frac{1}{2\pi\sqrt{LC_{eq}}}$$

where $1/C_{eq} = 1/C_1 + 1/C_2$. The value of R_E can be adjusted to alter the quiescent collector current until the resonator starts to oscillate.

7.1.3 Quartz-crystal oscillators

The symbol for a piezoelectric crystal is shown in figure 7.8a. The mechanical vibrations of a piezoelectric quartz crystal produce an alternating voltage on opposite sides of the crystal and vice versa. The frequencies of the natural vibrational modes of the crystal depend on its elastic constants which are relatively independent of temperature.

The quartz crystal provides a very stable frequency, though one that cannot be varied much other than by changing the crystal. The frequency stability depends partly on the supply voltage (about 0.1 ppm/V) and partly on the temperature (about 0.001 ppm/K). The equivalent circuit of the crystal is basically just a series RLC circuit in parallel with a capacitor, C_0, as shown in figure 7.8b, and so with a few more components a very stable reference frequency is achievable. Because $C_0 \gg C$, the resonant angular frequency, ω_0, is almost $1/\sqrt{(LC)}$; and because $R \ll 1/\omega_0 L$ the Q-factor is very large — usually greater than 10 000.

Figure 7.8

(a) Symbol for a piezo-
    electric crystal
(b) Its equivalent circuit
(c) The Pierce oscillator

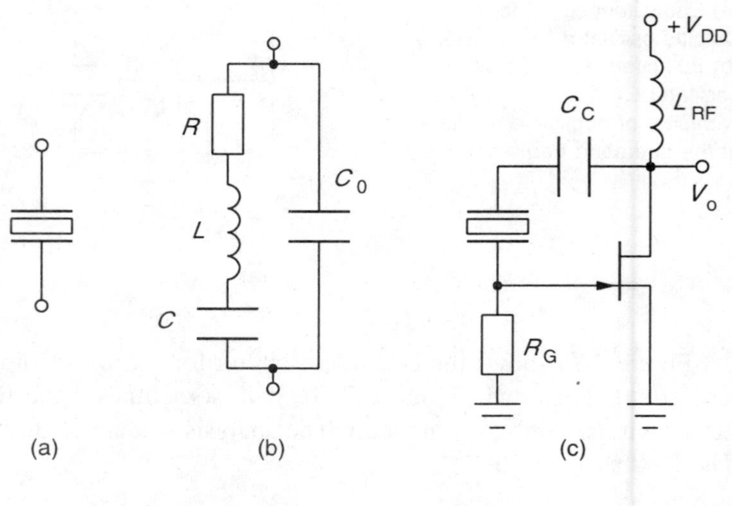

(a)   (b)   (c)

There are numerous crystal oscillator circuits in use, including the Hartley and Colpitts oscillators. The one shown in figure 7.8c is known as the Pierce oscillator and is the simplest (C_C is only a coupling capacitor). Quartz crystals are available from general suppliers of electronic components in a limited range of frequencies but specialist suppliers stock many more and can cut crystals to almost any desired frequency. For example many stockists supply 32.768 kHz (2^{15} Hz) and 4.194304 MHz (10^{22} Hz) crystals which can readily be divided down to any multiple or submultiple of two. Crystals operating at popular frequencies in the range from 2 MHz to 20 MHz cost only £1-2 each and CMOS packages with an 8-pin DIL 'footprint' with frequencies ranging from 4 MHz to 50 MHz can be bought for £3 each.

7.2 Square-wave generators

By 'square' one usually means a waveform such as that shown in figure 7.9a, where the positive-going half is of the same duration as the negative and the positive and negative excursions are equal in amplitude. In popular parlance, however, a 'square' wave can be one with any ratio of positive time to negative (or *mark-space ratio*) and with any DC offset: these should be called *rectangular* waves. For example, in figure 7.9b, the mark-space ratio, $T_+/T_- = 3$ and there is some positive DC bias. Sometimes *duty cycle* is used to describe the ratio of the ON (positive-going) time to the total time, so if the mark-space ratio is 1:1, the duty cycle is $\frac{1}{2}$ or 50%.

If both high and low output states are quasi-stable, that is one will spontaneously change to the other, then the device is *astable*. If only one state is stable, that is it persists unless some external input causes it to change, then the device is *monostable*. If both states are stable the device is *bistable*, and is usually called a *flip-flop*. Flip-flops are important components of logic circuits and as such are fully discussed in chapter 10.

Square waves can be readily generated from a sinewave in many ways, two such being a Schmitt trigger and a logic circuit, but if we start with only a DC supply a convenient way of generating a square wave is by means of an *astable multivibrator*.

7.2.1 The BJT astable multivibrator

Figure 7.9c shows a BJT astable with a variable mark-space ratio. The analysis of this circuit assumes that the transistors are identical and that when they are turned on they are in saturation. When Q1 has just turned from off to on V_{CE1} changes from V_{CC} to V_{CEsat} which is a large, almost instantaneous change of $-V_{CC} + V_{CEsat}$. C_2 cannot change its voltage instantaneously and the result is that V_{BE2} must change from +0.6 V (the forward voltage drop across the base-emitter junction) to $-V_{CC} + V_{CEsat} + 0.6$ V, which is an amount exactly equal to the change in V_{CE1}. However, C_2 begins to charge exponentially up to V_{CC} via R_{B2} with time constant $C_2 R_{B2}$. When its voltage reaches +0.6 V, Q1 turns on.

The same process now takes place with C_1. But C_2 now has to charge through R_{C1} for V_{CE1} to reach V_{CC}; thus the change from V_{CEsat} to V_{CC} as the transistors are turned off cannot be instantaneous: the time constants are $C_2 R_{C1}$ for Q1 and $C_1 R_{C2}$ for Q2.

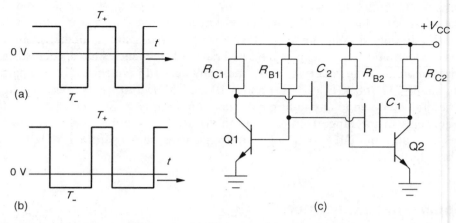

Figure 7.9 (a) A symmetrical square wave (b) a general square wave (c) a BJT astable multivibrator

Let us consider the voltage, V_{BE2}, across the base-emitter junction of Q2. When Q2 has just turned off and Q1 on, this voltage is

$$V_{BE2}(0) = -V_{CC} + V_{CEsat} + 0.6 \text{ V}$$

and C_2 is charging up through R_{B2} towards an ultimate voltage of V_{CC}. But when $V_{BE2} = +0.6$ V, Q2 turns on and Q1 off and the cycle repeats through C_1. The time constant is C_2R_{B2} and thus V_{BE2} as a function of time can be written

$$V_{BE2}(t) = A\exp\left(-t/C_2R_{B2}\right) + B$$

where A and B are constants found from the conditions that $V_{BE2}(0) = A + B = -V_{CC} + V_{CEsat} + 0.6$ and $V_2(\infty) = B = V_{CC}$, so that $A = 0.6 + V_{CEsat} - 2V_{CC}$. We also know that $V_{BE2}(T_2) = +0.6$, which means

$$V_{BE2}(T_2) = 0.6 = (0.6 + V_{CEsat} - 2V_{CC})\exp(-T_2/C_2R_{B2}) + V_{CC}$$

This equation is graphed in figure 7.10b, and from it we find T_2, the time that Q2 is OFF:

$$T_2 = C_2R_{B2}\ln\left(\frac{2V_{CC} - V_{CEsat} - 0.6}{V_{CE} - 0.6}\right) = C_2R_{B2}\ln\gamma \qquad (7.1)$$

The time when Q1 is off is given by the same expression as for Q2 but with a time constant of C_1R_{B1}:

$$T_1 = C_1R_{B1}\ln\gamma$$

Thus the mark-space ratio is T_2/T_1, when the output is taken from the collector of Q1, which is C_2R_{B2}/C_1R_{B1}, regardless of the value of the argument of the natural logarithm in equation 7.1.

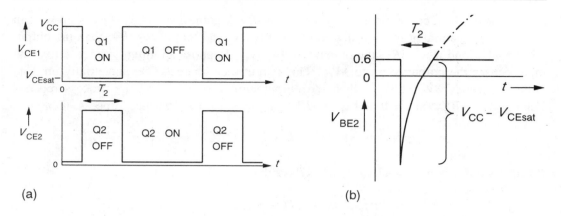

Figure 7.10 (a) Switching times for the transistors of figure 7.9c (b) Graph of V_{BE2} against t

Example 7.2

Design a square-wave generator using a BJT astable multivibrator that has a mark-space ratio of 1:99, works from a +5V supply and has a period of 60 ms. (Take $V_{CEsat} = 0.25$ V and $V_{BE}(on) = 0.6$ V.)

The timing periods are $T_1 = 0.6$ ms and $T_2 = 59.4$ ms. The argument to the natural log in equation 7.1 is

$$\frac{2V_{CC} - V_{BE} - V_{CEsat}}{V_{CC} - V_{BE}} = \frac{10 - 0.6 - 0.25}{5 - 0.6} = 2.0795$$

Then $\ln 2.0795 = 0.7321$ and equation 7.1 gives

$$C_1 R_{B1} = T_1/0.7321 = 0.6/0.7321 = 0.82 \text{ ms}$$

If we take $R_{B1} = 8.2$ kΩ, then $C_1 = 0.1$ μF. Similarly we find $C_2 R_{B2} = 81.14$ ms, so we can make $R_{B2} = 8.114$ kΩ and $C_2 = 10$ μF. We must now consider the time constant for turning off Q1, which is $C_2 R_{C1} \ll 0.6$ ms and this means $R_{C1} \ll 60$ Ω, which is too small for a small transistor. We can increase R_{B2} to 81.14 kΩ thus reducing C_2 to 1 μF and increasing R_{C1} by a factor of ten, which would probably suffice. Choosing $R_{C1} = 100$ Ω would give a time constant of 0.1 ms for switching Q1 off. The time constant for turning Q2 off is not a limitation; however, we assume that Q2 is in saturation when on and that means that $R_{C2} \approx 0.1 R_{B2} = 8.1$ kΩ. A substantially smaller value might result in Q2's not being saturated, depending on Q2's h_{fe} (or β). We see that when the mark-space ratio is very different from unity the design becomes more difficult.

7.2.2 The relaxation oscillator

The circuit of figure 7.11a is sometimes called a relaxation oscillator. It is basically a comparator circuit which outputs a square wave going from $+V_{CC}$ to $-V_{CC}$, and the op amp should ideally be a comparator. If we let the feedback gain, $B = R_2/(R_1 + R_2)$ then we can see that the potential at the non-inverting input is BV_o and that if $v_C < BV_o$ then the output will be $+V_{CC}$ and if $v_C > BV_o$ the output will be $-V_{CC}$. Suppose an initial voltage across the input causes the output to go to $+V_{CC}$. This output then charges C exponentially until its voltage exceeds $BV_o = BV_{CC}$ when the output jumps to $-V_{CC}$ and the process repeats. Taking $v_C = -BV_{CC}$ at $t = 0$ and $v_C = +BV_{CC}$ at $t = \infty$ for the charging process, then

$$v_C(t) = -V_{CC}(1 + B)\exp(-t/RC) + V_{CC}$$

But this stops when $v_C = +BV_{CC}$ at $t = \frac{1}{2}T$, so that

$$v_C\left(\frac{1}{2}T\right) = BV_{CC} = -V_{CC}(1 + B)\exp\left(-\frac{1}{2}T/RC\right) + V_{CC}$$

And from this we deduce that

$$T = 2RC\ln\left(\frac{1 - B}{1 + B}\right)$$

The discharging process starts at $t = \frac{1}{2}T$ and takes until $t = T$ for the voltage to reach $-BV_{CC}$. Figure 7.11b shows the waveforms.

The mark-space ratio is unity and the frequency of the square wave output is $1/T$. The feedback gain, B, serves only to change the frequency of the waveform. If we wish to alter the mark-space ratio then we must make the charging and discharging time constants different (see problem 7.7). A reasonable sinewave can be produced from the output with a lowpass filter (see problem 7.8).

Example 7.3

Design a relaxation oscillator to produce square waves over a frequency range from 10 Hz to 10 kHz.

We first find a comparator that will do (see section 5.6 for details of comparators). The LM311 is cheap and has a slew rate of 0.2 μV/s, which should be adequate. Let us suppose we are going to use variable resistors for R and R_2 in figure 7.11a and keep C and R_1 fixed.

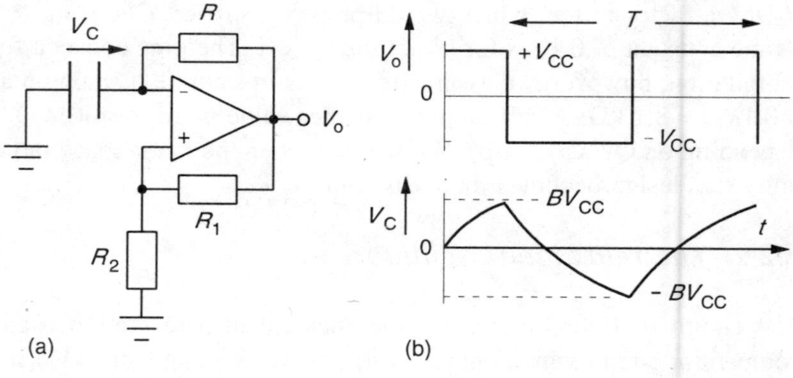

Figure 7.11

(a) A square-wave
    RC oscillator
(b) Its waveforms

(a)

(b)

The variation in frequency is over a range of 1000, but it is mostly controlled by RC as the ln term involving B cannot vary over more than a range of from ln1.2 to ln21 or from about 0.2 to 3, roughly a decade of frequency. The other resistance then has to cover two decades, which is reasonable if it is made of two variable resistors in series. Thus to obtain 10 Hz we could use a resistance, $R = 100$ kΩ and $C = 0.22$ μF, so that $2RC = 0.044$. Then we require

$$\ln\left(\frac{1+B}{1-B}\right) = \frac{1}{2RCf} = \frac{1}{0.44} = 2.27 \;\Rightarrow\; B = 0.813$$

that is

$$\frac{R_2}{R_1 + R_2} = \frac{1}{1 + R_1/R_2} = B = 0.813 \;\Rightarrow\; \frac{R_1}{R_2} = 0.23$$

so we could make $R_1 = 10$ kΩ and $R_2 = 43$ kΩ. At the high-frequency end of the range, we must make B as small as possible, say $B = 0.1$ and then

$$2RC\ln\left(\frac{1+B}{1-B}\right) = 2R \times 0.22 \times 10^{-6} \times \ln 1.222 = \frac{1}{f} = 10^{-4}$$

This gives $R = 1.1$ kΩ, and to make $B = 0.1$ we might try $R_1 = 10$ kΩ and $R_2 = 1.1$ kΩ. Thus R_1 is kept fixed at 10 kΩ while the range required for R_2 is easily accommodated with single variable resistor. R could be made from two variable resistors of 91 kΩ and 9.1 kΩ in series. In practice both the gain resistors, R_1 and R_2, are best left as fixed values.

7.2.3 *The 555 timer*

The versatile 555 timer IC is often used to provide square waves of variable frequency and variable mark-space ratio. Its stability is worse than most RC op amp square-wave generators — about 2% — but is adequate for many purposes. Besides operating as an astable multivibrator, the 555 timer can operate as a monostable (or one shot) and then gives out a single pulse, whose length can be set by external components, on receipt of a triggering input. For the astable multivibrator, two external resistors and a capacitor must be connected to the IC externally.

Figure 7.12a shows the BJT version of the 555 timer circuit (there is also a TTL-compatible CMOS version) which is contained in an 8-pin DIL package shown in figure 7.12c. The connections and external components required to establish the required output in astable mode are shown in figure 7.12b.

The three resistances, R, provide the threshold voltage of $\frac{2}{3}V_{CC}$ and the trigger voltage of $\frac{1}{3}V_{CC}$. When any voltage $< \frac{1}{3}V_{CC}$ is applied to the trigger terminal, the trigger comparator goes high ($\approx V_{CC}$) and sets the flip-flop (that is Q goes high and $\overline{Q}$ low). The output transistor then turns off and the external capacitor, C, charges up to $\frac{2}{3}V_{CC}$ at which point the threshold comparator resets the flip-flop (that is Q goes low and $\overline{Q}$ high). This turns on the output transistor and the capacitor discharges through R_2. The capacitor charges through R_1 and R_2 and so the voltage across C is

$$v_C(t) = V_{CC} - \tfrac{2}{3}V_{CC}\exp\left(\frac{-t}{C(R_1 + R_2)}\right)$$

because at $t = 0$ (the commencement of charging), $v_C = \tfrac{1}{3}V_{CC}$. If the output is high until v_C reaches $\tfrac{2}{3}V_{CC}$, which is at time $t = T_1$, then

$$\tfrac{2}{3}V_{CC} = V_{CC} - \tfrac{2}{3}V_{CC}\exp\left(\frac{-T_1}{C(R_1 + R_2)}\right) \;\Rightarrow\; T_1 = C(R_1 + R_2)\ln 2$$

During discharge through R_2, the capacitor voltage drops from $\tfrac{2}{3}V_{CC}$ to $\tfrac{1}{3}V_{CC}$, during which time, T_2, the output is low (that is ≈ 0 V). The capacitor's voltage during discharge is given by

$$v_C(t) = \tfrac{2}{3}V_{CC}\exp\left(\frac{-t}{CR_2}\right)$$

And so $T_2 = CR_2\ln 2$.

Thus the frequency of the output pulse train is

$$f = \frac{1}{T_1 + T_2} = \frac{1}{C(R_1 + 2R_2)\ln 2} \approx \frac{1.4}{C(R_1 + 2R_2)}$$

When R_1 and R_2 lie between 1 kΩ and 50 kΩ and $C = 0.1$ μF the timing error is about 2.25%. The frequencies of best accuracy lie between 100 Hz and 5 kHz, but the extreme range is from 0.2 Hz to 500 kHz.

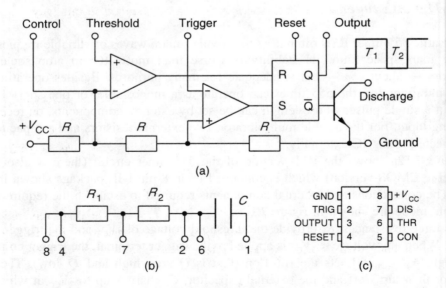

Figure 7.12 (a) Internal components of the BJT 555 timer (b) External components for astable operation (c) 8-pin DIL package

The mark-space ratio

The mark-space ratio, $T_1/T_2 = 1 + R_1/R_2 > 1$, during operation as described above. However, R_2 may be short-circuited with a diode and then the mark-space ratio can be found as follows. Looking at figure 7.13 we see that during discharge R_2 operates as normal and the discharge time is unchanged at $T_2 = CR_2\ln 2$. But during charging the capacitor charges according to

$$v_C(t) = A\exp\left(\frac{-t}{CR_1}\right) + B$$

since the charging current now passes only through R_1. Once more A and B are constants to find. When $t = \infty$ the voltage across C would be $V_{CC} - V_{AK}$, so that $B = V_{CC} - V_{AK}$. And when $t = 0$, $v_C = \frac{1}{3}V_{CC}$, making $A = -\frac{2}{3}V_{CC} + V_{AK}$. Hence

$$v_C(t) = \left(V_{AK} - \tfrac{2}{3}V_{CC}\right)\exp\left(\frac{-t}{CR_1}\right) + V_{CC} - V_{AK}$$

and this must be equal to $\frac{2}{3}V_{CC}$ at $t = T_1$. Substituting $V_C = \frac{2}{3}V_{CC}$ and $t = T_1$ into the equation above gives

$$T_1 = CR_1\ln\left(\frac{2V_{CC} - 3V_{AK}}{V_{CC} - 3V_{AK}}\right)$$

Problem 7.9 indicates how the mark-space ratio can be changed in this way.

Figure 7.13

Using a diode to reduce the mark-space ratio

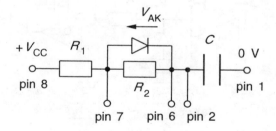

The control pin 5

When a voltage, $V_{CON} < V_{CC}$, is applied to the control pin the thresholds of the comparators become V_{CON} and $\frac{1}{2}V_{CON}$ instead of $\frac{2}{3}V_{CC}$ and $\frac{1}{3}V_{CC}$. Then the capacitor charges from $\frac{1}{2}V_{CON}$ to V_{CON} and discharges from V_{CON} to $\frac{1}{2}V_{CON}$. The charging time is found from

$$v_C(t) = A\exp\left(\frac{-t}{C(R_1 + R_2)}\right) + B$$

Now $v_C(t) = V_{CC}$ when $t = \infty$ and $v_C(t) = \frac{1}{2}V_{CON}$ when $t = 0$, so that $B = V_{CC}$ and $A = \frac{1}{2}V_{CON} - V_{CC}$. Then when $t = T_{CON1}$, $V_C = V_{CON}$, so that

$$T_{\text{CON1}} = C(R_1 + R_2)\ln\left(\frac{V_{\text{CC}} - \frac{1}{2}V_{\text{CON}}}{V_{\text{CC}} - V_{\text{CON}}}\right)$$

And the discharge time remains at $T_2 = CR_2\ln 2$. We can therefore also vary the mark-space ratio using the control pin. However, the minimum value permitted for V_{CON} is $0.35V_{\text{CC}}$, making the minimum value of $T_{\text{CON1}} = 0.24C(R_1 + R_2)$. It is best to derive V_{CON} from V_{CC} so that variations in supply voltage will not affect the timing.

Though cheap in itself (about 25p), the 555 timer requires a good low-leakage capacitor for reasonably accurate performance and these are relatively expensive when values are above 1-2 µF. The precision timer ZN1034E is ten times the price of the 555, but may be worth the extra money.

7.3 Triangular-wave generators

Once a square wave has been generated, a triangular wave can be obtained from it by integration. For the output to fall back to zero at the end of each pulse the square wave must have any DC component removed with a capacitor, or the integrating capacitor must be zeroed (with a FET switch for example) at the end of each cycle. Triangular waves can also be generated by using a comparator (OA1) with negative feedback from an RC integrator (OA2), as shown in figure 7.14a.

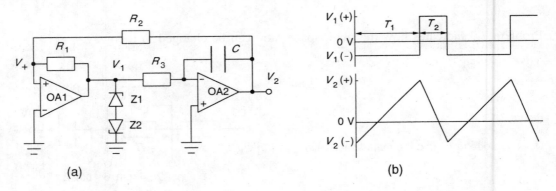

Figure 7.14 (a) A triangular wave generator (b) Its voltage waveforms

The two Zeners, Z1 and Z2, on the comparator's output control the positive and negative slopes of the integrator's ramp, while the frequency can be varied by varying R_3. The output's magnitude, V_2, is controlled by the gain resistors, R_1 and R_2, of the comparator, which has positive and negative output voltages of

$$V_1(+) = V_{Z1} + V_{AK2} \quad \text{and} \quad V_1(-) = -V_{Z2} - V_{AK1} \tag{7.2}$$

where $V_{AK2} = V_{AK1} = 0.7$ V, the forward drop of the Zeners.

The integrator will be ramping up when V_1 is negative and down when V_1 is positive and will switch from up to down mode when $V_+ = 0$. Let the time for ramping up be T_1 seconds and let T_2 be the time for ramping down. The peak voltage after ramping up (see figure 7.14b) will be

$$V_2(+) = V_2(-) - \frac{T_1 V_1(-)}{CR_3} \tag{7.3}$$

and a similar expression for the ramping-down time. Now since there is no current into the comparator, at the moment when $V_+ = 0$

$$V_2(+) = \frac{-R_2 V_1(-)}{R_1} \quad \text{and} \quad V_2(-) = \frac{-R_2 V_1(+)}{R_1} \tag{7.4}$$

and substituting these into equation 7.3 leads to

$$T_1 = \frac{CR_3 R_2[V_1(-) - V_1(+)]}{R_1 V_1(-)} \tag{7.5}$$

And in a like manner we can find the ramping-down time to be

$$T_2 = \frac{CR_3 R_2[V_1(+) - V_1(-)]}{R_1 V_1(+)} \tag{7.6}$$

Example 7.4

Design a triangular-wave generator as in figure 7.14a to give a frequency of 1 kHz with a positive ramp that is twice as long as the negative ramp and that has a peak-to-peak output of 2 V.

From equations 7.5 and 7.6 we see that the ramping time ratio is just

$$\frac{T_1}{T_2} = \frac{V_1(+)}{V_1(-)}$$

And these voltages are given by equation 7.2 so that

$$\frac{T_1}{T_2} = 2 = \frac{V_{Z1} + V_{AK2}}{V_{Z2} + V_{AK1}} = \frac{V_{Z1} + 0.7}{V_{Z2} + 0.7}$$

assuming forward drops of 0.7 V for silicon Zeners. If we choose $V_{Z2} = 4.7$ V, then $V_{Z1} = 10.1$ V. The peak-to-peak voltage required is 2 V and this is $V_2(+) - V_2(-)$. But

$$V_2(+) - V_2(-) = \frac{R_2[V_1(+) - V_1(-)]}{R_1} = \frac{R_2[10.8 - (-5.4)]}{R_1} = 2$$

Hence $R_1/R_2 = 8.1$ and we can choose $R_1 = 27$ kΩ and $R_2 = 3.3$ kΩ, both preferred values.
  The frequency of the wave is given by

$$f = \frac{1}{T_1 + T_2} = \frac{1}{3T_2} = 1000 \text{ Hz}$$

And then $T_2 = 1/3000 = 0.333$ ms. From equation 7.6 we have

$$T_2 = \frac{R_2}{R_1} \frac{V_1(+) - V_1(-)}{V_1(+)} CR_3 = \frac{2CR_3}{10.8} = 0.333 \text{ ms}$$

from which $CR_3 = 1.8$ ms. Choosing $C = 0.22$ μF makes $R_3 = 8.2$ kΩ.

7.3.1 A BJT sawtooth sweep generator

A sawtooth wave is just a triangular wave with as brief a down-ramp ('flyback') time as possible. In CROs the electron beam is swept by deflector plates connected to a sawtooth wave generator which is often a BJT circuit such as shown in figure 7.15a.

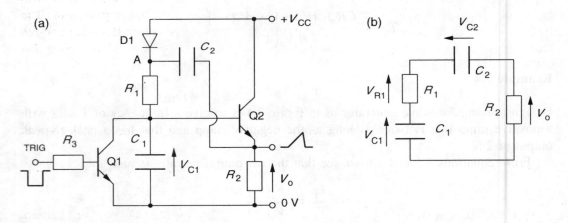

Figure 7.15 (a) A BJT sawtooth ramp generator (b) A loop from part of the circuit

  The emitter follower's output, v_o, follows v_{C1} as the capacitance, C_1, charges up when the input to Q1's base (TRIG) goes low and turns off Q1. This charging would normally result in an exponential rather than the linear change in voltage which is required. But the rise in voltage across R_2 causes the potential at A to rise and turns off the diode. C_1 is then charged from C_2 ($\gg C_1$) whose voltage does stay almost constant. If we examine the loop formed by R_1, C_1, R_2 and C_2 as in figure 7.15b, we see that by KVL

$$v_{C1} + v_{R1} = v_{C2} + v_o$$

But as $C_2 \gg C_1$ its voltage hardly changes while charging C_1, and as $v_o \approx v_{C1}$ we have $v_{R1} = v_{C2} =$ constant, too. Now the voltage across C_2 will be within a diode's forward drop of V_{CC}, hence the current, I_1, through R_1 is constant and nearly equal to V_{CC}/R_1. Therefore

The circuit produces a linear, ramp output as required. When the input to Q1 goes high,

$$v_{C1}(t) \approx v_o(t) = \int (I_1/C_1)\,dt = V_{CC}t/C_1R_1 \qquad (7.7)$$

C_1 is rapidly discharged through Q1 and C_2 is recharged through D1 and R_2.

Example 7.5

Design a ramp generator as in figure 7.15a to provide a ramp of 5 V from a 12 V supply with a ramp time of 10 ms and a flyback time of 0.5 ms.
From equation 7.7 we find

$$5 = 12 \times 10/R_1C_1 \quad \Rightarrow \quad R_1C_1 = 24 \text{ ms}$$

so if $R_1 = 5.1$ kΩ, $C_1 = 4.7$ µF. Making $C_2 = 470$ µF means it will discharge only by 50 mV when C_1 is charging. To choose R_2 we must consider that C_1 can lose charge though the base of Q2 and R_2 and the effective discharging resistance is $h_{fe}R_2$ ($= \beta R_2$) for an emitter follower, and the time constant is $h_{fe}C_1R_2$. Making this time constant equal to $5T_{ON}$, where T_{ON} = ramp time = 10 ms, will result in a loss in voltage of < 0.7%. Thus the requirement is that $h_{fe}R_2 = 5 \times 10 \times 10^{-3}/C_1 = 10$ kΩ. It is easy to find a transistor with $h_{fe} > 100$, and so we can make $R_2 = 100$ Ω and more than fulfil the requirement. The flyback is caused by discharging C_1 through Q1. This means that we must supply enough base current to allow the collector current to discharge C_1 in time $T_F = 0.5$ ms. Now this collector current must be

$$I_{C1} = \frac{C_1V_{C1}}{T_F} = \frac{4.7 \times 10^{-6} \times 5}{0.5 \times 10^{-3}} = 47 \text{ mA}$$

The base current must therefore be 0.47 mA ($= I_{C1}/h_{fe}$). If the trigger input is 5 V, then R_3 should be

$$R_3 = \frac{5 - V_{BE1}}{I_{B1}} = \frac{4.3}{0.47 \times 10^{-3}} = 9 \text{ k}\Omega$$

C_2 recharges during flyback via R_2. A little reflection will tell us that the charging current must be the same as the discharging current for C_1 which is 47 mA, and that means the voltage across R_2 must be 4.7 V, far too much. We can either replace Q2 with a source-follower FET — which also eliminates the problem of discharging C_1 during ramping up — and then we can make R_2 much smaller, or we can add a diode with its anode connected to the emitter of Q2 and its cathode to the collector of Q1 so that C_2 charges via Q1 rather than R_2. The latter solution will double the collector current in Q1 and therefore R_3 must be reduced to 4.5 kΩ.

7.3.2 A staircase ramp generator

In some cases a ramp voltage is required which increases in small, equal-height steps: in other words a staircase, as shown in figure 7.16b. This can be derived from the diode pump circuit of figure 7.16a.

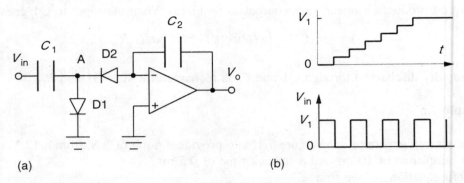

Figure 7.16 (a) A diode pump circuit (b) Input and output waveforms

The square-wave input almost instantaneously charges C_1 up to V_1, the peak input voltage, through the forward-biased diode, D1. During this process D2 is reverse biased and so the output voltage remains at zero. When the input voltage falls to zero, the voltage across C_1 cannot fall immediately and so node A becomes negative for an instant at a potential of $-V_1$. This forward biases D2 and reverse biases D1. Current therefore flows instantaneously through C_2 and charges it up, and through C_1 to discharge it. Since these currents are the same the charge transferred is the same

$$Q_1 = Q_2 \;\Rightarrow\; C_1 V_1 = C_2 V_o$$

and the output voltage is positive and equal to $V_1 C_1 / C_2$ after the first positive cycle of V_{in}. During the next positive input half cycle the charge on C_2 stays constant as D2 is reverse biased, but C_1 charges up again to the same voltage as before. Then on the zero-voltage half of the input cycle, C_2 can charge up while C_1 discharges as before. Thus after n cycles the voltage on C_2 is

$$V_o = n V_1 C_1 / C_2$$

Eventually V_o will rise to V_1 and the pumping ceases until C_2 is zeroed, for example by a FET placed in parallel with it. If the ramp is slow the capacitors and diodes must be low-leakage components ('computer grade'). With some loss in accuracy (see problem 7.12) the op amp can be omitted.

7.4 Waveform-generator ICs

There are several integrated circuits available which will produce square, sine and triangular waves from their outputs. Until recently these were of relatively high cost for modest performance but now there are several inexpensive ICs available with excellent performances, especially in view of their low cost. Two examples from are the XR-8038A from Exar and the ICL8038 from Harris, both in 14-pin DIL form (shown in figure 7.17) and costing from £3 to £5 in single quantities.

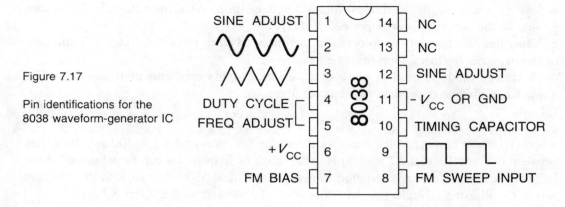

Figure 7.17

Pin identifications for the
8038 waveform-generator IC

The frequency ranges from 0.001 Hz to 200 kHz — 300 kHz for the Harris chip —
can be fully or partly swept on all three outputs in FM mode. Table 7.1 gives a few of the
chip specifications for the XR-8038AC.

Table 7.1 *8038 Waveform-generator specifications*

Parameter	min	typical	max
Single supply voltage	10	-	30
Dual supply voltage	±5	-	±15
Supply current mA	-	12	20
Frequency	0.001 Hz	-	200 kHz
Temperature stability	-	50 ppm/°C	-
Supply stability	-	0.05%/V	-
Square wave p-p	-	$0.98V_{CC}$	-
Duty cycle	2%	-	98%
Triangle/sawtooth/ramp p-p	-	$0.33V_{CC}$	-
Sinewave p-p	-	$0.22V_{CC}$	-
Distortion	-	0.3%	-

Problems

1 A Wien-bridge oscillator as in figure 7.2b is stabilised by placing an NTC thermistor
in series with R_4. The thermistor's resistance is given by

$$R_T = -125V_T + 5000 \ \Omega$$

where V_T is the voltage across the thermistor. If $R_3 = 2.7$ kΩ and $R_4 = 1$ kΩ, what is the
r.m.s. output voltage of the oscillator? *[8.84 V]*

2 The circuit of figure P7.2 is a Wien-bridge oscillator which is stabilised by the diodes placed across R_2. Show that, if $R_1 = 22$ kΩ and $R_2 = 5$ kΩ, the maximum output voltage is $5.4V_{AK}$, where V_{AK} is the diodes' forward voltage drop. What must the resistance from ground to the potentiometer wiper be? *[9 kΩ]*

3 Show that, for oscillations to occur in the phase advance circuit of figure 7.3, the gain of the negative feedback loop must be exactly −29.

4 A varicap diode is used in place of C_2 in the Colpitts oscillator of figure 7.7b. The diode has a voltage-dependent capacitance given by

$$C_D = C_2 = C_0 V_j^{-\frac{1}{2}}$$

where V_j is the reverse voltage across the diode junction and $C_0 = 400$ pF. If V_j lies between 6 V and 12 V and $L = 600$ µH, what range of frequencies can be achieved? What DC bias across the diode is required to give an output at 550 kHz? How is the DC bias across C_D adjusted? (Neglect C_1 which is $\gg C_2$.) *[508 kHz to 605 kHz; 8.2 V]*

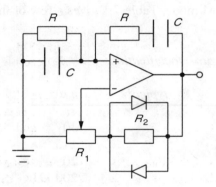

Figure P7.2

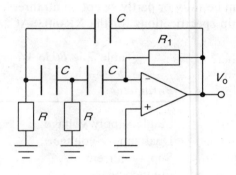

Figure P7.6

5 A Colpitts oscillator is built as in figure 7.7b with $C_2/C_1 = 80$, $V_{CC} = 15$ V and a load of 600 Ω. All the other reactive components can be neglected. If the bias resistor, $R_B = 750$ kΩ, the transistor has $h_{fe} = 200$, $h_{re} = h_{oe} = 0$, and $V_{CEsat} = 0.2$ V, what must R_E be for the circuit to oscillate at 300 K? What is the peak-to-peak load voltage? (Take $V_{th} = kT/q$ and $V_{BEQ} = 0.65$ V.) *[422 Ω, 8.5 V]*

6 The circuit of figure P7.6 is a phase-shift oscillator. Show that the circuit has an output only if $R_1 = 12R$ and its angular frequency must be $0.577/RC$.

7 Ignoring the diodes' forward voltage drops, what is the output frequency and the mark-space ratio of the circuit of figure P7.7, given $R_1 = R_4 = 10$ kΩ, $R_2 = R_3 = 20$ kΩ and $C = 0.22$ µF? If the diodes have forward drops of 0.7 V and V_o swings between +5 V and −5 V, what are the output frequency and mark-space ratio? *[219 Hz, 0.5; 195 Hz, 0.5]*

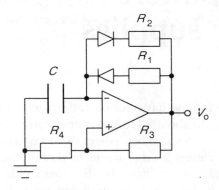

Figure P7.7

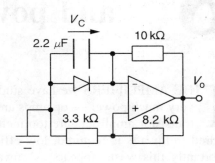

Figure P7.11

8 A relaxation oscillator has a square-wave output which has a mark-space ratio of unity and a frequency of 2 kHz. This is fed into a second-order Butterworth lowpass filter whose cut-off is at 2 kHz. Graph the output from the filter, using the fact that a square wave of angular frequency ω can be written as Fourier series $A(\sin\omega t + \frac{1}{3}\sin3\omega t + \frac{1}{5}\sin5\omega t + ...)$ where A is a scaling constant. Compare your result with a 2kHz sinewave. (You need only consider the first three terms in the Fourier series.)

9 What is the mark-space ratio for a 555 timer connected as in figure 7.13, where $V_{CC} = 5$ V, $V_{AK} = 0.7$ V, $R_1 = 6.2$ kΩ and $R_2 = 18$ kΩ? What must C be to make the frequency 5 kHz? By what percentage do these values change when V_{CC} increases to 6 V and the component values stay the same? *[0.5; 10.7 nF; −7%; +2.4%]*

10 What value of V_{CON} is required for a 555 timer in astable mode to produce square waves of mark-space ratio equal to unity if $R_1 = R_2 = 10$ kΩ and $V_{CC} = 12$ V? Suppose this value of V_{CON} is independent of V_{CC} which changes to 11.5 V, what is the mark-space ratio now? If $C = 3.6$ μF, what are the output frequencies? *[5.44 V; 1.07:1; 20 Hz; 19.4 Hz]*

11 If the capacitor voltage in figure P7.11 at $t = 0$ is +10 V, graph the output voltage as a function of time given $v_o(0) = +10$ V.

12 Show that the diode pump circuit of figure 7.16a will have an output after j cycles of

$$V_{oj} = \frac{C_1 V_1 + C_2 V_{o(j-1)}}{C_1 + C_2}$$

when the op amp is removed and the right-hand terminal of C_2 is connected to the cathode of D1. ($V_{o(j-1)}$ is the output after $j-1$ cycles). What are the first five step heights when $C_1 = 10$ nF, $C_2 = 2.2$ μF and $V_1 = 12$ V? *[54.3 mV, 54.1 mV, 53.8 mV, 53.6 mV, 53.3 mV]*

8 Power amplifiers and power supplies

THE AMPLIFIERS we have studied up to now have not been designed to supply very much power — op amps are limited to about 20 mA in output current and the transistor amplifiers are too inefficient to do so. When power of more than a few hundred milliwatts is required, low efficiency causes a large waste of power, and more importantly this waste appears as unwanted heat which must be disposed of. Indeed a major part of power engineering is concerned with the minimisation and removal of the heat produced by wasted power. Amplifiers are assumed to be driven by DC sources and amplify AC sources to deliver AC power to their loads. The efficiency is defined by

$$\eta = \frac{\text{AC power in load}}{\text{power from DC source}}$$

Power amplifiers are classified by letters according to the amount of time the active device is on while power is delivered to a load, and this largely determines the efficiency.

8.1 Class-A amplifiers

In these the active device, a BJT or FET, is on for the whole of the time that power is being delivered to the load; alternatively one could say that the device's duty cycle is 100%, the maximum possible. We shall consider only BJT common-emitter class-A amplifiers as the same considerations apply to FET class-A amplifiers and the maximum attainable efficiencies are the same in both. There are three ways of connecting a load to the CE amplifier and this results in three different maximum efficiencies.

8.1.1 The series-fed CE amplifier

This is the simplest way of using the CE amplifier, in which the collector resistor is made the load as shown in figure 8.1a. The disadvantage of this method of load coupling is the DC bias current flowing in the load as well as the amplified AC signal. Figure 8.1b shows the load's current waveform which is a single-frequency sinusoid (the amplified signal) imposed on the DC bias, that is I_{CQ}. The total current in the load is given by

$$i_L = I_{CQ} + I_m \sin \omega t$$

And the instantaneous power from the source is then given by

$$p_S = V_{CC} i_L = V_{CC}(I_{CQ} + I_m \sin \omega t)$$

The average power over one cycle from the source is

$$P_S = \frac{1}{T}\int_0^T p_S \, dt = \frac{1}{T}\int_0^T (V_{CC}I_{CQ} + I_m \sin \omega t) \, dt = V_{CC}I_{CQ}$$

since the average of a sinewave over one cycle is zero.

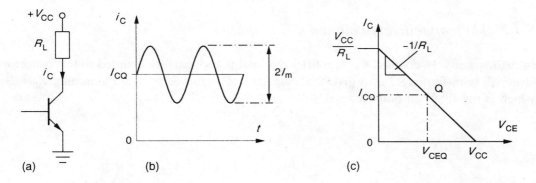

Figure 8.1 (a) Series-connected load (b) Current waveform (c) Load line

The average signal (AC) power in the load is just $I_{rms}^2 R_L = \frac{1}{2}I_m^2 R_L$. If the signal is not to be distorted then the maximum value of I_m is I_{CQ} and then the maximum signal power in the load is $\frac{1}{2}I_{CQ}^2 R_L$. Now the value of I_{CQ} which gives maximum undistorted load power is one situated midway between saturation (where $I_C = V_{CC}/R_L$, ignoring V_{CEsat}) and cut-off (where $I_C = 0$), as shown in figure 8.1c. So $I_{CQPmax} = \frac{1}{2}V_{CC}/R_L$ and the maximum efficiency is given by

$$\eta_{max} = \frac{\frac{1}{2}I_{CQPmax}^2 R_L}{V_{CC}I_{CQPmax}} = \frac{I_{CQPmax}R_L}{2V_{CC}} = \frac{(V_{CC}/2R_L)R_L}{2V_{CC}} = \frac{1}{4}$$

The maximum efficiency is therefore only 25%, assuming that the transistor has zero saturation voltage, that there is no loss of power in any biasing resistances and that the transistor is optimally biased.

The other 75% of the power is lost as heat in the transistor and the load. The DC load power is $I_{CQPmax}^2 R_L = \frac{1}{2}P_S$ (and is counted as lost, since it is not derived from the input signal) and the power lost in the transistor is

$$I_{CQPmax}V_{CEQ} = I_{CQPmax} \times \frac{1}{2}V_{CC} = \frac{1}{4}P_S$$

Example 8.1

A CE amplifier with a 500Ω series-connected load is operated from a 24V supply. The transistor's h_{fe} is 80 and it is biased at $I_{CQ} = 20$ mA. What is the efficiency if the input signal is 20 mV (r.m.s.) and losses in the bias resistors are negligible? (Take $V_{th} = 30$ mV.)

The CE amplifier's voltage gain is

$$A_v = \frac{h_{fe}R_L}{h_{ie}} = \frac{h_{fe}R_L}{V_{th}/I_{BQ}} = \frac{I_{CQ}R_L}{V_{th}}$$

which works out as 333, so that the AC load voltage is 6.67 V and the load power is $6.67^2/500 = 89$ mW. The power drawn from the supply is $V_{CC}I_{CQ} = 24 \times 20 = 480$ mW, so the efficiency is $89/480 = 18.5\%$. The biasing is a little suboptimal and the input signal a little below the maximum possible without clipping, producing an efficiency under 25%.

8.1.2 The capacitively-coupled CE amplifier

In the capacitively-coupled CE amplifier the load is capacitively coupled to the collector terminal as in figure 8.2a, to give the load lines of figure 8.2b, again assuming $R_E = 0$, which is not likely in practice.

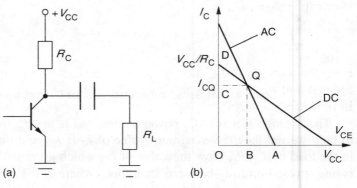

Figure 8.2

(a) Capacitively-coupled load
(b) Its load lines

For maximum efficiency we require maximum peak-to-peak load voltage and current. This means that we require in figure 8.2b that OB = BA. Now the slope of the AC load line is $-1/R_o$ where $R_o = R_C \,/\!/\, R_L$, so that BA $= I_{CQ}R_o$. The slope of the DC load line is

$$\text{CD/OB} = 1/R_C \quad \Rightarrow \quad \text{OB} = \text{CD} \times R_C = (V_{CC}/R_C - I_{CQ})R_C$$

Therefore, if OB = BA,

$$(V_{CC}/R_C - I_{CQ})R_C = I_{CQ}R_o$$

This expression gives I_{CQ} as

$$I_{CQ} = \frac{V_{CC}}{R_C + R_o} \tag{8.1}$$

The maximum voltage across the load is OB = BA $= V_m = I_{CQ}R_o$ and the r.m.s. load voltage is $V_m/\sqrt{2}$ and the maximum load power is

$$P_{Lmax} = \frac{1}{2}\frac{V_m^2}{R_L} = \frac{1}{2}\frac{(I_{CQ}R_o)^2}{R_L}$$

The power drawn from the DC supply is again $V_{CC}I_{CQ}$, making the efficiency

$$\eta = \frac{(I_{CQ}R_o)^2}{2R_L \times V_{CC}I_{CQ}} = \frac{I_{CQ}R_o^2}{2V_{CC}R_L}$$

But I_{CQ} is given by equation 8.1 and hence η is

$$\eta = \frac{V_{CC}R_o^2}{2V_{CC}R_L(R_C + R_o)} = \frac{R_C R_L}{2(R_C + R_L)(R_C + 2R_L)} \tag{8.2}$$

where we have substituted $R_C R_L/(R_C + R_L)$ for R_o. Equation 8.2 can be differentiated with respect to R_L and the result set equal to zero for maximum efficiency (see problem 8.1) and the result is $R_L = R_C/\sqrt{2}$. When this value for R_L is substituted into equation 8.2 it yields for the maximum efficiency

$$\eta_{max} = \frac{1/\sqrt{2}}{2(1 + 1/\sqrt{2})(1 + 2/\sqrt{2})} = 0.086$$

Even this very low efficiency is seldom approached in practice, since the biasing is rarely optimal and the load voltage is normally kept below the maximum to reduce distortion. In these circumstances efficiencies of 2-3% are all that can be attained.

8.1.3 The transformer-coupled CE amplifier

Of the CE amplifiers, the highest efficiency is found in the transformer-coupled version, yet it is scarcely used because it is bulky, expensive and less efficient than a class-B amplifier. In figure 8.3a the transformer has a primary-to-secondary turns ratio of

$$n = \frac{N_1}{N_2}$$

so that the load in the collector circuit is

$$R'_L = n^2 R_L$$

As far as DC is concerned the transformer primary represents a short circuit, so the DC load line is vertical in figure 8.3b, while the AC load line has a slope of $-1/R'_L$. The transformer primary voltage range is $\pm V_{CC}$, so that the maximum collector voltage is $2V_{CC}$ when the transformer primary is at a potential of $-V_{CC}$ and 0 V when the primary is at $+V_{CC}$. The AC load line goes from $I_C = 0$ and $V_{CE} = 2V_{CC}$ to $I_C = 2V_{CC}/R'_L$ and $V_{CE} = 0$. All of the load power is AC and is given by

$$P_L = \frac{V_{CC}^2}{2R'_L}$$

Therefore the efficiency is

$$\eta = \frac{V_{CC}^2/2R'_L}{V_{CC}I_{CQ}} = \frac{V_{CC}}{2I_{CQ}R'_L} = \frac{1}{2}$$

because $I_{CQ} = V_{CC}/R'_L$. The other half of the power is consumed by the transistor.

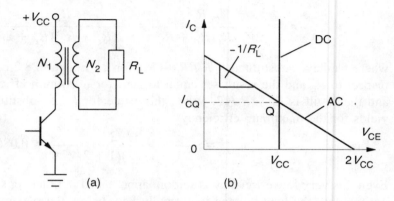

Figure 8.3

(a) A load transformer-coupled to the collector
(b) The corresponding load line with an ideal, loss-less transformer

Note that the maximum voltage across the transistor is $2V_{CC}$ not V_{CC} and that the transformer must be able to take the quiescent collector current without magnetically saturating. However, the load can be almost anything we like since the turns ratio can be adjusted to give maximum efficiency.

Example 8.2

A 24V DC supply is to drive a 500Ω load at maximum power using a silicon npn transistor which has a maximum for V_{CEO} of 60 V and a maximum power dissipation of 40 W. What turns ratio, voltage and power rating are required for a transformer-coupled CE amplifier?

The transistor will have a maximum V_{CE} of 48 V: the rating of 60 V is sufficient. The maximum power dissipation is 40 W in the transistor, so we could aim at 30 W in the load and leave a 10W safety margin. The total power from the supply will then be 60 W, implying $I_{CQ} = I_1 = P_s/V_{CC} = 60/24 = 2.5$ A. To develop 30 W in a 500Ω load means that the load current is

$$I_L = I_2 = \sqrt{\frac{P_L}{R_L}} = \sqrt{\frac{30}{500}} = 0.245 \text{ A}$$

The AC primary current has a peak value of $I_{CQ} = 2.5$ A, so its r.m.s. value is $2.5/\sqrt{2} = 1.77$ A. Thus the current ratio is

$$\frac{I_1}{I_2} = \frac{1.77}{0.245} = 7.2 = \frac{N_2}{N_1}$$

There are more turns on the low-current, high-voltage secondary. The voltage on the secondary is $\pm24 \times 7.2 = \pm173$ V $= 122$ V r.m.s. The transformer rating is 30 VA. If the frequency of the signal is in the audio range then an audio transformer could probably found with a specification near this, though it would be working in reverse, since the normal supply voltage would be high and the output voltage low.

8.2 Class-B amplifiers

In class-B operation the active device conducts during only 50% of the input cycle. We could make a class-B amplifier from a CE amplifier biased at cut-off, that is with I_{CQ} zero, as in figure 8.4a, whose load is shown in figure 8.4b.

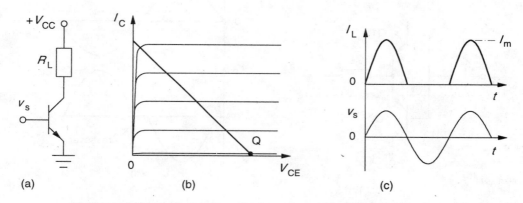

Figure 8.4 (a) A CE amplifier with Q-point at cut-off (b) The load line (c) The load current waveform

When the input signal to the base is positive, the device conducts and current flows through the load, but when the input goes negative, the device is cut off and no current flows. Ignoring any base-emitter voltage drop, we can see from figure 8.4c that the average current flowing is $2I_m/\pi$, the average value of the positive half cycle of a sinewave. The power drawn from the supply is therefore

$$P_S = 2V_{CC}I_m/\pi$$

The power in the load is $I_L^2 R_L = \frac{1}{2}I_m^2 R_L$, since the waveform is sinusoidal. But $I_m = V_{CC}/R_L$, assuming V_{CEsat} is negligible. Thus the maximum efficiency becomes

$$\eta_{max} = \frac{\frac{1}{2}I_m^2 R_L}{2V_{CC}I_m/\pi} = \frac{\pi I_m R_L}{4V_{CC}} = \frac{\pi}{4} = 0.785$$

The remaining 21.5% of the power must be dissipated in the transistor, though in practice the figure is usually higher. Nevertheless the efficiency is markedly better than in a class-A amplifier.

8.2.1 The class-B push-pull amplifier

It is obviously sensible to make use of both halves of the input waveform and the most convenient circuit for this is that of figure 8.5a, which contains a matched pair of complementary (npn and pnp) transistors driving a load with a bipolar DC power supply. For audio amplifiers — the commonest use for this amplifier — the transistors must have identical characteristics (apart from polarity) so that both halves of the input waveform are amplified identically. For other purposes this might not be necessary, but if fidelity is not required a more efficient class of power amplifier should be used. The amplifier is often called a push-pull amplifier because the current through the load alternates and can be said to be pushed through by Q1 and pulled through by Q2.

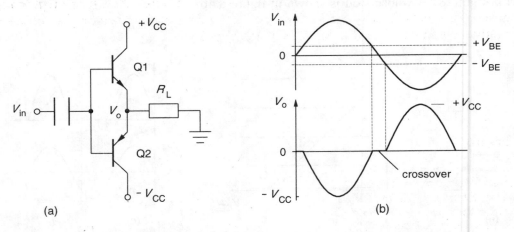

Figure 8.5 (a) A simplified class-B push-pull amplifier (b) Voltage waveforms

8.2.2 Class AB operation

The amplifier shown in figure 8.5a suffers from a number of drawbacks one of which is shown in figure 8.5b, namely *crossover distortion*. The load voltage waveform is non-sinusoidal with a sinusoidal input because of the need for a bias of $\pm V_{BE}$ at the transistors' bases before they can conduct, so giving the output waveform of the lower part of figure 8.5b. The easiest way to eliminate this is to place two diodes in series across the bases of the transistors so that they are just biased into conduction. Since the duty cycle of the transistors in now over 50% but less than 100% this mode of operation is known as class AB: it differs little from class B, however, in efficiency. A further drawback is the need for a bipolar supply. The circuit of figure 8.6 eliminates both of these problems. The additional emitter resistances stabilise the emitter current and ensure that only one transistor conducts at any time. Q1 and Q2 are normally matched power Darlingtons as the h_{fe} values of power transistors tends to be small. D1 and D2 can be mounted on the same heat sink as Q1 and Q2 so that temperature changes are accurately tracked. Alternatively a resistor can be used instead of the two diodes[1] and its value adjusted to give a voltage drop of $2V_{BEQ}$. Using two diodes ensures that the quiescent collector current is small, so that the standby power consumption (when no signal is being input) is minimised. The coupling capacitor, C_2 has to have a small reactance compared to R_L at the lowest frequency of operation. Neglecting V_{CEsat}, the load voltage has a maximum of $+\frac{1}{2}V_{CC}$ and a minimum of $-\frac{1}{2}V_{CC}$. If absolute fidelity is the goal the input signal can be fed directly to both transistor bases, as in figure 8.6, to avoid the small AC voltage drop across the diodes.

A disadvantage of the circuit shown in figure 8.6 is its low input resistance (unless Darlingtons are used), but this can be increased by using an emitter follower buffer on the input in place of R_{B1} or by using a CE amplifier in place of R_{B2}, or using a combination of both. There are numerous other refinements to the circuit that have been proposed and implemented, but the principal features are those of figure 8.6, including an efficiency slightly reduced from the maximum 78.5% of class B.

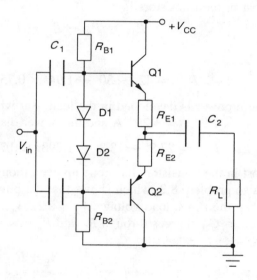

Figure 8.6

A class-AB push-pull amplifier. Crossover distortion is removed by biasing with D1 and D2. The transistors are then just on the verge of conduction with no input

Example 8.3

If $V_{CC} = 48$ V, $R_L = 8$ Ω and $R_E = 0.75$ Ω in figure 8.6, what is the maximum average power developed in the load? What is the efficiency? What is the efficiency if the peak load voltage is only two-thirds of the maximum possible, that is $V_{Lmax} = \frac{1}{3}V_{CC}$? What is the power lost in the transistors? What must C_2 be if the lowest frequency to be amplified is 20 Hz? What must R_{B1} and R_{B2} be to make I_{CQ} equal to 20 mA?

We assume that the r.m.s. output voltage is $V_{CC}/2\sqrt{2}$ which gives a load voltage of

$$V_L = \frac{R_L V_{CC}}{(R_E + R_L)2\sqrt{2}} = \frac{8 \times 48}{(0.75 + 8)2\sqrt{2}} = 15.5 \text{ V}$$

---

[1] The number of diodes should be increased to four if Q1 and Q2 are Darlingtons, because the effective V_{BE} of a Darlington is twice that of a single transistor.

The load power is $15.5^2/8 = 30$ W. The maximum current drawn from the supply is $I_L\sqrt{2}$ and since $I_L = V_L/R_L = 15.5/8 = 1.94$ A, then $I_{Smax} = 1.94\sqrt{2} = 2.74$ A. The mean current drawn from the supply is

$$I_S = I_{Smax}/\pi = 2.28/\pi = 0.872 \text{ A}$$

Hence the power drawn from the DC supply is $V_{CC}I_S = 48 \times 0.872 = 42$ W giving an efficiency of

$$\eta = P_L/P_S = 30/42 = 0.714 = 71.4\%$$

When the peak load voltage is $\frac{1}{3}V_{CC}$ the load power is reduced to $(\frac{2}{3})^2 \times 30 = 13.3$ W. The average current from the supply is also reduced by a factor of two-thirds to 0.581 A and so the power supplied is $V_{CC}I_S = 48 \times 0.581 = 27.9$ W, making the efficiency

$$\eta = 13.3/27.9 = 0.477 = 47.7\%$$

The power lost in the transistors is

$$P_{Tx} = P_S - P_L - I_L^2 R_E$$

which is

$$P_{Tx} = 42 - 30 - 1.94^2 \times 0.75 = 9.2 \text{ W}$$

when maximum power is developed in the load. But when the load power is reduced, the load current is $\frac{2}{3} \times 1.94$ A $= 1.293$ A and the power dissipated in the transistors becomes

$$P_{Tx} = 27.9 - 13.3 - 1.293^2 \times 0.75 = 13.35 \text{ W}$$

The power lost in the transistors has gone up even though the power in the load has gone down! (See also problem 8.2, which shows that the power dissipation in the transistors is maximum when the r.m.s. load voltage, $V_L = 0.225V_{CC}$, when R_E is negligible.)

The capacitor, C_2, charges through R_L and R_E in series so that its reactance must be

$$X_{C2} = \frac{1}{2\pi f C_2} << R_E + R_L = 8.75 \ \Omega$$

that is

$$C_2 >> \frac{1}{2\pi \times 20 \times 8.75} = 909 \ \mu\text{F}$$

A value of 2200 μF would probably suffice.

The diodes make the circuit effectively a current mirror, so that the current through R_{B1}, D1, D2 and R_{B2} is nearly I_{CQ}. Neglecting the base currents and the quiescent voltage drops across the emitter resistances we have

$$V_{CC} = 2I_{CQ}R_B + 2V_{AK} \Rightarrow R_B = \frac{48 - 1.4}{2 \times 20} = 1.165 \ \text{k}\Omega$$

where $R_B = R_{B1} = R_{B2}$ and we have taken the diodes' forward voltage drop to be 0.7 V. If we take R_E into account, then we see that the total voltage drop in the emitter resistors is $2 \times 0.75 \times 20 = 30$ mV. This is the extra voltage that has to be produced by an increased current through D1 and D2. A tenfold increase in current through a silicon diode gives a 60 mV larger voltage (see chapter two, example 2.7), so we require 15 mV per diode or a quarter of a decade increase in current. A quarter of a decade is $10^{0.25} = 1.78$, so we must increase the current by this. Thus the base-bias resistors must be reduced by a factor of 1.78, that is $R_B = 650\ \Omega$.

8.3 Class-C amplifiers

Class-C amplifiers have their active devices in operation for less than 50% of the time that the signal is applied. Their efficiency is therefore higher than that of class-B amplifiers, ranging from 78.5% to nearly 100%. The usual form of the class-C amplifier is that of the tuned circuit shown in figure 8.7a, in which the collector is connected to an LC resonant circuit (a *tank circuit*) which supplies power to the load during the time that the transistor is off. The base of the transistor is reverse biased by V_{BB} which is decoupled from the AC circuit by a choke inductance. The transistor is turned on when the input voltage rises above $V_{BE} - V_{BB}$ (V_{BB} is negative) and during the time that it conducts it replenishes the tank circuit current. The frequency is limited to the resonant frequency of the tank circuit as the Q-factor is large. The input and base bias are adjusted so that at the maximum input voltage the transistor is just driven into saturation.

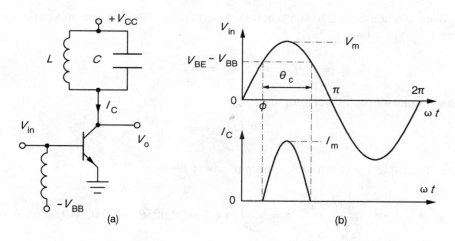

Figure 8.7 (a) A class-C amplifier (b) The input voltage and the transistor current waveforms

The transistor turns on during the input cycle at an angle given by

$$\sin\phi = \frac{V_{BE} - V_{BB}}{V_m} \tag{8.3}$$

It then conducts until the phase angle of the input is $\pi - \phi$ radians, that is the transistor conducts for a total of θ_c radians per cycle, where

$$\theta_c = \pi - 2\phi \tag{8.4}$$

as shown in figure 8.7b. This result can be expressed as

$$\theta_c = 2\cos^{-1}\left(\frac{V_{BE} - V_{BB}}{V_m}\right) \qquad (8.5)$$

Because the tank circuit oscillates from $+V_{CC}$ to $-V_{CC}$ the peak current flowing in the load is V_{CC}/R_C, where R_C is the total equivalent resistance in the collector circuit. This is made up of the load resistance and the equivalent parallel loss resistance of the coil ($= Q_cX_L = Q_c^2r$) in parallel. The tank circuit filters the transistor waveform of the lower part of figure 8.7b to give an almost perfect sinewave; that is to say only the fundamental component at ω_0 remains in the output, where ω_0 is also the resonant angular frequency of the tank circuit, $1/\sqrt{(LC)}$.

The useful power output of the amplifier depends on the ratio, γ, of the peak of the fundamental component (I_{fm}) to the peak of the transistor waveform (I_m), which is approximately

$$\gamma \approx 0.2\theta_c + 0.001\theta_c^2 - 0.00446\theta_c^3$$

when $0 < \theta_c < \pi$ radians. Then the maximum average load power is

$$P_L = \tfrac{1}{2}I_{fm}V_{CC} = \tfrac{1}{2}\gamma I_m V_{CC}$$

γ ranges from 0 when $\theta_c = 0$ to 0.5 when $\theta_c = \pi$ radians. The average power drawn from the source is $I_{av}V_{CC}$ where I_{av} is

$$I_{av} = \frac{I_m}{\pi}\frac{\theta_c}{\pi} = \frac{I_m\theta_c}{\pi^2}$$

Thus the average source power output is

$$P_S = \frac{\theta_c I_m V_{CC}}{\pi^2} \qquad (8.6)$$

Therefore the efficiency at maximum output power becomes

$$\eta = \frac{P_L}{P_S} = \frac{\tfrac{1}{2}\gamma I_m V_{CC}}{\theta_c I_m V_{CC}/\pi^2} = \frac{\pi^2\gamma}{2\theta_c} \approx 1 + 0.005\theta_c - 0.0234\theta_c^2 \qquad (8.7)$$

When θ_c is maximum, that is π radians, the efficiency at maximum output is 0.785, the same as the maximum efficiency of class B. This result is not surprising as the duty cycle of the transistor is 50% and the amplifier is then class B.

Loads can be matched to the collector circuit by using a transformer in place of the inductor as in figure 8.8a. In this case the effective collector resistance is $Q_1^2R_1 \,/\!/\, n^2R_L$ where n is the turns ratio of the transformer, Q_1 is the Q-factor of the primary and R_1 is the primary resistance, which is in series with X_L though not shown in the diagram. The use of base bias, V_{BB}, is not essential as the input can be capacitively coupled, as figure 8.8a also shows; in this case the base-emitter junction acts as a clamping diode and shifts the AC input down to one diode drop above ground. The input need not be at the resonant frequency of the tank circuit, but can be at a sub-multiple of it, so that the output is a higher harmonic of the input. This will only work for harmonics of an order much less than the Q-factor of the tank circuit plus load.

The class-C amplifier can easily be adapted for amplitude modulation using the circuit of figure 8.8b, which is very similar to that of figure 8.8a except that the transformer in the collector circuit is used to modulate the carrier signal which is input to the base of the transistor, and the output is at the collector.

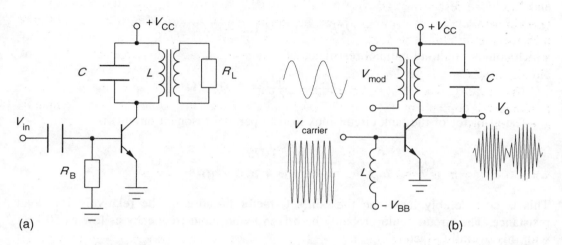

Figure 8.8 (a) Class-C amplifier with transformer-coupled load (b) Class-C amplitude modulator

Example 8.4

A class-C amplifier like that of figure 8.8a is used to drive a load of 50 Ω via a 5:1 step-down transformer, which has a primary resistance of 1.2 Ω and a Q-factor of 80 at the circuit's resonant frequency of 200 kHz. What is the capacitance of the tank circuit? What is the effective Q-factor of the collector circuit? If the input is a sinusoidal voltage of r.m.s. value 5 V, what is the duty cycle of the transistor? What is the conduction angle, θ_c? Find the efficiency according to equation 8.7. If $V_{CC} = 30$ V, calculate the load power. What is the peak collector voltage? If the secondary losses are negligible find the losses in the transformer. Given the transistor losses are 5.5 mW calculate the total losses and the efficiency. Why does this not agree with the previous calculation using equation 8.7? What is the peak current in the transistor collector? What is the AC current at 200 kHz in the equivalent resistance of the collector circuit? (Take $V_{BE} = 0.7$ V.)

The resonant frequency is 200 kHz, so the resonant angular frequency is 400π krad/s. The Q-factor of the primary coil is given by

$$Q_1 = \omega_0 L / R_1$$

where $Q_1 = 80$ and $R_1 = 1.2$ Ω. Thus the primary coil's inductance is

$$L_1 = \frac{Q_1 R_1}{\omega_0} = \frac{80 \times 1.2}{4\pi \times 10^5} = 76.4 \ \mu\text{H}$$

The resonant frequency is $\omega_0 = 1/\sqrt{(LC)}$ and so

$$C = \frac{1}{\omega_0^2 L} = \frac{1}{(4\pi \times 10^5)^2 \times 76.4 \times 10^{-6}} = 8.29 \text{ nF}$$

The equivalent resistances in parallel with L_1 and C are

$$R'_L = n^2 R_L = 5^2 \times 50 = 1.25 \text{ k}\Omega$$

and

$$R_{P1} = Q_1^2 R_1 = 80^2 \times 1.2 = 7.68 \text{ k}\Omega$$

which comes to a total parallel resistance of

$$R_p = \frac{1.25 \times 7.68}{1.25 + 7.68} = 1.075 \text{ k}\Omega$$

The Q-factor of the tank circuit plus load (a parallel resonant circuit) is

$$Q_p = \frac{R_p}{\omega_0 L} = \frac{1075}{4\pi \times 10^5 \times 76.4 \times 10^{-6}} = 11.2$$

This is considerably less than the coil's Q-factor because of the relatively low load resistance. The circuit could probably be driven by an input frequency as low as 50 kHz with this overall Q-factor.

The input voltage has a peak value of $5\sqrt{2} = 7.07$ V. The input to the transistor's base is therefore a sinusoid of the same peak-to-peak value as the input, but whose maximum is clamped to +0.7 V and whose minimum is −13.44 V. Thus the transistor is turned on at a phase angle of

$$\phi = \sin^{-1}\left(\frac{5\sqrt{2} - 0.7}{5\sqrt{2}}\right) = 64.3°$$

Then $\theta_c = 180° - 2\phi = 51.4° = 0.897$ radians. The transistor's duty cycle is therefore

$$D = \theta_c/2\pi = 0.897/2\pi = 0.143 = 14.3\%$$

Substituting this value of θ_c into equation 8.6 leads to $\eta = 98.6\%$. The primary voltage will oscillate between $+2V_{CC}$ and 0 V, just as it does in the transformer-coupled class-A amplifier. Thus the r.m.s. primary voltage, V_1, is $V_{CC}/\sqrt{2} = 21.2$ V and the r.m.s. load voltage will be 21.2/5 = 4.24 V. The load power, P_L, is then $4.24^2/50 = 360$ mW.

The peak collector voltage is $2V_{CC} = 60$ V. The power lost in the primary is V_1^2/R_{P1} where $R_{P1} = 7.68$ kΩ, the equivalent parallel resistance of the primary. The primary losses then are $21.2^2/7680 = 58.5$ mW and these are the total transformer losses. Adding them to the transistor losses of 5.5 mW gives a total losses of 64 mW. The power supplied becomes 64 + 360 = 424 mW and the efficiency is

$$\eta = \frac{\text{load power}}{\text{power supplied}} = \frac{360}{424} = 0.849 = 84.9\%$$

The reason this is not 98.6%, as calculated from equation 8.6, is that this equation — like all the maximum efficiency calculations — assumes there are no losses in the reactances of the tank circuit, but only in the transistor.

The peak collector current can be found from equation 8.5, which can be rewritten as

$$I_m = \frac{\pi^2 P_S}{\theta_C V_{CC}} = \frac{0.424\pi^2}{0.897 \times 30} = 0.156 \text{ A}$$

The peak AC current in the equivalent collector resistance is

$$I_{fm} = \gamma I_m = 0.177 \times 0.156 = 0.0276 \text{ A} = 27.6 \text{ mA}$$

which is 19.5 mA r.m.s. This result is more readily obtained from the r.m.s. load voltage of 21.2 V and the collector resistance of 1.075 kΩ by Ohm's law, $I = 21.2/1.075 = 19.7$ mA. The small (about 1%) disagreement reflects the approximation in γ.

8.4 Class-D amplifiers

Class-D (and class-E etc.) amplifiers are the most efficient of all because they operate with the active device either cut-off or in saturation, so that its power consumption is virtually nil. The active device merely switches the supply voltage on and off. The load voltage waveform may therefore be far from sinusoidal and more like a square wave, but with suitable filtering a sinusoidal output can be obtained if desired. Reactive loads often provide all the filtering necessary in any event. The power lost in the BJTs used in class-D amplifiers is

$$P_{BJT} = V_{CEsat} I_C$$

If I_C is also the load current, then the load power is $P_L = I_C^2 R_L$ and I_C is given by

$$I_C = \frac{V_{CC} - V_{CEsat}}{R_L}$$

and the efficiency is

$$\eta = \frac{P_L}{P_L + P_{BJT}} = \frac{I_C^2 R_L}{I_C^2 R_L + I_C V_{CEsat}} = \frac{I_C R_L}{I_C R_L + V_{CEsat}}$$

$$= \frac{V_{CC} - V_{CEsat}}{V_{CC}} = 1 - \frac{V_{CEsat}}{V_{CC}}$$

Thus when $V_{CC} = 12$ V and $V_{CEsat} = 0.3$ V, the efficiency is 97.5%.

Most class-D amplifiers use MOSFETs instead of BJTs because the input resistance is very high and the switching speeds are high also, even when the load current is high. BJTs carrying large currents cannot be switched as rapidly as MOSFETs. But ordinary MOSFETs have a relatively large on resistance, R_{DS}(on), which means that the power losses can be higher than with BJTs; BIFET power devices attempt to combine the better features of BJTs and FETs. A great deal of effort has been put into reducing the drain-source channel resistance (see for example figure 4.32) and hence $R_{DS}(on)$. The power lost in the MOSFET is $I_D{}^2 R_{DS}(on)$ so the efficiency of a class-D MOSFET amplifier approximates to

$$\eta = 1 - \frac{R_{DS}(on)}{R_L}$$

If $R_{DS}(on) = 0.2\ \Omega$ and $R_L = 10\ \Omega$, $\eta = 98\%$.

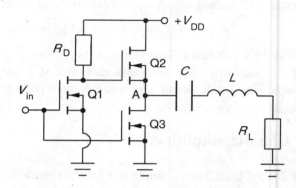

Figure 8.9

A class-D amplifier driving a series-resonant circuit with n-channel, enhancement-mode MOSFETs

Figure 8.9 shows an arrangement, known as a totem pole, often used in class-D MOSFET amplifiers. The input controls the gate of an n-channel enhancement-mode MOSFET, Q1, and the lower of the output transistors, Q3. When the input is high Q1 and Q3 are turned on and node A is at 0 V. When the input is low, Q1 and Q3 are off and Q2 is on and node A is at a potential of almost $+V_{DD}$. The load circuit connected to A is a series resonant circuit with $\omega_0 = 1/\sqrt{(LC)}$ which acts as a filter so that the load voltage is almost sinusoidal. If the input is a square wave then the voltage at A can be expressed as

$$v_A = \left(\tfrac{2}{\pi} V_{DD}\right)\left(\sin \omega t + \tfrac{1}{3}\sin 3\omega t + \tfrac{1}{5}\sin 5\omega t \dots\right)$$

At a frequency, $\omega = \omega_0$, the r.m.s. load current is almost entirely due to the fundamental, which is

$$I_L = \frac{2V_{DD}}{R_L \pi \sqrt{2}} = \frac{0.45 V_{DD}}{R_L}$$

since third and higher harmonics are filtered out by the load circuit. It can be shown (see problem 8.5) that the load current due to the third harmonic is only

$$I_L(3) = \frac{0.15}{1 + 7.11 Q^2} \frac{V_{DD}}{R_L} \tag{8.8}$$

where $Q = \omega_0 L / R_L$.

8.4.1 Pulse-width modulation

One of the chief uses of class-D etc. amplifiers is in industrial motor control wherein waveforms are synthesised from a series of pulses. The input waveform is sampled by a comparator having a sawtooth wave as its reference voltage as in figure 8.10a. The output of the comparator is a series of pulses as shown in figure 8.10b, whose width is proportional to the amplitude of the input. These pulses can then be used to control the output stages of a class-D amplifier. If the load is inductive the load current is

$$I_L = \frac{1}{L}\int V_o \, \mathrm{d}t$$

and so its waveform will approximate to the original input to the comparator, V_{in}. Speed control of induction motors can be achieved over a wide range by varying the frequency of the input to the comparator.

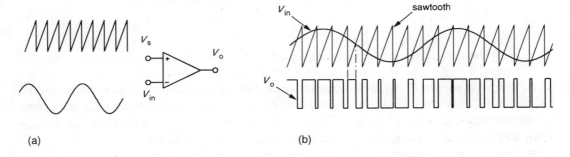

Figure 8.10 (a) The inputs to a PWM comparator (b) The output is a series of rectangular pulses

8.5 Power supplies

By the term 'power supply' we mean a mains-derived, DC power source which has appropriate characteristics for its purpose, which normally implies a constant voltage — within certain limits — over the full range of specified currents drawn from it. Figure 8.11 shows a typical power supply suitable for low-power applications. The starting point for such a power supply is a transformer of appropriate rating which will give one or more AC voltages of the right magnitude. The transformer output voltage is then rectified and smoothed to give DC. The rectification at mains frequencies is usually performed by a bridge rectifier in a single package. A large smoothing capacitor is required if the output ripple voltage must be small and the output current large. For safety a bleeder resistor of a few kΩ is placed across the smoothing capacitor so that no charge can remain on it at switch-off. The circuit must be protected by several devices: a fuse near the input is essential to avoid transformer burn out if the output is short-circuited, or the current drain excessive for any reason. The fuse rating should have a margin of about 20% over the largest current normally taken, and it should preferably be a slow-blowing fuse[2] to absorb the current surge on staring up. The varistor (voltage-variable resistor) across the input prevents any large voltage spikes from damaging the circuit by shunting them to ground. The snubber circuit placed across the primary of the transformer absorbs the energy in the circuit when it is broken or switched off while on load. The snubber capacitor should be a few nF and the resistor a few kΩ.

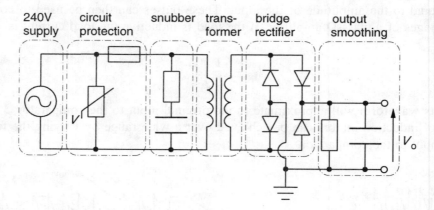

Figure 8.11 A small, unregulated, power supply with simple capacitive smoothing on the output

If a bipolar supply is required a derivative of the voltage doubler circuit of figure 2.17 can be used as shown in figure 8.12. Here C_1 is a small capacitor and C_2 is a large smoothing capacitor. By grounding one end of the secondary the output voltage swings from $+V_1$ to $-V_1$ instead of being doubled as in the voltage doubler.

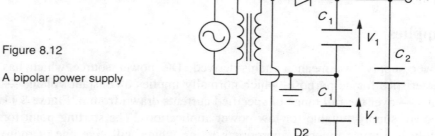

Figure 8.12

A bipolar power supply

Example 8.5

Design an unregulated power supply to deliver a maximum of 1.25 A into an 8Ω load with a peak-to-peak ripple of 0.2 V. What is the open-circuit voltage of your supply?

'Unregulated' means that the output voltage will vary according to the output current.

---

[2] These are sometimes listed as 'anti-surge' or 'semi-delay' or 'mid-delay' or 'T' fuses.

In section 2.2.7 we saw that the peak-to-peak ripple in a half-wave rectifier was given by

$$V_{rpp} = \frac{V_p}{RCf} = \frac{I_p}{Cf}$$

where R is the load resistance, V_p is the peak load voltage and I_p the peak load current. In a full-wave rectifier the ripple is half of this, and so

$$C = I_p/2V_r f = 1.25/2 \times 0.2 \times 50 = 0.0625 \text{ F}$$

This is a large capacitance for a small power supply, but the low voltage means that a 16V cylindrical electrolytic of 68 000 μF can be used with dimensions of 50 mm (L) by 35 mm (Ø), suitable for PCB mounting, and costing about £6. The bleed resistor should be chosen to give a discharging time constant of about a minute, which means that

$$R = T/C = 60/0.068 = 880 \ \Omega$$

A preferred value of 920 Ω will do, and as this is over a hundred times the load resistance its current consumption of 11 mA is negligible.

The peak load voltage is 10 V, so if we use a diode bridge rectifier the maximum on the transformer secondary is $10 + 0.7 + 0.7 = 11.4$ V, since there are two diode drops to take into account. A packaged bridge rectifier will use diodes of lower voltage drop than this: the usual voltage drop is only about 1 V at rated maximum current. The primary peak voltage is $240\sqrt{2} = 339$ V, so the transformer turns ratio is $339/11 = 31{:}1$. It is possible that a 20VA (the maximum load power is 12.5 W) off-the-shelf transformer will suffice. One with a 2×6V, 1.67A rating has two secondary coils which will give a 12V output when connected in series. However, if the load voltage has to be exactly 10 V at full current, then we can include some series resistance and put up with lower efficiency. The transformer costs about £4, but its *regulation* will be poor. The regulation of a transformer is expressed as

$$\%reg = \frac{100(V_{NL} - V_{FL})}{V_{NL}}$$

where V_{NL} is the secondary voltage on no-load (open-circuit output) and V_{FL} is the voltage on full load (the rated voltage). The regulation of a 20VA mains-input transformer is about 12%, which means that its open-circuit output voltage is

$$V_{NL} = \frac{V_{FL}}{1 - \%reg/100} = \frac{12}{1 - 0.12} = 13.6 \text{ V}$$

This is because the output impedance is a substantial fraction of the full-load resistance. Poor regulation is a common feature of all small mains-input transformers. In conjunction with poor regulation goes poor efficiency which in this case is about 80%, implying that the transformer consumes about 5 W on full load, and about 1 W on no-load or standby.

Though of no economic significance in themselves, these losses must be allowed for when the equipment is built to prevent an excessive rise in temperature. The input power on full load is then 25 W and the r.m.s. input current is 100 mA. In many cases PCB transformers are fully protected against short circuits etc. and require no additional components on the input side, so keeping costs to a minimum. If a fuse is required it should have a rating of 130 mA: these cost about 8p.

Finally a bridge rectifier must be selected which will be capable of delivering 1.25 A, a very modest requirement: a 2A bridge rectifier costs only 40p. The total cost of components for this small power supply is therefore only about £11, and that is as a prototype, or 'one-off': the cost when manufacturing large numbers would be about halved.

8.5.1 The series voltage regulator

As we have seen in the previous example, the regulation provided by small power supplies is poor because the transformers have poor regulation. Using larger transformers is not an economical way of improving the regulation, though it will certainly do that. The most common way of improving the regulation in small power supplies is to use a transistorised regulator, such as shown in figure 8.13, which is known as a series regulator because the output current goes through the pass transistor and load in series. Voltage regulators are available in IC form with numerous variations; they are the second most popular IC that is bought from the electronic component distributors — only op amps sell better. The reason is that they take a potentially highly variable supply voltage with a good deal of ripple and convert it into almost ripple-free DC at a very constant voltage from zero to maximum rated current. Table 8.1 gives a few of the commoner voltage regulators and their salient characteristics.

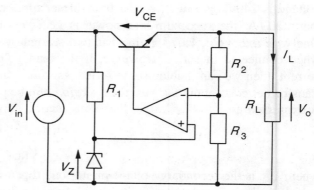

Figure 8.13

A voltage regulator controlled by an op amp. There is no limitation on current here

The key component is the Zener diode which sets the output voltage to

$$V_o = \left(1 + \frac{R_2}{R_3}\right)V_Z$$

By choosing V_Z to be between 4 V and 6 V, the tempco is kept down to about +0.5 mV/K with a dynamic resistance of about 75 Ω at a Zener current of 10 mA.

There are several problems with the simple regulator shown in figure 8.13, the first being the power lost in the output transistor, which is $I_L V_{CE} = I_L(V_{in} - V_o)$, so that if $V_{in} =$ 12 V and $V_o = 6$ V and $I_L = 1$ A, the transistor must dissipate 6 W, the same as the load!

The second problem is that if the load is short-circuited the transistor will try to supply a very large current and so burn out. For this reason nearly all voltage regulators have short-circuit protection and usually a thermal cut-out as well which comes into operation when the transistor reaches a preset temperature. There is also a minimum voltage across the output transistor (the 'drop-out'), so that the minimum input voltage is higher than the regulated output voltage. For various reasons connected with the details of the construction of the IC this drop-out can be as high as 3.6 V (see table 8.1).

Example 8.6

In the voltage regulator shown in figure 8.13 the resistance values are $R_1 = 1$ kΩ, $R_2 = 16$ kΩ, $R_3 = 20$ kΩ and $R_L = 80$ Ω, with $V_Z = 5$ V. If the input voltage varies from 11 V to 16 V what is the maximum power lost in the transistor? The Zener has a tempco of +2 mV/K and a dynamic resistance of 60 Ω at a current of 8 mA, where $V_Z = 5$ V at 25°C. By how much does the output voltage vary if the Zener temperature increases by 0.3 K/mW and the ambient temperature is 25°C?

The output voltage is

$$V_o = \left(1 + \frac{R_2}{R_3}\right) \times V_Z = \left(1 + \frac{16}{20}\right) \times 5 = 9 \text{ V}$$

so if the input voltage is 16 V the collector-emitter voltage is 7 V. The load current is 9/80 = 0.1125 A, and ignoring the voltage divider whose current consumption is only 0.25 mA, the transistor's maximum power consumption is $7 \times 0.1125 = 0.788$ W.

The Zener voltage extrapolated to zero current is

$$V_{Z0} = V_Z - r_z I_Z = 5 - 60 \times 8 \times 10^{-3} = 4.52 \text{ V}$$

At the lowest input voltage the voltage across R_1 is $11 - 5 = 6$ V, and therefore the current through it is 6 mA, and the actual Zener voltage will be

$$V_Z(\text{min}) = V_{Z0} + r_z I_Z = 4.52 + 60 \times 6 \times 10^{-3} = 4.88 \text{ V}$$

The Zener's power dissipation is $4.88 \times 6 = 29.3$ mW, so its temperature rise is $29.3 \times 0.3 = 8.8$ K. The Zener voltage then increases by $2 \times 8.8 = 17.6$ mV to 4.90 V. The output voltage is therefore $1.8 \times 4.9 = 8.82$ V.

At the highest input voltage the voltage across R_1 is $16 - 5 = 11$ V and the Zener current is therefore 11 mA. By the same procedure as before we find the Zener voltage to be 5.18 V and the power dissipation to be 57 mW, which produces a temperature rise of 17.1 K and a Zener voltage increase of 34.2 mV. The eventual Zener voltage is thus 5.21 V and the output voltage is 9.38 V.

With these Zener parameters the regulation is very poor, amounting to about 11% of the input voltage variation. The biggest source of variation is the Zener's dynamic resistance, which can be reduced by increasing the Zener current by reducing R_1, though this will lead to a greater voltage variation through temperature changes. In practice the Zener is connected to a constant-current source and is accurately temperature compensated, so that the output voltage is far less variable than suggested in this example.

Current limiting

Current limiting is essential for a series voltage regulator and a simple method of obtaining it is to place another transistor, Q2, and a small current-sensing resistor, R_4, in the output circuit as shown in figure 8.14. When the voltage across R_4 rises to $V_{BE} \approx 0.7$ V, Q2 turns on and diverts current from the base of Q1, thereby lowering the output current to a maximum of

$$I_L(\text{max}) = V_{BE}/R_4 \approx 0.7/R_4$$

If current limiting is used, the output voltage will of course fall when the load resistance falls below the point where it draws maximum current (see problem 8.7).

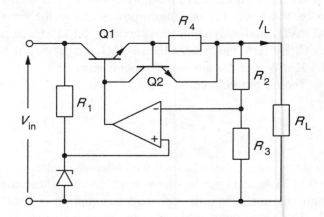

Figure 8.14

A series voltage regulator with current limiting by Q2. R_4 is the current sensing resistor

8.5.2 Practical voltage regulators

Because voltage regulators are readily available as cheap ICs and with remarkably good specifications, nobody needs to design one. A few practical points should, however, be noted. All the devices will dissipate power and the rise in temperature is determined by the thermal impedance of the package. The TO92 package is not designed to dissipate much power and in free air (no heat sink, no fan) will have a junction temperature rise of about 200 K/W, so at maximum power the rise above ambient will be about 120 K. The TO220 and TO3 packages have relatively low thermal impedances provided they are properly connected to appropriate heat sinks. If the device is improperly heat sunk and rises in temperature excessively, it will automatically shut down when its thermal overload protection is activated. This can cause unexpected and intermittent equipment outages.

The ripple rejection is in dBs, so that if the input peak-peak ripple is 1 V and the ripple rejection is 60 dB the output ripple will be only 1 mV. The input needs relatively little smoothing and the capacitance can be much smaller than those in unregulated supplies. Several figures are given for regulation given in data sheets: the line or voltage regulation is that defined for the transformer previously, and often called the 'input regulation'. Unfortunately there are many different ways of stating this. Some manufacturers give a percentage figure which covers the full range of input (and output voltages), others give a %/V figure which is the percentage change in output for each volt change in input, and some give mV/V, and others give just mV, which means the change in output over the full range in input voltage. The load (or output) regulation is normally stated as well as the line regulation. This is defined as

$$\%reg(\text{load}) = 100\Delta V_{L}/\Delta I_{L}$$

Thus if the load regulation is 0.2% and the load current changes by 3 A, the load voltage change will be $3 \times 0.2/100 = 0.006$ V $= 6$ mV. More important than either can be the variation in output voltage with temperature, which may be as much as 1% for a junction temperature change of 100 K. This figure is not always given.

Another point to bear in mind is that the regulator will only work above a certain minimum current, usually about 0.5-1% of the rated maximum. If a regulator is used to supply CMOS op amps, for example, in a precision circuit which consumes little current, it is possible for the output to fall out of regulation. If in doubt, assume the 1% minimum figure, which for a 3A maximum-current regulator is a rather large 30 mA.

When a bipolar supply is needed a pair of regulators must be chosen, with the appropriate input voltages, or a bipolar regulator such as the RC4195N should be used. The table lists only an arbitrary selection of devices: there are hundreds of voltage regulators on the market and in crucial applications the manufacturers' data sheets must be consulted.

Table 8.1 *Voltage regulator ICs*

Part No.	Pack	V_{in}[A]	V_{o}[B]	I_{L}[C]	drop[D]	RRF[E]	P[F]	Price[G]
LM79L05ACZ	TO92	−7/−35	−5	0.1	2	50	0.6	40p
MC78L12ACP	TO92	14/35	12	0.1	1.7	62	0.6	60p
RC4195N[α]	8-DIL	±30	±15	0.1	3	75	0.6	£1.50
LM309K	TO3	7/35	5	1	2.4	72	30	£3
MC7824CT	TO220	27/40	24	1	2	68	20	50p
LM2940CT-5	TO220	5.5/26	5	1	0.5	70	30	£1.50
LM317T[β]	TO220	4.2/77	1.2/37	1.5	2.3	80	20	80p
MC78T15CT	TO220	18/40	15	3	3	65	40	£1
IP3R18K-12	TO3	15/35	12	5	3	72	50	£8
LT1584CT[γ]	TO220	5/7	1.5/3	7.1	1.25	85	30	£8

Notes: All devices have internal current limiting and thermal protection
[A] Input voltage range
[B] Output voltage (fixed) or range (adjustable)
[C] Max load current in A
[D] Min dropout voltage
[E] Typical ripple reduction factor at 120 Hz in dB, worst case may be 10-15 dB lower
[F] Max device power dissipation in W, determined by the package
[G] Single quantities
[α] Bipolar regulator for op amp power supply in 8-pin DIL package
[β] Adjustable output
[γ] Adjustable output, Pentium supply, low-voltage logic supply

Example 8.7

Design a power supply which is to deliver a maximum of 1.25 A with an output voltage of 10 V with a maximum ripple of 2 mV peak-to-peak. Estimate its efficiency on full load.

The voltage is not a standard one of 5, 9, 12 or 15 V, so the adjustable regulator LM317T in table 8.1 must be used. Assuming a worst case for the ripple reduction factor of 80 − 15 = 65 dB, the input ripple should be not more than

$$V_{rpp} = 2 \times 10^{-3} \times 10^{65/20} = 3.6 \text{ V}$$

The drop out for the LM317T is given as 2.3 V, minimum: it would be prudent to allow 3 V, or even 4 V for this. Thus the input voltage must be at least 13 V, rising to a maximum of 16.6 V and we then must allow another 1.4 V for the bridge rectifier, which leads to a transformer secondary peak output of 18 V. The tolerance on off-the-shelf transformers is quite high — at least 5%, so it would be wisest to opt for a higher secondary voltage than 18 V. The next highest standard voltage is 24 V and the rating required is then at least 24 V × 1.25 A = 30 VA. This is not a standard value and we have to use a chassis-mounted 50VA transformer, which although quite cheap (about £6) is rather large and weighs about 1 kg.

The bridge rectifier can be a 2A one with a forward voltage drop of only 1 V, costing 40p. Thus the transformer and rectifier combination can supply a current of up to 2 A, instead of the 1.25 A required.

The capacitor size can be calculated from the usual formula to be

$$C = \frac{I}{2V_{rpp}f} = \frac{1.25}{2 \times 3.6 \times 50} = 3500 \ \mu\text{F}$$

though the ripple is now so large that the approximation is only rough, but it errs on the safe side: the actual ripple will be less than 3.6 V. But the tolerances on electrolytic capacitors are ±20%, so it would be best to opt for less ripple and use a smaller transformer. There are cylindrical 6 800μF electrolytic capacitors available rated at 25 V, costing only £1.50, which are small: 35 mm (L) × 22 mm (Ø). This size of capacitor will reduce the ripple to about 1.8 V_{pp} and thus the transformer secondary voltage need only be 16.2 V: an 18V transformer can be used rated at 18 × 1.25 = 22.5 VA, which means we still require a 50VA transformer.

If the transformer secondary voltage is higher than necessary it will increase the power lost in the regulator: taking the mean rectified voltage, 16 V, for the regulator input voltage, the drop across the regulator is 6 V producing a regulator dissipation of 6 × 1.25 = 7.5 W at maximum output current, which is tolerable. The regulator will still require a good heat sink: a 6.8°C/W TO220 heat sink costs about £1.50 and will produce a temperature rise in the TO220 case of 7.5 × 6.8 = 51°C at maximum power.

The LM317T needs additional resistors to enable the output voltage to be adjusted, as shown in figure 8.15, which shows the TO220 package and the circuit used. The regulated output voltage is given by

$$V_o = V_{ref}(1 + R_2/R_1) = 1.25(1 + R_2/240) = 10 \text{ V}$$

since V_{ref} for the LM317T is 1.25 V and R_1 is specified by the makers to be 240 Ω. This gives $R_2 = 1.68$ kΩ, so a 4.7kΩ potentiometer costing 15p would do for R_2. The additional capacitors shown in figure 8.15b are used if better transient suppression is called for (C_2) or if the regulator is somewhat remote from the smoothing capacitor (C_1). There may also be a need for a small series surge-suppression resistor before the smoothing capacitor. The surge current is $I_S = V_{Cmax}/R_S$ where V_{Cmax} is 17 V and the surge rating of the rectifier is 60 A, making $R_S = 0.3$ Ω. This will hardly affect the voltages calculated previously and consumes only 0.5 W. The circuit is shown in figure 8.16.

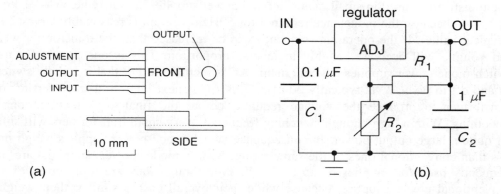

Figure 8.15 (a) LM317 voltage regulator in a TO220 package (b) Output voltage adjustment

At full load the regulator will consume 7.5 W. The bridge rectifier will dissipate about 1.25 W, taking the average current to be 1.25 A and the average forward voltage drop to be 1 V. Thus the power output from the transformer is

$$P_{Tx} = P_L + P_{reg} + P_R + P_{rect} = 12.5 + 7.5 + 0.5 + 1.25 \approx 22 \text{ W}$$

If the transformer is 75% efficient then the power input is 22/0.75 = 29 W. The overall efficiency is therefore

$$\eta_{max} = P_{Lmax}/P_{in} = 12.5/29 = 0.43$$

Small power supplies are always rather inefficient, but the losses are small anyway.

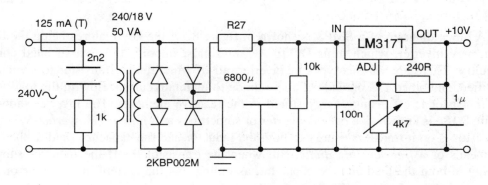

Figure 8.16 The power supply for example 8.7

The cost of the power supply's components comes to about £10.50 (excluding the additional capacitors shown in figure 8.16), which is almost the same as the unregulated one in example 8.5, though the efficiency is slightly lower and the total power dissipation a little more.

8.6 Switch-mode power supplies (SMPS)

The transformers used in small mains-derived power supplies are fairly heavy — 1 kg per 50 VA — because the inductance required for 50Hz operation is reasonably large (2 H is a typical value). If the operating frequency can be increased then the transformer weight and volume will decrease roughly in inverse proportion. This is one reason for using switch-mode power supplies wherein mains AC is rectified to DC that is rapidly switched on and off to give AC at typically 20 kHz. This AC is next input to a small ferrite-cored transformer to produce the voltage required before the final-stage rectification and smoothing. With a high-enough switching frequency, very small transformers will suffice for quite a large output power. In consequence switch-mode power supplies weigh much less than conventional ones of the same rating. Switch-mode power supplies are now frequently used for supplies of up to 3 kW, even though they are more complex than conventional power supplies, because while their overall cost is similar, their weight is considerably lessened.

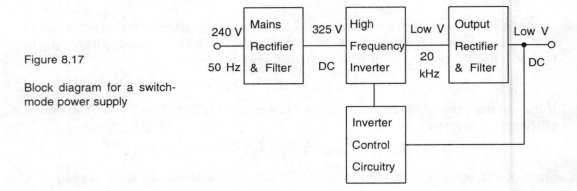

Figure 8.17

Block diagram for a switch-mode power supply

A block diagram of a SMPS is shown in figure 8.17, the key features being the high-frequency inverter, which turns DC to AC, and most importantly its associated control circuitry. There are various types of inverter, all of which use a transistor to switch the rectified, smoothed DC on and off and a diode to maintain current flow in the load when the transistor is off, using energy stored in the reactive elements. The inverter shown in figure 8.18a is known as a *buck converter* or sometimes as a *forward converter*. When the transistor is on it transfers energy to both the inductor and the capacitor (and to the load) by means of a *freewheeling diode*. But when the transistor is off the inductor supplies energy to both the load and the capacitor, assuming that the current in the inductor is at all times greater than zero. The input voltage, V_{in}, from the rectifier and the load voltage,

V_o, are essentially constant if the supply is to be DC. The resulting current and voltage waveforms for the diode and the inductor are shown in figure 8.18b.

When the transistor is ON, the voltage across the inductor is $V_{in} - V_o$ and so the change in the inductor current if the ON time is T_{ON} is

$$\Delta I_L = \frac{(V_{in} - V_o)T_{ON}}{L} \tag{8.9}$$

using $V = L\Delta I/\Delta t$. When the transistor is OFF the voltage across the inductor must be V_o (apart from a diode drop) and so if the OFF time is T_{OFF} the current change is

$$\Delta I_L = V_o T_{OFF}/L$$

Equating the two expressions for ΔI_L leads to

$$V_o = \frac{V_{in}T_{ON}}{T_{ON} + T_{OFF}} = \frac{V_{in}T_{ON}}{T} = V_{in}\delta \tag{8.10}$$

where the duty cycle, $\delta = T_{ON}/T$, and $T_{ON} + T_{OFF} = T$. Thus the output (load) voltage is proportional to the ON time of the transistor and must always be less than V_{in}.

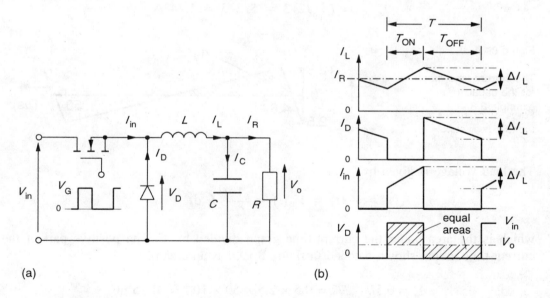

(a)   (b)

Figure 8.18 (a) A buck converter (b) Approximate voltage and current waveforms

Example 8.8

A buck converter like that of figure 8.18a is to operate at a switching frequency of 20 kHz with an input voltage of 325 V and an output voltage of 30 V across a 2Ω load. What is the ON time for the transistor? What is the value of inductance required if the current through it changes by 5 A during charge and discharge? Draw the capacitor current as a function of time. What is its r.m.s. value? What must C be if the peak-to-peak load voltage ripple is to be 1%? If the duty cycle can vary from 0-100%, when is the change in inductor current greatest? What is the load voltage and its ripple then?

The period of the transistor switching waveform is $T = 1/f = 1/20000 = 50$ μs. Thus the ON time for the transistor is, using equation 8.10,

$$T_{ON} = V_o T/V_{in} = 30 \times 50/325 = 4.6 \text{ μs}$$

The change in inductor current, by equation 8.9, is

$$\Delta I_L = \frac{(V_{in} - V_o)T_{ON}}{L} = \frac{(325 - 30) \times 4.6 \times 10^{-6}}{L} = 5 \text{ A}$$

whence $L = 0.272$ mH.

If the load voltage is constant, so is the load current, I_R, and then by KCL at the top right-hand node of figure 8.18a,

$$I_C = I_R - I_L$$

And thus the capacitor current-time graph is as in figure 8.19, a triangular wave-form whose r.m.s. value is

$$I_{Crms} = I_{Cpp}/2\sqrt{3} = 5/2\sqrt{3} = 1.44 \text{ A}$$

Figure 8.19

The current-time graph for the capacitor in example 8.8

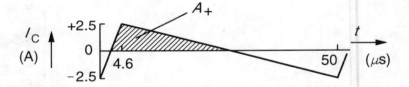

The load voltage is given by

$$V_C(t) = V_R(t) = \frac{1}{C}\int_0^t I_C \, dt$$

which is the area under the current-time graph divided by C. The positive part of the current-time graph, shown hatched in figure 8.19, has an area of

$$A_+ = 0.5 I_{Cmax} T/2 = 0.5 \times 2.5 \times 50 \times 10^{-6} = 31.25 \text{ μC}$$

because the current is positive for half a cycle and negative for the other half. The ripple voltage required is $0.01 \times 30 = 0.3$ V. And the increase in voltage during the charging of C is half the peak-to-peak ripple voltage, hence

$$0.5 V_{pp} = A_+/C \quad \Rightarrow \quad C = 2A_+/V_{rpp} = 2 \times 31.25 \times 10^{-6}/0.3 = 208 \text{ μF}$$

The inductor current as a function of T_{ON} is found by substituting equation 8.9 into equation 8.10 to obtain

$$\Delta I_L = \frac{V_{in}(1 - \delta)T_{ON}}{L} = \frac{V_{in}(1 - \delta)T\delta}{L}$$

This is greatest when $\delta = 0.5$ or $T_{ON} = T/2$ and then

$$\Delta I_{Lmax} = V_{in}T/4L = 325/(20\,000 \times 4 \times 0.27 \times 10^{-3}) = 15 \text{ A}$$

The load voltage is $V_{in}\delta = 325 \times 0.5 = 162.5$ V and the average load current must be $162.5/2 = 81.25$ A. Since the inductor current change has increased by a factor of three and the capacitor current now goes from $+7.5$ A to -7.5 A, the ripple voltage has increased by a factor of three to 0.9 V, which is $0.9 \times 100/162.5 = 0.55\%$. The percentage change in load voltage has decreased with the increased duty cycle (see also problem 8.9).

8.6.1 The switching regulator

The switching regulator shown in figure 8.20 makes use of a PWM comparator to control the switching transistor.

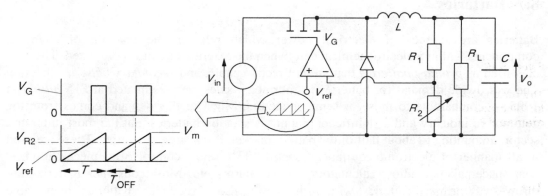

Figure 8.20 A switching voltage regulator using a PWM comparator to drive the MOSFET gate

The reference voltage into the comparator is a sawtooth wave whose peak value is V_m. The comparator output goes high and switches on the transistor when $V_{ref}(max) > V_{R2}$ and so the OFF time is TV_{R2}/V_m and the ON time is therefore $(1 - V_{R2}/V_m)T$. The duty cycle is

$$\delta = \frac{T_{ON}}{T} = 1 - \frac{V_{R2}}{V_m}$$

Then the output voltage is

$$V_o = V_{in}\delta = \left(1 - \frac{V_{R2}}{V_m}\right)V_{in}$$

wherein

$$V_{R2} = \frac{V_o R_2}{R_1 + R_2} = \frac{V_o}{1 + K}$$

and $K = 1 + R_1/R_2$. Substituting for V_{R2} in the expression for V_o leads to

$$V_o = \left(1 - \frac{V_o}{(1 + K)V_m}\right)V_{in}$$

This gives the output voltage as

$$V_o = \frac{V_{in}}{1 + \dfrac{V_{in}}{(1 + K)V_m}} = \frac{(1 + K)V_m}{(1 + K)V_m/V_{in} + 1}$$

If $V_m = 10$ V and $V_{in} = 200$ V, then $V_o = 18$ V when $K = 1$ rising to 167 V when $K = 100$.

8.7 Batteries

Batteries are sources of electrical power which rely on the reaction of chemical components within a electrochemical cell[3] when the external circuit is completed. They are classified as *primary batteries* that are not rechargeable, and *secondary batteries* that are. The worldwide demand for batteries is enormous: sales were estimated at £15 *billion* in 1995 — comparable to those of semiconductors. Much of this demand comes from the automotive industry and a significant amount is standby batteries, but the most important sector, amounting to about half of the battery market, is consumer batteries. These are used in all manner of electronic equipment, such as TV remote controls, timepieces, torches, heart pacemakers, radios, calculators, smoke alarms etc. Most consumer batteries are throw-away primary batteries, with an increasing use of secondary batteries for portable tools and other equipment driven by electrical motors in which the power demand is high for relatively short periods. The discussion here is mainly restricted to consumer batteries.

8.7.1 Types of battery

Batteries can be classified according to the chemical system which provides the electrical power. The negative electrode (from which electrons flow into the external circuit) is invariably a metal such as zinc (Zn), cadmium (Cd) or lithium (Li), while the positive electrode (the source of conventional current in the external circuit) is an oxidant such as manganese dioxide (MnO_2). The two electrodes are immersed in a conducting electrolyte which in the so-called 'dry' cell is actually a paste. The chemical system is then written

---

[3] Strictly speaking a battery comprises two or more cells connected in series or parallel or both, while a cell has just two terminals and cannot be split up further; but in common usage this distinction has long gone.

with the negative electrode material first, then the electrolyte, then the positive electrode material as in $Zn/ZnCl_2/MnO_2$. Often the electrolyte is omitted as in the Zn/MnO_2 system, which is popularly called zinc-carbon. Most of the commonly-used chemical systems were discovered long ago; the Leclanché cell (Zn/MnO_2) was first described in 1868 and the Planté (or lead-acid, Pb/PbO_2) cell in 1859. Table 8.2 lists some of the most popular consumer batteries.

Table 8.2 *A selection of commonly-used consumer batteries*

Type[A]	Size[B]	Voltage[C]	Capacity[D]	Price[E]	p/Wh[F]
[G]Zn/MnO_2	D	1.5	3.75 (2 Ω)	£0.84	15
[H]$Zn/ZnCl_2/MnO_2$	D	1.5	3.8 (2 Ω)	£1.10	19
Alkaline Zn/MnO_2	D	1.5	7.5 (2 Ω)	£1.40	12
			15.6 (10 Ω)	£1.30	6
[K]$Cd/NiO.OH$*	D	1.2	4	£5.40	113
computer Ni/Cd	Non-std pack	4.8	5	£60	250
camcorder Ni/Cd	Pack	6	2.4	£40	278
mobile phone Ni/Cd	various	7.2	0.7	£20	397
[L]Ni/MH*	AA	1.2	1.1	£4	303
[M]Pb/PbO_2*	51 × 43 × 50H	6	2 (60 Ω)	£8	67
	67H × 36Ø	2	2.5 (20 Ω)	£4	80
car battery[N]	200 × 165 × 170H	12	45 (5 Ω)	£30	5.6
'Lithium' batteries					
Li/V_2O_5*	3.2H × 30Ø	3	0.1	£4.25	1417
$Li/(CF)_x$	3H × 23Ø	3	0.26	£1.40	179
Li/FeS_2	AA	1.5	2.57 (3.9 Ω)	£2.60	67
Li/MnO_2	PP3	9	1.2 (75 Ω)	£6.30	58
$Li/SOCl_2$	AA	3.5	2 (75 Ω)	£4.25	61
	D	3.5	16.5 (25 Ω)	£14	24
Zn/air	5.4H × 8Ø	1.4	0.21 (250 Ω)	£0.75	255
	26 × 17 × 45H	8.4	1.5 (500 Ω)	£7	56
Zn/AgO	5.4H × 11.6Ø	1.5	0.15	£0.84	373

Notes: * = Rechargeable

[A] −ve electrode active material/+ve electrode active material or −/electrolyte/+

[B] Sizes in mm. H = height, Ø = diameter of cylinder. The standard D (R20) cell is 57.2H × 31.8ϕ and the AA (R6) cell is 47.8H × 13.5Ø; the PP3 is 26.5 × 17.5 × 48.5H.

[C] The working voltage may be considerably less [D] In Ah with load in brackets

[E] For single quantities [F] Price ÷ (Ah capacity × nominal voltage)

[G] Zinc-carbon or Leclanché cells [H] Chloride battery [K] Ni/Cd or Nicad [L] MH = metal hydride [M] Lead-acid [N] For comparison, not strictly a 'consumer' battery

The development of the Leclanché cell has been continuous since its invention in 1868. In fact the D-sized zinc-carbon battery of today has about eight times the capacity it had in 1920 and the alkaline-manganese battery has doubled its capacity in the last 12 years. A great deal of money is now being spent on the development of rechargeable batteries as portable, battery-powered equipment is becoming much more widely used, especially mobile telephones, television remote controls, camcorders and power tools. Compared to semiconductor devices the pace of improvement of batteries is slow but steady.

Batteries are an expensive way to provide power unless they are very large and can be repeatedly recharged many times. The choice of battery depends on how and where it is used: the maximum power and current, the required voltage, the duty cycle, the service temperature and the shelf life. A heavy-duty battery should not be used where a light-duty one would suffice, since it will cost more and will offer little better performance if lightly discharged. For example alkaline 'heavy-duty' batteries will repay their extra cost when delivering 1 A continuously, but not when delivering 50 mA intermittently. Rechargeable batteries are especially dear and should only be used when frequently-repeated recharging is required, as in portable power tools and shavers. Remember that the cost of the recharger has to be recovered and that the rechargeable batteries themselves can only undergo a certain number of recharge cycles. As a rough guide, only if you need to replace a battery at least once a month and at least twenty times is it worthwhile buying rechargeable batteries.

The most recent commercial development has been the used of nickel-metal hydride rechargeable batteries instead of nickel-cadmium. Not only does this replace cadmium — a potentially-harmful heavy metal pollutant — with a safe alternative, but it also offers a 50% increase in capacity for a given size of battery. Lead-acid batteries, though rechargeable, are less robust than Nicad batteries, and will not last long if deeply discharged, which is why they have been less popular as consumer batteries.

In many systems the battery e.m.f. declines continuously during discharge; for example the Leclanché cell's voltage can fall steadily to zero, but is reckoned to have reached its practical discharged state when its e.m.f. has fallen to 0.9 V. Where constancy of e.m.f. is necessary one must choose lithium, zinc-air or zinc-silver oxide batteries. Batteries employing lithium (Li) as the negative electrode have become popular since their introduction in about 1970. There are many different materials used for the positive electrode in lithium batteries and none seems to have achieved commercial dominance. Lithium batteries are generally only used if space is at a premium, since they are relatively expensive. If the battery has to be held in reserve for an emergency it must have a good shelf life: zinc-air batteries can be stored indefinitely without loss of capacity since they only work when exposed to air. Some batteries have rather short shelf lives, especially in hot climates.

8.7.2 Battery characteristics

Among the many characteristics of batteries are the voltage-time discharge curves as a function of temperature and discharge rate, the charge retention as a function of time and temperature and the charging characteristics. All of these vary, even from battery to battery of the same type, though modern production methods have reduced this variation considerably. Some performance curves for the more popular batteries are shown in figure 8.21, about which a word of caution: they are only a very small sample of huge amounts of manufacturers' data, which should always be consulted when designing battery-powered equipment of any kind. The figures in Ah are the battery's capacity when discharged through the load indicated in ohms or at the rate indicated in amps.

Figures 8.21a and 8.21b show clearly one major disadvantage of using zinc-carbon or alkaline manganese batteries for powering electronic equipment: their terminal voltages drop continuously during discharge. For this reason they are normally reserved for driving small motors and lamps. Figure 8.21b shows the advantage of the alkaline battery over the zinc-carbon battery when the current drain is heavy, an advantage that is almost entirely absent when the current drain is light.

Nicad batteries maintain their voltages much better on discharge than zinc-carbon batteries, as figure 8.21c shows, but they have only about a quarter of the capacity of even ordinary zinc-carbon cells. The great advantage of Nicad batteries is their rechargeability. Figure 8.21d shows the superior capacity of rechargeable nickel/metal-hydride batteries (NiMH) compared to Nicads, and this figure is out of date by now, given the relatively rapid improvement in NiMH batteries.

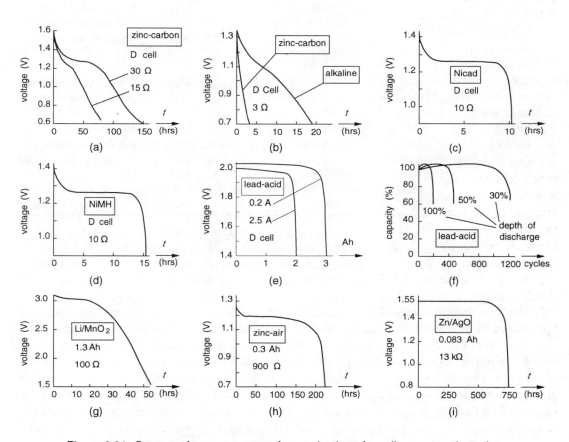

Figure 8.21 Some performance curves for a selection of small consumer batteries

The other common rechargeable battery is the lead-acid ($Pb/H_2SO_4/PbO_2$) battery, which is now available in sealed form owing to the development of jellied electrolyte. Lead-acid batteries have much greater capacity than Nicads and work at the much higher cell voltage of 2 V; they are also capable of very high discharge rates: a single D-cell can deliver 25 A for a minute or two. Unfortunately, figure 8.21f shows that if the depth of discharge exceeds about 30%, then the battery life is much reduced: deep (50-100%) discharges must be avoided. Figure 8.21 does not show a further drawback of lead-acid batteries, which is their inability to remain discharged for long without suffering permanent reduction in capacity, in contrast to Nicads which suffer little loss on being left discharged. Lead-acid batteries should therefore always be connected to a charger; their characteristics are well suited to car ignition systems.

Table 8.3 *Consumer batteries compared*

System	Shelf life[A]	Service[B] temperature	Emf/cell[C]	Cost[D]	Typical use
Zn/MnO_2	2	−7°C to +50°C	1.5 (1.2)	low	toys, clocks, alarms
$Zn/ZnCl_2/MnO_2$	2	−20°C to +70°C	1.5 (1.2)	low	tape recorders, radios
$Zn/alkaline MnO_2$	4	−30°C to +54°C	1.5 (1.25)	low	toys, torches, motors
$Cd/NiO.OH$ ('Nicad')	recharge	−40°C to +50°C	1.35 (1.2)	high	tools, camcorders
NiMH	recharge	−40°C to +50°C	1.4 (1.25)	high	laptops, mobile phones
$Pb/H_2SO_4/PbO_2$ (lead-acid)	recharge	−45°C to +60°C	2.1 (2.0)	high	standby power & lighting
$Li/SOCl_2$	10	−55°C to +85°C	3.6 (3.3)	medium	remote data collection
Li/MnO_2	10	−40°C to +70°C	3.2 (2.7)	medium	sonobuoys, mines
Li/V_2O_5	recharge	−55°C to +70°C	3.4 (2.5)	high	backup power on PCBs
Li/CuO	10	−40°C to +60°C	2.3 (1.5)	high	oil-well loggers
Zn-air	indefinite	−20°C to +60°C	1.45 (1.25)	medium	hearing aids, watches
Zn/AgO	2	−40°C to +75°C	1.6 (1.5)	med/high	instruments, calculators

Notes: [A] In years at 20°C; lower temperatures will increase and higher diminish this figure
[B] The capacity may be considerably reduced at the lower temperatures
[C] Open-circuit e.m.f. with typical on-load voltage in parentheses
[D] Initial cost for batteries of identical size, in appropriate service

The development of MOS devices for performing logical functions has led to a proliferation of small battery-powered devices for all manner of purposes. The advantage of the CMOS family for example, is its very small current consumption and above all, its wide operating voltage range, typically from 3-18 V, which makes it ideal for battery operation. The advent of low-voltage CMOS capable of operating for considerable periods from a single 1.5V cell has accelerated this trend. Well-designed CMOS circuits are so economical that the frequency of changing batteries is more dependent on their shelf life than their capacity. A small digital barometer designed by a group of students has operated in my office now for several years on battery power, and I have no idea of what the battery is or where to find it, though it is probably a zinc-carbon PP3. The semiconductor strain-gauge pressure sensor uses little current and the rest of the circuitry is CMOS, with an LCD display which also uses minimal current.

Lithium batteries, exemplified by the Li/MnO_2 system in figure 8.21g, have high terminal voltages and reasonably flat discharge characteristics. The figure does not show their superior capacity compared to all the other batteries. A lithium D-cell has a capacity of about 16 Ah, four times that of a zinc-carbon D cell, and at twice the voltage, so the capacity in Wh is actually eight times greater (see also problem 8.9). Lithium batteries also have very long shelf lives (up to ten years) because their self-discharging rates are very low, but they suffer from an inability to supply large currents. Zinc-air and zinc-silver batteries are used where constancy of voltage or long shelf life are of paramount importance and are usually supplied as small button cells, but they have been made in very large sizes by specialist manufacturers for standby service. In terms of shelf life the zinc-air battery is unsurpassed since it can be kept indefinitely by simply blocking the air-holes.

Service temperatures have not been mentioned since it is assumed that most electronic equipment will be used at room temperature, but when it must be used outside there are serious problems with reduced battery power at low temperatures. Table 8.3 summarises the chief properties of these battery systems.

8.8 Cooling

The heat dissipated in electrical machines and circuits must ultimately be removed, or they will suffer breakdown through excessive temperature rise. This is particularly important for devices based on p-n junctions since all the heat is effectively produced at the junction, whose temperature can therefore rise to the point that either the device fails through excessive leakage current, or is destroyed by thermal runaway. In the case of silicon devices the junction temperature is limited to about 150°C. The heat produced at the junction is removed by conduction through the silicon to the case, which is usually made of a highly thermally-conductive material in power devices. When the heat reaches the outside of the case it can be lost to the air by convection and to remote objects by thermal radiation. Unless the heat lost by these processes equals the heat produced in the junction, the junction temperature will rise until a balance is achieved.

The rate of loss of heat is proportional to the temperature difference between the junction and case, which can be found from the thermal resistance specified by the manufacturer. Thermal resistance, θ, is defined by

$$\theta = \Delta T/P$$

where ΔT is the rise in temperature produced by the power dissipated, P. Thus if the thermal resistance from junction to ambient, θ_{JA}, of a small transistor is given as 200°C/W, it means that a power dissipation of 100 mW will produce a rise in the junction temperature of

$$\Delta T_J = \theta_{JA}P = 200 \times 0.1 = 20°C$$

Then if we know the ambient air temperature to be 35°C, we can see that the junction temperature is

$$T_J = T_A + \Delta T_J = 35 + 20 = 55°C$$

The rate of heat removal from the case may be insufficient without the use of a heat sink, which is a piece of metal that is stuck onto the case to increase its surface area. This increases the convection and radiation in proportion to the increased surface area, provided the heat sink has effectively infinite thermal conductivity. Aluminium is the preferred material for heat sinks because its thermal conductivity is very high and it is much cheaper than other high-conductivity metals. The thermal resistance of heat sinks is always quoted by the suppliers and must be added to that of the device from its junction to its case, as in the next example.

Example 8.9

A silicon transistor has $\theta_{JC} = 10°C/W$ and has a thermal resistance from case to ambient, $\theta_{CA} = 30°C/W$. What is the maximum power dissipation permitted if the ambient is at 30°C and the junction limited to 150°C? If it is fitted with a heat sink with a thermal resistance, $\theta_{HA} = 7°C/W$, what is the junction temperature when the power dissipated is 5 W?

The maximum temperature difference from junction to ambient, $\Delta T_{JA} = 150 - 30 = 120°C$ and so the maximum power dissipation is

$$P_{max} = \Delta T_{JA}/\theta_{JA} = \Delta T_{JA}/(\theta_{JC} + \theta_{CA}) = 120/(10 + 30) = 3 \text{ W}$$

When the heat sink is fitted θ_{CA} is replaced by θ_{HA}, the thermal resistance from heat sink to ambient. The maximum power dissipation then becomes

$$P_{max} = \Delta T_{JA}/(\theta_{JC} + \theta_{HA}) = 120/(10 + 7) = 7 \text{ W}$$

When the power is 5 W, the temperature rise is

$$T_J = T_A + \Delta T_{JA} = T_A + (\theta_{JC} + \theta_{HA})P = 30 + 17 \times 5 = 115°C$$

The thermal contact between case and heat sink must be of low thermal resistance and a thermally conducting paste should be used to bond the heat sink to the device. Without this a small air gap of very high thermal resistance will exist between case and heat sink and the heat transfer will be poor. Take note that nothing can be done to reduce the thermal resistance from junction to case as this is determined by the way the device is made. In the above example even if the heat sink had zero thermal resistance the maximum permitted dissipation would be restricted to $120/\theta_{JC} = 120/10 = 12$ W.

Heat sinks are made in many shapes and sizes and it is necessary to choose with care because an over-specified heat sink can waste a lot of money. For example a TO3 heat sink with $\theta_{HA} = 7°C/W$ costs only 40p, while one with $\theta_{HA} = 4°C/W$ costs £2. Large heat sinks with very low thermal resistances can be used when many devices require cooling. The one shown in cross-section in figure 8.22 for example costs £14, has a thermal resistance of 0.5°C/W and can accommodate a dozen TO3 packages. Heat sinks are often painted matt black to aid radiative heat transfer.

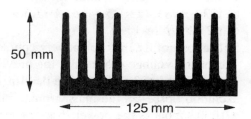

Figure 8.22

An extruded aluminium heat sink. With L = 150 mm, θ_{HA} = 0.5°C/W (L is the dimension normal to the page)

50 mm

125 mm

If the heat sink gets too hot it may be necessary to use forced convection, that is, to cool it by a fan. Fans are capable of improving the dissipation of a heat sink by a factor of two or three and may be worthwhile economically, since a smaller heat sink will be cheaper and will occupy less space perhaps than a small fan. A mains operated fan of 80mm × 80mm × 30mm dimensions costs only about £6, can blow 13 l/s of air and consumes 13 W. Axial fans are now available with thermistor sensors for speed control so that the air temperature remains almost constant. This cuts down both the noise and the power consumption. Peltier-effect heat pumps can also be used to cool or control the temperature of especially sensitive devices, but they are not cheap: about £1 per watt extracted. When optimally adjusted a temperature difference of 65°C can be maintained across the devices, which are only 3-5 mm thick. Again a heat sink, and possibly a fan too, is needed on the hotter face of the heat pump.

Suggestions for further reading

Switched mode power supplies: design and construction by H W Whittington, B W Flynn and D E MacPherson (Research Studies Press, 1992) Despite the publisher's name, a very practical book.
Rechargeable batteries applications handbook by Technical marketing staff of Gates Energy Products, Inc. (Butterworth-Heinemann, 1992)
Battery reference book by T R Crompton (Butterworth-Heinemann, 2nd ed., 1995)

Problems

1 Show that the maximum efficiency of a capacitively-coupled common-emitter amplifier occurs when the load resistance is equal to the collector resistance divided by $\sqrt{2}$.
2 Calculate the maximum efficiency of the CE amplifier of figure P8.2, including all the biasing components, if R_L = 1 kΩ. Assume the maximum peak-to-peak output voltage is sinusoidal and the smaller of $2V_{CEQ}$ or $2I_{CQ}(R_C \mathbin{/\mkern-5mu/} R_L)$. Take V_{BEQ} = 0.7 V and $h_{fe} = \beta = 50$, but do not assume $I_1 \gg I_{BQ}$. *[4.15%]*
3 Show that maximum power is dissipated in the transistors of a class-B push-pull amplifier when the r.m.s. load voltage, $V_L = 0.225V_{CC}$.
4 In the circuit of figure P8.4 Q3 and Q4 are a matched complementary pair, V_{CC} = +36 V, R_1 = 10 kΩ, R_2 = 4.7 kΩ, $R_3 = R_4$ = 1 kΩ, R_5 = 12 kΩ and $R_6 = R_7$ = 680 Ω.

Neglecting any base currents, what must R_8 be to give a quiescent voltage at the output of 18 V? What then is I_{CQ} in Q2? Also neglecting the base currents, what are all the quiescent voltages at the lettered nodes in figure P8.3? (Take all the diode drops and V_{BEQ} values to be 0.7 V.) *[48.4 Ω; 25.4 mA; A 11.5 V, B 25.2 V, C 10.8 V, D 1.93 V, E 18.6 V, F 17.3 V, G 1.23 V, H 18 V]*

5 If a load of 8 Ω is capacitively coupled to the output of figure P8.4 with the resistance values and voltage drops as in the previous problem, what is the maximum possible peak load voltage if Q3 and Q4 have β = 100? What are then the load power and the efficiency, excluding losses in the biasing/preamplifying network? What are these three values if β = 1000? What is the power loss in the biasing/preamplifying network?
[9.4 V; 5.5 W; 41%; 16 V, 16 W, 70%; 1.5 W]

6 Derive equation 8.8, assuming the components of figure 8.9 are ideal. The load of a class-D amplifier as in figure 8.9 is a series-resonant RLC circuit with $L = 1.2$ mH, $C = 56$ nF and $R_L = 10$ Ω. When the supply voltage, $V_{DD} = 60$ V, what power is developed in the load, assuming the components are ideal and that it is driven at the circuit's resonant frequency? How many dB down on the fundamental is the third harmonic in the load? Repeat these calculations when the inductor's Q-factor is 50 at the circuit's resonant frequency. *[72.9 W, 73.2 dB; 43.6 W, 68.7 dB]*

7 For a slow-blowing fuse, the minimum time to blow in seconds is given by

$$t_B = \frac{3000}{\exp[1.4(I/I_0 - 1)] - 1}$$

where I_0 is the rated fuse current and I is the r.m.s. current carried by the fuse. If a ramp current, given by $i = 0.1I_0 t$ passes through the fuse, at what time and current will it blow? How long will this take? If $I_0 = 2$ A and $I = 21\exp(-10t)$ A when will the fuse blow? What value of I_m will just cause the fuse to blow if $I = I_m\exp(-0.5t)$? *[$t_B = 65$ s and $i = 6.5I_0$; $t = 11$ ms; $I_m = 16.25$ A]*

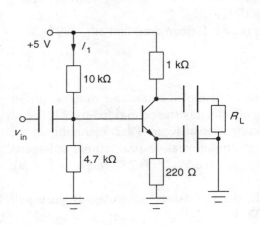

Figure P8.2

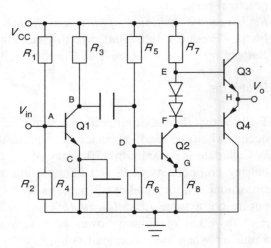

Figure P8.4

8 A current-limited series voltage regulator as in figure 8.14 has $R_1 = 470\ \Omega$, $R_2 = 30\ k\Omega$, $R_3 = 24\ k\Omega$ and $R_4 = 1.2\ \Omega$. If $V_Z = 5.3$ V and $V_{BE2} = 0.6$ V, what will be the load currents and output voltages when $R_L = 0\ \Omega$, $10\ \Omega$, $25\ \Omega$ and $50\ \Omega$?

9 The input to a series voltage regulator is subject to a variation in DC level from 12 V to 14 V and superimposed on this is a 100 Hz peak-to-peak ripple of 5%. The load voltage is to be maintained at 9 V with currents ranging from 0.1 A to 3 A. If the dropout of the regulator is 1.5 V and it has a ripple rejection at 100 Hz of 40 dB, what is the maximum voltage ripple across the load? What is the maximum power lost in the regulator?

10 A series voltage regulator has an output voltage of 6 V and a dropout of 1 V. It feeds a load of 90 Ω and is supplied by 6×1.5 V carbon-zinc D-cells. These have the characteristics shown in figure 8.21a. How long will they last? How many Wh will they have delivered to the load? If they cost 85p each, how much is the cost/Wh? If 3×3 V Li/MnO_2 D-cells are used instead with the characteristics of figure 8.21g, but with the timescale multiplied by 6, how long will they last and at what cost/Wh, given that they are £12 each? Which is the better choice? *[35 h, 14 Wh, 36p/Wh; 250 h, 100 Wh, 36p/Wh]*

11 Show that the load voltage regulation for the forward converter of figure 8.18 is given by

$$\%reg = \frac{100(1 - \delta)}{4LCf^2}$$

where $f = 1/T$ and $\delta = T_{ON}/T$.

12 What are the output and ripple voltages for the circuits of figure P8.12a and P8.12b, assuming the diodes all have forward drops of 0.8 V? *[1.354 kV, 13.54 V; 40 V, 0.125 V]*

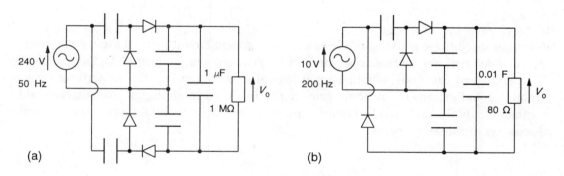

Figure P8.12

13 A power supply is built using 8 output transistors each of which has $\theta_{JC} = 5°C/W$ and must dissipate on average 5 W. They are all mounted on a heat sink which has $\theta_{HA} = 1.7°C/W$ using a bonding paste with a thermal resistance of 0.2°C/W. If the ambient temperature is 30°C, what is the temperature of the case and of the junction? It is necessary to increase the transistors' power output to 10 W each, and a fan is used to force cool the heat sink. If the fan increases the heat sink's dissipation by a factor of 2.5 what is the junction temperature now? What is the maximum allowable power dissipation for each transistor now if the junction temperature is limited to 150°C?
[99°C, 124°C; 136°C, 11.3 W]

9 Combinational logic

C OMBINATIONAL LOGIC in electronic engineering enables the processing of information that is effectively in the form of binary numbers, since the inputs and output(s) of a logical circuit can be in only one of two states. The output state is a function of the input states, and can be used to control some other device or machine or circuit, or can just be used as information. Figure 9.1 shows schematically what combinational logic does. Two-state inputs can be expressed as two-state variables, or Boolean[1] variables, which can be combined according to the rules of Boolean algebra. In formal terms we can write

$$Q = f(A, B, C, D...)$$

where Q is the output and A, B, C, D etc. are the inputs.

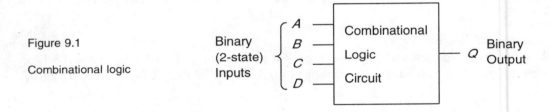

Figure 9.1

Combinational logic

Logic circuits are made up of elementary component circuits, called *logic gates* or just *gates*, which perform the basic operations of Boolean algebra, such as AND, OR and NOT. Besides logical functions, combinational logical circuits can be used to perform binary arithmetic and extremely fast logic gates are in fact used in all digital computers to add, subtract, multiply and divide binary numbers. We shall therefore start with a brief discussion of binary numbers.

9.1 Binary and hexadecimal numbers

Normally one counts and performs arithmetic in decimal numbers made up of the set of digits $\{0,1,2,3,4,5,6,7,8,9\}$, that is numbers are represented as powers of 10. Decimal numbers are said to be to *base* 10. For example the decimal number 11.25_{10} represents

---

[1] George Boole (1815-64) was born in Lincoln and taught himself mathematics. He was a schoolmaster in that city when he devised what he called *The Laws of Thought* in the years 1847-54. He became Professor of Mathematics at Queen's College, Cork, where he died. His school can still be seen close by the cathedral.

$$10^1 + 10^0 + 2 \times 10^{-1} + 5 \times 10^{-2}$$

The subscript 10 on 11.25 indicates the base. Computers and calculators perform arithmetic with the set {0,1} of *binary* numbers, to base 2. The advantage of using binary numbers is that a digit can be either a 1 or a 0 and nothing else so that two-state devices can be used to store and operate on them.

A binary number such as 1011.01_2 represents

$$\begin{aligned}
&1 \times 2^3 + 0 \times 2^2 + 1 \times 2^1 + 1 \times 2^0 + 0 \times 2^{-1} + 1 \times 2^{-2} \\
&= 8 + 0 + 2 + 1 + 0 + 0.25 \\
&= 11.25_{10}
\end{aligned}$$

Whatever the base, the point indicates where the powers of the base change from 0 to −1.

9.1.1 Conversion from decimal to binary and vice versa

Conversion from integer decimal numbers to binary is accomplished by dividing successively by 2 and storing the remainders, for example 23_{10} can be converted as follows:

$$\begin{aligned}
23 \div 2 &= 11, && \text{remainder 1 (= LSB)} \\
11 \div 2 &= 5, && \text{remainder 1} \\
5 \div 2 &= 2, && \text{remainder 1} \\
2 \div 2 &= 1, && \text{remainder 0} \\
1 \div 2 &= 0, && \text{remainder 1 (= MSB)}
\end{aligned}$$

The remainders written down from the bottom (MSB, most significant bit) up to the top (LSB, least significant bit) give the binary form as 10111_2.

Fractional decimal numbers such as 11.25_{10} must be split into the integral part, 11_{10}, which is converted to binary as above, and the fractional part 0.25_{10}, which is converted by multiplying successive fractional parts by 2 until no fractional part remains:

$$\begin{aligned}
0.25 \times 2 &= 0.5 \text{ integral part } 0 = \text{MSB} \\
0.5 \times 2 &= 1.0 \text{ integral part } 1 = \text{LSB}
\end{aligned}$$

Thus 0.25_{10} is 0.01_2 and 11.25_{10} is 1011.01_2; however, most decimal fractional numbers give recurring binary fractional numbers. For example, 0.2_{10} gives

$$\begin{aligned}
0.2 \times 2 &= 0.4, \text{ integral part } 0 = \text{MSB} \\
0.4 \times 2 &= 0.8, \text{ integral part } 0 \\
0.8 \times 2 &= 1.6, \text{ integral part } 1 \\
0.6 \times 2 &= 1.2, \text{ integral part } 1 \\
0.2 \times 2 &= 0.4, \text{ integral part } 0
\end{aligned}$$

The digits 0011 are recurring so that 0.2_{10} is $0.00\dot{1}\dot{1}_2$

9.1.2 Binary arithmetic

Addition is performed on two binary numbers at a time and if three or more binary numbers are to be added the third is added to the sum of the first two and so on, to avoid carries of more than 1. For example 1001_2 is added to 1101_2 as follows:

$$\begin{array}{r} 1001 \\ +\ 1101 \\ \hline 10110 \end{array}$$

since $1 + 1 = 0$, carry 1. Subtraction is similar, with a borrow if needed. Multiplication and division are the same as for decimal numbers, but in computers these operations are done by shifting the binary point.

Subtraction is performed in computers by addition of the complement of the *subtrahend* (the number that is subtracted). Thus instead of performing the subtraction $X - Y$, the computer adds: $X + (-Y)$. Two types of complement can be used, the 1's complement and the 2's complement. The 1's complement of an n-digit binary number, N, is defined as $(2^n - 1) - N$, and the 2's complement is defined as $2^n - N$, which is the 1's complement plus 1. For example the 1's complement of 101100_2 is

$$(2^6 - 1) - 101100_2 = 111111_2 - 101100_2 = 010011_2$$

We can see that forming a 1's complement is easy, just change all the 0s to 1s and vice versa. The 2's complement of 101100_2 is

$$2^6 - 101100_2 = 1000000_2 - 101100_2 = 010100_2$$

But it is easier to add one to the 1's complement:

$$010011_2 + 1 = 010100_2$$

To subtract 101100_2 from 1000101_2 directly we have

$$\begin{array}{r} 1000101 \\ -\ 101100 \\ \hline 11001 \end{array}$$

The borrowing is complicated here! Using 2's complement addition is much easier:

$$\begin{array}{r} 1000101 \\ +\ 010100 \\ \hline 1011001 \end{array}$$

The leading 1 is then dropped to give the correct result, 11001_2. What has been done is like saying $100_{10} - 89_{10} = 100_{10} + 11_{10} - 100_{10}$.

Figure 9.2 shows the process of machine multiplication of binary numbers. It is accomplished by shifting the multiplicand (the number being multiplied by another) to the left and adding the shifted parts together. For example $110_2 \times 101_2$ would be interpreted by a machine firstly as placing 110_2 in a multiplicand *register* (a place to store a number) and then adding it to the contents of another register called an *accumulator*, which is initially cleared to zero. This is the action $110_2 \times 1_2$, that is multiplication by the LSB of the multiplier. Secondly, since the next multiplier digit is zero, nothing need be done. Thirdly, the final multiplier digit is 1, so the contents of the multiplicand register are shifted two places left (equivalent to multiplying by 100_2) and added to the accumulator. Division is similar, but involves shifting right and subtraction.

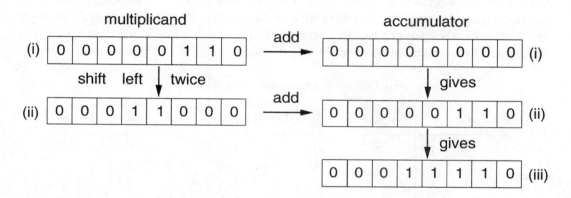

Figure 9.2 Successive stages of multiplicand register and accumulator are denoted (i), (ii), (iii).

9.1.3 Hexadecimal numbers

Hexadecimal numbers have 16_{10} as a base and the set $\{0,1,2,3,4,5,6,7,8,9,A,B,C,D,E,F\}$ as digits, where the hexadecimal numbers A, B, C, D, E and F represent the decimal numbers 10, 11, 12, 13, 14 and 15. They can be formed from binary numbers by grouping them in fours, for example

$$10110011_2 = |1011|_2|0011|_2 = |11|_{10}|3|_{10} = B3_{16}$$

9.1.4 Binary-coded decimal (BCD) numbers

When numbers are to be output to a display or other device as decimal numbers, it is convenient to represent each decimal digit as a 4-bit binary number. This is said to be *binary-coded decimal* or BCD form. For example the decimal number 395 is coded in BCD as

$$\begin{array}{ccc} 0011' & 1001' & 0101 \\ 3 & 9 & 5 \end{array}$$

The binary numbers 1010, 1011, 1100, 1101, 1110 and 1111 are meaningless in BCD.

9.2 Logic functions and Boolean algebra

Boolean algebra was first applied to logical switching circuits by Shannon[2] (who also made great contributions to information theory. Consider the variables A and B each of which can be in one of two states, and suppose we wish to combine them according to just two rules symbolised by '+' and '•', which bear a resemblance to 'ordinary' addition and multiplication, but are called OR and AND in Boolean algebra. What does '$A + B$' (A OR B) mean? We can only tell by looking at all possible outcomes of that operation; fortunately there are only four combinations, shown in Table 9.1. There are three ways we can fill in the truth table depending on how we wish to represent the states of the variables. In the first the two states are called 'true' (T) and 'false' (F) as in symbolic logic, in the second they are denoted ON and OFF corresponding to the switching circuit of figure 9.3b, and in the third they are given binary values 1 and 0. It is not important what we call the two states: if the operation obeys table 9.1, it must be the OR function.

Table 9.1 *Truth tables for A+B*

A	B	$A + B$	A	B	$A + B$	A	B	$A + B$
T	T	T	ON	ON	ON	1	1	1
T	F	T	ON	OFF	ON	1	0	1
F	T	T	OFF	ON	ON	0	1	1
F	F	F	OFF	OFF	OFF	0	0	0

These tables may be taken as a *definition* of the operation $A + B$ (A OR B). When the logic is implemented with logic ICs (or 'gates'), the ON or TRUE state is associated with a HIGH voltage (often +5 V), and the OFF or FALSE state with a LOW voltage (usually ground or 0 V) in the positive logic convention normally used. The logic circuit symbol for an OR gate is shown in figure 9.3a, together with a parallel switching arrangement which will turn the lamp ON in agreement with table 9.1.

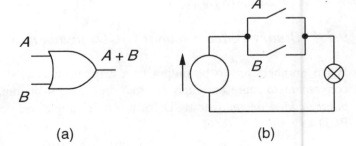

Figure 9.3

(a) The logic symbol for OR
(b) A switching circuit for OR

(a)   (b)

[2] C E Shannon, *A symbolic analysis of relay and switching circuits* Trans. AIEE **57** 723ff (1938)

The second operation we wish to make use of is AND: $A{\cdot}B$, defined in table 9.2, with its logic circuit symbol in figure 9.4a and a switching circuit in figure 9.4b. By examining the binary forms of tables 9.1 and 9.2, it can be seen that the AND operation looks very much the same as ordinary multiplication, while OR is like ordinary addition except that 1 OR 1 (1 + 1) is defined as 1, since only the states 0 and 1 exist. We shall see a little later that the truth table where this last outcome is defined as 0 also exists and is called the *exclusive OR* or XOR, denoted $A \oplus B$.

Table 9.2 *Truth tables for AND*

A	B	$A{\cdot}B$	A	B	$A{\cdot}B$	A	B	$A{\cdot}B$
T	T	T	ON	ON	ON	1	1	1
T	F	F	ON	OFF	OFF	1	0	0
F	T	F	OFF	ON	OFF	0	1	0
F	F	F	OFF	OFF	OFF	0	0	0

Figure 9.4

(a) The logic circuit symbol for AND
(b) A switching circuit for AND

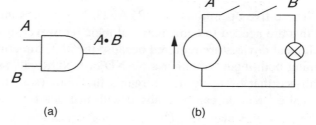

(a)   (b)

As well as AND and OR it is convenient to use the NOT function, also called *inversion* or *complementation*, whereby A becomes NOT-A or $\overline{A}$ or A' (the inverse or complement of A), AND becomes NAND and OR becomes NOR. Table 9.3 is the very brief truth table for NOT and the full logic circuit symbol is shown in figure 9.5, though whenever it is possible the inversion operation will be shown by a circle placed on the appropriate input or output of the logic gate.

Table 9.3 *Truth tables for NOT*   Figure 9.5 The circuit symbol for NOT

A	A'	A	A'
T	F	1	0
F	T	0	1

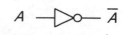

For example, table 9.4 shows the inversion of OR and AND, denoted $\overline{A + B}$, or $(A+B)'$ or A NOR B, and $\overline{A \cdot B}$ or $(A{\cdot}B)'$ or A NAND B, respectively. So the circuit symbols for these devices, shown in figure 9.6, have circles on their outputs to indicate inversion. In most of what follows we will use only the binary form of the truth tables.

Table 9.4 *Truth tables for NOR and NAND*

A	B	$A + B$	$(A + B)'$	$A \cdot B$	$(A \cdot B)'$
1	1	1	0	1	0
1	0	1	0	0	1
0	1	1	0	0	1
0	0	0	1	0	1

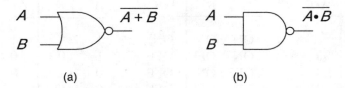

(a)   (b)

Figure 9.6 (a) Circuit symbol for a NOR gate (b) Circuit symbol for a NAND gate

These five operations — OR, AND, NOT, NOR and NAND — turn out to be almost all that are needed in combinational logic. However, fewer than five functions in the form of logical devices are required because the NOT function, for example, can be realised by joining both inputs of either a NAND or a NOR gate as shown in figure 9.7. Later on we shall see that by using de Morgan's theorems (sometimes called de Morgan's law) any logical expression can be realised with just *one* type of gate.

Figure 9.7

Inversion using NOR and NAND gates

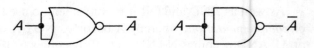

Having looked at some operations in Boolean algebra, its laws can be formally stated:

$$A + 0 = A$$
$$A \cdot 0 = 0$$
$$A + 1 = 1$$
$$A \cdot 1 = A$$

These define an element, 0, much like zero in 'normal' algebra, and another element, 1, that behaves in some ways like unity does normally. The rules for the complement of A are:

$$A + \overline{A} = 1$$
$$A \cdot \overline{A} = 0$$

The *idempotency* (from the Latin *idem* = the same and *potens* = power) laws are:

$$A + A = A$$
$$A \cdot A = A$$

The laws of *association*:

$$A + (B + C) = (A + B) + C = (A + C) + B$$

and

$$A \cdot (B \cdot C) = (A \cdot B) \cdot C = (A \cdot C) \cdot B$$

And finally the *distributive* laws:

$$A \cdot (B + C) = A \cdot B + A \cdot C$$

$$(A + B) \cdot (A + C) = A + B \cdot C$$

The last seems a little strange until it is realised that

$$(A + B)(A + C) = AA + AB + AC + BC$$

$$= A(1 + B + C) + BC$$

$$= A + BC$$

since $1 + B + C = 1$. Note that the dot is omitted by custom between ANDed variables. Note that in the absence of parentheses AND takes precedence over OR. All of the above can be proved using truth tables and the definitions of AND, OR and NOT.

Example 9.1

Prove that $A + A'B = A + B$.

Since $A = A(1 + B) = A + AB$, the equation above gives

$$A + AB + A'B = A + B(A + A') = A + B$$

which is the desired result. Table 9.5 is a truth table which verifies the algebra above.

Table 9.5 *The truth table for $A + A'B$*

A	B	A'B	A + A'B
1	1	0	1
1	0	0	1
0	1	1	1
0	0	0	0

9.3 Logic ICs

Devices that implement binary logical operations are made from transistors, diodes and resistors forming an integrated circuit (IC) and are called logic gates or just gates. These operate at voltage levels which are either HIGH (anything from about +3 V to +18 V) or LOW (usually ground or 0 V, but possibly as high as +2 V). In the positive logic convention, HIGH corresponds to 'true' or 1 and LOW to 'false' or 0. Usually several gates are put on a single chip which is then packaged in plastic with externally available tinned copper leads. An example is the quad, dual-input NAND, which has four 2-input gates in a 14-pin DIL (Dual In Line) package. There are many different ways in which a particular function can be realised, not just different in circuit design, but also in mode of IC fabrication. In consequence a number of *logic families* have been marketed, such as the various 74 series shown in table 9.6, which are discussed more fully in section 9.11.

Table 9.6 *Logic families*

Family	Type	Gate Delay[a]	Power[b]	Price[c]	Year[d]
74TTL[e]	5V TTL	10	10	15/34	1964
74LS	5V Low-power Schottky	8	2	10/22	1971
4000	3-18 V CMOS	50	0.001	11/16	1972
74ALS	Advanced 74LS	6	1	28/46	1980
74FAST	5V High-current TTL	3	6	10/22	1982
74HC	2-6V CMOS	8	0.001	10/22	1983
74HCT	5V CMOS, TTL-compatible	8	0.001	11/20	1983
74AC	2-6V Advanced CMOS	2	0.001	16/40	1986
74ACT	5V TTL-compatible 74AC	2	0.001	18/40	1986
74VHC	2-5V fast CMOS	4	0.001	20/40	1993
74LV	3.3V HCMOS	18	0.001	20/40	1994
74LVX	5V input-tolerant 74LV	12	0.001	30/60	1994

Notes: [a] Propagation delay/gate in ns [b] No-load power in mW [c] Price of quad NAND in pence for 100+ and single quantities [d] Year of introduction in original form. FAST is a trademark of Philips Gloeilampenfabrieken [e] Transistor-transistor logic

The 74 series is the cheapest and most popular today, and exists in many forms. The oldest family listed is the 74TTL series, which originated as long ago as 1964, in a rather different form. It operates from a +5V supply, is slow and uses lots of power, mainly because of its input transistors (TTL stands for transistor-transistor logic, an advance on the early DTL, diode-transistor logic). CMOS stands for Complementary Metal Oxide Semiconductor, which uses pairs of *n*-channel and *p*-channel transistors. Stand-by power consumption is very low in CMOS ICs because the inputs are to high-impedance, insulated, FET gates.

The 14-pin DIL packages contain four, two-input, single-output gates (taking 12 pins) together with a ground pin and a positive supply pin. The external appearance of a 74LS quad 2-input NAND is shown in figure 9.8b, while the pin arrangements of the gates are shown in figure 9.8a. Logic devices are oriented with pin 1 at the bottom left of a DIL package, marked by a circle or dot on the top of the device (as in figure 9.8b), or with an indentation on the left-hand side viewed from the top (as in figure 9.8a). The DIL package must be the right way round to read the numbers stamped on it. Then the pin numbering runs from 1 at the bottom left to 7 (nearly always ground) at the bottom right, then 8 at the top right to 14 (nearly always $+V_{CC}$) at the top left. Inverters are single-input devices, so six of them can be arranged in one 14-pin DIL package. It is also possible to buy multi-input devices, such as the triple, 3-input, NAND from Harris Semiconductor, CD74HCT10E, and the MC74LS21N dual, 4-input AND from Motorola Semiconductors. 74HCT and 74LS are the logic families to which the devices belong and 10 and 21 are the device code numbers. Letters preceding the logic family are the manufacturer's identification code, and letters following the device number are a code for the type of package or encapsulation process. Any alphanumerics following on subsequent lines are the batch and date codes stamped on by the manufacturer, so that subsequent failures can be identified.

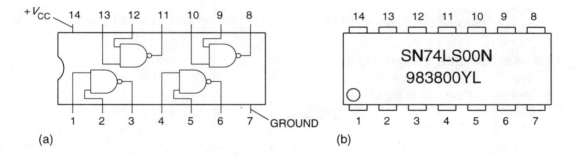

Figure 9.8 (a) Pin-outs a quad 2-input NAND IC (b) Top view of DIL package

9.3.1 *The exclusive OR (XOR)*

As an example of the use of these logic gates in combination, let us look at the truth table given in table 9.7, defining the XOR function, which is written $A \oplus B$. Calling the output function Q (that is $Q \equiv A \oplus B$), it is apparent that when $A = 1$ *and* $B = 1$, $Q = 0$ (and $Q' = 1$); *or* when $A = 0$ *and* $B = 0$, then $Q = 0$ (and $Q' = 1$). Thus the logical expression for Q' must include the expression $AB + A'B'$. Now both AB and $A'B'$ will be zero if $A \neq B$, therefore the truth table for $AB + A'B'$ is as shown in table 9.8 and we see that $Q' = A \oplus B$. The XOR function is often used, for example as a controlled inverter, since when $A = 1$ the output, $Q = B'$, and when $A = 0$, $Q = B$, as can be seen from the truth table.

Table 9.7 *The truth table for A⊕B*   Table 9.8 *The truth table for AB + A′B′*

A	B	A⊕B
0	0	0
0	1	1
1	0	1
1	1	0

A	B	AB	A′B′	Q	Q′
0	0	0	1	1	0
0	1	0	0	0	1
1	0	0	0	0	1
1	1	1	0	1	0

Since $A \oplus B = AB + A'B'$, the logic circuit for $A \oplus B$ will be that of figure 9.9a. In the next section it will be seen that this can be simplified by using de Morgan's theorems. The XOR function is often required so it has been given its own circuit symbol (figure 9.9b) and is available as one quarter of the 7486, a quad XOR chip.

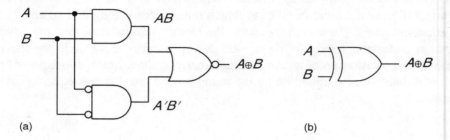

(a)   (b)

Figure 9.9 (a) A logic circuit for $A \oplus B$ (b) The circuit symbol for an XOR gate

9.4 de Morgan's theorems[3]

de Morgan's theorems state:

$$\overline{A + B} = \overline{A} \cdot \overline{B}$$

$$\overline{A \cdot B} = \overline{A} + \overline{B}$$

Sometimes these equations are called de Morgan's law, best remembered as '*change the sign and break the line*'. In the left-hand side of the first of the equations, if the + sign is changed to a • and the line above is broken, so that half appears over the A and half over the B, then the right-hand side results. In the left-hand side of the second, the • is replaced by a + and the line broken, then the right-hand side follows. De Morgan's law enables us to manipulate logical expressions into more convenient forms. The first of the theorems shows that the NOR operation on A and B is identical to performing an AND operation on their complements A' and B', while the second shows that NANDing is identical to ORing the complements. Usually the dot is omitted from ANDed variables and must be (mentally) inserted when using de Morgan's law.

---

[3] Augustus de Morgan (1806–1871) was Professor of Mathematics at the University of London. He was a great logician and champion of George Boole.

With de Morgan's law it must be remembered that the line to be broken is the one *immediately* above the sign that is altered and that lines crossing a bracket, or an implied bracket, must first be broken outside the bracket. For example, in the expression

$$Q = \overline{AB + A'B'}$$

the terms AB and $A'B'$ have implied brackets round them, so the line must be broken at the + to give

$$Q = \overline{AB}\ \overline{A'B'}$$

The lines above AB and $A'B'$ can then be broken further, but now we must insert brackets:

$$Q = (A' + B')(\overline{A'} + \overline{B'}) = (A' + B')(A + B)$$

Complementing a variable twice leaves it unchanged: $(A')' = A$.

Example 9.2

The circuit of figure 9.10a uses NOR gates to form the logic function W, given by

$$W = \overline{A + B} + \overline{A' + C'}$$

Use de Morgan's law to turn it into a form suitable for implementing solely in NAND gates.

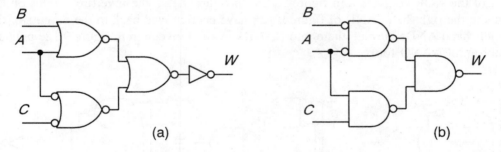

Figure 9.10 (a) The all-NOR circuit for W (b) The all-NAND circuit for W

By de Morgan's law

$$\overline{A + B} + \overline{A' + C'} = A' \cdot B' + \overline{A'} \cdot \overline{C'} = A'B' + AC$$

Then we can invert the result twice, leaving its value unchanged and apply de Morgan's law to the result:

$$W = \overline{W'} = \overline{\overline{A'B' + AC}}$$

$$= \overline{\overline{(A'B')} \cdot \overline{(AC)}}$$

This expression is in all-NAND form, and leads directly to the circuit of figure 9.10b.

9.4.1 Redrawing circuits with de Morgan's theorems

To restate, de Morgan's theorems are

$$\overline{A + B} = A' \cdot B' \quad \text{and} \quad \overline{A \cdot B} = A' + B'$$

In the first, a NOR operation becomes an AND with inverted inputs and in the second a NAND operation becomes an OR with inverted inputs. Thus the gates in figure 9.11 are equivalent, and circuits containing one may be directly redrawn using its equivalent.

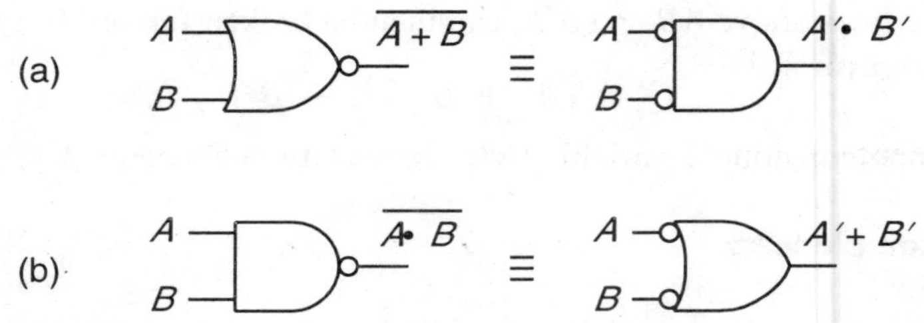

Figure 9.11

(a) A NOR gate and its equivalent
(b) A NAND gate and its equivalent

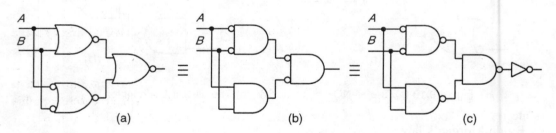

For example consider the circuit of figure 9.12a contains only NOR gates, which can be replaced with their equivalent from figure 9.11a, to give the circuit of figure 9.12b. The inverted inputs to the lower NOR of figure 9.12a have cancelled out the inversions on the input of the NOR equivalent in figure 9.12b. In figure 9.12c, the inverting circles on the inputs to the right-hand AND of figure 9.12b have been moved back to the outputs of the two left-hand ANDs, turning them into NANDs. A final inversion in figure 9.12c puts the circuit in all-NAND form.

Figure 9.12 (a) An all-NOR circuit (b) Each NOR is replaced by its equivalent (c) The all-NAND form

The resulting circuit can be verified by Boolean algebra as follows. The output of figure 9.12a can be written as

$$Q = \overline{\overline{A + B} + \overline{A' + B'}}$$

By using de Morgan's law thrice this can be turned into

$$Q = \overline{\overline{A' \cdot B'} \cdot \overline{A \cdot B}}$$

which is the output of figure 9.12c.

9.5 Minterms and maxterms

An expression for the output of a logic circuit can be deduced from its truth table. Here we shall explain how this may be done, but first we must introduce two new words: *minterm* and *maxterm*. Suppose we have a function of three variables: $W = f(A,B,C)$ (we could have had as many variables as we liked, but three will suffice); then the minterms are the result of ANDing each of the three variables or its complement, while the maxterms are the result of ORing the same. Thus a minterm is an expression such as ABC, or $A'B'C$, and a maxterm is an expression such as $A' + B + C$, or $A' + B' + C$. There are 2^3 or eight possible minterms and eight possible maxterms for W.

Example 9.3

Deduce a logical function for W from table 9.9, the truth table for W in example 9.2.

Table 9.9 *The truth table for W*

A	B	C	W
0	0	0	1
0	0	1	1
0	1	0	0
0	1	1	0
1	0	0	0
1	0	1	1
1	1	0	0
1	1	1	1

We note that there are four entries in the output column where $W = 1$. The first of these has for inputs $A = 0$, $B = 0$ and $C = 0$, so that the minterm $A'B'C' = 1$. If any other combination of values for A, B or C is put into this minterm, such as $A = 0$, $B = 1$, $C = 1$, it will be zero. The next corresponds to $A = 0$, $B = 0$ and $C = 1$, so that the minterm $A'B'C = 1$; again, any other combination of values for A, B or C will give zero for this minterm. The other two minterms corresponding to $W = 1$ are $AB'C = 1$ and $ABC = 1$. If we now form the sum of these four minterms, we obtain

$$f(A,B,C) = ABC + AB'C + A'B'C + A'B'C'$$

We can see that $f(A,B,C)$ is the required function, since each term will be zero except for the sole combination of values of A, B and C that is required for W to be 1. For example, if $A = 1$, $B = 0$ and $C = 0$, each of the four terms is zero (and $W = 0$); while if $A = 0$, $B = 0$ and $C = 1$, the first two terms and the last term of $f(A,B,C)$ are 0, while the third term is 1 — as it must be since $W = 1$ for this combination. W has thus been expressed as a *sum of minterms* (also known as a *sum of products* or *SOP*). The number of minterms is the same as the number of times $W = 1$ in the truth table.

The SOP can be simplified with Boolean algebra:

$$f(A,B,C) = ABC + AB'C + A'B'C + A'B'C'$$

$$= AC(B + B') + A'B'(C + C') = AC + A'B'$$

By using de Morgan's law this can be put in all-NAND form. Complementing it twice produces:

$$f(A,B,C) = W = \overline{\overline{W}} = \overline{\overline{AC + A'B'}}$$

$$= \overline{\overline{(A\,C)}\cdot\overline{(A'B')}}$$

This W is identical to that of example 9.2.

W can also be expressed as a *product of maxterms*, sometimes called a *product of sums* or *POS*. To do this we look for the combinations producing a zero for W. The first of these is $A = 0$, $B = 1$ and $C = 0$, so the OR function $(A + B' + C)$ will be 0 for this combination, but 1 for *all of the rest*. The next time $W = 0$ occurs when $A = 0$, $B = 1$ and $C = 1$, then $A + B' + C' = 0$ for this combination only. The other two maxterms formed in this way are $(A' + B + C)$ and $(A' + B' + C)$. The product of these four maxterms is:

$$F(A,B,C) = (A + B' + C)(A + B' + C')(A' + B + C)(A' + B' + C)$$

and it can be seen that $F(A,B,C)$ is a valid expression for W, since each maxterm is 0 only at one combination of values for A, B and C, and that is the combination for which $W = 0$. The number of maxterms must therefore be the same as the number of zero entries for W in the truth table.

The POS expression above for $F(A,B,C)$ could also have been obtained from the SOP or minterm expression for W', which is

$$W' = A'BC' + A'BC + AB'C' + ABC'$$

Then by de Morgan's law

$$F(A,B,C) = \overline{W'} = \overline{A'BC' + A'BC + AB'C' + ABC'}$$

$$= \overline{A'BC'} + \overline{A'BC} + \overline{AB'C'} + \overline{ABC'}$$

$$= (A + B' + C)(A + B' + C')(A' + B + C)(A' + B' + C)$$

$F(A,B,C)$ can also be simplified by Boolean algebra:

$$(A + B' + C)(A + B' + C') = A + B'$$

and

$$(A' + B + C)(A' + B' + C) = A' + C$$

so that

$$F(A,B,C) = (A + B')(A' + C) = A'B' + AC + B'C$$

Perhaps surprisingly this expression for W differs from the one found from the sum of minterms, yet it is undoubtedly an expression which obeys the truth table for W, as can quickly be checked. The extra term $B'C$ is redundant and can logically be omitted. That is not to say that $B'C = 0$, but that $B'C = 0$ when $A'B' + AC = 0$. We shall see later in section 9.10.4, that in practice the redundant term may be necessary to make the circuit work properly.

9.6 Karnaugh mapping and circuit minimisation

Karnaugh mapping is a means by which examination of a truth table can produce an efficient, or minimised, logical expression for a function of several variables. First let us examine a very simple specimen, the Karnaugh map for the function $Z = XY + XY' + X'Y'$, in table 9.10.

Table 9.10 *The Karnaugh map for $Z = XY + XY' + X'Y'$*

Along the top row are the values for X (just two, of course) and down the first column the two values for Y. The four cells forming the Karnaugh map contain the values of Z corresponding to all four combinations of X and Y. We can circle a horizontal group of cells containing ones (1-cells) and a vertical group[4] of 1-cells also. Since each cell corresponds to a minterm such as XY, the circled horizontal group corresponds to $XY' + X'Y'$, which is just Y'. The circled vertical subcube corresponds to $XY + XY'$, which is just X. Thus the logical expression for Z must be $Z = X + Y'$. In the horizontal subcube, X changes value, and therefore must give rise to a term in the logical expression for Z which does *not* contain X. As this occurs when $Y = 0$, that term can only be Y'.

Similarly for the vertical subcube, when $X = 1$, Y is both 0 and 1, so this term must be independent of Y, that is just X. There are then only two terms in the expression for Z: $Z = X + Y'$. The same result could have been found by manipulating the original expression for Z, but we might have overlooked that $X + X'Y'$ was the same as $X + Y'$.

To be sure, in such a simple Karnaugh map we can see immediately that only one entry is zero, when $X = 0$ and $Y = 1$, so that $Z' = X'Y$ and then $Z = (X'Y)' = X + Y'$ — a much quicker solution.

Look now at the Karnaugh map for $S = ABC' + A'B'C + A'BC + A'BC'$ in table 9.11: note that because there are three variables, we have had to group two together (A with B) along the top row. In this row only one variable is different in adjacent cells. Then we proceed to circle the three groups 1-cells. The top horizontal subcube where $C = 1$, $A = 0$ and

---

[4] These groups are known as *subcubes* though they are not cubes, squares or even rectangles.

B varies, must correspond to the term $A'C$. The bottom horizontal subcube, wherein A varies, must correspond to the term BC'. The vertical subcube corresponds to $A'B'$, so the expression for S must be $A'C + BC' + A'B$. But the vertical subcube overlaps both the others and is therefore redundant, hence the minimised expression for S is $S = A'C + BC'$.

Table 9.11   *The Karnaugh map for S*

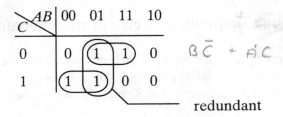

We can now see from a Karnaugh map of the function, W, whose truth table is given in table 9.9, how the redundant term in the SOP analysis arose. Table 9.12 shows two versions of the Karnaugh map for W. Looking at the left-hand table, it is clear that two terms suffice: the horizontal subcube with $A = 1$ and $C = 1$, and the vertical subcube with $A = 0$, $B = 0$. However there is a third subcube with $B = 0$, $C = 1$ that overlaps completely the other two, corresponding to the redundant term, $B'C$, the other two terms give the correct minimisation for W, $W = AC + A'B'$.

There are no edges to Karnaugh maps, so that the column furthest to the left, headed 00, is adjacent to the 10 column. This follows, since the numbers in the top row could have been written just as easily in the order 11, 10, 00, 01 — as in the right-hand version of the map. The only rule is: adjacent cells differ in only one variable — you may not put 00 next to 11 or 01 next to 10. If there were four variables in an expression we should split them two horizontally and two vertically.

Table 9.12   *Equivalent versions of the Karnaugh map for W*

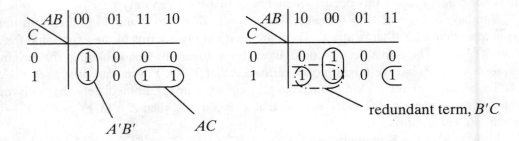

Example 9.4

Suppose we have a lamp operated by three switches as shown in figure 9.13. What is the function which describes the operation of the circuit?

First we make up a truth table: when switch A is closed we shall set $A = 1$ and when it is open we shall put $A = 0$ and the same for the other switches. When the light is on the output function $Q = 1$. The result is the truth table of table 9.13.

Figure 9.13

A switching circuit for a lamp

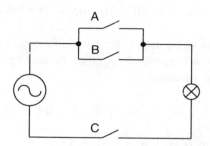

Table 9.13 *The truth table for the lamp-switching circuit*

A	B	C	Q
0	0	0	0
0	0	1	0
0	1	0	0
0	1	1	1
1	0	0	0
1	0	1	1
1	1	0	0
1	1	1	1

This is not particularly helpful as it stands, so the Karnaugh map for Q, table 9.14, must be examined.

Table 9.14 *The Karnaugh map for the lamp-switching function*

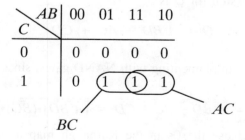

At once it is clear that the combinations which switch the light on are those of the subcubes where $C = 1$, $AB = 01,11$ and $C = 1$, $AB = 11,10$. The first of these corresponds to $BC = 1$ (as A can be 0 or 1) and the other corresponds to $AC = 1$ (since B can be 0 or 1). Thus the light switching function is $Q = BC + AC$. It is hardly surprising that C is part of each term in Q, since it is the switch controlling the only return path.

9.6.1 Karnaugh maps with four input variables

The principles are the same as for three variables, except that the variables are grouped in pairs. The like cells are grouped together in as large blocks as possible. Consider for example the Boolean function

$$Q = (A'B + C)(BC + D)\overline{AD'}$$

whose Karnaugh map is shown in table 9.15.

Table 9.15 *The Karnaugh map for Q, leading to equation 9.1*

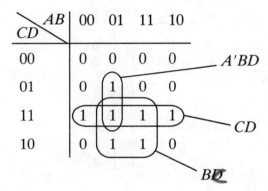

We can make a grouping of the four 1s in a square block, corresponding to the term minterm BC, and a group of two 1s vertically which correspond to the minterm $A'BD$. Finally we can make a group of the four 1-cells running right across the table, which corresponds to the minterm CD, as it is independent of the variables A and B. Thus the minimal minterm expression for Q is

$$Q = A'BD + BC + CD \tag{9.1}$$

This form is suitable for implementing with NAND gates, since by complementing twice we find

$$Q = \overline{\overline{A'BD + BC + CD}} = \overline{\overline{(A'BD) \cdot \overline{(BC)} \cdot \overline{(CD)}}}$$

Alternatively, we could use the 0s in the Karnaugh map as in table 9.16 and form a maxterm expression for Q.

Table 9.16 *The Karnaugh map for Q, leading to equation 9.2*

Now there is a square group of four 0s corresponding to the maxterm $A' + C$ and a line of four 0s corresponding to $C + D$. The 0 in the cell where $AB = 00$ and $CD = 01$ can be joined to three other 0s to form a block of four corresponding to the maxterm $B + C$. The two remaining 0s in the bottom corners can be joined to those in the two top corners to give a further block of four corresponding to the maxterm $B + D$. Thus the minimal maxterm expression for Q is

$$Q = (A' + C)(C + D)(B + C)(B + D) \qquad (9.2)$$

This can be readily implemented in all-NOR form by complementing twice.

9.6.2 Don't cares

There are cases when the output or switching function does not exist for some combination of the input variables, or the input variables cannot be set to a particular combination (such as 1100 in BCD coding), so we don't care whether it is 0 or 1. In these cases the *don't care* combinations are marked with a cross on the Karnaugh map and we are at liberty to set these cells to 0 or 1 as convenient. As an example look at the Karnaugh map of table 9.17.

Table 9.17 *Don't cares*

The simplest solution is to call all the don't care cells 0 except the 0101 cell, giving the subcube of 1s shown. In this subcube $C = 0$ and $B = 1$, while A and B change; so the switching function is $S = BC'$. When Karnaugh maps contain more 1s than 0s the switching function may be more simply expressed as a *POS* or in terms of its inverse and the *SOP*.

9.7 Practical examples

We shall give just a few examples of circuits using combinational logic, starting with some of the basic building blocks in the arithmetic unit of digital computers.

9.7.1 The half adder

The half adder circuit's function is to add two binary digits, A and B, which gives the truth table as in table 9.18. There are two outputs: S, which is the least significant digit in the sum of the two added digits and C, the carry which must be added into the sum of the next, more significant, digit. S is seen to be just the XOR function while C is the AND function.

Table 9.18 *The truth table for a half adder*

A	B	S	C
0	0	0	0
0	1	1	0
1	0	1	0
1	1	0	1

The Karnaugh map for S leads to no simplification, so the *SOP* expression is, as we have seen before, $S = AB' + A'B$. Looking for the *POS* expression we find

$$S = (A + B)(A' + B')$$

so that

$$S' = \overline{(A + B)(A' + B')} = \overline{A + B} + \overline{A' + B'}$$

by de Morgan's law. Inverting this produces

$$\overline{S'} = S = \overline{\overline{A + B} + \overline{A' + B'}}$$

which is in all-NOR form. The advantage of this representation is that it gives the carry, C, which is AB. By de Morgan's law once more

$$C = \overline{C'} = \overline{\overline{AB}} = \overline{A' + B'}$$

and this is one of the terms in the all-NOR form of S. Hence the circuit for the half adder with carry is as in figure 9.14a, and requires three dual-input gates. If an XOR gate is used to form S and an AND gate to form C, the circuit of figure 9.14b results.

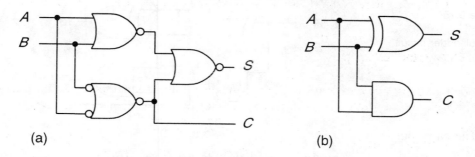

(a)   (b)

Figure 9.14 Half-adder circuit with carry using (a) NOR gates and inverters (b) XOR and AND gates

9.7.2 The full adder

The half adder is not much use on its own as it cannot cope with a carry in from a previous stage, C_i. Thus three inputs are required for a full adder, as shown in the truth table of table 9.19.

Table 9.19 *The truth table for a full adder*

A	B	C_i	S	C_o
0	0	0	0	0
0	1	0	1	0
1	0	0	1	0
1	1	0	0	1
0	0	1	1	0
0	1	1	0	1
1	0	1	0	1
1	1	1	1	1

The outputs from the full adder are a sum, S, and a carry out, C_o. From the truth table we can form the SOP for S:

$$S = A'BC_i' + AB'C_i' + A'B'C_i + ABC_i$$

which rearranges to

$$S = C_i'(A'B + AB') + C_i(A'B' + AB)$$

$$= C_i'(A \oplus B) + C_i(\overline{A \oplus B}) = (A \oplus B) \oplus C_i$$

The circuit for S therefore requires two XOR gates, as in figure 9.15.

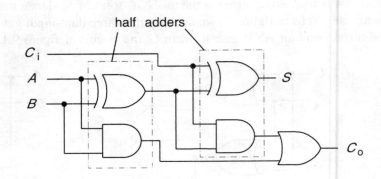

Figure 9.15

The full adder circuit

The carry out is

$$C_o = ABC_i' + A'BC_i + AB'C_i + ABC_i$$

$$= C_i(A'B + AB') + AB(C_i' + C_i)$$

$$= C_i(A \oplus B) + AB$$

This expression for C_o can be obtained by ANDing C_i and the output of one XOR, $A \oplus B$, and ORing this with A AND B. The circuit of figure 9.15 performs the logic for S and C_o and comprises two half adders with an OR gate to form C_o from the carry outs.

9.7.3 Encoders and decoders

It is often necessary to encode and decode data: for example one might wish to code keyboard alphanumericals into ASCII binary code prior to transmission, or to encode decimal digits into BCD form. One can then decode transmitted data back into the form in which it is required. As a simple example, suppose we wished to encode the ten decimal digits into XS3 binary, that is binary plus 11_2, a form which makes for easier subtraction. The block diagram for the coder/decoder is shown in figure 9.16 and the truth table is that of table 9.20.

Figure 9.16

Block diagram for an encoder/decoder

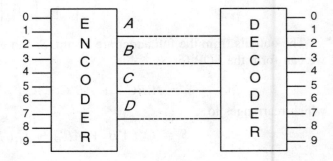

Table 9.20 *Decimal to excess-3*

Decimal	XS3 Binary			
	A	B	C	D
D_0	0	0	1	1
D_1	0	1	0	0
D_2	0	1	0	1
D_3	0	1	1	0
D_4	0	1	1	1
D_5	1	0	0	0
D_6	1	0	0	1
D_7	1	0	1	0
D_8	1	0	1	1
D_9	1	1	0	0

The logic can be derived by simply ORing the entries with 1. For example $A = D_5 + D_6 + D_7 + D_8 + D_9$, where D_5 is the input activated by pressing the key marked '5' etc. The decoding is done by Karnaugh mapping the decimal numbers as in table 9.21, which is the Karnaugh map for D_0, the unused binary numbers being entered as don't cares. It is of course these don't cares which lead to logical simplification and here we see that $D_0 = A'B'$. The full circuit for the XS3 to decimal decoder is shown in figure 9.17. This is easily performed by a programmable logic array (PLA) — see section 9.9.

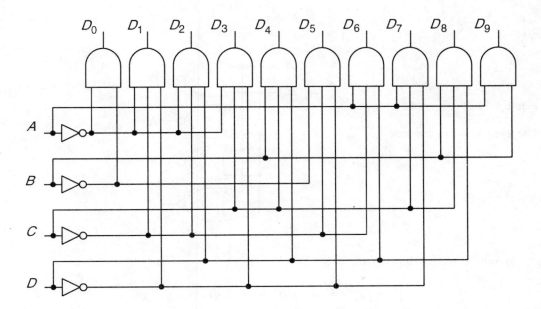

Figure 9.17 The binary XS3 to decimal decoder

Table 9.21 *The K. M. for D_0*

CD\AB	00	01	11	10
00	×	0	0	0
01	×	0	×	0
11	1	0	×	0
10	×	0	×	0

The 2-line-to-4-line decoder

An *n*-bit binary number can convey up to 2^n different messages. A decoder takes an *n*-bit binary number at its input and decodes it into one of up to 2^n outputs. Thus a 3-bit decoder can have 8 outputs (or fewer) and a 2-bit decoder four. The truth table for the 2-line-to-4-line decoder is shown in table 9.22.

Table 9.22 *Truth table for a 2-line-to-four-line decoder*

A	B	W	X	Y	Z
0	0	1	0	0	0
0	1	0	1	0	0
1	0	0	0	1	0
1	1	0	0	0	1

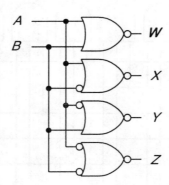

Figure 9.18

The 2-line-to-4-line decoder

From the truth table we see that

$$W = A'B' = \overline{A + B}$$

$$X = A'B = \overline{A + B'}$$

$$Y = A B' = \overline{A' + B}$$

$$Z = A'B' = \overline{A + B}$$

Hence the logic for *W*, *X*, *Y* and *Z* requires only the four two-input NOR gates and two inverters shown in figure 9.18.

9.7.4 BCD to 7-segment display

In section 9.1.4 we saw how decimal numbers could be conveniently represented in binary form by encoding them digit by digit in BCD form. In devices like calculators decimal numbers are displayed by suitably selecting segments from a 7-segment array, as in figure 9.17a. If we take the decimal number, 6 (0110 in binary), as an example, we can see that segments *c*, *d*, *e*, *f* and *g* are to be lit, while *a* and *b* are not. Considering all the numbers for which a particular segment, such as *d*, must be lit, a truth table can be drawn up as in table 9.23, where *W* is the MSB and *Z* is the LSB.

Table 9.23 *The truth table for segment 'd'*

Decimal number	Binary				d
	W	X	Y	Z	
0	0	0	0	0	1
1	0	0	0	1	0
2	0	0	1	0	1
3	0	0	1	1	1
4	0	1	0	0	0
5	0	1	0	1	1
6	0	1	1	0	1
7	0	1	1	1	0
8	1	0	0	0	1
9	1	0	0	1	0

Table 9.24 *The Karnaugh map for segment 'd'*

WX \ YZ	01	11	10	00
01	1	×	0	0
11	0	×	×	1
10	1	×	×	1
00	0	×	1	1

The Karnaugh map for d in table 9.24 shows that we can make three POS terms by lumping some of the don't cares in with the 0s, none of which includes W. The POS for d is then

$$d = (X + Y + Z')(X' + Y' + Z')(X' + Y + Z)$$

We could AND three 3-input ORs with a 3-input AND gate, or we could turn it into all-NOR form by de Morgan's law:

$$d = \overline{d'} = \overline{\overline{X + Y + Z'} + \overline{X' + Y' + Z'} + \overline{X' + Y + Z}}$$

This can be achieved with four 3-input NORs as shown in figure 9.19b, and does not require an input from W, the most significant BCD bit. We could use, for example, the 74ALS27 device, a triple 3-input NOR, to implement the circuit, or a pair of dual 4-input NORs such as the 74HC4002. Each IC package costs about 50p in single quantities. In the 74HC4002 the spare inputs to the 4-input NORs should be tied to logical 0, in case they drift to 1. Likewise, the spare inputs of AND or NAND gates should be tied to logical 1 in case they become grounded. The LEDs themselves will sink quite a lot of current and it will be necessary to use the logic output to turn on a transistor to drive the display. Then all that remains is the logic design for the other six segments. We can best accomplish this kind of task with a programmable logic array (PLA), explained next, though in practice a BCD 7-segment LED decoder/driver chip would be used.

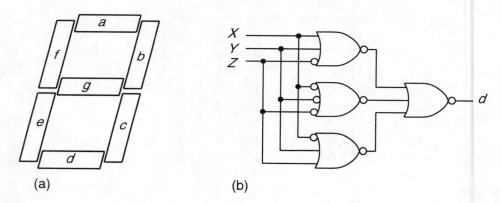

Figure 9.19 (a) 7-segment LED (b) Logic for segment 'd'

9.7.5 Gray-to-binary decoder

Gray code is used to convert decimal numbers into a form of binary number in which only neighbouring digits change in counting up or down. For example the three-bit binary counting sequence is 001, 010, 011, 100, 101 etc. and the count from 3_{10} to 4_{10} involves the change of all three binary digits. In Gray code the same decimal sequence would become 001, 011, 010, 110, 111 etc. so that no more than one digit changes at a time. The advantage of the Gray code is that errors in one bit of an encoded decimal number can only affect the decimal count by one, which would not be the case with unencoded binary representation. The conversion from Gray to binary is made by a combinational logic circuit. Consider the 3-digit Gray code and its binary equivalent in table 9.25, from which we can draw up three Karnaugh maps — one for each binary bit — as in tables 9.26a-c.

Table 9.25 *Gray code and binary*

No.	G_1 G_2 G_3	B_1 B_2 B_3
0	0 0 0	0 0 0
1	0 0 1	0 0 1
2	0 1 1	0 1 0
3	0 1 0	0 1 1
4	1 1 0	1 0 0
5	1 1 1	1 0 1
6	1 0 1	1 1 0
7	1 0 0	1 1 1

Table 9.26 *(a) The KM for B_1 (b) The KM for B_2 (c) The KM for B_3*

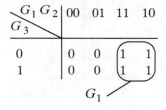 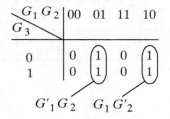

The Karnaugh map in table 9.26a shows that $B_1 = G_1$ and table 9.26b that

$$B_2 = G'_1G_2 + G_1G'_2 = G_1 \oplus G_2$$

Table 9.26c has no groupings of 1s or 0s, but the symmetry suggests an XOR function for B_3 which is

$$B_3 = G_1 \oplus G_2 \oplus G_3$$

Therefore the logic circuit for the 3-bit Gray-to-binary decoder is that of figure 9.20 below.

Figure 9.20

Gray-to-binary decoder

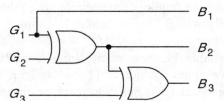

9.8 Programmable logic arrays

Programmable logic arrays (PLAs), programmable logic devices (PLDs), or gate-array logic devices (GALs) can be purchased in IC form. They are programmed by the user to perform a variety of logic functions that might have used many logic ICs of the NAND, NOR and NOT type. Figure 9.21 shows the arrangement of the connections in a PLA which has programmable input connections to AND gates and programmable OR outputs.

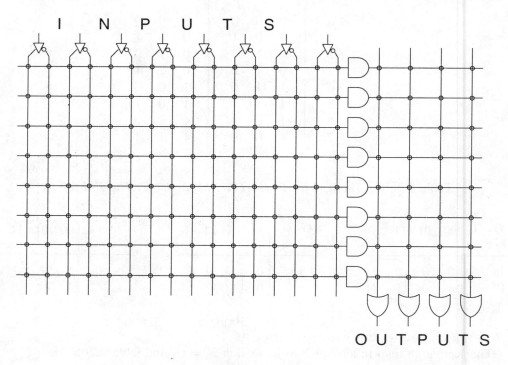

Figure 9.21 A programmable logic array

The inputs can be connected to all the AND gates in both complemented and uncomplemented form. Figure 9.22 indicates the actual connection symbolised in figure 9.21. The cross-links are connected where desired by a suitable PLA programming device, which can be instructed by a microcomputer. Some PLAs are one-time programmable (OTP) and some may be reprogrammed when desired, like an erasable programmable read-only memory (EPROM), by exposing them to ultra-violet light, and some are electrically erasable as in EEPROMs. There are programs which will convert a logical expression into instructions to make the required connections, but otherwise one has to tell the programming device exactly which links to make. In figure 9.21 the connections are shown for forming the logical expression $Q = ABC' + A'BC$, as a simple example.

Commonly-available PLAs have anything from 10 to 22 inputs and from 8 to 10 outputs. Some have outputs which can be reconfigured as inputs. The cheaper PALs are in 20, 24 or 28-pin packages. Some have programmable OR outputs as in figure 9.21, while others are hard wired. Prices range from about £1 for the GAL16V8B-25LP device

(16 inputs, 8 programmable outputs, 25 ns propagation delay, 20-pin package) up to £10 for a GAL22V10B-7LP device (22-input, 10-programmable output, 7.5 ns propagation delay, 24-pin package).

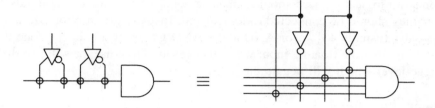

Figure 9.22 The actual configuration of the circuit in figure 9.21

Figure 9.23

The links that must be programmed to make the output of the PAL $Q = ABC' + A'BC$

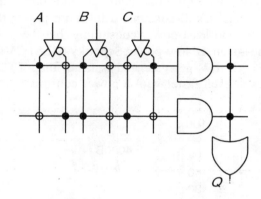

9.9 Practical aspects of logic circuits

Logic circuits are made from transistors, diodes and resistors in the form of ICs. Besides the differences in the methods of construction (MOS, bipolar, Schottky etc.) there is an inherent variability in the manufacturing processes which leads to variations in properties. These inevitable variations can be made worse by operation at elevated (or lowered temperatures) or by employing a supply voltage at the limits of the specification. We shall briefly consider a few important practical points.

9.9.1 Logic voltage levels: noise immunity

Logic devices need a certain minimum input voltage before they will register a logical 1, and a certain maximum input voltage below which they will register a 0. These are specified by the manufacturers, but a standard +5V TTL or TTL-compatible device will normally operate on an input 1-level (one-level) of +2.0 V and an input 0-level (zero-level) of about +0.8 V. The *noise immunity* specified by the manufacturers is 0.4 V, which means that 0.4 V can be added to, or subtracted from, the output levels before they produce malfunctions in subsequent logic. Thus the 0-level output voltage must then be +0.4 V or less and the 1-level output voltage 2.4 V or more (0.8 − 0.4 V and 2.0 + 0.4 V). The 1-level is usually determined by the supply voltage, which must therefore be not less than that specified by the data sheets. CMOS has a greater noise immunity than TTL and ECL has a lower noise immunity.

Spare inputs to logical devices should either be connected to one of the inputs that *is* connected, or they should be connected to logical zero (usually ground, for OR and NOR gates) or logical 1 (usually $+V_{CC}$, for AND and NAND gates). If this is not done there is a strong possibility that the floating input will either respond to noise, or will just drift and go to a voltage level which will negate the valid inputs.

9.9.2 *Power and speed*

The power consumed by logic gates depends on whether their inputs and outputs are HIGH or LOW, but more importantly depends on the device family being used. If the IC is a fast ECL type, it will consume a lot more power than a slower TTL type. CMOS has been slowest and least power consuming, but the position has changed since the introductions of the advanced low-power Schottky (ALS) CMOS, which has an average power consumption of 1 mW per gate and is faster than TTL. The AC (advanced low-power CMOS) series is even faster with a power consumption of 1 µW/gate.

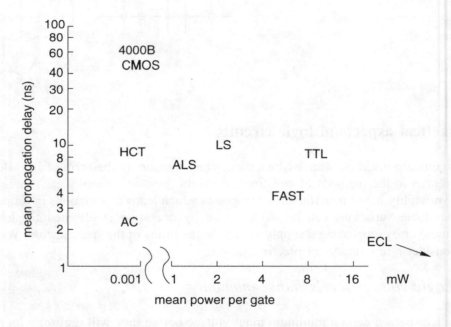

Figure 9.24 Average propagation delays and power consumption/gate for some logic families

The speed of logic devices is expressed as *propagation delay*, the time taken for the signal to pass from input to output at some percentage (usually 50%) of the maximum signal. Figure 9.24 shows propagation delays and power consumptions for some logic families, mostly 74-series. See also table 9.2. The speed-power product is often used as a figure of merit (the lower the better) and this is simply the product of average power/gate and propagation delay (see section 9.10.4). The advent of CMOS and Schottky transistors has resulted in a considerable reduction in speed-power product — from around 100 pJ for standard TTL to around 5-10 pJ for advanced, low-power, Schottky (ALS) devices.

9.9.3 Fan-out

Logic devices require current to operate. The current required depends on the device type and the logic level. The amount of current they can supply from their outputs is also a function of device type and logic level. If a logic gate can supply sufficient current under worst operating conditions to a maximum of ten inputs to gates of the same logic family (logic families should not be mixed), then it is said to have a fan-out of ten. Fan-outs are typically ten or more.

   If a logic circuit is interfaced with analogue circuitry the current drawn by, or supplied by, that circuit from the digital part must be less than the maximum specified. When the logic devices are sourcing current, they can supply typically from 10 mA to 65 mA, and when sinking current from −10 mA to −25 mA, depending on the logic family.

9.9.4 Propagation delay

For each logic gate a finite time, known as the propagation delay, elapses before the output changes in response to a change of input. This delay can have serious consequences if the designer of a circuit assumes that the output is at 1 (or 0) when it may for a short time be at 0 (or 1) instead. In the case of the now-obsolescent, slow, 4000B MOS family, the typical propagation delay for a NAND gate transition from LOW to HIGH, t_{PLH}, is 40 ns at a supply voltage of +15 V, rising to 125 ns with a +5V supply. These times should be doubled to allow for worst-case devices. The 74 series has a typical propagation delay of 11 ns, while the 74HC and 74HCT series are a little faster. Fastest at the moment is the 74ABTC (advanced BiCMOS) family with $t_{\text{PLH}} = 3.6$ ns, maximum. See also figure 9.24.

9.9.5 Three-state logic

The word 'three-state' is slightly misleading for there are still only two logic states, 1 and 0, as in normal logic, but three output states: HIGH, LOW and open circuit. The reason for this is that some types of TTL gate cannot be connected together, because when one is HIGH and the other LOW they try to pull each other into their state: the outputs are said to be in contention. Three-state gates connected to a common bus are enabled one at a time, the others being disabled by an enabling input so that they present open circuits to the bus. Three-state logic allows the use of the same devices to transmit or receive, or multiplexers to multiplex or demultiplex according to the state of the enabling input.

   Figure 9.25 shows some three-state logic gates which are enabled when $E = 0$ and disabled when $E = 1$, and a three-state bus with data transmitting (Tx) and receiving (Rx) lines controlled by three-state inverters.

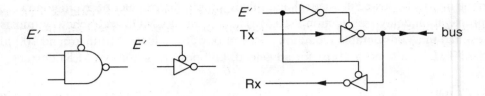

Figure 9.25 Three-state gate symbols and a two-way three-state bus

9.9.6 *Hazards*

Hazards are states caused by propagation delays which arise at the output of a logical circuit which should not be present according to the truth table. Consider the circuit of figure 9.26a which has a logical output of $Q = AB' + BC$ and whose Karnaugh map is shown in table 9.27.

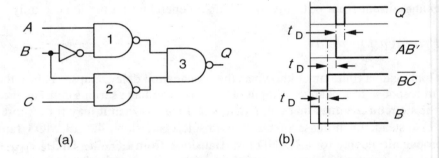

(a)   (b)

Figure 9.26 (a) Circuit which produces a hazard condition (b) timing diagram for (a)

Table 9.27 *The KM for the circuit of figure 9.26a*

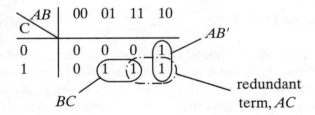

There is a propagation delay caused by the inverter which produces B' at the input of NAND1. If the inputs are initially $A = 1$, $C = 1$ and $B = 1$, then when B switches from 1 to 0 the output at NAND2 switches from 0 to 1 after one gate propagation delay, t_D. But the output of NAND1 at $t = t_D$ is 1, and so at $t = 2t_D$ the output of NAND 3 is 0, not 1 as suggested by the KM. After $3t_D$ the output at NAND3 will be 1 as required. The timing diagram for this is shown in figure 9.26b. While the output is 'wrong' a problem may have arisen (lift doors closed, alarm bells rung, sprinklers turned on and so forth).

The changing variable which causes the hazard is that in the redundant loop corresponding to the term AC, and it can be overcome quite simply by including the redundant term in the circuit, that is we form

$$Q = AB' + BC + AC$$

as in figure 9.27. We can see from this figure that the output must remain at 1 when B switches from 0 to 1 while A and C are 1. The term AC, which is 1, ensures that the output of NAND3 is 0 during the time B switches and so the output of NAND5 stays at 1. Hazards arise whenever adjacent 1s in a Karnaugh map are not looped together. Logically-redundant terms might not be redundant practically.

Figure 9.27

The circuit for Q with the redundant term included to eliminate the switching hazard

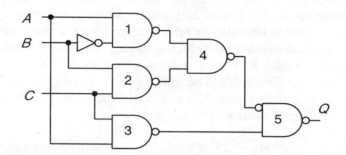

9.10 Multiplexers and demultiplexers

A multiplexer is a device with a number of data inputs which can be selected one at a time as output. The selection is performed by address (or data selection) inputs in accordance with the desired output sequence. The output can then be transmitted and demultiplexed into the original inputs.

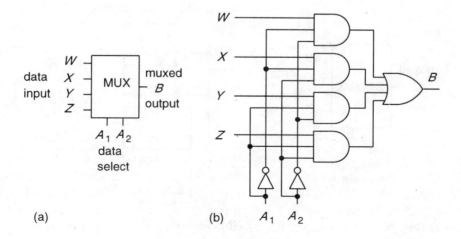

Figure 9.28 (a) A 4-input multiplexer (MUX) (b) The logic for data selection

Figure 9.28a shows a multiplexer with four inputs, W, X, Y and Z, an output, B, and two address inputs A_1 and A_2. If W is selected by $A'_1A'_2$, X by $A_1A'_2$, Y by A'_1A_2 and Z by A_1A_2 then the logic for the output, B, is

$$B = WA_1'A_2' + XA_1'A_2 + YA_1A_2' + ZA_1A_2$$

and this can be realised with the circuit of figure 9.28b.

It is clear that if there are 2^N data lines then N address (data-selection) lines are needed. If the address inputs are switched so that the data on the data inputs stays constant during one complete address cycle, then the data is presented to the output in multiplexed bit form, and parallel-to-serial conversion has been effected. If the address cycle time is a multiple of the cycle time for bit multiplexing then the multiplexed data stream comprises several bits of each data input. Thus the data can also be multiplexed in bytes (normally of eight bits) or nibbles (half a byte) or half nibbles.

The demultiplexer (which is usually the same IC as the multiplexer) takes the multiplexed serial data stream and converts it to the component data streams once more as shown in figure 9.29a. The truth table to demultiplex the data from the circuit of figure 9.28b is given in table 9.28 and is like that of the 2-line-to-4-line decoder. The logic for this is shown in figure 9.29b.

Table 9.28　*The truth table for the demultiplexer of figure 9.29*

A_1	A_2	W	X	Y	Z
0	0	B	0	0	0
0	1	0	B	0	0
1	0	0	0	B	0
1	1	0	0	0	B

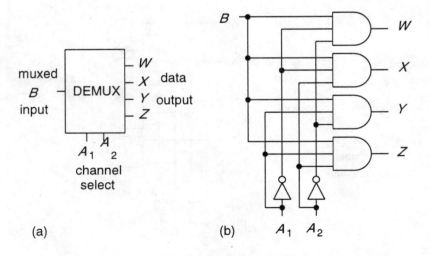

(a)　　　　　　　　　　　　　　(b)

Figure 9.29　(a) A 4-channel demultiplexer (b) The logic for (a)

9.10.1 *Boolean functions using multiplexers*

By using the data input and data select lines it is possible to form Boolean functions at the output of a multiplexer. Almost any combination of three variables is possible using a 4-channel multiplexer. For example consider the truth table of table 9.29, which can be expressed in SOP form as

$$Q = A'B'C' + A'B'C + A'BC' + AB'C + ABC$$

If we make A and B the address lines of a 4-to-1 multiplexer, then the first two terms are

$$A'B'C' + A'B'C = A'B'1 = A_1'A_2'W$$

Thus the W-input must be connected to logical 1. The third term is $A'BC' \equiv A_1'A_2X$ so that $X = C'$. The fourth and fifth terms give $Y = Z = C$. The multiplexer must then be configured as shown in figure 9.30 to give Q.

Table 9.29 *The truth table for Q*

A	B	C	Q
0	0	0	1
0	0	1	1
0	1	0	1
0	1	1	0
1	0	0	0
1	0	1	1
1	1	0	0
1	1	1	1

Figure 9.30

A MUX configured for a Boolean function

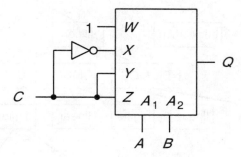

9.11 Logic families

Integrated circuits or ICs can be made in a variety of ways as shown in figure 9.31, which shows not only ICs made of semiconductors but also those made from thick and thin films on insulating substrates. The latter are not considered here as they are not used for logic ICs. Gallium arsenide (GaAs) ICs are restricted to a few specialised uses, mainly at very high frequencies: all the ICs supplied by popular stockists are made from silicon.

We have mentioned previously various types of logic: TTL, CMOS, ECL and versions of these. Broadly speaking there are two types of constructions for logic circuits: those using MOS (metal-oxide semiconductor) technology and those using bipolar (BJT) technology. Recently BiCMOS has been introduced which uses both and this may prove to be the most popular type in the future; certainly a great deal has been invested in this process. The chart does not show any subdivisons of MOS ICs, mainly to save space. Originally these were all made from PMOS (p-channel MOS) transistors, which were superseded by NMOS (n-channel MOS), but now CMOS (which uses complementary pairs of NMOS and PMOS transistors) is completely dominant, especially in logic devices. One advantage of an MOS-based process is that the standby power consumption is low, and in CMOS it is very low indeed, while that of the bipolar-based circuits is much larger. Another advantage lies with packing density: MOS transistors are inherently smaller than BJTs. The drawback of CMOS is reduced speed of operation, though this is becoming less of a problem with high speed CMOS of the 74HCT series.

The packing density of ICs is an important consideration in VLSI (very-large-scale integrated circuits) which may have up to ten million transistors on a single chip. VLSI circuits used NMOS technology for this reason, but now even in the largest circuits CMOS has taken over.

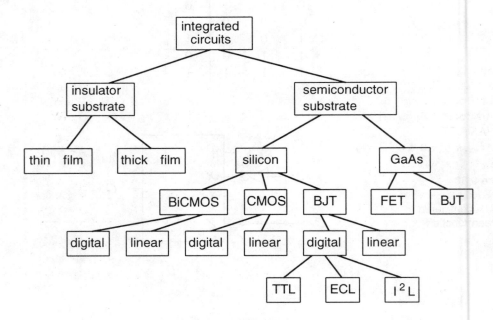

Figure 9.31 Integrated circuit types

The oldest of the present-day digital logic families is bipolar TTL (transistor-transistor logic, a successor to resistor-transistor logic, RTL, and diode-transistor logic, DTL). ECL (emitter-coupled logic) gates are very fast, but consume much more power than other types of logic and are largely restricted to special purposes such as high-speed computing. I²L (integrated injection logic) was an attempt to increase the packing density and speed of TTL which has faded like the NMOS and PMOS precursors of CMOS. Virtually all readily-available logic ICs now are 74 series TTL, CMOS or BiCMOS and we shall restrict our discussion to these.

9.11.1 TTL

TTL is still widely used because it is cheap, operates at well-defined and well-known voltage levels and is robust. It has been undergoing continuous development for 25 years in an effort to reduce its power consumption and increase its speed. It suffers from a larger power consumption than CMOS but it is normally faster. Both of these aspects of performance have been enhanced by the advent of Schottky TTL, but probably not by enough to ensure TTL's ultimate survival. Drop-in CMOS replacements for advanced low-power Schottky TTL have been available for some time, offering greatly reduced power consumption at comparable speeds.

Unlike many CMOS logic device families, TTL has standard operating voltages given in table 9.30 along with typical NAND propagation delays. The voltage requirements mean that all TTL logic devices, no matter where made, can be used interchangeably.

Table 9.30 *74 series TTL and TTL-compatible CMOS parameters*

74 family	NAND No.	V_{CC}	V_{IL}	V_{IH}	t_{pd}	P
Standard TTL	7400	4.75-5.25	0.8	2.0	9	10 mW
Schottky TTL	74S00	4.75-5.25	0.8	2.0	3	25 mW
Low-power Schottky TTL	74LS00	4.75-5.25	0.8	2.0	16	2 mW
Advanced Schottky TTL	74AS00	4.5-5.5	0.8	2.0	2	5 mW
FAST TTL	74F00	4.5-5.5	0.8	2.0	2.5	5 mW
Advanced LS TTL	74ALS00	4.5-5.5	0.8	2.0	5	1.25 mW
Low-voltage TTL	74LV00	2.7-3.6	0.8	2.0	9	0.75 mW
Hi-speed CMOS	74HCT00	4.5-5.5	0.8	2.0	10	< 1 μW
Advanced CMOS	74ACT00	4.5-5.5	0.8	2.0	2	< 1 μW
Low-voltage BiCMOS	74LVT00	2.7-3.6	0.8	2.0	3	< 60 μW

Notes: NAND = quad 2-input NAND V_{CC} = supply voltage V_{IL} = maximum input voltage for LOW level transition V_{IH} = min. input voltage for HIGH level transition t_{pd} = typical propagation delay in ns for a quad 2-input NAND P = quiescent power.

TTL output stages

Bipolar TTL outputs are of three kinds: open collector, totem pole and three state[5]; section 9.10.5 discusses three-state logic. The open-collector output is used when external loads are to be driven, which become the collector pull-up resistor as in figure 9.32.

We can see in figure 9.32 that when either or both of the inputs, A and B, are LOW the input transistor turns on and draws base current from the supply and reverse biases the output transistor, whose output therefore HIGH. Only when both inputs are HIGH can current flow through the input transistor into the base of the output transistor to turn it on, and so send the output LOW. The logic for the circuit is therefore $Q = 0$ when A AND B are 0, that is $Q' = AB$ or $Q = (AB)'$: a NAND gate.

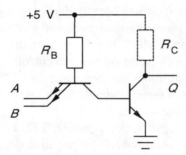

Figure 9.32

Open-collector TTL NAND

When two or more devices are connected together (a wired-OR connection) at their collector outputs, as in figure 9.33, the output will be pulled LOW (it is said to be *active LOW*) when any output is LOW. Thus the circuit of figure 9.33 performs the operation

$$Q' = AB + CD \quad \Rightarrow \quad Q = \overline{AB + CD} = \overline{AB} \cdot \overline{CD}$$

and should be called a wired AND rather than a wired OR. If the circuits had been equipped with pull-down outputs, that is active HIGH, they would truly be wired-OR circuits.

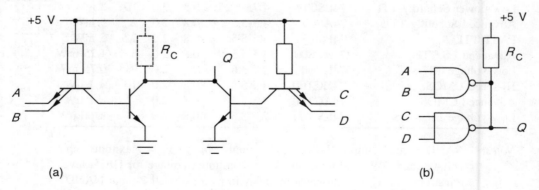

(a)   (b)

Figure 9.33 (a) Wired-OR connection (b) The equivalent circuit

---

[5] The term tri-state, commonly used, is a registered trademark of the National Semiconductor Corporation.

The usual output stage for TTL is that of figure 9.34, known as a totem-pole output, and in fact very like a push-pull amplifier. From this diagram, which is that of a NAND gate, we can understand the logic levels in TTL. Table 9.31 gives the voltages at the numbered nodes when the output is high and when it is low, assuming 0.65 V for V_{BE} and the diode's forward drop, and 0.2 V for V_{CEsat} and the LOW input voltage. These are easily found, for when the inputs are both HIGH, Q1 supplies base current to Q2 and turns it and Q4 on. Q4 is saturated so the output voltage is 0.2 V. When one of the inputs is LOW (0.2 V), Q2 and Q4 are off while Q3 turns on. The output will be 5 V with no load and 3.7 V when current is drawn through Q3 and the diode. When the output is changing state, both Q3 and Q4 will conduct and so a 120Ω resistor is placed in the collector of Q3 to limit the switching current from the supply, which would otherwise cause considerable noise problems.

The logic transition voltages can be worked out from figure 9.34. Suppose one input is LOW and the other HIGH and then the LOW input is increased from 0.2 V, that is the circuit is going to switch output from HIGH to LOW. The potential must rise above 1.3 V at node 2 for Q2 and Q3 to conduct, and then there is a further 0.2 V for V_{CEsat} of Q1, for a total of 1.5 V at the input. But Q1 could have $V_{CEsat} = 0.4$ V and the V_{BE} values could be as high as 0.75 V, and then the minimum input which would cause the output to go LOW becomes 1.9 V.

The push-pull type of output has important consequences for speed of operation, which is largely dependent on the charging and discharging of stray capacitance at the output terminal. When the output switches from HIGH to LOW, this capacitance can discharge via Q4, which offers a low-resistance path to ground and therefore a short transition time. When the output switches from LOW to HIGH, Q3 provides a charging path from the supply via the 120Ω resistor. If the stray capacitance is of the order of 10 pF (a large value compared to most stray logic IC capacitances) the charging time is still only about 1 ns.

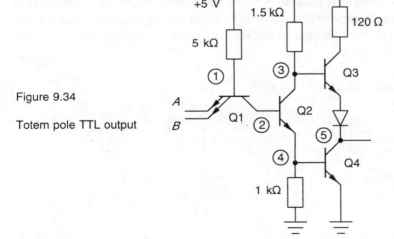

Figure 9.34

Totem pole TTL output

Table 9.31 *Circuit voltages*

node	output HIGH	LOW
1	0.85	1.95
2	0.4	1.3
3	5	0.85
4	0	0.65
5	5 ‡	0.2

‡ 3.7 V with load

9.11.2 Schottky TTL logic

In Schottky TTL logic the transistors are replaced by Schottky transistors, whose symbol is shown in figure 9.35a. These have Schottky diodes (see section 2.4) placed across their base-collector junctions to reduce the base current flowing when the device is in saturation, as shown in figure 9.35b.

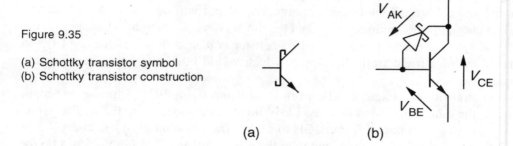

Figure 9.35

(a) Schottky transistor symbol
(b) Schottky transistor construction

(a)   (b)

The Schottky diode has a low forward voltage drop of about 0.3 V and ensures that $V_{CE}(\min) = V_{BE} - V_{AK} \approx 0.4$ V. The transistor cannot be taken fully into saturation, thereby reducing the diffusion capacitance of the transistor and hence the time taken to charge when switched. The result is a considerable improvement in speed.

9.11.3 CMOS

Besides TTL-compatible CMOS families there are a considerable number of other CMOS logic families, some of which are shown in table 9.32. We can see from this table that the CMOS switching voltages depend on the supply voltage, and the present standard for CMOS is that $V_{IL} = 0.2V_{DD}$ and $V_{IH} = 0.7V_{DD}$ with the output HIGH and LOW voltages being 0.1 V from V_{DD} and 0 V respectively.

CMOS, and MOS ICs generally, used to be very susceptible to malfunction because the thin gate insulation is easily broken down by the voltages arising from static electricity. Handling therefore required the handler to be earthed until the device was properly connected to its circuit.

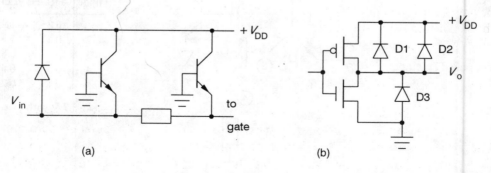

(a)   (b)

Figure 9.36 Protection against static electricity (a) Input (b) Output

The problem has now largely been overcome by incorporating protection diodes into the gate during manufacture. Figure 9.36 shows a refinement of this used by Texas Instruments which prevents any damaging voltages from building up or damaging currents from flowing.

Table 9.32 *Some CMOS logic families*

Family	NAND No.	V_{DD}	V_{IL}	V_{IH}	t_{pd}
high speed	74HC00	2-6			
		6	1.2	4.2	8
		4.5	0.9	3.5	9
		2	0.3	1.5	45
advanced	74AC00	3-5.5			
		5.5	1.65	3.85	3
		4.5	1.35	3.15	
		3	0.9	2.1	
very high speed	74VHC00	2-5.5			
		5	1.0	3.5	4.7
4000B	4011B	3-15			
		5	-	-	60
		10	-	-	30
		15	-	-	25
low voltage	74LVC00	2.7-3.6	0.8	2.0	4
advanced low voltage	74ALVC00	2.7-3.6	0.8	2.0	2

On the input protection side positive voltages are limited to within one diode drop of the supply, and negative inputs cannot produce destructive currents because these are limited by the transistor/resistor network. On the output side, D1 and D3 are parasitic diodes (produced willy-nilly by the production process) and D2 is specially added as a protection diode to improve the discharge capability for positive output voltages.

The most basic logical device is the inverter or NOT gate, discussed in detail in section 4.7.3, but we will recapitulate a little. The CMOS inverter is shown in figures 9.37a and 9.37b, the former being the normal circuit symbol and the latter one more suited to logic circuits. Q1 is a p-channel transistor which turns OFF when

$$V_{in} > V_{DD} - |V_{TP}|$$

where $-|V_{TP}|$ is the threshold voltage of Q1, written like this to make the sign explicit. And Q2 is an n-channel transistor which turns ON when $V_{in} > V_{TN}$, the threshold voltage for Q2. In this case the output is pulled down to ground, or logical 0. Then when $V_{in} < V_{TN} < V_{DD} - |V_{TP}|$ Q1 is ON and Q2 OFF and the output voltage is pulled up to V_{DD}. The transfer characteristic is that of figure 9.37b, which shows that the output voltage swing is from 0 V (logical 0) to V_{DD} (logical 1). The threshold voltage for changes in logical state is almost exactly $\frac{1}{2}V_{DD}$ which give maximum noise immunity, a great boon when the supply voltage is low (as little as 3 V is acceptable). The gate's ability to source or sink current is only limited by R_{DS}(on), a relatively small resistance. And the only time that current can flow from the supply to ground through Q1 and Q2 is when the output is in transition. Thus in the quiescent state the device dissipates virtually no power, which is the reason for CMOS's very low average power consumption.

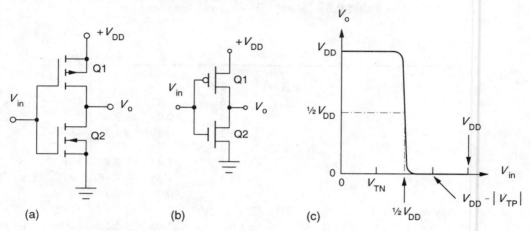

Figure 9.37 CMOS inverter (a) Circuit symbol (b) Alternative symbols for logic (c) Transfer characteristic

CMOS logic gates based on the push-pull configuration of the inverter in figure 9.37 can be assembled into logical circuits with ideal transfer characteristics, illustrated by the cascaded inverters of figure 9.38a. If the input to the cascade is not zero, but still less than V_{TN} (and in practice only a bit less than $\frac{1}{2}V_{DD}$), the output of NOT1 is V_{DD} (logical 1) and this is the input to NOT2, which is therefore driven hard to 0 V (logical 0). Similarly when the input to the cascade is $> V_{DD} - |V_{TP}|$ (and in practice only just greater than $\frac{1}{2}V_{DD}$) the output of NOT1 is 0 V, and this drives NOT2 hard into logical 1. The logic levels in CMOS are therefore always ideal, with $V_{DD} \equiv 1$ and $0 \text{ V} \equiv 0$. Other gate designs can be based on the inverter of figure 9.37 quite readily; we shall look at the CMOS NOR and the CMOS NAND gates to show how.

Figure 9.38

(a) An inverter cascade
(b) Switch model of inverter

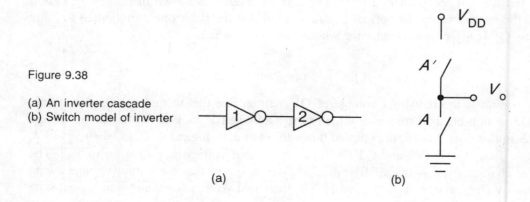

The CMOS NOT gate can be modelled as a pair of switches, labelled A and A', as in figure 9.38b, in place of Q2 and Q1 respectively in figure 9.37. Then when $A = 1$, switch A closes and switch A' opens to give an output of 0, and vice versa when $A = 0$. Using this switch model we can construct a 2-input NOR gate as in figure 9.39a, from which we see that when $AB = 11$, 01 or 10 the output is 0 as either A' or B' is open, and the output is 1 only when $AB = 00$, the logic for a NOR gate indeed.

Figure 9.39

A CMOS NOR gate
(a) The switching equivalent circuit
(b) The transistor circuit

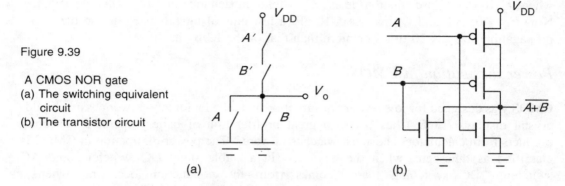

The CMOS NAND gate is soon derived using a switching equivalent circuit as in figure 9.40a, leading to the CMOS circuit of 9.40b. It is readily seen that when $AB = 11$ the output is 0 and that all other input combinations give an output of 1, which is the logic of a NAND gate. Additional inputs can be provided by adding more complementary transistor pairs with the PMOSFETs in parallel with each other and the NMOSFETs in series with each other. The same goes for multi-input NOR gates except that here the PMOSFETs are in series and the NMOSFETs in parallel.

Figure 9.40

A CMOS NAND gate
(a) The switching equivalent circuit
(b) The transistor circuit

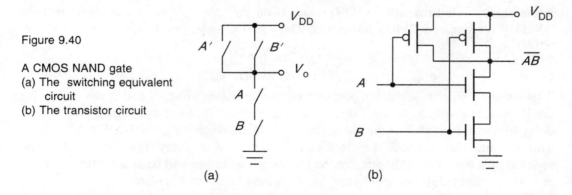

We can see from these examples of CMOS logic gate design that CMOS logic gates require two transistors for each input, which is twice that required for non-complementary logic gates, and this can only be avoided at the expense of greater quiescent power dissipation. Given the very small size of MOSFETs in logic ICs (about 2 μm × 2 μm) the penalty paid in increased size is less than the reward of low power consumption. There is, however, another penalty to pay: increased capacitance and so increased propagation delay times — about 50% more than in non-complementary MOS logic.

But in practice propagation delays are more the result of interconnections between logic ICs and the rest of the circuitry than internal delays within the ICs, since the total propagation delay time is roughly

$$t_{pd}(\text{tot}) = t_{pd}(\text{IC})C_T/C_{IC}$$

where $t_{pd}(\text{IC})$ = IC propagation delay, C_T = total capacitance and C_{IC} = IC capacitance. Now $C_{IC} \approx 0.3$ pF and C_T may be 5-10 pF (a few cms of interconnection), so the actual propagation delay is 20 times or more than that of the IC by itself.

Power dissipation in CMOS

CMOS was designed for low power consumption and allied with low-power passive liquid-crystal displays (LCDs) has led to a great proliferation of battery-operated consumer products from calculators and wrist watches onwards. The power dissipation in CMOS is classified as quiescent, when the device is in a stable state, DC switching and AC switching. DC switching losses comes from the conduction occurring when a complementary pair switches state. AC switching losses come from the need for charging capacitance when transistors change state. As mentioned before, the quiescent losses are negligible compared to the self-discharge of batteries.

The DC switching power losses are proportional to the propagation delay of the gate and are given in the worst case by

$$P_{DC} = \frac{V_{DD}It_{pd}(\text{G})}{T} = \frac{V_{DD}{}^2t_{pd}(\text{G})}{3R_{DS}(\text{ON})T}$$

assuming three transistors between supply and ground, as in a NOR or NAND circuit. T is the clock cycle time and $R_{DS}(\text{ON})$ depends on the gate voltage and is approximately 40 kΩ/V for standard CMOS transistors. Thus the DC switching losses are at most

$$P_{DC} = \frac{V_{DD}{}^3t_{pd}(\text{G})}{120\,000T}$$

This equation emphasises the importance of low operating voltages and is the reason for developing low-voltage logic. Taking $V_{DD} = 5$ V, the TTL level and a gate propagation delay of about 1 ns, then if the clock rate is 10 MHz we find $P_{DC} = 10$ µW/NAND gate. This is quite high because the clock rate is high, and in many applications could be reduced by a large factor. In practice the DC switching losses will be smaller than this and are often neglected, leaving only the AC switching losses to calculate.

The AC switching losses are caused by capacitance charging and do not depend on the circuit resistance, since the charging and discharging losses are $\frac{1}{2}CV^2$ per cycle each, making the loss power is CV^2/T in total. The capacitance is largely interconnection capacitance, about 0.3 pF, so the AC losses are

$$P_{AC} = \frac{CV_{DD}{}^2}{T} = \frac{0.3 \times 10^{-12} \times 5^2}{10^{-7}} = 27 \ \mu\text{W}$$

also per NAND (or NOR) gate. Thus the AC losses are more significant than the DC (more so than shown here, since the DC losses are exaggerated), and are directly proportional to capacitance. The total losses of about 40 µW/NAND gate are proportional to the clock frequency. If this can be reduced without loss of accuracy to 1 MHz for example, the losses can be reduced to 4 µW/NAND. With losses of 4 µW/NAND a small battery-driven circuit containing the equivalent of 15 NAND gates could run for about a year, but only about 1 month at a 10 MHz clock rate. No account has been taken of any loads driven by the logic of course.

9.11.4 Low-voltage logic

Because the overall power dissipation in CMOS goes as $(V_{DD})^n$, where $3 > n > 2$, there is a great incentive to reduce V_{DD}, the supply voltage. There has been a great proliferation of low-voltage logic families in the last few years, such as the 74LV, 74LVX, 74LCX, 74LVC, 74LVCT and HLL series, most of which require supply voltage in the range from 2.7 V to 3.6 V.

Low-voltage ICs are intended for use with batteries and both regulated and unregulated supplies. Several of these families will operate at voltages down to 1.2 V, though at a lower speed, and these can be powered by a single cell. Several low-voltage families now operate on TTL logic levels, though with a reduced supply voltage. The standard for full interfacing and compatibility with TTL logic is still under discussion. Operation at lower supply voltages mean that the switching power consumption is reduced, but the need for tight parameter control and hence tight process control is accentuated. Another advantage of low-voltage operation is reduced EMI (electromagnetic interference) because of much smaller switching transients. Low supply voltages will cause an increase in propagation delay unless device sizes can be reduced and so the MOS channel length on low-cost standard ICs is now only about 1 µm and the BJT basewidth of BiCMOS devices is about 0.1 µm, compared to values three times as great five years ago. The ultra-fast HLL series has multiple supply and grounding points to increase speed. There seems to be little doubt that low-voltage CMOS will become standardised and further developments will be in the direction of lower voltages yet, perhaps under 1 V.

Figure 9.41

A low-voltage, TTL-compatible, BiCMOS output stage

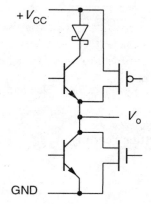

The LVCT CMOS versions actually have BiCMOS construction, which utilises BJTs for speed and CMOS output stages for reduced power consumption. Figure 9.41 shows a typical low-voltage, TTL-compatible, BiCMOS output stage, which combines low-current CMOS rail-to-rail switching and high-current BJT, which can drive loads down to 35 Ω at currents of up to 60 mA.

The need for small propagation delays has meant that the package size must be reduced, so that of the main low-voltage logic series on the slowest, the LV series, has logic gates in standard DIL packages; all the others use the smaller SO (Small Outline) SSOP (Shrink Small Outline Package) or the TSSOP (Thin SSOP) packages. Figure 9.42 shows the 14-pin plastic DIL versions of these packages: the SOT27 is the standard package, the SOT108 is the plastic small outline package, the SOT337is the plastic shrink small outline package and the SO402 is the plastic thin shrink small outline package. The latter three are surface mount devices and are drawn to twice the scale of the standard DIL package.

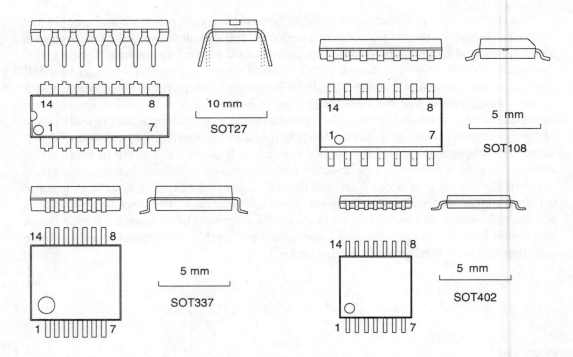

Figure 9.42 Standard 14-pin package styles

Suggestions for further reading

See chapter 10

Problems

1 Convert the hexadecimal number B2E7 to binary and thence to decimal form. *[1011001011100111$_2$, 45799$_{10}$]*

2 Use the 2's complement to subtract 11010110.111$_2$ from 11100101.100$_2$. *[1110.101]*

3 Add in binary arithmetic 1000111$_2$, 110011.11$_2$ and 11101.101$_2$. *[1011000.011]*

4 A BCD number is 100101001.01110101, what is its decimal form? *[129.76]*

5 Convert 38.7$_{10}$ to binary form. *[100110.101100110011...]*

6 Simplify $F = AB + AB' + A'B'$

7 Find the simplest Boolean expression for Y if $Y' = ABC' + AB'C' + A'B'C + ABC + AB'C$, firstly by Boolean algebra and then by using a Karnaugh map. Use de Morgan's law to derive an all-NAND expression, and hence a logic circuit, for Y that uses only 2-input NAND gates. Repeat the last part using only NOR gates. *[Y = A'B' + A'B]*

8 Show by Boolean algebra and truth tables that equations 9.1 and 9.2 are the same.

9 Draw up a truth table for the expression

$$f = \overline{BD'}(AB' + C)(A + B'C)$$

then draw a Karnaugh map to find the minimal expression for f. Draw a logic circuit for f which uses only NOR gates or only NAND gates, whichever is simplest.
[f = B'(A + C) + ACD]

10 The truth table for a Boolean function, Q, is given in table P9.10. Write down a SOP expression for Q and simplify by using Boolean algebra. Use a Karnaugh map to find the minimal logical expression for Q. From this draw the logic diagram for Q. Then use de Morgan's law to find an all-NOR expression and a logic circuit for Q that uses only 2-input NOR gates. *[Q = A + B'C]*

11 Draw up a truth table for the logic circuit of figure P9.11 and then make a Karnaugh map for Q. Hence draw logic circuits which use solely NAND or solely NOR gates to perform the same function as the circuit of figure P9.11.

12 Show that the XOR function obeys the law of association, that is $(A{\oplus}B){\oplus}C = A{\oplus}(B{\oplus}C) = (A{\oplus}C){\oplus}B$

Table P9.10

A	B	C	Q
0	0	0	0
0	0	1	1
0	1	0	0
0	1	1	0
1	0	0	1
1	0	1	1
1	1	0	1
1	1	1	1

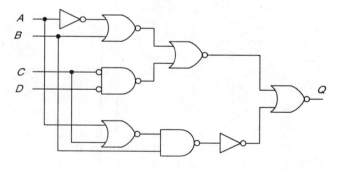

Figure P9.9

13 Show that

$$\overline{(A \oplus B) \oplus (C \oplus D)} = (A \oplus B) \oplus \overline{(C \oplus D)} = \overline{(A \oplus B)} \oplus (C \oplus D)$$

14 The Boolean burglary squad of a certain police force notices that in one neighbourhood houses are burgled ($B = 1$) at any time when a door is left open ($D = 1$), but only at night ($N = 1$) if a window is left open ($W = 1$), and never if a dog is in the house ($H = 1$). Draw up a truth table and a Karnaugh map for B. Derive a Boolean expression for the burgling function in terms of D, H, N and W, then draw a logic circuit for B using only 2-input NOR gates. [$B = H'(D + NW)$]

15 The inputs, A and B, in the circuit of figure P9.15 are either 0 V or +5 V. Make out a truth table for the output, C. What sort of logic gate is it?

16 For the circuit of figure P9.16, the inputs, A and B, are 0 V or +5 V. Draw up a truth table for the output, C. What sort of logic gate is it?

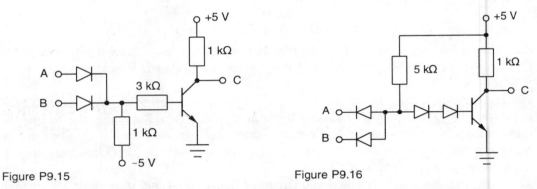

Figure P9.15   Figure P9.16

17 Deduce a logical function for Q from its truth table below, and implement it with XOR gates only.

A	B	C	D	Q
0	0	0	0	0
0	0	0	1	1
0	0	1	0	1
0	0	1	1	0
0	1	0	0	1
0	1	0	1	0
0	1	1	0	0
0	1	1	1	1
1	0	0	0	1
1	0	0	1	0
1	0	1	0	0
1	0	1	1	1
1	1	0	0	0
1	1	0	1	1
1	1	1	0	1
1	1	1	1	0

18 Draw up the truth table for lighting segment *g* in the BCD LED of figure 9.16a and from it deduce the simplest Boolean expression for lighting segment *g* and then draw the logic diagram for this. Draw a circuit for lighting segment *g* which uses only 2-input NAND gates.

19 Deduce which linkages need to be made in a PLA such as that of figure 9.17 so that all seven segments of a BCD LED can be selected, assuming there are seven output ORs and sufficient ANDs.

20 Draw up the Karnaugh map for the function, *Q*, whose truth table is given below. Draw a logic circuit for *Q* which eliminates the hazard present and uses fewest gates all of the same type.

A	B	C	Q
0	0	0	1
0	0	1	1
0	1	0	1
0	1	1	0
1	0	0	0
1	0	1	1
1	1	0	0
1	1	1	0

21 The 4-bit XS3 code for the decimal numbers 0-9 in sequence is 0011, 0100, 0101, 0110, 0111, 1000, 1001, 1010, 1011, 1100. Design a converter to turn this into binary.

10 Sequential logic

B OTH SEQUENTIAL and combinational logic circuits operate on binary data, but whereas in combinational logic the output is a function only of the inputs, in sequential logic circuits it depends also on past outputs as well as present inputs. A sequential logic circuit therefore can be considered to have memory, since it 'remembers' the previous outputs and uses them to determine the new output. This is achieved by using positive feedback from the output terminal(s) to form one or more of the inputs. Sequential logic, as its name implies, is used to generate sequences of states in response to a succession of inputs (asynchronous mode) or clock pulses (synchronous mode). These sequences can be used to store or read information, or to count.

10.1 Flip-flops and latches

The basic building block of many sequential logic circuits is the bistable multivibrator (usually called just a bistable), of which there are several types with different charact-eristics. As its name implies, a bistable has two stable output states — HIGH and LOW, or 1 and 0 — in contrast to the monostable and astable multivibrators, which were discussed in section 7.2. Latches and flip-flops are both bistables, the difference being that a flip-flop will only respond to changes in inputs if it is enabled by the right type of clock pulse. In common parlance both are called flip-flops, which becomes just an alternative name for a bistable.

10.1.1 The SR latch

Figure 10.1 shows and SR latch made from a pair of NAND gates where the output of NAND1 is fed back to the input of NAND2 and vice versa. The other two inputs are known as set (S) and reset (R). The two outputs, Q_1 and Q_2, are normally complements, Q and Q', as we shall see.

Figure 10.1

An SR latch made from two NANDs

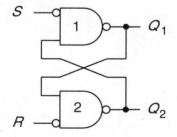

Suppose that $Q_1 = 1$ and $Q_2 = 0$, while $S = 0$ and $R = 0$. The inputs to NAND1 are then 1 and 0, so the output is $Q_1 = 1$, while the inputs to NAND2 are 1 and 1, so the output is $Q_2 = 0$. The configuration is stable, and clearly if the initial state were $Q_1 = 0$ and $Q_2 = 1$ it would also be stable. With these values of S and R the latch is storing the state of Q_1 (and Q_2) before S and R became 0.

Now if $SR = 10$ and $Q_1 = 0$, the inputs to NAND1 are 0,1 and its output changes to $Q = 1$, while the inputs to NAND2 are 1,0 so that its output changes to $Q_2 = 0$. Had the initial state been $Q_1 = 1$, nothing would have happened. When $SR = 01$ and $Q_1 = 1$, the output of NAND2 changes to $Q_2 = 1$, which causes the output of NAND1 to change to $Q_1 = 0$. If $Q_1 = 0$ initially, then it stays unchanged. The combination $SR = 10$ is said to *set* the latch to 1, while the combination $SR = 01$ is said to *reset* the latch to 0.

What if $SR = 11$? It is clear that both outputs will go to 1, which is undesirable as they should be complementary. The combination is forbidden. Further, if now $SR = 00$ the outputs will settle at 01 or 10 depending which was first set to zero. If S and R are simultaneously zeroed, the actual output will depend on the zero levels of the individual gates, but in any case the output state will not be the same as the previous state, which invalidates the first row of the truth table. This is said to be a *critical race*; we shall return to the subject in section 10.3.1 which deals with races.

We can see that for the SR latch the condition $SR = 00$ is equivalent to storing the previous value of the output, while the conditions $SR = 10$ and $SR = 01$ are equivalent to writing 1 and 0 to the output respectively. Table 10.1 summarises the properties of the SR latch. In the table q is the previous state of Q and asterisked inputs will lead to unpredictable states when SR is changed to 00.

Table 10.1   *The truth table for the SR latch of figure 10.1*

S	R	Q
0	0	q
1	0	1
0	1	0
1*	1*	1

Like any truth table, the one above can be expressed as a Karnaugh map with $SR = 11$ entered as don't cares. Table 10.2 is such a Karnaugh map.

Table 10.2   *The Karnaugh map for the SR latch*

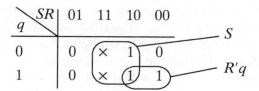

It leads to a logical function for Q:

$$Q = S + R'q$$

The present state of the output, Q, is thus a function of its previous state, q — the essential feature of all sequential logic.

SR latches can also be made from NOR gates connected in the same way as figure 10.1 (see problem 10.1); they behave slightly differently when made this way. In the 74 series of logic devices the SR latch is available in quad form as the device numbered 74279 shown in figure 10.2 and costs about 65p. In the 74279 SR latch the set and reset inputs are not inverted as in figure 10.1, that is the device is set when $SR = 01$ and reset when $SR = 10$ and are therefore written in the data sheet as S' and R'. Sometimes this latch is known as an $S'R'$ latch. There is no complementary output in the commercial package. Two of the devices have two S' inputs instead of one and these operate as in the notes to the truth table which is table 10.3.

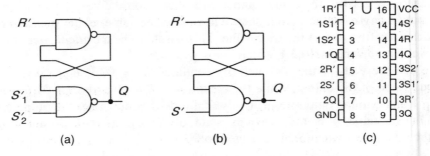

Figure 10.2 The 74279 quad SR latch (a) Latches 1 and 3 (b) Latches 2 and 4 (c) Pin identifications

Table 10.3 *The truth table for the 74279 SR latch*

	S'†	R'	Q
stores q →	1	1	q
writes 1 →	0	1	1
writes 0 →	1	0	0
not allowed →	0	0	1*

Notes: * = Data not valid.

† For latches with two S' inputs (1 & 3), $S' = 1$ means both inputs are 1 and $S' = 0$ means one or both inputs are 0.

10.1.2 *The clocked SR flip-flop*

The unclocked SR latch is prone to interference and transients caused by adjacent devices switching, as well as uncertainty regarding the exact time that it changes state. These problems can be overcome by the use of a clock as shown in figure 10.3a, where the inputs are controlled by NAND gates which disable the inputs when CK (the clock) is LOW. When the clock is HIGH the flip-flop works just like an SR latch and follows table 10.1. The transition from one state to another will occur when the clock pulse goes HIGH and so the flip-flop of figure 10.3a is called *positive-edge triggered*. This is also shown by an upward-pointing arrow on the leading edge of the clock pulse.

Figure 10.3b shows the symbol for a positive-edge triggered SR flip-flop and figure 10.3c that for a negative-edge triggered; the triangle on the CK input indicates edge triggering and the additional circle negative-edge triggering. The actual devices contain further flip-flops, not shown in figure 10.3a, which ensure that only one state transition can occur on each clock pulse.

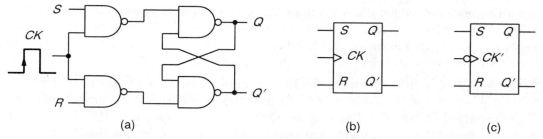

Figure 10.3 (a) Clocked SR flip-flop and circuit symbols (b) positive-edge (b) negative-edge triggered

10.1.3 The JK flip-flop

The JK flip-flop's inputs are termed J and K rather than S and R, but it is just like the SR flip-flop except that the input condition 11 is allowed and causes the output to invert or *toggle*. This is achieved by feeding back the output and its complement to the clocking NANDs as shown in figure 10.4a and then the truth table is as in table 10.4.

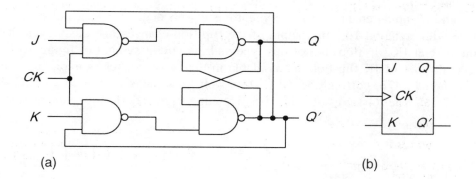

Figure 10.4 (a) The circuit for a JK flip-flop (b) Its circuit symbol

Table 10.4 *The truth table for a JK flip-flop*

J	K	Q
0	0	q
1	0	1
0	1	0
1	1	q'

Figure 10.4b shows the circuit symbol for a positive-edge triggered JK flip-flop. JK flip-flops are usually preferred to SR flip-flops because of the toggling feature and because there are no disallowed input combinations. When the Karnaugh map is drawn for table 10.4 (see problem 10.2), it shows that the logic function describing the JK flip-flop's operation is

$$Q = Jq' + K'q \qquad (10.1)$$

and we see that the output at any given time is a function of the previous output.

The master-slave JK flip-flop

The circuit given in figure 10.4a for the JK flip-flop is unsatisfactory because when $JK = 11$ (that is when it is in toggle mode) and the clock is HIGH, the output can and will toggle rapidly from 0 to 1 and back again. Only when the clock pulses HIGH for a very short time would the output toggle just once, and then there is the risk that it might not toggle at all. The solution is to use two SR flip-flops with feedback from the outputs of the second (the *slave*) to the inputs of the first (the *master*). The master's outputs are connected to the slave's inputs and the slave's clock is inverted, as shown in figure 10.5a. Together the pair acts as a single JK flip-flop. Now when $JK = 11$ and the master's clock is HIGH, the slave is disabled by its clock being LOW. The master's output can then toggle, but will not be transferred to the slave's output until the slave's clock pulse has gone HIGH and the master's LOW. Data is thus transferred only when the master's clock goes from HIGH to LOW. The device is thus a negative-edge triggered JK flip-flop, as shown in the timing diagram of figure 10.5b.

The J and K inputs change state in synchronism with negative clock pulses, and in accordance with equation 10.1 the output of the flip-flop changes one clock cycle later. Most commercial JK flip-flops are of this type and the circuit symbol of figure 10.4b is used. Several types of JK flip-flop are available in the 74 series, for example, as a dual device numbered 7476, costing about £1.50 as well as variants numbered 7472, 7473, 7478, 74107 etc. The 74 series also has positive-edge triggered JK flip-flops numbered 7470 and 74109.

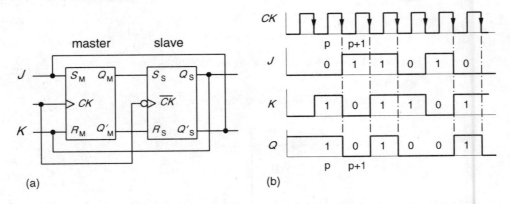

Figure 10.5 (a) A master-slave JK flip-flop (b) The timing diagram for (a)

Preset and clear

Some JK flip-flops have preset (*PR*) and clear (*CR*) inputs as well as the normal data pins. These are additional inputs to NANDs 3 and 4 of the master flip-flop, shown in figure 10.6a.

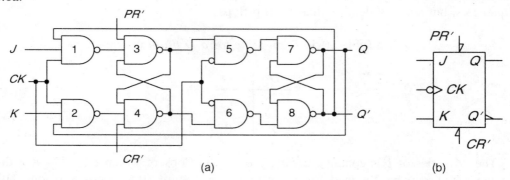

Figure 10.6 (a) A master-slave JK flip-flop with preset and clear (b) IEEE/CIE circuit symbol

When $PR = 0$, $Q_M = 1$ before any clock transitions occur, so that when the clock next goes negative, the outputs of the slave (and hence the entire flip-flop) become $Q_S = 1$ and $Q'_S = 0$. And when $CR = 0$, $Q'_M = 1$, also before any clock transitions occur, and then the outputs of the JK flip-flop become $Q_S = 0$ and $Q'_S = 1$ when the next negative clock transition occurs. Since the preset and clear inputs are active when LOW, they are usually written as PR' and CR'.

In many data books the IEEE (US Institution of Electronic and Electrical Engineers) convention is used whereby a small flag or half arrowhead is placed on the active-low input as in figure 10.6b, which also indicates the inverted output with a half arrowhead pointing away from the device.

10.1.4 The D-type flip-flop

The D-type is a special type of RS or JK flip-flop in which the inputs are complementary: $J = K'$ as shown in figure 10.7a. Only one input — by custom called D — is necessary as shown in the circuit symbol of figure 10.7b.

Figure 10.7

(a) A D-type flip-flop made from a JK flip-flop
(b) The circuit symbol

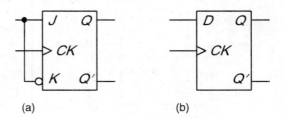

Substituting $J = K'$ into equation 10.1 gives for the logic of a D-type flip-flop the expression

$$Q = J(q' + q) = J = D$$

Thus the D flip-flop transfers the input to the output with a delay of one clock cycle (if clocked), hence the name. The truth table is that of table 10.5, a particularly simple one. Despite this simplicity D-type flip-flops can be used to make complex counters and sequencers, but are less flexible than JK flip-flops.

Table 10.5 *Truth table for a D flip-flop*

D	Q
1	1
0	0

The 74 series of ICs contains a variety of D flip-flops of which the 7474 is a dual version which can be obtained for as little as 35p. Figure 10.8 shows a 14-pin, DIL-packaged, advanced, low-power, Schottky (74ALS74) device which has preset (*PR*) and clear (*CR*) on LOW, together with the IEEE circuit symbol for one of the flip-flops. The truth table for this device, given in table 10.6, is complicated by the preset and clear inputs, but is the same as table 10.5 when these are held HIGH.

Figure 10.8

The 74ALS74 D flip-flop with preset and clear

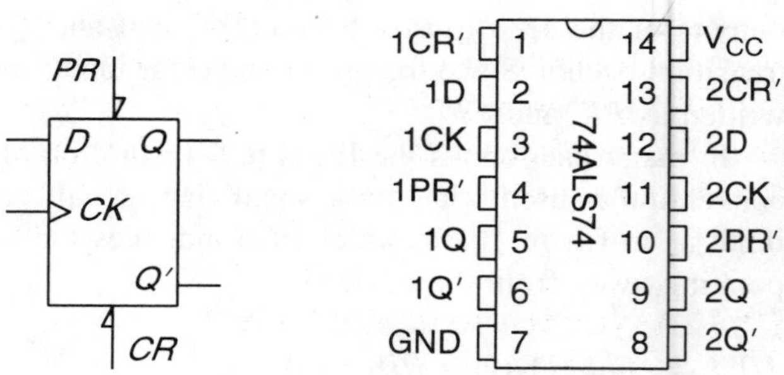

Table 10.6 *The truth table for the 7474 D flip-flop*

PR	CR	CK	D	Q	Q'
0	1	×	×	1	0
1	0	×	×	0	1
0	0	×	×	1*	1*
1	1	↑	1	1	0
1	1	↑	0	0	1
1	1	0	×	q	q'

Notes: × = don't care * unstable ↑ positive-going

The data latch

The D flip-flop of figure 10.7 has an output which merely follows the input with a delay of one clock cycle.

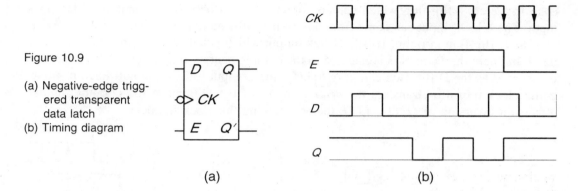

Figure 10.9

(a) Negative-edge trigg-
    ered transparent
    data latch
(b) Timing diagram

(a)   (b)

A transparent data latch can be made from an SR flip-flop with an enabling input instead of (or as well as) a clock and then the output is held ('latched') when the enabling input is LOW and can be written to when the enabling input is HIGH and the output is said to be *transparent* to the input. Figure 10.9a shows a negative-edge triggered, transparent, data latch and figure 10.9b its clock, enable, input and output waveforms. The output was initially HIGH and, when the enable is HIGH and the clock goes from HIGH to LOW, data immediately prior to the clock transition is shifted to the output. The last bit of data is latched because the enable pulse was HIGH immediately prior to the clock's negative transition.

10.2 Asynchronous counters

Asynchronous counters move through a succession of states in response to changes of input, some of which are stable and can persist when the inputs are stationary, whereas some are unstable and will not persist, even when the inputs are static. In synchronous circuits the input variables are changed only when the circuit is stable to prevent the possibility of malfunction. In asynchronous counters the clock inputs of the flip-flops are not connected to a common clock but to some other input or the output of a preceding flip-flop. The stable output state of an asynchronous counter is preceded by one or more short-lived unstable states as the flip-flops change state. When the number of flip-flops is not small the overall delay between stable states may be significant. Nevertheless the speed of asynchronous counters is generally superior to that of synchronous counters because they do not have to wait for the right clock pulse to be registered before changing state; the speed is limited only by the propagation delays of the flip-flops.

Unless the circuit is particularly simple, however, or there is an exceptional need for speed, it is preferable to use clocked flip-flops.

10.2.1 Divide-by-two counter

Sometimes called a 'scale of two counter', this circuit uses just a single flip-flop, which can be either a JK flip-flop in toggle mode (sometimes called a T flip-flop) or a D flip-flop. It is the basic building block for all binary counters. Figure 10.10a shows a positive-edge triggered T flip-flop which has an output which is half the frequency of the clock input. The 'clock' input can come from any source; it need not be a regular timing device.

The D flip-flop in figure 10.10b has an output which is half the frequency of the clock input, but here the D input is connected to the complementary output, Q'. This connection is essential as the D flip-flop can only transfer the D input to its Q output. Each positive-going clock transition transfers the *prior* contents of the D input to the Q output and so effects an inversion of Q, Q' and D, thereby halving the clock frequency.

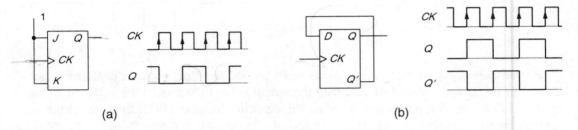

(a)   (b)

Figure 10.10 Divide-by-two counters using (a) a T flip-flop (b) a D flip-flop

10.2.2 Asynchronous binary counters

By adding further stages to the single divide-by-two counter a binary counter can be constructed with any desired range. Each flip-flop registers a binary digit so that the count in decimal notation goes from 0 to 2^{n-1}, where n is the number of flip-flops. Figure 10.11 shows such a counter, often called a *ripple counter*, which uses three T flip-flops. The output of the least-significant bit register is fed into the clock of the next, more-significant bit register. With positive-edge triggering the counter counts down; counting up is achieved by using the complementary (Q') outputs, or negative-edge triggered flip-flops.

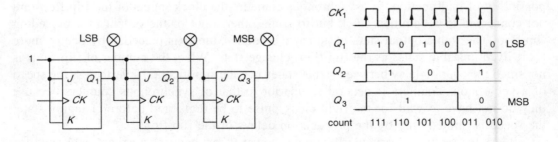

Figure 10.11 A three-bit down counter using T flip-flops

10.2.3 An asynchronous up/down counter

With the addition of some logic circuitry it is possible to make the counter in figure 10.11 into a bidirectional (up/down) counter. Suppose we wish to control the direction of the count with a switch, S, counting up when $S = 1$ and down when $S = 0$. This means inverting Q_1 and Q_2 before connecting them to the next flip-flop's clock. The truth tables are therefore

Q_1	S	CK_2
0	0	0
1	0	1
0	1	1
1	1	0

Q_2	S	CK_3
0	0	0
1	0	1
0	1	1
1	1	0

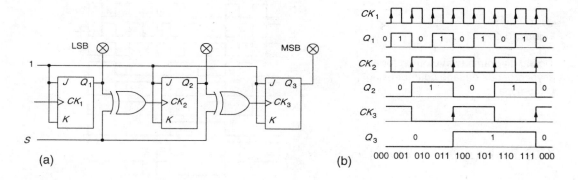

(a)   (b)

Figure 10.12 (a) An asynchronous 3-bit up/down counter (b) Timing waveforms for $S = 1$

Thus the switching functions are $CK_2 = S \oplus Q_1$ and $CK_3 = S \oplus Q_2$ and the logic circuit is that of figure 10.12a. When $S = 1$ the circuit counts up as shown in figure 10.12b, which is the timing diagram. The XOR gates here are acting as controlled inverters, since they will invert the other input when $S = 1$ and leave it unchanged when $S = 0$.

10.2.4 Asynchronous decade counters

A decade counter counts from 0_{10} to 9_{10} (0_2 to 1001_2) and then resets to zero. It therefore requires four flip-flops and some means to detecting the count of 10_{10} or 1010_2 and using that to reset all the flip-flops. If T flip-flops are used which have a clearing input (usually CR', so that the flip-flop clears on zero), then we can NAND the outputs of the first and third flip-flops to produce a zero and clear all of them. The circuit is shown in figure 10.13, but it is unsatisfactory insofar as the unwanted count of ten (1010) will be registered very briefly until the flip-flops can be cleared. This will not matter in an asynchronous circuit as there will be transitory states anyway between the proper counts.

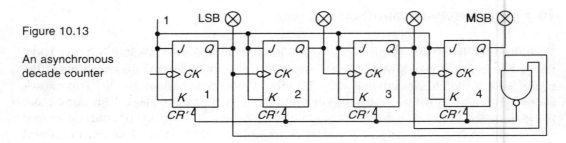

Figure 10.13

An asynchronous
decade counter

If we wish to be sure that the unwanted count does not appear even for an instant, then we must find another way of resetting the flip-flops. We will use positive-edge triggered D flip-flops connected as in figure 10.10b, the divide-by-two counter, the clocks being connected to the previous stage's complementary output.

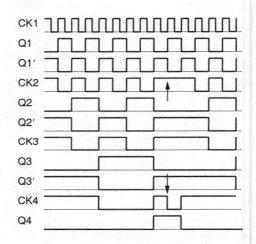

Figure 10.14

Timing diagrams (waveforms) for an
asynchronous decade counter using
D flip-flops

The desired waveforms of the outputs, Q1-Q4 are first drawn (see figure 10.14) and then the clock waveforms necessary to achieve these. At the points arrowed the required clock waveforms (CK2 and CK4) differ from the complementary outputs (Q1' and Q3') connected to those clocks. These occur after a count of 8_{10} is registered, which is 1000_2, so that ANDing Q4 and Q1' will give a 1 to control an XOR inverter as in the previous counter. The circuit is shown in figure 10.15.

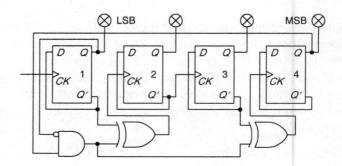

Figure 10.15

An asynchronous decade counter
using D flip-flops

10.3 Synchronous counters

In asynchronous circuits the state of the circuit at any time depends on the state of the inputs and any outputs that are fed back as input. Changes of state in the circuit depend on changes in the inputs, but are under no other form of control so that if the feedback contains unstable or transient states these can affect the eventual output of the circuit and cause its malfunction. Synchronous circuits also respond to their inputs of course, but they only change state at a definite time, that is on receipt of a clock pulse. Thus the initial and final state of each flip-flop in a circuit is better defined and erratic behaviour is much less likely.

Synchronous counters and shift registers can be constructed from JK, T or D flip-flops. Straightforward up (or down) counters, which count up to 11...1 in binary and then reset to 00...0 or vice versa, require the least logic between stages. The standard 4-bit binary counter constructed from T flip-flops, shown in figure 10.16, has a circuit which is readily derived by the method outlined later. (See also problem 10.1.)

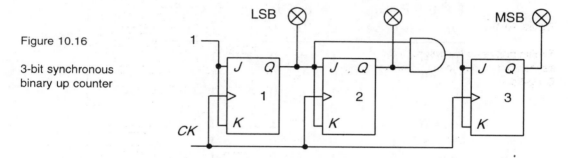

Figure 10.16

3-bit synchronous
binary up counter

The additional logic required for a particular counter or sequence generator can be less with JK flip-flops because the output state is controlled by two separate inputs rather than one as in D or T flip-flops, but then the actual process of design is more complicated. The twisted ring counter shown in figure 10.17 uses JK flip-flops and requires no additional logic at all. It generates the Johnson code sequence: 000, 100, 110, 111, 011, 001, 000, ... (See problem 10.2.)

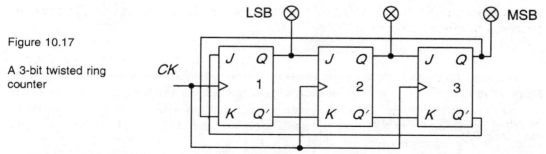

Figure 10.17

A 3-bit twisted ring
counter

Whatever flip-flop is chosen the same method must be used to design the counter. First a table is drawn up of present outputs and next outputs on receipt of a clock pulse. Then from this a logical circuit can be derived to drive the flip-flops into the required states. The method is the same for any counter or sequence generator and can be demonstrated by examples using three types of flip-flop for identical sequences. Further logic can then be used on the outputs to drive other circuitry to turn things on or off, or display the time of day, the atmospheric pressure — any of a host of functions.

Example 10.1

As an example let us design a counter to count from 0_{10} to 5_{10} and then return to 0, that is a scale-of-six counter, using T flip-flops, which are JK flip-flops with $J = K$. The largest count in binary is 101, so that three flip-flops are needed and the state tables are shown as table 10.7 below. Whenever the present state changes in the next state (that is after a clock pulse) the toggle state must be 1. The LSB is subscripted 1 and the MSB 3; present outputs are lower case, next upper.

Table 10.7 *Present (q) and next (Q) output states, and present (T) toggle state*

PRESENT			NEXT			PRESENT		
q_1	q_2	q_3	Q_1	Q_2	Q_3	T_1	T_2	T_3
0	0	0	1	0	0	1	0	0
1	0	0	0	1	0	1	1	0
0	1	0	1	1	0	1	0	0
1	1	0	0	0	1	1	1	1
0	0	1	1	0	1	1	0	0
1	0	1	0	0	0	1	0	1
0	0	0	1	0.	0	1	0	0

Table 10.8 *Karnaugh maps for toggling (a) T_2 (b) T_3*

(a)   (b)

The logic required to produce the states of the three toggles is derived from Karnaugh maps of present states, one map for each toggle, but we see at once that $T_1 = 1$ and only T_2 and T_3 require mapping. These are given in table 10.8, in which the numbers not in the required sequence — in this case 6_{10} and 7_{10} or 101_2 and 111_2 — are entered as don't cares. The logical expressions are thus

$$T_1' = 1, \quad T_2 = q_1 q_3' \quad \text{and} \quad T_3 = q_1 q_2 + q_1 q_3 = q_1(q_2 + q_3).$$

Hence the circuit which fulfils the requirements is that of figure 10.18.

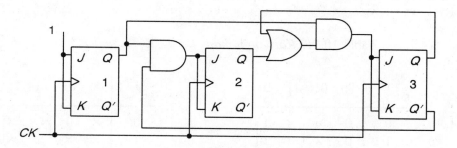

Figure 10.18 A scale-of-six counter made from T flip-flops

Now let us repeat the exercise using JK flip-flops.

Example 10.2

Design a scale-of-six binary counter with JK flip flops.

The transitions for a JK flip-flop are given in table 10.9 below.

Table 10.9 *Transitions for a JK flip-flop*

$q \quad Q$	J	K
$0 \to 0$	0	×
$0 \to 1$	1	×
$1 \to 0$	×	1
$1 \to 1$	×	0

Here we must put in the J and K values required for the next state and enter a don't care when the J or K value can be either 0 or 1. Table 10.10 shows the present and next output states and the J and K values. Again MSBs are given the subscript 3 and LSBs the subscript 1.

Table 10.10 *Present and next states for a JK flip-flop scale-of-six counter*

PRESENT			NEXT			PRESENT					
q_1	q_2	q_3	Q_1	Q_2	Q_3	J_1	K_1	J_2	K_2	J_3	K_3
0	0	0	1	0	0	1	×	0	×	0	×
1	0	0	0	1	0	×	1	1	×	0	×
0	1	0	1	1	0	1	×	×	0	0	×
1	1	0	0	0	1	×	1	×	1	1	×
0	0	1	1	0	1	1	×	0	×	×	0
1	0	1	0	0	0	×	1	1	×	×	1
0	0	0	1	0	0	1	×	0	×	0	×

Examination of table 10.10 shows at once that $J_1 = K_1 = 1$ and so four Karnaugh maps are needed for J_2, K_2, J_3 and K_3 as in table 10.11 below.

Table 10.11 *Karnaugh maps for J_2, K_2, J_3 and K_3*

J_2

q_3 \ q_1q_2	00	01	11	10
0	0	×	×	1
1	0	×	×	1

K_2

q_3 \ q_1q_2	00	01	11	10
0	×	0	1	×
1	×	×	×	×

J_3

q_3 \ q_1q_2	00	01	11	10
0	0	0	1	0
1	×	×	×	×

K_3

q_3 \ q_1q_2	00	01	11	10
0	×	×	×	×
1	0	×	×	1

These maps give $J_2 = K_2 = q_1$, $J_3 = q_1q_2$ and $K_3 = q_1$, which leads to the circuit of figure 10.19. This is substantially simpler than the toggle flip-flop circuit of figure 10.15 as would be expected if the full versatility of the JK flip-flop is utilised. Despite this example, the JK flip-flop design may not always work out simpler (though it cannot be more complicated) than one using T flip-flops. The design process is much simpler with T flip-flops and therefore quicker and less error-prone.

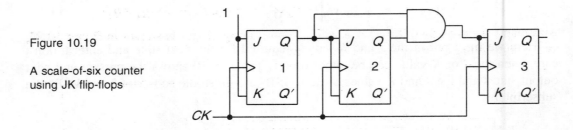

Figure 10.19

A scale-of-six counter
using JK flip-flops

Example 10.3

Repeat the scale-of-six design in D flip-flops.

One might predict this design to have the most complicated logic of the three since the D flip-flop is the least versatile. The D flip-flop transfers the data at the D input to the Q output, and in complementary form to its Q' output, on receipt of a clock pulse. Thus the next state of Q has to be the present state of D as in table 10.12 below.

Table 10.12 *Present and next states for a D flip-flop scale-of-six counter*

PRESENT			NEXT			PRESENT		
q_1	q_2	q_3	Q_1	Q_2	Q_3	D_1	D_2	D_3
0	0	0	1	0	0	1	0	0
1	0	0	0	1	0	0	1	0
0	1	0	1	1	0	1	1	0
1	1	0	0	0	1	0	0	1
0	0	1	1	0	1	1	0	1
1	0	1	0	0	0	0	0	0
0	0	0	1	0	1	1	0	0

The resulting Karnaugh maps are shown in table 10.13 below.

Table 10.13 *Karnaugh maps for a D flip-flop scale-of-six counter*

$q_1 q_2$ / q_3	00	01	11	10
0	1	1	0	0
1	1	×	×	0

(a)

$q_1 q_2$ / q_3	00	01	11	10
0	0	1	0	1
1	0	×	×	0

(b)

$q_1 q_2$ / q_3	00	01	11	10
0	0	0	1	0
1	1	×	×	0

(c)

Map (a) shows that $D_1 = q_1'$ which could have been seen by inspecting table 10.12 cursorily. Map (b) shows that

$$D_2 = q_1'q_2q_3' + q_1q_2'q_3' = q_3'(q_1'q_2 + q_1q_2') = q_3'(q_1 \oplus q_2)$$

And map (c) shows that $D_3 = q_1q_2 + q_1'q_3$. The resulting circuit is shown in figure 10.20, and is indeed the most complicated of the three.

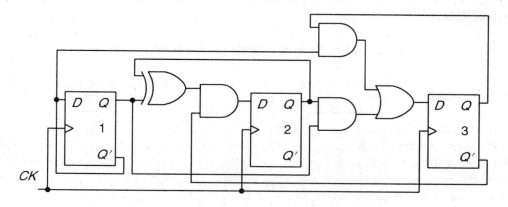

Figure 10.20 Scale-of-six counter using D flip-flops

It is possible that the counter will be in a state which is not in the required sequence when powered up and on clocking may never enter the desired sequence. This problem can be overcome by additional logic, but it is better to use the preset and/or clear inputs to start the counter at the right place in the sequence.

Sequence generators can be used to produce any arbitrary sequence of binary numbers, but they are in all respects the same as counters except the sequence is not necessarily in numerical order. Counters can be used to control switching sequences, provided the count is large enough and the clock cycle time appropriate.

Example 10.4

As an example consider the traffic-light switching sequence of table 10.14. With a clock cycle time of 5 s (0.2 Hz), the counts are as shown in the table and it can be seen that a scale-of-twelve counter, going up to 11_{10} (1011_2) and resetting, is needed.

Table 10.14 *Traffic-light switching sequence*

Light(s)	time	count	cumulative
RED	20 s	4	4
RED & AMBER	5 s	1	5
GREEN	30 s	6	11
AMBER	5 s	1	12
RED	20 s	4	4

Some output logic is needed to recognise the count and activate the relays which switch the lights on and off. We shall assume that a logical 1 at the appropriate output will suffice for this. The counter requires four flip-flops and if we use T flip-flops the transition table is that of table 10.15, which gives $T_1 = 1$ and $T_2 = q_1$ immediately. Then the Karnaugh maps for T_3 and T_4 in table 10.16, with unused counts entered as don't cares, give $T_3 = q_1 q_2$ and $T_4 = q_1 q_2 q_3$.

Table 10.15 *Transition table for the traffic lights*

P R E S E N T				N E X T				P R E S E N T			
q_1	q_2	q_3	q_4	Q_1	Q_2	Q_3	Q_4	T_1	T_2	T_3	T_4
0	0	0	0	1	0	0	0	1	0	0	0
1	0	0	0	0	1	0	0	1	1	0	0
0	1	0	0	1	1	0	0	1	0	0	0
1	1	0	0	0	0	1	0	1	1	1	0
0	0	1	0	1	0	1	0	1	0	0	0
1	0	1	0	0	1	1	0	1	1	0	0
0	1	1	0	1	1	1	0	1	0	0	0

Table 10.15 (cont.)

1	1	1	0	0	0	0	1	1	1	1	1
0	0	0	1	1	0	0	1	1	0	0	0
1	0	0	1	0	1	0	1	1	1	0	0
0	1	0	1	1	1	0	1	1	0	0	0
1	1	0	1	0	0	0	0	1	1	0	1

Table 10.16 *The Karnaugh maps for T_3 and T_4*

T_3 $q_3 q_4$ \ $q_1 q_2$	00	01	11	10
00	0	0	1	0
01	0	0	1	0
11	×	×	×	×
10	0	0	1	0

T_4 $q_3 q_4$ \ $q_1 q_2$	00	01	11	10
00	0	0	0	0
01	0	0	0	0
11	×	×	×	×
10	0	0	1	0

The counter can now be constructed, then the output logic to drive the relays on and off must be designed. It is probably more efficient in ICs to do this by using logic to drive JK flip-flops rather than just combinational logic, and for this we need another transition table, that of table 10.17, which gives the present states of the counter and the desired states of the JK flip-flops, one for each colour.

Table 10.17 *Transition table for the traffic lights*

q_1	q_2	q_3	q_4	R	J_R	K_R	A	J_A	K_A	G	J_G	K_G
0	0	0	0	1	1	×	0	×	1	0	0	×
1	0	0	0	1	×	0	0	0	×	0	0	×
0	1	0	0	1	×	0	0	0	×	0	0	×
1	1	0	0	1	×	0	0	0	×	0	0	×
0	0	1	0	1	×	0	1	1	×	0	0	×
1	0	1	0	0	×	1	0	×	1	1	1	×
0	1	1	0	0	0	×	0	0	×	1	×	0
1	1	1	0	0	0	×	0	0	×	1	×	0
0	0	0	1	0	0	×	0	0	×	1	×	0
1	0	0	1	0	0	×	0	0	×	1	×	0
0	1	0	1	0	0	×	0	0	×	1	×	0
1	1	0	1	0	0	×	1	1	×	0	×	1
0	0	0	0	1	1	×	0	×	1	0	0	×

Table 10.18 shows the six Karnaugh maps which are needed for the JK light drivers. These lead to $J_R = q_1 q_2$, $K_R = q_1 q_3$, $J_A = q_1 q_2 q_4 + q_2' q_3$, $K_A = 1$, $J_G = q_1 q_3$ and $K_G = q_1 q_2 q_4$.

Table 10.18 *Karnaugh maps for the light-switching flip-flops*

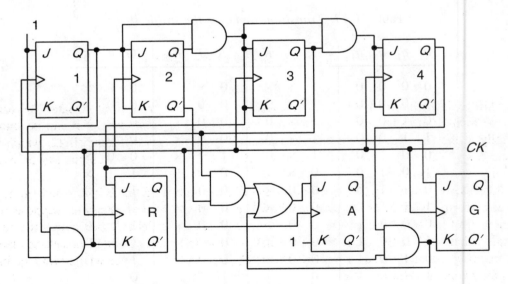

Figure 10.21 shows a circuit for the traffic light switching.

Figure 10.21 The traffic-light sequencer

10.4 Shift registers

Shift registers are used to store and move data, generally from one bit register to the next, from left to right or from right to left. All those here are of shift-right variety. They can be used to multiply numbers in the manner indicated in section 9.1.2. 8-bit shift registers are available in 14-pin DIL packages in the 74 series, for example the 74HC164M1R device, costing only 40p. Large shift registers can be made from these by connecting the serial output of one to the serial input of the next. Shift registers can be classified according to whether the data is read in or out in serial or in parallel, so that one can have serial in/parallel out (SIPO), or parallel in/serial out (PISO) or serial in/serial out (SISO), or parallel in/parallel out (PIPO) shift registers, though the PIPO 'shift register' is really just a clocked latch and does not shift. 'Serial' means that the bits are read one at a time, 'parallel' means all at once.

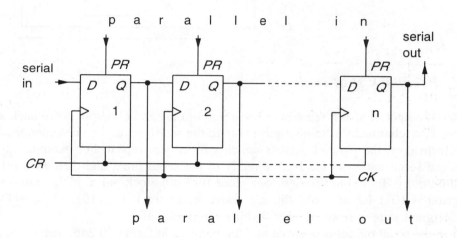

Figure 10.22 An n-bit shift register

Shift registers can be made from clocked flip-flops, one for each bit of data. The data can be input in parallel form by using the preset and clear functions, or it can be input serially at the first flip-flop and clocked through to the last as in figure 10.22, which is made from D flip-flops; JK could equally well have been used. The output can be read serially or in parallel.

Alternatively data can be loaded in series or in parallel using the D inputs and suitable logic performs loading or shifting, as shown in figure 10.23. For example, suppose we want to read in data to the shift register of figure 10.22 in parallel and then output the data serially, that is a PISO shift register: the data is read in by clearing and presetting; then on clocking the bits appear in reverse order, Q_n first and Q_1 last, the whole word taking n clock cycles to read out.

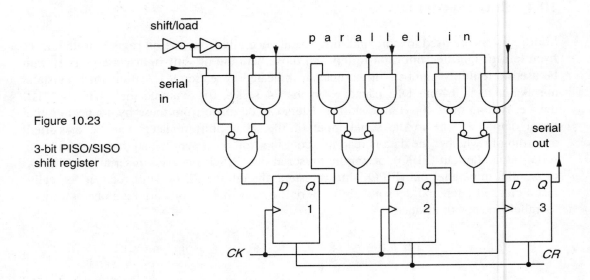

Figure 10.23

3-bit PISO/SISO
shift register

10.4.1 *Feedback shift registers*

If the serial output of a shift register or some part of the parallel output is fed back directly or inverted, or combined in some logical way to the serial input, then a sequence of states can be formed on the parallel outputs which will in time repeat. For example, in figure 10.24a, the serial output is fed back directly to the serial input and then each bit of the parallel output is shifted to the right one place for each clock pulse. If the initial state of the register is 1011 for example, the successive states are 1011, 1101, 1110, 0111, 1011 ... The original state is restored after 4 shifts (or clock pulses).

When the serial output is inverted and fed back, as in figure 10.24b, then the sequence is 1011, 0101, 0010, 1001, 0100, 1010, 1101, 0110, 1011 ... and the original state is restored after 8 shifts. Combinations of the parallel outputs can be used as in figure 10.24c, where the two furthest right outputs, Q_4 and Q_3 (the MSB and the next MSB), are XORed and fed back to the serial input. If the initial state of the register was 1011 the sequence generated is the longest possible, 15, as shown in table 10.19. The device is a pseudo-random-number generator, since the probability of any bit being 1 (or 0) is 50%. The maximum sequence is always 2^n, where n is the number of bits, though in the case of 8, 16 and 32-bit registers three bits must be combined to produce the maximum sequence length (see problem 10.10). If the clock rate is 1 MHz and the shift register has 32 bits, then the maximum sequence requires more than an hour for its generation. This can be made the key to a code generator in which only the initial state of the register need be known for decryption. It can be shown that if the generator has n bits and the kth bit and the last produce a maximum-length sequence when XORed, then so also will the $(n - k)$th and the last.

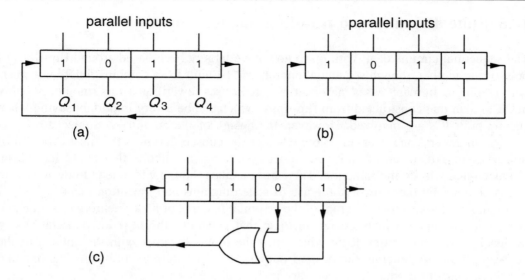

Figure 10.24   Feedback shift registers (a) Directly fed-back serial output (b) Fed-back, inverted, serial output (c) A pseudo-random-number generator with combinational feedback

Table 10.19   *Truth table for a pseudo-random-number generator*

Q_1	Q_2	Q_3	Q_4	$Q_3 \oplus Q_4$	Decimal No.
1	0	1	1	0	13
0	1	0	1	1	10
1	0	1	0	1	5
1	1	0	1	1	11
1	1	1	0	1	7
1	1	1	1	0	15
0	1	1	1	0	14
0	0	1	1	0	12
0	0	0	1	1	8
1	0	0	0	0	1
0	1	0	0	0	2
0	0	1	0	1	4
1	0	0	1	1	9
1	1	0	0	0	3
0	1	1	0	1	6
1	0	1	1	0	13

10.5 State diagrams and transition maps

The sequential circuits dealt with up to now have largely been based around flip-flops of one sort or another, though sequential circuits can be built from combinational logic gates with feedback. In many cases the resulting circuit uses fewer transistors (though possibly not ICs) than that constructed from flip-flops, which can be a very important saving when custom-built ICs are involved. However, the design of the circuitry is subject to uncertainties: in general one does not know whether the chosen design is the most economical in gates, since there are usually a number of equivalent circuits that could have been produced that will do the same job. There are methods though which lead to designs with few problems and these involve the use of state diagrams and transition maps.

Figure 10.25 is a block diagram of the essential features of an asynchronous sequential circuit; it has inputs which are fed to logic that generates the next internal state using feedback, and it has output logic which uses the inputs together with the outputs of the next-state logic to generate the desired outputs. It is the purpose of state diagrams and transition maps to produce the necessary logic equations.

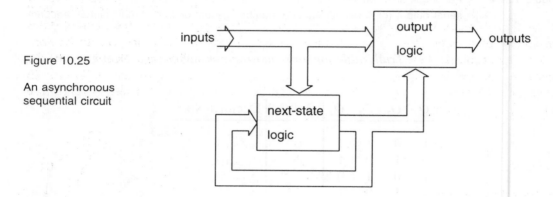

Figure 10.25

An asynchronous
sequential circuit

10.5.1 Stable and unstable states: races

The major problem with asynchronous sequential circuits is the possibility of malfunction because of transient unstable states, and the design therefore must be done very carefully. As an example examine the asynchronous 2-bit counter of figure 10.26a which is constructed from two JK flip-flops with $J = K = 1$, that is in toggle mode.

If the initial count is $Q_1 = 0$, $Q_2 = 0$, then on receiving a clock pulse, FF1 changes state from 0 to 1, which constitutes a rising clock pulse for FF2, which then also changes state, but with a delay equal to the propagation delay of the flip-flop. Thus before the stable state $Q_1 = 1$, $Q_2 = 1$ (or 11) is established, an unstable state $Q_1 = 1$, $Q_2 = 0$ (or 10) is gone through because Q_1 changes state before Q_2. And on receipt of the next clock pulse, Q_1 will change to 0 (a negative clock-going, and therefore unoperational, transition in FF2) and Q_2 will stay on 1, so the state 01 is stable. The next clock pulse causes FF1 to change to $Q_1 = 1$ and so Q_2 will change to 0, passing through the unstable state 11 en route to the stable state 10. The next clock pulse will return the system to the stable state 00. The state-transition diagram is given in figure 10.26b, where stable states are ringed. The unstable states have arisen whenever the output state has required a change in both Q_1 and Q_2.

Figure 10.26

(a) An asynchronous
    sequential circuit
(b) State transitions for (a)

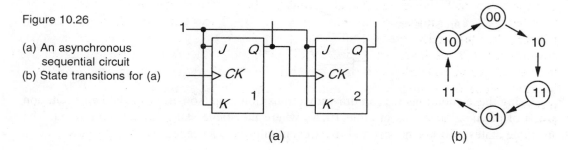

(a)   (b)

Consider now the synchronous circuit of figure 10.27a, which has exactly the same output sequence as that of figure 10.26b, but a different state-transition diagram, shown in figure 10.27b. Now when both output variables change state we cannot say what intermediate state is followed, but if FF1 is faster in operation than FF2 then the sequences 00, 10, 11 and 01, 11, 10 will be followed rather than 00, 01, 11 and 01, 00, 10. These are said to be *races* because there is a race between flip-flops to change their output states on receipt of a clock pulse. The two races shown in figure 10.27b are said to be *non-critical races*, because the final state achieved is the same in either case. Note also that the unstable intermediate state is only registered for a time equal to the *difference* in propagation delays of the two flip-flops, and not the propagation delay itself, as is the case with asynchronous operation.

Figure 10.27

(a) A synchronous circuit
(b) State-transitions for (a)

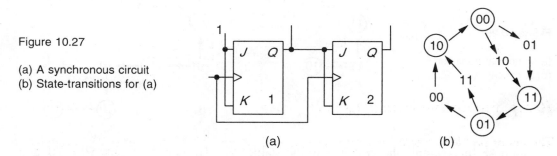

(a)   (b)

Critical races

Critical races must be eliminated in sequential circuits because they will lead to unpredictable behaviour, which must never happen in good circuit design. To see how critical races arise, let us return to the SR latch of figure 10.28, whose outputs are given by the Boolean expressions

$$Q_1 = \overline{S'q_2} = S + q'_2 \quad \text{and} \quad Q_2 = \overline{R'q_1} = R + q'_1$$

where the lower case is used for previous output states.

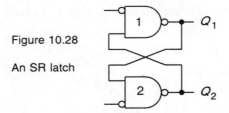

Figure 10.28

An SR latch

Using these equations we can draw up a transition map (sometimes called an excitation map) of the next states, as in table 10.20, where the stable states have been circled. The unstable states will not persist and must eventually be succeeded by stable states.

Table 10.20 *Transition map for the SR latch*

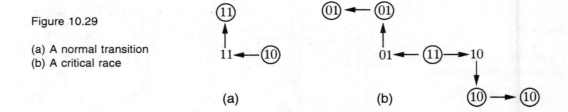

Figure 10.29

(a) A normal transition
(b) A critical race

(a)   (b)

Suppose for example the circuit is in the stable state 10 with the input, $SR = 10$, and we change this to $SR = 11$. The circuit state moves to the left by one column to enter the transitional state 11, and then moves up to the stable state above, where the present output is $Q_1Q_2 = 11$ with $q_1q_2 = 11$ (the previous output) and $SR = 11$, as shown in figure 10.29a. But now if both S and R are switched to 0 the eventual state of the circuit is uncertain. If NAND1 is faster the circuit state moves left to the cell with unstable state 01, then up to stable state 01 until R changes to 0 and the circuit state moves left to the stable state 01 (see figure 10.29b). Had NAND2 been faster, the circuit state would have moved right to unstable state 10, then down to stable state 10 until S switched to 0 and the circuit moved further right (over the edge of the map) to stable state 10. The outcome of the race between the two states is uncertain: we can never know which state the circuit will end up in and the race is termed critical.

Critical races are avoided by ensuring that the inputs change singly (a *unit distance* change), by changing the stable output state in only one variable (that is the stable output states are *logically adjacent*) and by changing inputs only when the circuit is in a stable state. The latter means waiting for the circuit to stabilise for a time longer than the longest cumulative propagation delay time.

Example 10.5

As an example of asynchronous sequential design consider the waveforms of figure 10.30, for which a sequential logic circuit is required. The inputs, *A* and *B*, are constrained so that only one can change state at a time, thereby reducing the likelihood of a critical race.

Figure 10.30

Input and output waveforms and the corresponding state assignments

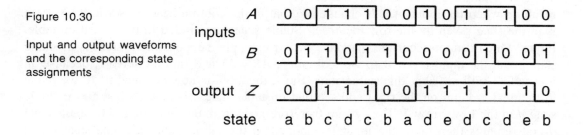

The states of the system can be given letters according to the states of the inputs and outputs, for example when *A* = 0, *B* = 0 and *Z* = 0, the system is in state *a*. After a time no new states appear and in this case it can be seen that there are just five, lettered from *a* to *e*. The states can be now be placed on a state diagram, as in figure 10.31, where arrows indicate transitions from one state to another.

Figure 10.31

The state diagram corresponding to the waveforms of figure 10.30

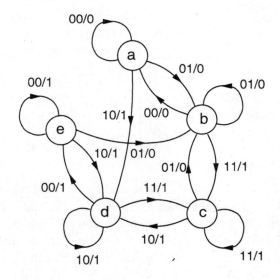

The numbers next to the arrows indicate the input values producing the transition and the output value when the transition is complete, for example the transition from state *b* to state *c* is accomplished by inputting 11 and the output is then 1, so 11/1 is placed next to that transition. Every state in figure 10.31 is a stable state, which is indicated by a *sling*, that is a transition arrow which starts and ends on the same state, so that state *d*, for example, has a sling with 10/1 written next to it because the state will remain unchanged with those inputs and output. Because of the lack of a clock, which means that the output is determined solely by the inputs, asynchronous sequential systems cannot distinguish between successive identical inputs, that is a sequence such as 00,01,11,10 cannot be distinguished from 00,01,11,11,11,10,10.

The next step is to draw up a primitive transition map as in table 10.21, where the inputs are entered as in a Karnaugh map along the column heading and the internal states are given numbers used as row identifiers. The inputs to the table are the *next* state resulting when the input, *AB*, is changed to that in the column heading while the system is in the state given by the row identifier. Stable states are circled in the transition table. Inputs which cannot occur are marked '×'. For example, suppose the system is in stable state *c* and the input is 11, then we look under *AB* = 11 and find a circled *c*, because an input of 11 will not alter this stable state. But if the input is now changed to 10 we move to the next column on the right and see that unstable state *d* is entered just above stable state *d* in the lower row 4. The system therefore goes from unstable state *d* to stable state *d* and remains there until the input is changed. If the input while in stable state *c* had changed to 01 instead of 10, the system would have moved to the left one column and entered unstable state *b* prior to moving up to stable state *b* in that column. Since an input of 00 is not allowed when in stable state *c*, because the input then is 11 and two inputs cannot change at once, a cross is entered in column 00, row 3.

Table 10.21 *The primitive transition map derived from figure 10.30*

		AB		
present state	00	01	11	10
a	(a)	b	×	d
b	a	(b)	c	×
c	×	b	(c)	d
d	e	×	c	(d)
e	(e)	b	×	d

Table 10.21 can now be compressed by merging equivalent internal states (rows). For example, rows *a* and *b* can be merged since the states in each are identical (except for stability), provided we take the two disallowed × states as don't cares, and provided that the outputs of the stable states are the same. The row identifiers are now changed to the internal present-state variables, q_1 and q_2, which can be given arbitrary numerical values, in this case 00, 01, 11, 10, following the same order as in a Karnaugh map. This is *essential* to prevent critical races from occurring, since the internal state variables cannot

be controlled and transitions involving a change of both will *always* involve a race, though not necessarily a critical one.

Table 10.22 *Transition map from merging rows a and b of table 10.21*

next state ＼　　　　　*AB*

q_1q_2	00	01	11	10
00	(a)	(b)	c	d
01	×	b	(c)	d
11	e	×	c	(d)
10	(e)	b	×	d

Although this process of enumeration can be achieved in several ways, each leading to a different logical circuit, there is no way of deciding which is best, other than by ensuring that states with an output of 1 (*c*, *d* and *e*) are assigned to rows with identifiers containing 1, such as 11, 01 or 10, but not 00. Only by analysing alternative numbering sequences can the most efficient solution be found. In the present case the only alternative to the sequence for q_1q_2 is 00, 10, 11, 01. The choice in this instance is immaterial, since on further examination it can be seen that in table 10.22 rows $q_1q_2 = 01$ and $q_1q_2 = 11$ can be merged, as stable states *c* and *d* both have outputs of 1, and also rows $q_1q_2 = 00$ and $q_1q_2 = 10$ where all the stable states have outputs of 0. Table 10.23 is the result, where the auxiliary or internal present-state variables, q_1 and q_2, have been reduced to just one, *q*. The assignment of values to *q* is also arbitrary, but the output logic is simplified if *q* is made equal to the output of the stable states in its row, that is $q = 0$ for states *a* and *b*, and $q = 1$ for states *c*, *d* and *e*.

Table 10.23 *The final transition map*

next state ＼　　　　*AB*

q	00	01	11	10
0	(a)	(b)	c	d
1	(e)	b	(c)	(d)

Now the states in table 10.23 can be replaced by the value of *Q*, the next state of *q* for that state, which is in this case the same as the output state. For stable states in the row where $q = 0$ this must be $Q = 0$, and for stable states in the row where $q = 1$ this must be $Q = 1$. The unstable states are then replaced by the output of the corresponding stable state. Hence the final Karnaugh map for the internal state variable, *Q*, is that shown in table 10.24 below.

From table 10.24 we deduce that $Q = A + B'q$. The output state map will be the same as that for Q in this instance because we have arranged that $q = 1$ corresponds to the output, $Z = 1$.

Table 10.24 *The Karnaugh map for Q*

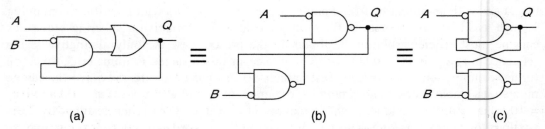

The logic for Q and therefore Z is shown in figure 10.32a, and that can be redrawn using only NAND gates as in figure 10.32b, whereat we recognise the SR flip-flop of figure 10.32c.

Figure 10.32 (a) Circuit for $Z = A + B'q$ (b) Circuit redrawn in all-NAND form (c) The SR flip-flop

In the usual form of a sequential design problem, waveforms are not given and the problem is stated in words as in the next example.

Example 10.6

A circuit has two inputs, A and B, of which only one can change at a time. The output of the circuit, $Z = 0$ when $A = 0$. If B changes while $A = 1$, then $Z = 1$ and stays at 1 until $A = 0$. Design an asynchronous sequential circuit to do this.

It is best to write down a sequence of inputs covering all the obvious possibilities and then find the corresponding output and draw up a state-transition diagram.

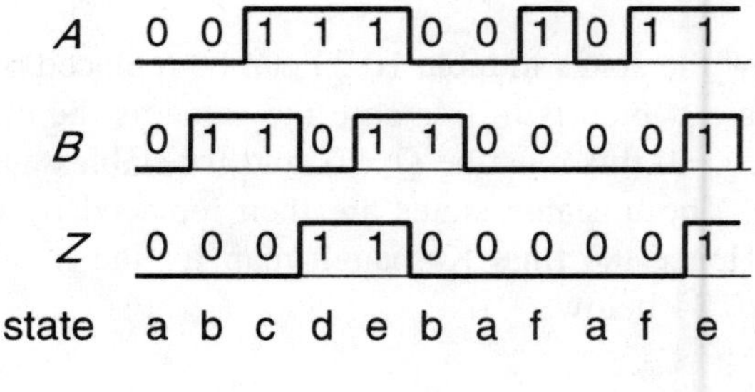

Figure 10.33

Test waveforms for example 10.6

A 0 0 1 1 1 0 0 1 0 1 1

B 0 1 1 0 1 1 0 0 0 0 1

Z 0 0 0 1 1 0 0 0 0 0 1

state a b c d e b a f a f e

Any gaps in this diagram can then be filled. The following cycle of inputs, for example, covers a lot of combinations: 00,01,11,10,11,01,00,10,00 and the waveforms for these input and the output are shown in figure 10.33 above.

From the waveform diagram of figure 10.33 a state diagram can be constructed as in figure 10.34, which has six stable states. These should therefore all have slings, but for simplicity they have been omitted from the diagram.

Figure 10.34

The state diagram for problem 10.6

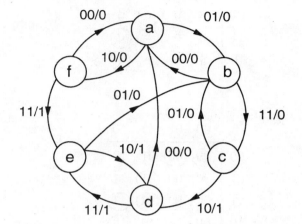

Taking note that states *d* and *e* have outputs of 1, table 10.25, the primitive transition map, is now drawn up and then compacted by merging rows *a* and *b*, and *d* and *e* to give table 10.26a, which has four rows and therefore requires two state variables, q_1 and q_2. The choice of rows to merge is arbitrary and we cannot at the outset say which choice will give the simplest logic. We could equally well merge rows *a* and *f*, and *b* and *c*, and *d* and *e*. By so doing we would get the 3-row compacted transition map of table 10.26b, where the unused row is entered as don't cares. The states can now be replaced by their state variable numbers, $q_1 q_2$, as in tables 10.26c and 10.26d. Here we replace states *d* and *e* (which have outputs of 1) with $q_1 q_2 = 11$, to try to simplify the output logic.

Table 10.25 *The primitive transition map for figure 10.34*

	next state \ *AB*	00	01	11	10
present state	a	(a)	b	×	f
	b	a	(b)	c	×
	c	×	b	(c)	d
	d	a	×	e	(d)
	e	×	b	(e)	d
	f	a	×	e	(f)

Table 10.26 *(a), (b) Compacted transition maps (c), (d) State variable assignments*

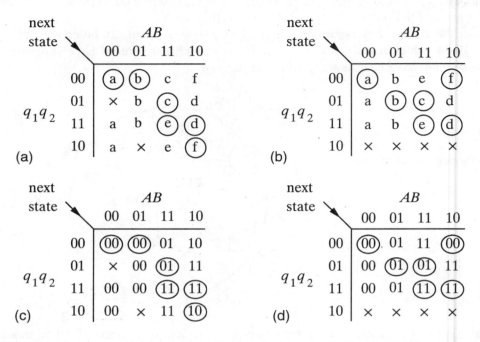

(a)   (b)   (c)   (d)

The transition map of table 10.26d is complicated rather than simplified by the spare row of don't cares, because there are four places possible for it. And because the stable state b becomes $Q_1Q_2 = 01$ in table 10.26d, the second column is brought into play, which complicates the expression for Q_2. In fact the simplest solutions for Q_1 and Q_2 emerge from table 10.26c. To determine Q_1 and Q_2 the Karnaugh maps in tables 10.27a and 10.27b, respectively, are drawn up using table 10.26c for the data. Examination of these Karnaugh maps shows that $Q_1 = Aq_1 + AB'$ and $Q_2 = Aq_2 + AB$.

Table 10.27 *The Karnaugh maps for Q_1 and Q_2 derived from table 10.26c*

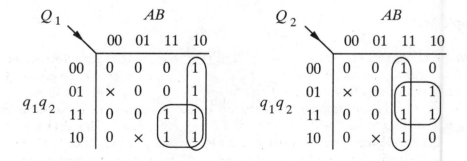

The Karnaugh map for the output, Z, in table 10.28 is constructed by entering 1s in the cells where the stable output is 1 (the unstable states with $Z = 1$ could have been used also, but they do not lead to simpler output logic). This map yields $Z = AQ_1Q_2$ and the complete circuit diagram is shown in figure 10.35.

Table 10.28   *The Karnaugh map for the output, Z*

$$
\begin{array}{cc|cccc}
& & \multicolumn{4}{c}{AB} \\
Z & & 00 & 01 & 11 & 10 \\
\hline
& 00 & 0 & 0 & 0 & 0 \\
& 01 & \times & 0 & 0 & 0 \\
q_1q_2 & 11 & 0 & 0 & 1 & 1 \\
& 10 & 0 & \times & 0 & 0 \\
\end{array}
$$

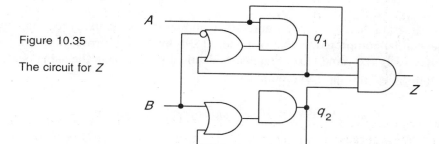

Figure 10.35

The circuit for Z

Both asynchronous and synchronous sequential circuits can be designed to recognise certain inputs or combinations of inputs and make a response. But if there is only one input, asynchronous sequential circuits are handicapped by their inability to distinguish 1, 0, 1 from 1, 0, 0, 1, that is a repeated input has no meaning. They are usually used then for cases where there is more than one input and then the design follows much the same path as example 10.6. Synchronous sequential circuits are required for sequence recognition when the duration of an input state matters, as in the next example.

Example 10.7

Design a synchronous sequential circuit to detect the sequence 011 anywhere in the single input data stream. The circuit will output a 1 on receiving the correct sequence (for just one clock cycle) and a 0 at all other times.

First we specify the problem in terms of a state diagram. Let us suppose that the system is in state a to start with and that the first bit input is a 0. This data is part of the correct sequence so the system must move on to another internal state, b. If the first bit had been a 1, not the first bit of the correct sequence, then the system would have stayed in state a, that is state a must have a sling labelled 1/0 (input 1, output 0). State a is thereby recognised as a stable state. Any further inputs of 1 will keep the system in state a. When in state b, if a 0 is received, not the second but the first bit of the correct sequence, the system must stay in state *b* waiting for a 1. If a 1 is received when in state *b* the system moves on to state *c*. When in state *c* if a 1 is received, the final bit of the sequence, the system moves to state *d* and outputs a 1. When in state *c* and a 0 is received, the system must return to state *b* and await a 1. The system never stays in state *c* longer than a clock pulse and no slings are needed. Neither does state *d* require a sling, since receipt of either 1 or 0 will cause transition to another state: a 0 sends the system to state *c* and a 1 to state *a*.

Figure 10.36

The state diagram for example 10.7

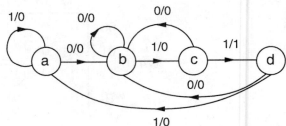

The state diagram then looks like figure 10.36 and produces the transition map of table 10.29a. This can be compacted to that of table 10.29b by merging rows *a* and *d*, since *a* and *d* are effectively the same state. States are identical if they have the same next states for either input value and the same output.

Table 10.29 *Transition maps for figure 10.36*

	next state	A 0	1
present state	a	b	a
	b	b	c
	c	b	d
	d	b	a

(a)

	next state	A 0	1
present state	a	b	a
	b	b	c
	c	b	a

(b)

A row of don't cares may then be inserted, as in table 10.30a, probably simplifying the logic. Two state variables, q_1 and q_2, and hence two flip-flops are still required, however, since there are still three internal states.

Table 10.30 *Transition maps for Q_1 and Q_2*

$Q_1 Q_2$		A 0	1
	00	01	00
$q_1 q_2$	01	01	11
	11	01	00
	10	×	×

Q_1		A 0	1
	00	0	0
$q_1 q_2$	01	0	1
	11	0	0
	10	×	×

Q_1		A 0	1
	00	1	0
$q_1 q_2$	01	1	1
	11	1	0
	10	×	×

From the transition maps for Q_1 and Q_2 we can draw up the states required for J_1 and K_1 should we choose to implement the design with JK flip-flops. These are derived from the present states of the variables q_1 and q_2 and the next states, Q_1 and Q_2 and hence the required states for J, K (as given in table 10.9).

Tables 10.31a and 10.31b give the Karnaugh maps for these and from them we can see that $J_1 = Aq_2$, $K_1 = 1$, $J_2 = A'$ and $K_2 = Aq_1$. Finally we must draw up a Karnaugh map for the output, Z, which is 1 only when the system is in state c and going to state d in table 10.29a, or state a as it becomes in table 10.29b, which corresponds to the cell $q_1q_2 = 11$, $A = 1$ in table 10.30. The Karnaugh map for Z is that of table 10.31c, which gives $Z = Aq_1$. The circuit that results is shown in figure 10.37.

Table 10.31 *Karnaugh maps for (a) J_1K_1 (b) J_2K_2 (c) Z*

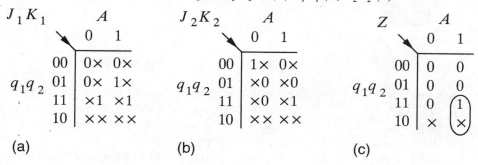

J_1K_1	A	
	0	1
00	0×	0×
01	0×	1×
11	×1	×1
10	××	××

(a)

J_2K_2	A	
	0	1
00	1×	0×
01	×0	×0
11	×0	×1
10	××	××

(b)

Z	A	
	0	1
00	0	0
01	0	0
11	0	1
10	×	×

(c)

(In maps (a), (b), (c) the row labels q_1q_2 appear at the left.)

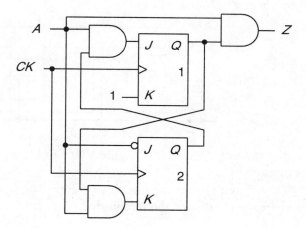

Figure 10.37

The circuit for example 10.7

Suggestions for further reading

Introductory digital design by M S Nixon (Macmillan 1995)
Digital fundamentals by T L Floyd (5th ed., Macmillan Publishing, New York, 1994)
Modern digital systems design by J Y Cheung and J G Bredeson (West Publishing, 1990)
Digital logic and state machine design by D J Cromer (3rd ed., Saunders College Publishing, 1995)

Problems

1 Show that the circuit of figure P10.1 will act as an SR flip-flop.

2 Show that the sequence generated by the twisted-ring counter of figure 10.17 is 000, 100, 110, 111, 011, 001, 000 ...

3 The D flip-flops of figure P10.3 are initially both set to 1. Draw the output waveforms at Q_1 and Q_2 when the circuit is clocked. Will the sequence of states be different if the initial state was 00 instead of 11?

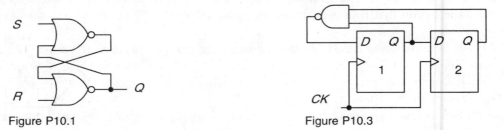

Figure P10.1   Figure P10.3

4 Design a counter using D flip-flops which will cycle through the successive states 000, 001, 011, 111, 110, 100, 000. Will the counter hang up if the initial state is one of those not in the sequence? (These missing states can be entered as don't cares in the K. M.)

5 Repeat the previous problem using JK flip-flops.

6 If the initial state of the counter in figure P10.6 is 1111, what will be the sequence generated by successive clock pulses?

Figure P10.6

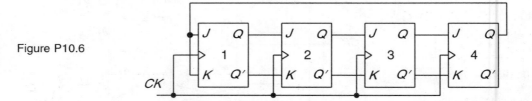

7 If the initial state of the circuit of figure P10.7 is 11, what will be the waveform at Q_2 when it is clocked?

8 If $Q_1 = Q_2 = 1$ initially in the circuit of figure P10.8, what will be the output at Q_2 when it is clocked. What does the circuit do?

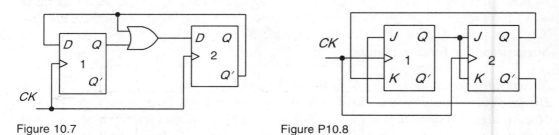

Figure 10.7   Figure P10.8

9 Design a scale-of-five counter using (a) D (b) T and (c) JK flip-flops.

10 Design a synchronous circuit using JK flip-flops which will produce the output sequence 000, 011, 101, 110, 001, 111, 100, 010, 000 ...

11 Show that an 8-bit shift register cannot produce a maximum length sequence when only two bits are XORed and fed back to produce a pseudo-random-number sequence. What is the longest sequence?

12 Produce a timing diagram such as that of figure 10.30 for a flip-flop which only changes state when the two inputs are 1s and hence derive a circuit for the flip-flop in terms of simple logic gates.

13 Repeat problem 10.12 but for a circuit whose output toggles every time the inputs are (a) different (b) the same.

14 A circuit has two inputs, X and Y, which can change only one at a time. The output goes HIGH when X goes LOW provided Y is LOW. The output goes LOW when Y goes HIGH. Design a circuit for this. (Use an SR flip-flop for the secondary variable.)

15 Design a circuit with two inputs, A and B, which can change only one at a time. The output must take on the value of B whenever $A = 0$, and when $A = 1$ the output remains in its previous state. Are any static hazards present? If so eliminate them.

16 Design a serial adder using JK flip-flops. (A serial adder adds the bits from two inputs and carries when necessary, so that the output sequence is the correct sum.)

17 Draw up an excitation map for the asynchronous circuit of figure P10.17, in which only one of the inputs, A and B, can change at any one time. From this construct the state diagram and deduce the output waveform for all possible input combinations.

18 Repeat problem 10.17 for the circuit of figure P10.18.

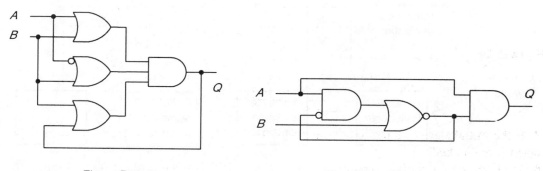

Figure P10.17   Figure P10.18

19 Design a circuit which has one input, Y, and an output, Z, which blanks alternate inputs as in figure P10.19; if there are static hazards eliminate them. (Hint: there are four states and you will need two secondary variables.)

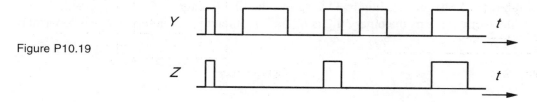

Figure P10.19

11 Data conversion

AT SOME POINT in much electronic equipment there is a need to convert analogue data to digital and vice versa. Transducers at the interface between the equipment and the outside world may be used to transform a physical property such as pressure, temperature, light intensity, colour, speed etc. into an analogue electrical signal, which may be a simple or complicated function of the physical parameter sought. This signal can be transformed into digital data by an analogue-to-digital converter (ADC), and then processed and displayed. In some cases the processed digital data is transformed afterwards back into an analogue signal via a digital-to-analogue converter (DAC), which produces a physical output in the outside world via another transducer or actuator. Figure 11.1 shows a block diagram of an electronic system of such a kind.

All digital multimeters and most digital instruments contain ADCs, which to a large extent determine their quality: a high-quality, high-resolution instrument with a good bandwidth requires a fast ADC with up to 20 bits — possibly more — and these are very expensive. Digital-to-analogue conversion is in general easier than the converse, and many ADCs contain DACs, so we shall first discuss DACs and then ADCs.

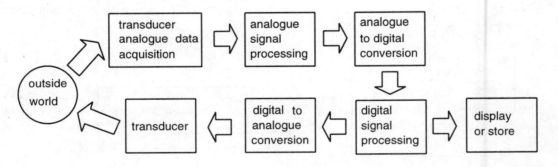

Figure 11.1 Block diagram of conversion process

11.1 Digital-to-analogue converters (DACs)

A DAC takes a binary word of n bits and converts it to a voltage which is proportional to a reference voltage, V_{ref}. Figure 11.2 shows the block diagram for this process.

If the word contains the binary digits b_1, b_2 ... b_n, then the output voltage can be written

$$V_o = GV_{ref}\left(\frac{b_1}{2^1} + \frac{b_2}{2^2} + ... + \frac{b_n}{2^n}\right)$$

414

where G is a constant scale factor. Putting this in summation notation

$$V_\text{o} = GV_\text{ref}\sum_{k=1}^{n} b_k 2^{-k} \tag{11.1}$$

Thus when the input word is 1101, say, and $V_\text{ref} = 2.5$ V and $G = 4$, the output voltage will be

$$V_\text{o} = 4 \times 2.5 \times \left(\frac{1}{2} + \frac{1}{4} + \frac{0}{8} + \frac{1}{16}\right) = 8.125 \text{ V}$$

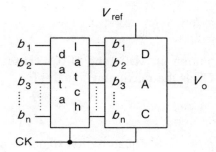

Figure 11.2

A clocked digital-to-analogue converter

11.1.1 Characteristics of DACs

Figure 11.3 shows the transfer function of an ideal 3-bit DAC, which has an output given by

$$V_\text{o} = FS\sum_{k=1}^{3} b_k 2^{-k} + \frac{FS}{2^4} \tag{11.2}$$

where FS = full-scale reading. The output is raised by one sixteenth of the full-scale reading above that given by equation 11.1, resulting in a smaller maximum error; then an input of 101 gives an output of $0.6875FS$ rather than $0.625FS$.

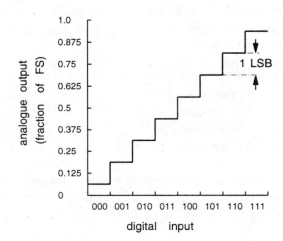

Figure 11.3

Ideal transfer characteristic of a 3-bit DAC

We see from figure 11.3 that the full-scale range (the difference between the biggest and the smallest output values of the DAC) is 1 LSB less than *FS*. The resolution of the DAC is the largest error which can occur in the output, and is not 1 LSB but $\pm\frac{1}{2}$ LSB at the point where the input changes value.

The dynamic range of a DAC is defined as the ratio of full-scale range divided by the resolution, that is

$$DR = \frac{FS - LSB}{0.5LSB} = \frac{2FS}{LSB} - 2 = 2^{n+1} - 2 = 2^n$$

Sometimes this is given in dB using

$$DR_{dB} = 20\log_{10}2^n = 20n\log_{10}2 = 6n \tag{11.3}$$

The finite resolution of ADCs and DACs causes *quantisation noise* which has an r.m.s. value, taking *FS* as the maximum signal, of

$$N_q = \frac{FS}{2^n\sqrt{12}} \tag{11.4}$$

and can be expressed as a quantisation signal-to-noise ratio, S/N_q or SNR_q, which is in dB

$$S/N_{qdB} = 20\log_{10}(FS/N_q) = 20\log_{10}(2^n\sqrt{12}) = 6n + 10.8 \tag{11.5}$$

The quantisation noise falls rapidly with *n*.

Example 11.1

A DAC is required which obeys equation 11.1 and has a dynamic range of 50 dB with a FS voltage output of 9 V. How many bits are required? What are the quantisation noise and the quantisation signal-to-noise ratio? What is the largest error, assuming only quantisation error is significant?

From equation 11.3 we see that $n = DR_{dB}/6 = 50/6 = 8.33$, but as *n* must be an integer, we require it to be 9. The quantisation noise is given by equation 11.4; taking *FS* = 9 V and *n* = 9, we find $N_q = 5$ mV. This is given in dB by equation 11.5, from which we find $N_{qdB} = 64.8$ dB. The largest error is $\pm\frac{1}{2}$ LSB and the LSB is $FS/(2^n - 1) = 9/511 = 17.6$ mV, so $\frac{1}{2}$ LSB = 8.8 mV.

Four types of error are generally recognised in DACs

> (1) Offset error
> (2) Gain error
> (3) Non-linearity
> (4) Non-monotonicity

These four errors are illustrated in figures 11.4 and 11.5, which refer to a 3-bit DAC for the sake of clarity. The quantisation error is so large in this that the non-linearity and non-monotonicity errors are much exaggerated. The gain error in figure 11.4b can be expressed as a percentage deviation. The non-linearity error in figure 11.5a can be expressed as a maximum deviation from the ideal output anywhere on the characteristic from zero to full scale. In the example shown this error is a maximum of +0.8 LSB at an input of 111 and −0.8 LSB at an input of 010. A monotonic DAC is one in which the output always rises when the input does. Non-monotonicity, shown in figure 11.5b, can only occur if the non-linearity at any point exceeds 1 LSB.

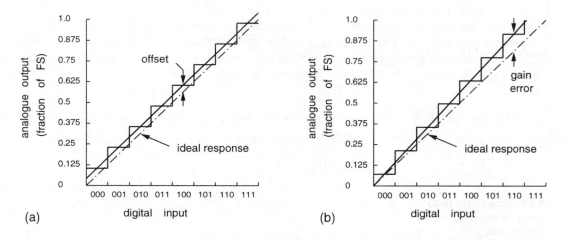

Figure 11.4 Errors in a 3-bit DAC (a) Offset error (b) Gain error

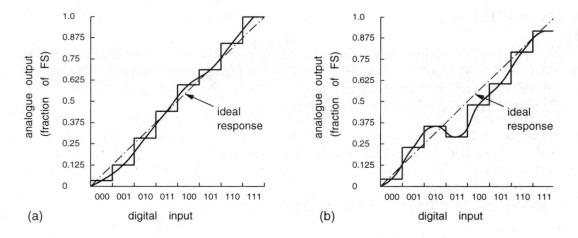

Figure 11.5 Errors in a 3-bit DAC (a) Non-linearity (b) Non-monotonicity

Example 11.2

An 8-bit DAC is just monotonic and obeys equation 11.2. What is the maximum error if the FS voltage is 10 V? What must the gain be if the reference voltage is 1.24 V and the FS output must be 12 V?

The maximum error is $\pm\frac{1}{2}$ LSB if we use equation 11.2, and the LSB = $10/(2^8 - 1)$ = $10/255$ = 39.2 mV, so the maximum error is 19.6 mV. The full-scale output is $GV_{ref} = FS$ = 12 V, so $G = 12/1.24 = 9.68$.

11.1.2 Current-scaling DACs

DACs can be of three basic types, with some hybridisation between them: current-scaling, voltage-scaling and charge-scaling DACs. Though it is possible to make serial DACs, all those discussed here are parallel DACs which are faster and simpler than serial DACs. Figure 11.6 shows a prototype current-scaling DAC which uses an op amp with a current-summing node (previously discussed in section 5.4.4).

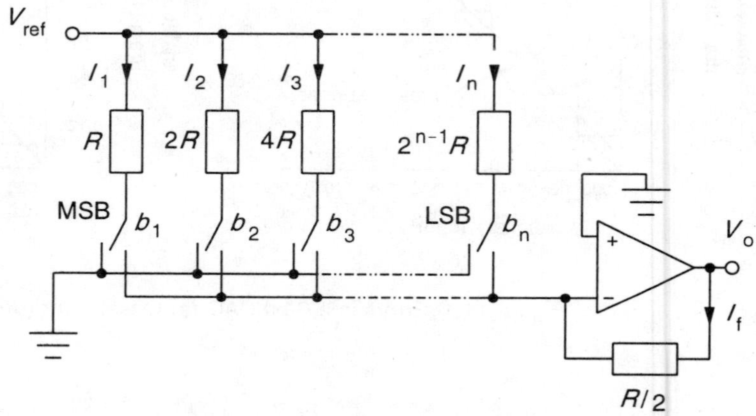

Figure 11.6

A current-scaling DAC using binary-weighted resistances

The switches connect the resistors to the op amp if $b = 1$ and to ground if $b = 0$. Switching in all DACs is done by MOSFETs but the op amps can be implemented in either bipolar or MOS technology.

The current fed back from the output is given by

$$I_f = \frac{2V_o}{R} = -(b_1 I_1 + b_2 I_2 + b_3 I_3 + ... + b_n I_n)$$

$$= -V_{ref}\left(\frac{b_1}{R} + \frac{b_2}{2}R + \frac{b_3}{4}R + ... + \frac{b_n}{2^{n-1}R}\right)$$

$$\Rightarrow \quad V_o = -V_{ref}\left(\frac{b_1}{2^1} + \frac{b_2}{2^2} + \frac{b_3}{2^3} + ... + \frac{b_n}{2^n}\right) = -V_{ref}\sum_{k=1}^{n} b_k 2^{-k}$$

If gain is needed the feedback resistance can provide it; when $R_f = R/2G$ then the output is scaled by G. This type of current scaling suffers from the range of resistance values needed: an 8-bit DAC covers a range of from R to $2^7 R = 128R$, and the tolerance required of the MSB resistance is that of the LSB/128. Then if the tolerance of the LSB resistance is ±20%, that of the MSB is 0.16%. Another problem is that of switch resistance, that is $R_{DS}(\text{on})$ for FETs, which may be several hundred ohms and is dependent on current. For accuracies of better than 1% or so the resistors require trimming. If thin-film resistors are used, fast, accurate laser trimming is possible, but the construction is complicated by the mix of thin-film and silicon technologies.

A large range of resistance values is also difficult to achieve in ICs and can be avoided by using an R-$2R$ ladder as shown in figure 11.7, where the resistance of the whole network is R and that of each branch $2R$, so that the outgoing current divides equally at each node.

Figure 11.7

A current-scaling DAC using an R-$2R$ ladder

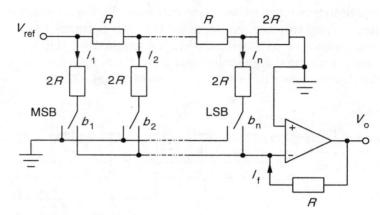

The current from the output is equal to the sum of the currents from the input:

$$I_f = \frac{V_o}{R} = -(b_1 I_1 + b_2 I_2 + b_3 I_3 + ... + b_n I_n)$$

$$= -V_{ref}\left(\frac{b_1}{2R} + \frac{b_2}{2^2 R} + \frac{b_3}{2^3 R} + ... + \frac{b_n}{2^n R}\right)$$

$$\Rightarrow \quad V_o = -V_{ref}\left(\frac{b_1}{2^1} + \frac{b_2}{2^2} + \frac{b_3}{2^3} + ... + \frac{b_n}{2^n}\right) = -V_{ref}\sum_{k=1}^{n} b_k 2^{-k}$$

Current-scaling DACs of this type with more than 8 bits become less accurate because the tolerance for the resistance associated with the MSB is still 2^{n-1} times less than that for the LSB. One way of avoiding this drawback is to use an attenuating network at the halfway point in the ladder and arrange the attenuation so that the currents in each half are equal when the input is maximum. The DAC is then two DACs with half the number of bits in cascade. This approach is discussed in section 11.1.4, which deals with capacitive-scaling, where similar problems arise.

11.1.3 Voltage-scaling DACs

Voltage-scaling DACs require a voltage-dividing resistor network with one resistor for every input state so that, for an n-bit DAC, 2^n resistors are required. Figure 11.8 shows a possible configuration for a 3-bit version of this type of DAC. The op amp acts merely as a buffer. To see how it works consider what happens when the input is 101. The ganged switches labelled b_1, b'_2 and b_3 close and b'_1, b_2 and b'_3 stay open. The resistance tap selected is therefore $5R$ counting up from ground, and the output voltage is $5V_{ref}/8$. The version shown in figure 11.8 has a direct connection to ground for 000, but if it is desired to have an output of $V_{ref}/2^4 = V_{ref}/16$ for 000 instead, the resistance, R, adjacent to V_{ref} can be split into two halves, with one half going between ground and the first switch and the other staying in place. An input of 111 then becomes $15V_{ref}/16$ instead of $7V_{ref}/8$. The large number of resistors and switches required for this type of DAC has made it less popular than the charge-scaling and current-scaling types, but it is used in hybrid voltage-and-charge-scaling or voltage-and-current-scaling DACs.

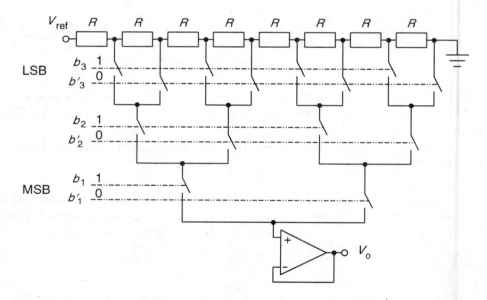

Figure 11.8 A 3-bit voltage-scaling DAC

Once more the accuracy required for the resistance values increases with the number of bits. Suppose for example the resistances are $R_0 \pm \delta R_0$, then in the worst case they could be arranged with a sequence of k minimum resistance values, R_1, with the remaining $2^n - k$ all at maximum resistance, R_2. The voltage output from the kth resistance is then

$$V_{k1} = \frac{kR_1}{(2^n - k)R_2 + kR_1}$$

compared to the ideal voltage, $V_{k0} = k/2^n$, taking $V_{ref} = 1$ V. The error in output is

$$V_{k1} - V_{k0} = \Delta V_k = \frac{kR_1}{(2^n - k)R_2 + kR_1} - \frac{k}{2^n} \qquad (11.6)$$

It can be shown (see problem 11.6) that the error is maximum when $k = 2^{n-1}$, that is halfway along the chain, and then

$$\Delta V_{max} = \frac{R_1}{R_2 + R_1} - \frac{1}{2} = \frac{R_0 - \delta R_0}{R_0 + \delta R_0 + R_0 - \delta R_0} - \frac{1}{2}$$

$$= \frac{R_0 - \delta R_0}{2R_0} - \frac{1}{2} = \frac{-\delta R_0}{R_0}$$

This is a worst case, so in general the voltage-scaling DAC will do better. For a minimal requirement of $\pm\frac{1}{2}$ LSB, the relative resistance error must be

$$\frac{\delta R_0}{R_0} = \frac{0.5}{2^n} = \frac{1}{2^{n+1}}$$

An 8-bit DAC would require a resistance tolerance of $\pm0.2\%$. If the resistance tolerance is 2% then five bits is the maximum. To build highly accurate DACs requires a different design to these.

11.1.4 Charge-scaling DACs

Charge-scaling DACs use capacitors to turn the bits of the input word into analogue form, and capacitors are generally more convenient for IC production than resistors. Figure 11.9 shows an array of capacitors in the familiar sequence of binary-scaled values.

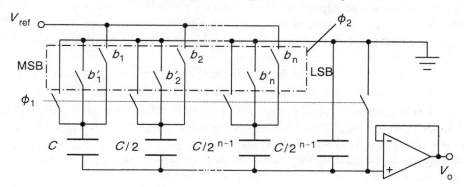

Figure 11.9 A charge-scaling DAC

Initially, during the first switching phase, ϕ_1, the capacitors are discharged. In the second switching phase the switches in the box marked ϕ_2 are operated in accordance with the values of the input bits, b_k closing if bit k is 1, b'_k if bit k is 0. Those connected to a 1-bit will charge up to a voltage of V_{ref} and those connected to a 0-bit will stay grounded. The capacitor furthest right, connected between the non-inverting terminal of the buffer and ground, is a terminating capacitor. During ϕ_2, the charging phase, the capacitors connected to V_{ref} are in parallel to give a total capacitance of ΣC_1, and those connected to ground are in parallel to give a total capacitance of ΣC_0.

The two groups are in series as shown in the equivalent circuit of figure 11.10, so that the charge divides between the two groups, the larger voltage being across the smaller capacitance. Therefore

$$V_o = \frac{\Sigma C_1 V_{ref}}{\Sigma C_1 + \Sigma C_0} = \frac{\Sigma C_1 V_{ref}}{C + C/2 + ... + C/(2^n - 1) + C/(2^n - 1)}$$

$$= \frac{V_{ref}(b_1 C + b_2 C/2 + ... + b_n C/2^{n-1})}{2C} = V_{ref}\left(\frac{b_1}{2^2} + \frac{b_2}{2^2} + ... + \frac{b_n}{2^n}\right)$$

$$= V_{ref}\sum_{k=1}^{n}\frac{b_k}{2^k}$$

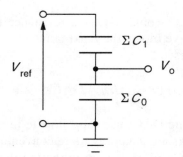

Figure 11.10

The equivalent circuit of figure 11.9

In IC manufacture the capacitor size is directly proportional to its area, which means that if the minimum capacitor area is 1 μm^2 an n-bit DAC will require an area of 2^{n-1} μm^2 for its largest capacitor. Not only that, but the tolerance for the MSB capacitance will be $1/2^{n-1}$ times that for the LSB capacitance.

The simplest way to overcome these difficulties is to cascade two DACs as shown in figure 11.11, where two 3-bit DACs are cascaded by the attenuating 8/7 pF capacitor. In practice cascading would only be used for 8-or-more-bit DACs. The circuit works in similar fashion to that of figure 11.9, except that the equivalent circuit is that of figure 11.12a, which reduces to that of figure 11.12b.

The charge on the side of the MSBs is shared between the capacitors of the 1-bits, ΣC_{M1}, and those of the 0-bits, ΣC_{M0}, to give a voltage of

$$V_M = \frac{V_{ref}\Sigma C_{M1}}{\Sigma C_{M0} + \Sigma C_{M1}} = \frac{V_{ref}\Sigma C_{M1}}{7} \qquad \text{(in V and pF)}$$

The same thing happens on the side of the LSBs and the voltage on that side of C is

$$V_L = \frac{V_{ref}\Sigma C_{L1}}{\Sigma C_{L1} + \Sigma C_L)} = \frac{V_{ref}\Sigma C_{L1}}{8}$$

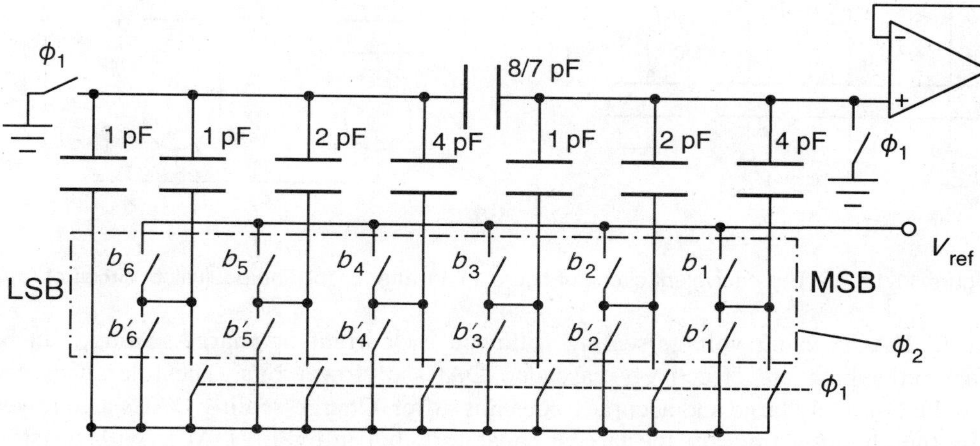

Figure 11.11 A six-bit, cascaded, current-scaling DAC

V_L and V_M are shown as voltage sources in figure 11.12b, which indicates that on the side of the LSBs the capacitances ΣC_L (= 8 pF) and C (= 8/7 pF) are in series so their equivalent capacitance is

$$C_{eq} = \frac{\Sigma C_L \times C}{\Sigma C_L + C} = \frac{8 \times 8/7}{8 + 8/7} = 1 \text{ pF} \tag{11.7}$$

The output voltage is therefore

$$V_o = \frac{V_L C_{eq} + V_M \Sigma C_M}{C_{eq} + \Sigma C_M} \tag{11.8}$$

$$= \frac{V_L + 7V_M}{8} = \frac{V_{ref}\Sigma C_{L1}}{8\Sigma C_L} + \frac{7V_{ref}\Sigma C_{M1}}{8\Sigma C_M}$$

Hence $\quad V_o = \dfrac{V_{REF}\Sigma C_{L1}}{64} + \dfrac{V_{REF}\Sigma C_{M1}}{8} = V_{REF}\left(\dfrac{\displaystyle\sum_{k=1}^{3} b_k C_k}{64} + \dfrac{\displaystyle\sum_{k=4}^{6} b_k C_k}{8} \right)$

The attenuating capacitor's value must be 8/7 pF to scale down the LSB capacitances by exactly 8. If the maximum allowable error is ± 0.5 LSB, then it can be shown (see problem 11.2) that the maximum tolerance for C is about $\pm 10\%$ in this example. Charge-scaling cascades are therefore relatively insensitive to errors in the attenuating capacitor.

With this approach it is possible to build 16-bit DACs with an accuracy of $\pm\frac{1}{2}$ LSB requiring a tolerance in the MSB capacitance of about $\pm0.4\%$.

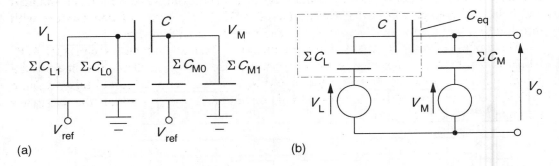

(a)   (b)

Figure 11.12 (a) The equivalent circuit of figure 11.11 after ϕ_2 (b) The equivalent circuit of (a)

Hybrid DACs, using voltage-scaling followed by current or charge-scaling, can be manufactured which are effectively cascaded DACs of fewer bits. The tolerances can therefore be relatively large and accuracy does not suffer. Charge-scaling DACs are slowed down by the charging time for the largest capacitors; but in hybrid DACs, with resistor-scaling followed by charge-scaling, this problem is reduced and the speed is greater.

11.2 Analogue-to-digital converters (ADCs or A/Ds)

Most physical data is collected in analogue form from sensors and transducers, which turn changing physical quantities such as mass, length, time, current etc. into time-varying electrical signals. The signals may be wanted in digital form for storage, display, transmission or further processing; and after filtering and scaling, they are fed into an analogue-to-digital converter to achieve this.

There are at least four main types of analogue-to-digital converter: serial, successive approximation, parallel and high performance. Serial and successive approximation ADCs are slower and cheaper, while parallel ('flash') are faster but use many more comparators. The fourth type — hybrid, high-performance ADCs — are becoming cheaper and more popular with the improvement in IC technology, especially switched-capacitor techniques.

It is not possible to convert a varying voltage into digital form without first sampling and holding it in some way, as shown in figure 11.13.

Figure 11.13

A sample-and-hold circuit

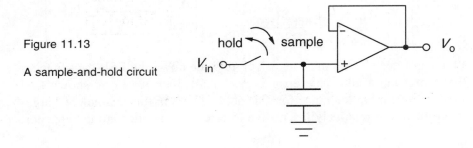

The sampling frequency determines the highest frequency that can be converted from the analogue waveform to a digital representation. In fact the sampling frequency must be at least twice the highest frequency component in the analogue waveform. Once sampled and held, the voltage can be fed to one (serial ADCs) or a series of comparators with threshold voltages set at levels suitable for the required digital output.

The process of sampling and holding produces a voltage at the input to the actual ADC which is constant for a definite period of time, denoted as V_{in}* to distinguish it from the actual analogue input, V_{in}. Because the acquisition of the analogue signal takes a finite time, t_a, and the output of the sample-and-hold circuit takes time to settle, t_s, the maximum sampling frequency, f_s, must be

$$f_s = \frac{1}{T_s} < \frac{1}{t_a + t_s}$$

And this frequency must be at least twice the highest frequency component of the input signal, f_{max}, for the sampled signal to be a valid approximation to the input, that is

$$f_s \geq 2f_{max}$$

This is known as the Nyquist criterion, and the critical sampling frequency, $2f_{max}$, the Nyquist frequency or Nyquist rate.

11.2.1 Serial A/D converters

There are two types of serial ADC, one using single-slope integration and the other dual-slope integration.

Single-slope integration

Figure 11.14 shows a single-slope integrator ADC.

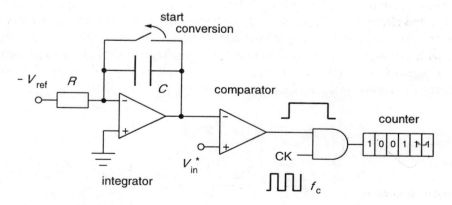

Figure 11.14 An ADC using single-slope integration

At the start of conversion the integrator output is a ramp voltage given by

$$V_o = \frac{V_{ref}t}{RC}$$

This is fed to the inverting input of the comparator whose non-inverting input is connected to the sampled signal, $V_{in}*$. The comparator output starts off HIGH and gates the clock pulses through to the counter via the AND. After a time the integrator's output voltage is equal to $V_{in}*$ and the comparator's output goes LOW. The AND gate then stops the clock pulses and the counter registers a count which is the digital representation of $V_{in}*$. If the input voltage is too large the counter will register a maximum count of $2^n - 1$, in an n-bit ADC, and hold it until the next conversion.

The time taken for the ramp to reach $V_{in}*$ is given by

$$V_{in}* = \frac{V_{ref}t}{RC} \quad \Rightarrow \quad t = \frac{RCV_{in}*}{V_{ref}}$$

If the clock frequency is f_c ($= 1/T_c$), the count reached will be

$$N = \frac{t}{T_c} = \frac{RCV_{in}*}{T_c V_{ref}} = \frac{RCf_c V_{in}*}{V_{ref}}$$

and the maximum count, N_{max}, should occur when $V_{in}* = V_{ref}$, that is

$$N_{max} = RCf_c = 2^n - 1$$

The time taken to ramp up to V_{ref} is

$$t_{max} = RC = \frac{N_{max}}{f_c} = \frac{2^n - 1}{f_c}$$

The maximum sampling rate must be less than $1/t_{max}$, that is

$$f_s < \frac{1}{t_{max}} = \frac{f_c}{2^n - 1}$$

If the number of bits is 12 and $f_c = 1$ MHz, $f_s < 244$ Hz, which is typical of serial ADC sampling rates.

The time constant of the integrator, RC, is given by

$$RC = t_{max} = \frac{N_{max}}{f_c} = \frac{2^n - 1}{f_c}$$

which is about 4 ms if $n = 12$ and $f_c = 1$ MHz. Hence if R is 1 MΩ, then C must be 4 nF, too large a capacitance to put in an IC.

In switched-capacitor ICs the resistance, R, is realised by a capacitor (see section 6.4.6), and is given by

$$R = \frac{1}{f_c C_R}$$

assuming the capacitor is switched at the clock frequency. C_R is the required capacitance value to simulate a resistance of R. Thus the time constant is

$$RC = \frac{C}{f_c C_R} = \frac{N_{max}}{f_c} \quad \Rightarrow \quad \frac{C}{C_R} = N_{max}$$

and if $N_{max} = 2^{12} - 1$, $C/C_R = 1000$, which is acceptable. Beyond 12 bits or thereabouts the capacitance ratio becomes too great for accuracy to be maintained at $\pm\frac{1}{2}$ LSB.

Single-slope integration relies on a constant, linear ramp from the integrator, which is a demanding requirement for the op amp. In dual-slope integrators the variable op amp parameters are largely cancelled out and their performance is better, though slower by a factor of two, than that of the single-slope integrator.

Dual-slope integration

Figure 11.15 shows a block diagram for a dual-slope ADC, in which the sample-and-hold voltage, $V_{in}*$ (assumed to be positive) is integrated for a set number of counts, N_{ref}, producing a negative ramp. Then a negative reference voltage, $-V_{ref}$, is applied to the integrator's input in place of $V_{in}*$. This causes the integrator to ramp up to its original level, set by the comparator, V_{TH}. The time taken for this to occur is proportional to $V_{in}*$ and is registered on a counter.

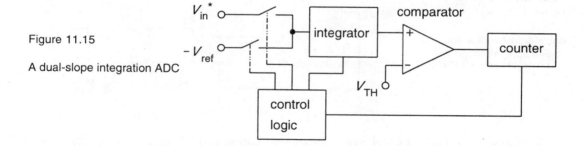

Figure 11.15

A dual-slope integration ADC

The integrator's output is shown in figure 11.16 for two different input voltages, $V_{in1}*$ and $V_{in2}*$. These produce counts, after zeroing the counter when the reference ramp starts, of N_1 and N_2, which are the digital representations of the sampled voltages. The analogue voltage, $V_{in1}*$, for example, is

$$V_{in1}* = \frac{N_1 V_{ref}}{N_{ref}}$$

Assuming that the offsets, drift etc. of the integrator remain fairly constant during the conversion, the error due to varying ramp rate is almost eliminated (at the price of halving the conversion rate) since the integrator ramps up as well as down. See also problem 11.7.

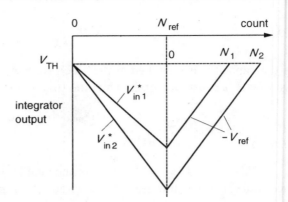

Figure 11.16

Integrator output for two
different input samples, with
the corresponding counts

11.2.2 Successive approximation ADCs

Successive approximation ADCs have a conversion time of n clock cycles, one for each
bit of the output word. Thus a 12-bit successive approximation ADC is $2^{12}/12 = 340$ times
faster than a serial ADC, which means it can sample at about 100 ksps (kilo samples per
second) instead of 200 sps and is therefore usable for converting signals with significant
components of up to about 40 kHz. They are thus medium-speed ADCs. Figure 11.17
shows a block diagram of a successive approximation DAC.

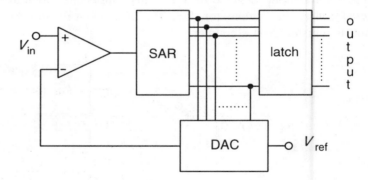

Figure 11.17

Block diagram of a succ-
essive approximation ADC

It comprises a DAC which converts an approximation to V_{in} from the successive approximation register (SAR) to analogue form and compares it to V_{in} with the comparator. If
the approximation is low the comparator's output is high and the approximation is increased. Suppose for example the correct digital output is 10110. The first digital approximation generated is 10000, which is low, so the next approximation tried is 11000, which
is high. The comparator output goes low when the approximation is high, which signals
that the next approximation should be 10100. This is low, so the next approximation must
be 10110, the correct output, which is held by the approximation register. The sequence
is shown in figure 11.18, which indicates all the possible states of the approximation
register. Each approximation determines one bit of the result, implying that the conversion
time is n cycles, the same as the number of bits in the result.

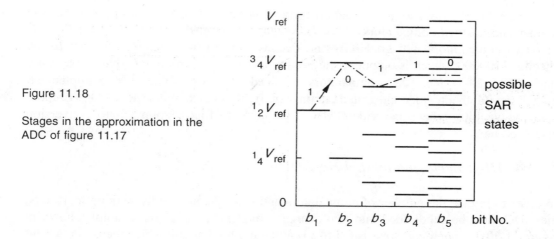

Figure 11.18

Stages in the approximation in the ADC of figure 11.17

11.2.3 Parallel ('flash') ADCs

Parallel ADCs operate in one clock cycle and are n times as fast as successive approximation ADCs. With sampling rates of 1 Msps and more, flash ADCs demand large numbers of op amps and are therefore much more expensive to manufacture than successive approximation or integration types. Their advantage is their high speed which enables high sampling rates so that high-frequency video waveforms can be adequately represented.

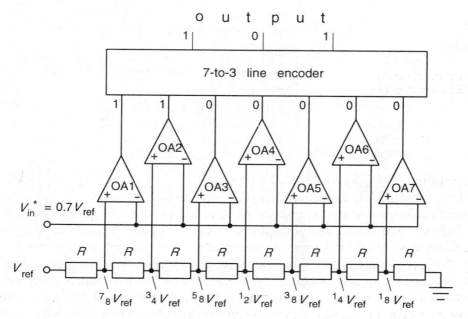

Figure 11.19 A 3-bit flash ADC

Figure 11.19 shows a block diagram for a 3-bit parallel ADC with an input voltage divider and a seven comparators (OA1-7). When the sampled input is $V_{in}^* = 0.7V_{ref}$ the first two comparators will go HIGH and the last five LOW, making the input to the encoder 1100000, which is output as the 3-bit word 101. The problem is that, for a 10-bit flash ADC, $2^{10} - 1 = 1023$ comparators are needed; so it is expensive to implement, though the Raytheon video integrated circuit TDC1020 is an example. Most ADCs of more than twelve bits, other than video ADCs, employ the fourth approach to conversion discussed next.

11.2.4 *High-performance hybrid ADCs*

In order to reduce the number of comparators and still operate rapidly with up to 18 bits, the ADC must be subdivided into two or more smaller ADCs. In the example shown in figure 11.20, the analogue input is fed to a sample-and-hold circuit which feeds into a 6-bit flash converter for the six MSBs. These are fed to an error-correcting logic block, to a 12-bit DAC and directly to the MSB pins of the output. The 12-bit DAC output is compared to the input signal with a comparator whose output adjusts the contents of a successive approximation register to generate the six LSBs, which are fed to the DAC and the LSB output pins. The device thus has the speed of a 6-bit successive approximation ADC instead of a 12-bit, but the main advantage is that its linearity is independent of that of the flash converter and its overall cost is much lower than for a 12-bit flash converter.

Table 11.1 gives some data for various commonly-used ADCs.

Table 11.1 *Selected data for A/D Converters*

Device ID	Bits	Speed	Type	Power	Voltage	Price
CA3306E	6	15M	F	1.5 mW†	3/7.5	£8
ADC0804LCN	8	10k	SA	6.5 mW†	5	£4
AD7821KN	8	1M	HP	75 mW	5	£15
LTC1098CN8	8	33k	SA	0.5 mW†	3-10	£5
ZN427E-8	8	100k	SA	125 mW	5	£12
AD773JD	10	10M	HP	1.5 W	±5	£80
ADC12032CIN	12	100k	SA	30 mW	5	£11
AD16071CIN	16	200k	HP	400 mW	5	£20
TSC850CPL	16	20	DS	1 mW†	+6/−9	£20
ADS7807	16	40k	SA	50 µW†	±10/5	£40

Notes: Speed is the sampling rate in samples/s. Type: DS = Dual-slope integration, SA = successive approximation, F = flash, HP = hybrid high performance. Power is the normal operating power, except † = standby power consumption. Voltage is the positive supply voltage, except where indicated otherwise. Price is for single quantities.

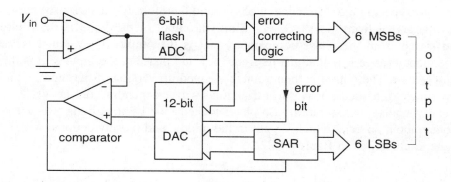

Figure 11.20 A hybrid, high-performance ADC combining a 6-bit flash ADC and a 6-bit SAR

11.3 Phase-locked loops

Phase-locked loops (PLLs), after a slow start[1], have lately become very widely used in IC form (for example, the CD4046 from National Semiconductor and its equivalent MC144046 from Motorola (which costs less than £1) the SDA2112-2 from Siemens, the SL650B, SL651B and SL652C from Plessey and the NE564 from Philips Semiconductors, which operates at up to 50 MHz). They are a very good means of extracting clean sine or other waves from noisy input signals, and they can be used also for AM and FM demodulation, tone decoding, frequency-shift keying, waveform generation, tracking filters, voltage-to-frequency conversion and motor speed control: the uses are limited only by the designers' imaginations, it seems. Many of these uses require specialised PLL constructional techniques and are beyond the capabilities of low-cost ICs. Figure 11.21 shows a block diagram of a PLL.

Figure 11.21

Block diagram of a phase-locked loop

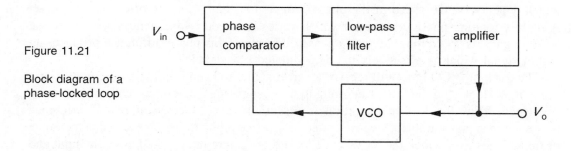

---

[1] The concept originated as long ago as 1922 and PLLs became widely used in specialised roles during the 1960s and 1970s, see for example *Phaselock Techniques* by F Gardner (John Wiley, 1966), 'Miniaturized RC filters using phase-locked loops' by G Moschytz, *Bell System Technical Journal*, May 1965 and *Phase-locked Loops* by A Blanchard (John Wiley, 1976). Philips Semiconductors' application notes AN177 and AN178 make a good introduction to PLLs.

The input signal's phase is compared to that generated by a voltage-controlled oscillator (VCO) and the output is low-pass filtered and amplified. The phase comparator's output depends on its type and the difference in phase (and therefore frequency also) between the input and the signal from the VCO, V_{ref}. Since the input of the VCO is the filtered and amplified signal from the phase comparator, that is the output, V_o, when it is within a certain frequency range it will adjust its frequency and phase to the input's. It is then said to have *locked on*. The output is then a much-cleaned-up version of the input, which may be a pure sinusoid, a square wave or a triangular wave, depending on the VCO.

One can visualise the operation of a phase-locked loop by imagining two heavy discs free to rotate about an axle, which is split in the middle and loosely coupled with a spring or a magnetic clutch, as in figure 11.22.

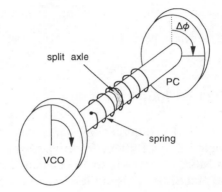

Figure 11.22

The mechanical analogue of a PLL

When one disc, the phase-comparator, is moved rapidly to a new position, the loose coupling means that initially the other disc, the VCO, remains stationary for a time before slowly accelerating to catch up with the first. When it does so it will overshoot and then stop and reverse direction. This will result in oscillations of the VCO disc about the mean position until it eventually comes to halt at a position slightly out of phase with the phase comparator, giving rise to a characteristic lock-in signal. If the phase-comparator disc moves slowly from its stationary position, the VCO disc will track it with a phase error which depends on the speed of the phase-comparator, that is its rate of change of phase. A rapidly-rotating phase comparator which varies its speed ($\equiv$ frequency) will cause the VCO to vary its speed too, but with a phase difference. This phase difference produces an output signal which is effectively the demodulated FM signal.

The tracking VCO must still receive an error signal to keep it in lock, otherwise it will fall out of lock and into its free-running frequency, f_O, sometimes called the centre frequency. Thus the lock-in condition is essentially one of constant phase difference between the signal and its locally-generated, noise-free replica. The *lock range* is the range of frequencies over which the VCO, once locked on, can remain locked onto the input and runs from $f_O - f_L$ to $f_O + f_L$, that is the lock range is $2f_L$. The lock range must be greater than the *capture range*, $2f_C$, which is the range of frequencies over which the VCO can lock onto the input, starting out of lock. The capture range is also centred on f_O. The time taken to lock on is the *lock time*, t_L which is inversely proportional to bandwidth. PLL ICs are readily available with free-running frequencies up to 50 MHz.

The loop gain, K_v, of the PLL is made up of three parts — the gains of the three component boxes in figure 11.21 — that is the phase comparator's conversion gain, K_D (in V/rad), the low-pass filter gain, K_{LP}, and the VCO conversion gain, K_O (in rad/s/V):

$$K_v = K_D K_{LP} K_O \quad \text{(in s}^{-1}\text{)} \tag{11.9}$$

The unusual units of the loop gain arise because the VCO's gain, K_O, is given by

$$\Delta \omega_O = \frac{d\Delta\phi}{dt} = K_O V_{LP}$$

$$\Rightarrow \quad \Delta\phi = \int K_O V_{LP} dt$$

since angular frequency is the rate of change of phase. If the output of the low-pass filter, V_{LP}, is of the form $V_m \sin \omega t$ then the phase output of the VCO is

$$\Delta\phi = \frac{K_O V_m (-\cos \omega t)}{\omega} = \frac{K_O V_m \sin(\omega t - 90°)}{\omega} = \frac{K_O V_m \sin \omega t}{j\omega}$$

It includes a $j\omega$ term in the denominator because the VCO has introduced a phase shift of $-90°$.

In designing a PLL it is necessary to write down the exact expression for the loop gain and then select a cut-off frequency (or frequencies) which will give unity loop gain at or a little above the frequency of interest. The VCO can be non-linear in output without seriously affecting the operation of the PLL, it just means that K_O is also a function of ω and the effective loop gain equation should take this into account. However, in most applications the variations in frequency are such that the VCO is very linear.

The damping factor, ζ, the stiffness of the spring in figure 11.22, is given by

$$\zeta = 1/\sqrt{4\pi K_v}$$

If the loop gain is large, the damping factor is small and the PLL will respond in an oscillatory manner to transients and is hardly stable. On the other hand, if the damping factor is large the response is slow. Ideally $0.2 < \zeta < 2$. Thus we should make K_v only as large as necessary in order to prevent excessive variations in the output of the PLL. This is largely a matter of adjusting the component values in the low-pass filter as the other terms are outside of our control, unless we are building the whole thing from scratch.

Figure 11.23 shows the phase-locked loop of National Semiconductor's CD4046BM CMOS IC (equivalent to Motorola's MC144046CP and SGS-Thomson's HCF4046BEY) in more detail. The inhibit input when HIGH causes the PLL to go into standby mode to reduce power consumption. One of two phase comparators can be selected; PC1 is a phase comparator which is just an XOR gate. Its output is a square wave — shown in figure 11.24a — when in the locked-on condition, since the VCO and the phase comparator will be out of phase by $90°$ because of the phase shift of the VCO (discussed later on). It is possible with this phase detector to get lock-in when the input is a harmonic of the VCO's centre frequency, f_O.

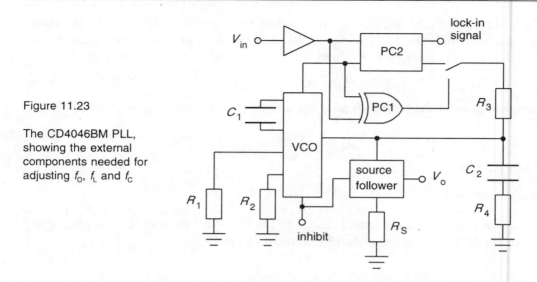

Figure 11.23

The CD4046BM PLL, showing the external components needed for adjusting f_0, f_L and f_c

Figure 11.24b shows the waveforms of PC2 when the VCO is locked-on. The slight residual phase difference between the output of the VCO and the input signal causes PC2 to output a small error voltage and a locked-in signal at the frequency of the input. Without an input signal PC1 causes the VCO to output its centre frequency, f_0, and if PC2 is used it outputs the minimum frequency, $f_0 - f_L$.

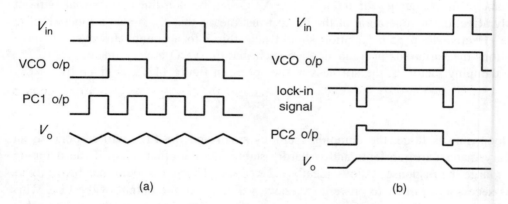

Figure 11.24 (a) Waveforms when using PC1 (b) Waveforms when using PC2

The adjustable components in figure 11.23 are the resistors R_1, R_2, R_3, R_4 and R_S, and the capacitors C_1 and C_2. In practice all the resistors should be above 10 kΩ and then R_S has no effect on the PLL. R_4 prevents the high-frequency gain of the low-pass filter from falling off at high frequencies, and can be omitted. R_3 and C_2 determine the cut-off of the low-pass filter. R_1 and C_1 determine the centre frequency of the VCO, f_0; while R_2 and C_1 determine the lock range, $2f_L$. Unfortunately these two frequencies are markedly dependent on the supply voltage and there may be considerable variation from chip to chip, so some experimentation is usually needed. The centre frequency with $R_2 = \infty$ is approximately

$$f_O \approx \frac{0.15 V_{DD}}{R_1 C_1}$$

Thus if $R_1 = 100$ kΩ, $C_1 = 1$ nF and $V_{DD} = 5$ V, $f_O = 7.5$ kHz. The practical limit for f_O is about 1 MHz, but may be only half of this. The lock range is given by

$$2f_L \approx \frac{0.5 V_{DD}}{R_2 C_1}$$

so if $C_1 = 1$ nF, $V_{DD} = 5$ V and $R_2 = 500$ kΩ, then $2f_L = 5$ kHz and the VCO can vary its output frequency from 5 kHz to 10 kHz.

The capture range is the same as the lock range when PC2 is used, but with the XOR detector, PC1, the capture range depends on the components of the low-pass filter and if $R_4 = 0$ it is

$$2f_C = \sqrt{\frac{2f_L}{\pi R_3 C_2}}$$

If $R_4 \neq 0$ the capture range is a more complicated function of the component values. If $f_L = 2.5$ kHz, $C_2 = 1$ nF and $R_3 = 220$ kΩ, then $2f_C = 2.69$ kHz $< 2f_L$ as it must be. The low-pass cut off is at $\omega_{LP} = 1/R_3 C_2 = 4.55$ krad/s or $f_{LP} = 723$ Hz. Thus at 10 kHz the low-pass gain is about -23 dB (at -20 dB/decade), that is $K_{LP} = 0.072$, which gives us a condition for the phase comparator's gain using equation 11.9

$$K_D = \frac{K_v}{K_O K_{LP}}$$

taking K_O, the VCO gain, as 6 krad/V, we find (if the loop gain, K_v is to be unity at 10 kHz) that $K_D \approx 2.3$ mV/rad.

Suggestions for further reading

VLSI design techniques for analogue and digital circuits by R L Geiger, P E Allen and N R Strader (McGraw-Hill, 1990) is useful for some aspects of ADC and DAC design.
Phase-locked loops: principles and practice by P V Brennan (Macmillan 1996) is an advanced text dealing with all aspects of PLL design.

Problems

1 How many binary digits are required of an ADC which has a resolution of 5 mV and a full-scale range of 6 V? If the input is a 0.5V sinusoid of frequency 400 Hz, what sampling rate is required for full resolution?

2 Show that the maximum error permissible in the attenuating capacitor (nominally of 8/7 pF capacitance) is about 10% if the maximum error in the conversion of a 6-bit input is to be $\pm\frac{1}{2}$ LSB. (Use equations 11.7 and 11.8.)

3 If the error in an n-bit DAC as in figure 11.7 is to be no more than $\pm\frac{1}{2}$ LSB, what is the permitted relative error in the first R resistance of the top row, the one nearest to V_{ref}? *[2/(2^n − 1)]*

4 A $4\frac{1}{2}$-digit DVM is designed to use a single-slope integration ADC to convert DC voltages up to a maximum of 1.999 V. How many bits are required? If the integrator used in the ADC has $R = 47$ kΩ and $C = 220$ nF, what clock frequency is required?

5 If a single-slope integrator, as in the previous problem, has a leaky capacitor with a parallel loss resistance of R_p show that for an accuracy of ± 1 LSB, $R_p > 2^{n-1}R$, where R is the input resistance for the reference voltage source.

6 Show that the error is a maximum in equation 11.6 when $k = 2^n - 1$.

7 Show that the maximum error in a dual-slope integration ADC, with an input offset voltage of V_{os}, is V_{os} when $V_{in}* = 0$, and is zero when $V_{in}* = V_{ref}$. By how much is the error caused by leakage resistance across the capacitor reduced?

8 A PLL is to be built using the CD4046 device of figure 11.23 with centre frequency of 500 Hz, a lock range of 500 Hz and a capture range of 300 Hz. Choose the components necessary for this assuming $V_{DD} = 9$ V, making sure all the resistances are greater than 10 kΩ. If $K_L = 1$ at 750 Hz, $K_O = 6$ krad/V and $K_D = 1$ mV/rad, what must the low-pass filter gain be at 750 Hz (taking $R_4 = 0$)? What component values are required for this?

12 Computers, microprocessors and microcontrollers

COMPUTERS ARE complete machines with a large number of input and output devices, or *peripherals*, and are capable of being programmed in a variety of high-level languages such as Fortran, Pascal, C, Ada, Modula2, etc. They also possess very large data and program storage capabilities involving a number of different media, such as magnetic disc, magnetic tape, CD ROM, semiconductor, etc. Microprocessors and microcontrollers on the other hand are computers on a chip, which require the user to supply operating instructions — usually in low-level or machine-coded form, though high-level language compilers for specific microcontrollers can be expected to become more common — as well as all the connections to peripheral devices necessary for performing the desired functions. Most have fairly small on-chip memories, but have the ability to address large amounts of external ROM (Read-Only Memory). Microprocessors and microcontrollers are now used almost interchangeably, but originally a microcontroller was a microprocessor with additional input and output capability for receiving status information and sending out control signals. Because the range of functions to be performed is very great, so are the numbers of different microcontrollers, as one would expect when the technology is in a state of rapid development. Eventually the numbers of different devices may decline as the market matures and the applications become more standardised.

A microcontroller is only a computer of limited capability on a single chip without peripherals, so we shall start by describing computers briefly and then go on to examine microcontrollers, with reference to one specific type which is possibly the most widely used 8-bit device family today.

12.1 Computers

Calculating machines are as old as history, and punched-card control of machines dates back to the early eighteenth century when they were used in France to control looms used for weaving elaborate patterns. In England, Charles Babbage[1] produced designs for a mechanical computer that were too far in advance of technology to produce a working machine. Babbage's proposed 'analytical engine' of c. 1840 had an arithmetic unit and a memory and used punched cards, not only for input and output, but also to store programs capable of performing iterative calculations with conditional branching. It was only when the Mark I computer at Manchester University began working in 1948 that Babbage's vision became reality. There is not time enough nor the space to go into the origins of the

---

[1] Charles Babbage (1792-1871). Lucasian Professor of Mathematics at Cambridge, 1828-39. M V Wilkes claims that Babbage invented microprogramming well over a hundred years before Wilkes himself reinvented it.

computer; those wishing to can try *The Origins of Digital Computers*, edited by B Randell, 3rd ed., Springer-Verlag (1982).

Hollerith used punched cards to tabulate the data from the US census of 1890 and a tabulator company was formed in 1896, which eventually became the International Business Machine (IBM) corporation in 1924. By about 1900 an adding facility was incorporated into the Hollerith tabulator to produce a punched-card-controlled calculator. Telephone relays were used in various calculating machines during the 1930s, but were eventually superseded by much faster thermionic valves (or tubes). However, the unreliability of vacuum tubes hampered any machine incorporating more than a few.

The beginning of the computer age can be dated to 1937 when a very remarkable paper[2] was published by a young English mathematician, A M Turing[3]. The paper presented a solution to a difficult problem in pure mathematics, though as fate decreed it had been solved by several others independently at about the same time. But what stayed forever in the minds of those that read it, was the technique by which Turing reached his conclusions — the conception of a universal computer, known as the Turing machine. Turing stated: 'It is possible to invent a single machine which can be used to compute any computable sequence.' Moreover any other machine could be emulated by his universal machine. It appears that Turing saw no essential difference between the calculating processes of a human computer and those of his hypothetical machine and that it would soon be possible to build machines that were, by all objective criteria, *intelligent*. To Turing the manipulation of numbers and the manipulation of non-numerical data were similar: a computing machine would play chess and write music as well as perform calculations accurately at lightning speed. This is the crucial distinction between a computer and a mere calculating machine.

The first general-purpose computer that used thermionic vacuum tubes in large numbers (about 19,000) to perform arithmetic and logical operations was ENIAC, built by Eckert and Mauchly at the University of Pennsylvania between 1943 and 1946. The number of tubes was such that under normal conditions a failure would have occurred long before any useful computation could have been completed. By careful selection of the tubes and by using them at minimal current the mean time between failures was increased enormously and computation became practical.

The first stored-program-controlled computer was designed and built at the University of Manchester by F C Williams and T Kilburn. It had a random-access store using cathode-ray tubes and ran its first program on 21st June 1948. The design was manufactured commercially by Ferranti as the Mark I computer. The first of these was delivered to a customer site in February 1952, beating UNIVAC in the USA by about five months. Turing, who had moved to the University of Manchester in September 1948, was one of

---

[2] 'On computable numbers, with an application to the Entscheidungsproblem', *Proc. Lond. Math. Soc.* **42**(2), 230-265 (1936-7). The paper was received in May, 1936. Turing was 23 when he wrote it.

[3] Alan Mathison Turing (1912-1954). BA in Mathematics 1934. Fellow of King's College, Cambridge 1935. Ph D from Princeton 1938. From 1939-1945 played an important part in breaking the German Naval (Enigma) codes using a series of Boolean electronic computers ('Colossi'). He worked on the ACE project while at the National Physical Laboratory 1946-1948. His ACE report was far ahead of its time. Turing became reader at Manchester University in 1949. He died in controversial circumstances. There is an excellent biography by Hodges, *Alan Turing: The Enigma*, Burnett Books (1983).

the first users of the computer and also wrote the first programming manual. In parallel with the work at Manchester, a group at Cambridge under M V Wilkes[4], designed the EDSAC I computer, using mercury delay lines as the storage medium. The system was completed in 1949 and was reliable enough to be used by non-specialists, who could be taught programming techniques. The first commercial manifestation of EDSAC was the Leo ('Lyons Electronic Office') computer system of December 1951. EDSAC I was retired in July, 1958 (and unfortunately destroyed), though by then EDSAC II had been in service some time. EDSAC II ceased working in 1965. From the mid 1950s the USA assumed a position of overwhelming dominance in computers, particularly in the business and commercial field, where IBM had a near monopoly in spite of anti-trust legislation and protracted litigation.

Computer development was greatly assisted by the invention of the transistor and later the integrated circuit, leading to decreased size and increased reliability and speed. The advent of the computer-on-a-chip has caused a major shift away from multi-user large computers towards small desk-top machines (with as much power as a mainframe of the 1960s) and embedded microcontrollers to give a degree of intelligence to many commonplace artefacts.

12.1.1 Computer architecture

The way in which the component parts of a computer (the hardware) interact with each other is called computer *architecture*. Computers require input and output ports, a *central processor unit* (CPU) and a memory, as figure 12.1 shows. The CPU carries out all the arithmetic, logic and control operations and is the fastest and most complex part of the hardware.

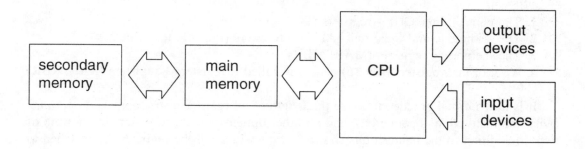

Figure 12.1 A block diagram of a computer system

The memory in figure 12.1 is shown in two parts, but often the secondary memory is itself split into several levels. The main or primary memory gives read and write access to random addresses. It is the fastest level, but also the most expensive and most limited

---

[4] Wilkes took a degree in mathematics at Cambridge the same year that Turing did. His very readable autobiography, *Memoirs of a Computer Pioneer* (MIT Press), came out in 1985.

in capacity. The secondary memory is generally divided into blocks of several thousand *bytes* (nearly always a group of 8 bits) with transfer of blocks between primary and secondary memories being managed by the CPU. The first level of secondary memory is normally magnetic discs and the second level magnetic tape. Secondary memory is slower, cheaper and of much higher capacity than primary memory. One of the most important tasks of a computer designer is to arrange for the fast memory to be used most effectively as the program is executed.

Input and output devices, or peripherals, are attached to the input and output ports. In some cases, such as a disc drive, the input and output device is the same, whereas others can only be used for input (such as a keyboard) or output (a printer, for example). Magnetic tape and disc can serve as memory (though of very slow access time) as well as input and output devices. The CPU must monitor the input port continually for *interrupt* or *wait* commands and from time to time send data to the output port as instructed by the program or the input port.

The memory stores two types of information: data and instructions. The computer carries out a sequence of instructions that forms a program. The data are manipulated in some way by the program to produce the output data required. The program may be to carry out a complex mathematical operation or may be simply for sorting a list into alphabetical order. The CPU must be able to select locations, or *addresses*, in memory and write to them or read and decode the information in them.

12.1.2 The CPU

The architecture of a typical CPU might look like that of figure 12.2. The CPU comprises four main parts each with associated *registers*, which are special-purpose stores containing one or two bytes:

1. The control unit (CU) + registers (PC, IR)
2. The arithmetic and logic unit (ALU) + registers (Fn, In1, In2, Out, SR)
3. The memory + registers (MAR, MBR)
4. A set of fast registers (say 16) generally called R registers + registers (RAR, RBR)

It is now normal practice to make the length of all register instructions and *operands* (what the instructions operate on), and all other dimensions of the system, as powers of two. In addition to the connections shown in figure 12.2, all the registers are attached to a highway, and data transfers between registers can be initiated by the CU.

The program counter (PC) contains the address of the next instruction to be executed. When an instruction is fetched into the instruction register, the PC is incremented by one so that it is always pointing to the next instruction. However, it is possible to take instructions out of order by issuing a command to *jump*, that is to move the pointer unconditionally up or down the memory by a certain number of places.

The instruction register is connected to the control unit, which decodes the instruction. Decoding derives the information and addresses needed to generate the sequence of register-to-register transfers required for executing the instruction.

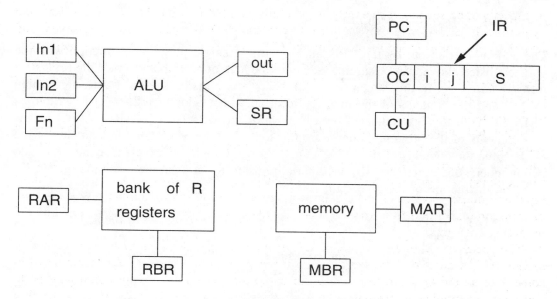

Figure 12.2 The structure of the central processor

The R registers are loaded from or stored to the memory by instructions of the type F,i,j,S. F is the operation code, often one byte of 8 bits, i and j (4 bits, a *nibble*, each) identify the registers involved and S is a 16-bit integer, giving a 32-bit instruction. In one mode of operation the j-field is not used and S is sent to the memory-address register (MAR) to select the appropriate memory location. In the second mode of operation the contents of register R_j are added to S and the result is sent to the MAR. This process is called address modification. The memory-buffer register (MBR), sometimes known as the memory-data register, temporarily holds data fetched from memory or data which is written to memory.

The CPU has a set of short, 16-bit instructions to perform arithmetic and logical operations using the ALU and two of the R registers, such that $R_i = F(R_i,R_j)$. Instructions to perform these operations have the format F,i,j. The operations are carried out in the ALU, which has a function register (Fn), two input registers (In1 and In2), one output register (Out) and a status register (SR). The SR has *flags*, which are bits that change according to whether the result of an operation is positive, negative or caused an *overflow*. Overflow means that during an arithmetic operation a number was produced that was too large to be stored in a register. The jump command referred to above can be arranged to jump conditionally, depending on the status of individual flags in the SR.

Most modern computers also have a floating-point arithmetic unit which interprets the contents of the R registers as being in two parts, a and b, representing the number $a.2^b$. The longer part, x, represents a binary fraction and the remaining digits, y, a binary integer. The range of numbers that can expressed is thereby greatly expanded. In many computers the R registers are of 32 bits and are used for integers; in this case a set of separate 64-bit X registers is used for floating-point numbers.

A typical instruction sequence

The actions carried out by the CPU, in executing a single program instruction with the registers listed above, can be summarised with the help of a symbolic code called *register-transfer language*. In this language, the contents of a register, say the MAR, are written [MAR] and the contents of the memory location whose address is [MAR] is written as [[MAR]]. Transfers are indicated by a backwards arrow, so that [MAR] ← [PC] means 'the contents of the program counter are transferred to the MAR' or 'the contents of the PC become the contents of the MAR'. The following might be a typical sequence to add two numbers, one in register, R_2, the other in register, R_3, and put the result in register, R_2, using an instruction of the type *F,i,j*:

(1) [MAR] ← [PC]		*the contents of the PC are moved to the MAR*
(2) [PC] ← [PC] + 1		*the program counter is incremented by one*
(3) [MBR] ← [[MAR]]		*the contents of the address in memory specified*
		by the PC are moved to the MBR

The time from loading the MAR in step 1 to the loading of the MBR in step 3 is very long compared to one cycle of the CPU clock, accordingly it is necessary for step one also to make the MBR busy, for step 3 to conclude by making the MBR free and for step 4 to wait for the MBR to be free. All that has happened until now is that a single instruction has been read to the IR. After step 3 the contents of the IR will be 'add,2,3' and the following sequence is required to complete the process:

(4) [RAR] ← [IR(R_i)]	*the first address of IR is transferred to RAR[5]*
(5) [RBR] ← [[RAR]]	*the contents of R_2 are transferred to RBR[6]*
(6) [In1] ← [RBR]	*the contents of RBR are transferred to ALU In1*
(7) [RAR] ← [IR(R_j)]	*the second address of IR is transferred to RAR*
(8) [RBR] ← [[RAR]]	*the contents of R_3 are transferred to RBR*
(9) [In2] ← [RBR]	*the contents of RBR are transferred to ALU In2*
(10) [Fn] ← [IR(OC)]	*the operation code, add, is transferred to ALU Fn*
(11) [RBR] ← [Out]	*when ALU has placed the result in the Out register, it is*
	then transferred to RBR
(12) [RAR] ← [IR(R_i)]	*the first address of IR is transferred to RAR*
(13) [[RAR]] ← [RBR]	*the contents of RBR are transferred to R_2*

Each of the thirteen steps above takes at least one cycle of the CPU clock. The operation codes are given mnemonics such as 'INC-PC', which means INCrement the contents of the PC by one, but the *machine code* is a binary word. If an *assembly-language* program of operation codes is written in mnemonics, it must be translated into machine code by an *assembler*. The complete set of operation codes is known as the *instruction set*

---

[5] RAR is the register-address register.

[6] RBR is the register buffer.

of the computer. For an operation code that is an eight-bit word there are $2^8 = 256$ possible codes, a number normally sufficient for microprocessors and microcomputers.

Machine code is said to be the *lowest-level* language, assembly language is a *low-level* language and Fortran, C, Basic, Cobol, etc. are said to be *high-level* languages. A high-level language may be translated into assembly language at first by a *compiler*, and the assembly language must then be translated into machine code. Some high-level languages such as Basic, Lisp and Prolog are not usually compiled but are *interpreted* step by step by an interpreter which generates the machine code for each step. High-level languages are intended to be easier to use than lower-level languages, but there are wide variations in their level of difficulty. We shall return later to programming.

12.1.3 Memory

In the UK the word 'storage' was used at first, but the American term 'memory' has replaced it in popular usage. Almost anything that can be changed from one stable state to other stable states can be used to store data, and there is no need to store numbers in binary if more than two states are available. However, it was soon realised that two-state stores were more readily constructed and easier to use, so they became standard. At first a motley assortment of stores was pressed into service, but soon magnetic storage of one kind or another was used by all commercial computers[7]. Magnetic stores were of three kinds: ferrite toroids (known as ferrite cores) were fastest, rapidly spinning magnetic drums were of intermediate speed (though orders of magnitude slower than ferrite cores), while magnetic tape was the slowest and mainly used for back-up and file storage. Ferrite cores started off as toroids about the size of necklace beads and the read and write wires were threaded through by hand just like stringing a necklace. As a consequence they were expensive and ferrite-core memory capacity was very small, even on the most advanced computers. Programmers had to know where their variables were stored so that those frequently used were placed in the ferrite-core memory. It could happen that a small program change would cause one or more variables to be moved from the fast core store to the slow drum store and the program's execution time would increase enormously.

Ferrite cores lasted a long time (about 25 years) because they developed steadily to beat off competition: their size fell from 10 mm down to 1 mm in diameter and they were eventually stamped out of ferrite-impregnated plastic sheets; machines did the wire threading and Mbit core stores could be made. However, their demise was inevitable; everyone in the industry knew that semiconductor memories were ordained to take over the task of fast-access storage; they had done so by about 1980.

Memory is organised into cells which are of a size in bits determined by the computer designer. Almost all modern machines use cells of 8 bits, called bytes, a term introduced by IBM in the 1960s. Each cell has associated with it a number, which is its unique address. As we have seen before, the address is quite distinct from its contents. While the contents of a cell may change many times during the execution of a program, its address is unchanged from program to program. *Read-only memories* (ROMs) have invariant contents as well as invariant addresses.

---

[7] In his autobiography, Wilkes says ferrite cores were mainly responsible for the computer's commercial success.

Memory hierarchy

The memory can be divided into two types according to the method of access:

(1) Random-access memory (RAM)
(2) Serial-access memory (SAM)

In random-access memory the time taken to locate any address is constant, no matter if it is the next cell to the one just accessed or one furthest away in terms of address. RAM is the fastest (shortest access time) memory; semiconductor memories are usually RAM. In serial-access memory the next address to be accessed is always next to the one just accessed. To access a cell M addresses from the one being accessed takes M times as long as accessing the next cell. Magnetic tapes are an example. In direct-access memory the accessing is a combination of random and serial and the access time for a particular address is not predictable (though the average access time might be). Magnetic discs and drums are examples. Drums and discs are divided up into circular tracks produced by *formatting*. Within the tracks, access is serial, but if each track has a read/write head associated with it, then the track access is random.

Memory is organised into a hierarchy, as in figure 12.3, because fast-access memory is much more expensive than slow. Registers are fewest and work fastest. Next in speed are *cache* memories[8], which are small blocks of very fast memory used to store frequently-accessed instructions and variables.

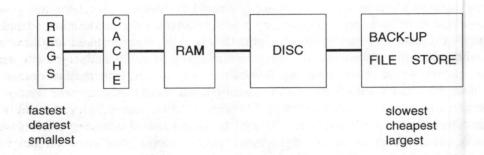

Figure 12.3 Memory hierarchy

Cache memories contain a small, continuously varying subset of the total addresses, together with the contents of those addresses; they are fast enough to be accessed in one or two CPU clock cycles. They are essential if the maximum computational speed of the CPU is to be realised. Management of cache memory is generally by hardware, as management by the operating system would be too slow. When an address leaves the CPU it goes first to the cache; if the address in the cache then the data access is very fast. If a match is not found in cache (a 'cache miss'), the address goes to the RAM and the

----

[8] The use of cache memories was first proposed by M V Wilkes in 1965.

contents of the required address and those of a few surrounding addresses, which form a block, are transferred to cache together with the block address. The reason for this is that neighbouring addresses are frequently required in succession. This process is slow, but subsequent accesses to addresses within the transferred block will of course be fast. When the cache is full, a block has to be removed before a new block can be accommodated, the choice being made by hardware, based on the activity of all the blocks in the cache. Cache misses cause a significant downgrading from peak performance attained when all data accesses are within the cache. If only 1% of accesses are not in the cache, performance is degraded by about 10%.

Most up-to-date machines operate with a virtual-address space which is larger than the real address space required to cover the whole of RAM and disc. Both of these memories are divided into pages of a fixed size of a few thousand bytes. The virtual address generated by the CPU is first presented to the cache as described above then, if not found in the cache, it goes to the RAM. If the real page containing the required virtual page is in RAM it is accessed, the matching of virtual and real pages being achieved by fast hardware. If the address is not in the RAM, the operating system finds the location on the disc of the real page containing the required virtual address and moves this page into an empty page of the RAM. The operating system maintains at least one empty page in the RAM by moving pages back to the disc when they are no longer actively used. The integration of RAM and disc into a 'one-level store' was first implemented at the University of Manchester in 1967[9].

Semiconductor memories

Semiconductor RAM and ROM are made from MOS devices, especially low-power CMOS. RAM may be *static* or *dynamic*: SRAM or DRAM. The information stored in dynamic RAM gradually leaks away and so the RAM must be refreshed every now and again, adding to the complexity of the controlling software. DRAMs typically have 'refresh' times of from 4 ms to 32 ms. SRAM does not lose its information so long as power is supplied, but a SRAM bit requires four components instead of the one needed by DRAM, increasing cost and size. Both SRAM and DRAM are *volatile*, that is they lose their contents when power is lost. All memory ICs have increased in size very rapidly and 2Mbit DRAM chips are available for the same price as a 2kbit chip a few years ago. A 128Mbit DRAM IC came out in 1996 and no doubt we shall soon see 1Gbit chips[10].

Table 12.1 shows a selection of the many hundreds of memory ICs now available. The prices are for single quantities and are many times the price paid by manufactures of large quantities of equipment. As can be seen DRAM is now very cheap at about £2 a Mbit, and EPROM — one-time programmable (OTP) ROM — is not much more expensive, but SRAM and E^{2}PROM (electrically-erasable PROM) are still fairly dear at about £30 and £15 per Mbit respectively.

---

[9] 'One-level storage system' by T Kilburn, D B G Edwards, M J Lanigan and F H Sumner, *IRE Trans. on Electronic Computers*, **EC-11**, No.2, 230-65 (1962). The first proposal for a virtual memory.

[10] Since the earliest DRAMs came out they have consistently quadrupled in size every three years, so a 1Gbit DRAM IC can be forecast for 1999 or 2000.

The capacity of a '1 Mbit' memory is actually $1\,048\,576$ (= 2^{20}) bits, arranged as $262\,144 \times 4$-bit, $131\,072 \times 8$-bit or $65\,536 \times 16$-bit words. Addressing as words rather than bits has the effect of speeding up the reading time by a factor comparable to the word length. Access time for the CX8MX32-60SMT memory IC is 60 ns, which is slow compared to fast CPU clocks. This chip requires 32 mA of standby current at V_{DD} = 5 V, so that a memory-back-up capacitor of 5 F will power it in emergencies for only a few minutes. Many CMOS ICs now require only 1 mA or less standby current and these will keep going for several hours with memory-back-up capacitors. A good proportion of SRAM chips are now non-volatile.

Table 12.1 *Semiconductor memory ICs*

Device No.	Type	Capacity in words × bits	Maker	Access time	Standby current†	Price (1-off)	Price /Mbit
TMS4256-10NL	DRAM	262144×1	TI	100 ns	4.5 mA	£5	£20
µPD42460LE-80	DRAM	262144×16	NEC	80 ns	2 mA	£24	£6
µPD424400	DRAM	1048576×4	NEC	60 ns	2 mA	£20	£5
TC5118160A-70	DRAM	1048576×16	Toshiba	70 ns	16 mA	£30	£1.88
CX1MX32-70	DRAM*	1048576×32	Centon	70 ns	16 mA	£50	£1.56
CX4MX36-70	DRAM*	4194304×36	Centon	70 ns	24 mA	£300	£2.08
NM24C03	E²PROM	256×8	NSC	–	60 µA	£1.70	£850
X24C04P	E²PROM	512×8	Xicor	–	50 µA	£1.50	£375
X24C08P	E²PROM	1024×8	Xicor	–	150 µA	£3	£375
AT29LV256-20JC	E²PROM	32768×8	Atmel	200 ns	1 mA	£14	£54
AT29C010-90JC	E²PROM	131072×8	Atmel	90 ns	0.1 mA	£30	£30
AT29LV040-20TC	E²PROM	524288×8	Atmel	200 ns	1 mA	£60	£15
M2732A-4F1	EPROM	4096×8	SGS-T	450 ns	35 mA	£10	£312
CY27C128-45PC	EPROM	16384×8	Cypress	45 ns	15 mA	£12	£93
AM27C256-90	EPROM	32768×8	AMD	90 ns	1 mA	£3.80	£15
AM27C020-120	EPROM	262144×8	AMD	120 ns	1 mA	£7	£3.50
AM27C040-120	EPROM	524288×8	AMD	120 ns	1 mA	£13	£3.25
CY7C128A-20PC	SRAM	2048×8	Cypress	20 ns	20 mA	£6.80	£415
STK12C68-S45	SRAM	8192×8	Simtek	45 ns	2 mA	£20	£305
CXK58257AP-70LL	SRAM	32768×8	Sony	70 ns	5 µA	£7	£28
GR51281-7	SRAM	524288×8	Greenwich	70 ns	1 mA	£160	£40

Notes: Most ICs require a 5V or 5.5V supply and will operate at voltages down to 2.7 V
† Mostly maximum values. The supply current when active will be 10-1000 times greater. * Modules made of 4 or more ICs.

Non-volatile ROM is available in various forms of PROM — programmable ROM. The one-time programmable PROM is now hardly used as EPROMs (erasable PROM) and E²PROMs become cheaper. EPROMs have a window on the back which can be exposed to ultra-violet light to erase the data stored. E²PROMS can be erased electrically and in this way are like SRAMs, but can be write-protected. Large computers now use much faster chips (1-2 ns access time) than those shown in table 12.1, but at greater expense and with larger standby currents. The power consumption of these is quite high, necessitating special cooling methods such as Peltier-effect heat pumps.

Figure 12.4 shows the block diagram for a 4-Mbit SRAM arranged as a memory matrix of 1024 rows by 512 columns of 8-bit words. This chip can hold roughly 100,000 words of English text. The control logic truth table is also shown in figure 12.4, in which CS = chip select, OE = output enable, WE = write enable.

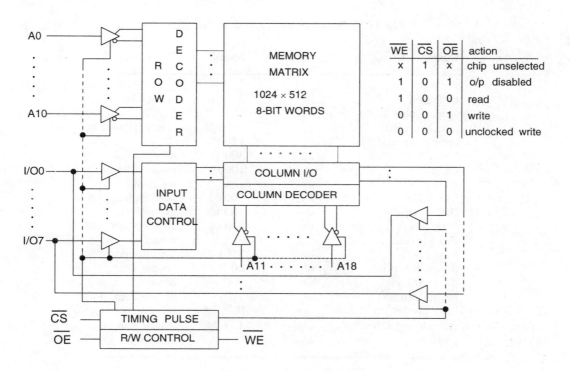

WE	CS	OE	action
x	1	x	chip unselected
1	0	1	o/p disabled
1	0	0	read
0	0	1	write
0	0	0	unclocked write

Figure 12.4 Block diagram of a 4Mbit SRAM chip arranged as 1024 × 512 × 8-bit words

The memory chips in a computer are connected to two *buses*: the data bus and the address bus, which are simply lines (usually several in each bus) to and from which messages and data are sent. The memory in figure 12.4 can only be addressed when the chip-select control is low so that the CPU can select one chip at a time for connection to the data or address buses. The connections from chip to bus are via three-state gates, explained in section 9.9.5. In so-called *von Neumann* architecture the data and instructions are stored in one block of RAM, so that a particular address may contain data or instructions. In *Harvard* architecture the RAM is divided into data RAM and instruction RAM so that a single address can refer to both data and instructions. The CPU has to decide from the context which is required when an address is specified.

12.1.4 Input and output devices

One reason why computers have so many uses is the steady increase in types of input and output device. Though the devices differ greatly in their methods of operation, the techniques by which they are connected (or *interfaced*) to computers are more-or-less standard. Table 12.2 lists some of the more common input and output devices.

Table 12.2 *Some input and output devices*

Device	Purpose[A]	Speed[B]	Comments
keyboard	i/p	5	echoed to CRT
'paper' tape	i/p	500	fragile
punched card	i/p	1000	easily damaged
optical bar code	i/p	1000	
MCR[C]	i/p	300	robust
OCR[D]	i/p	1000	easily misread
mouse	i/p	slow	for menus
CD ROM	i/p	500+ k	fast, high capacity
'large' tape[E]	i/o	100 k	expensive
'small' tape[F]	i/o	1-10 k	cheap when slow
flexible disc	i/o	35+ k	cheap
hard disc	i/o	350+ k	fast, not dear
dot-matrix printer	o/p	60	obsolescent
drum printer	o/p	500	obsolescent, noisy
ink-jet printer	o/p	120	obsolescent, quiet
laser printer	o/p	0.25-25 k	price $\propto$ speed
cathode-ray tube	o/p	1++ M[G]	fastest, not dear
plotter	o/p	-	slow

Notes:　[A] i/p = input, i/o = input and output, o/p = output
　　　　[B] very approximately, in characters per sec (cps), 100 k = 100,000 cps
　　　　[C] magnetic-character recognition
　　　　[D] optical-character recognition
　　　　[E] with tensioning arms or vacuum chambers
　　　　[F] without tensioning arms or vacuum chambers
　　　　[G] the limitation is the rate at which characters are sent to the CRT

The problem with peripheral devices is that they are not synchronised with the computer, and the computer may have no idea of how long a peripheral will take to perform a task. For this reason they are linked to the computer via interface units, as figure 12.5 shows. The organisation of the CPU, memory, interface units and buses can be arranged differently to that shown. But while the interface units may all differ when viewed from the peripheral device, they are all alike when viewed from the computer. Looking from the peripheral the interfaces must all be dissimilar, since the requirements of each type of device are quite distinct: the control commands for a laser printer are totally unlike those of a cathode-ray tube. Not all the interface registers in figure 12.5 may be needed for a particular peripheral.

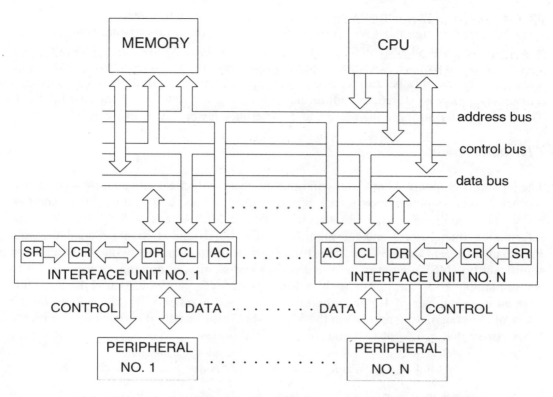

Figure 12.5   Computer with a common bus system for memory and peripheral interface units. AC = address comparator, CL = control logic unit, CR = control register, DR = data register, SR = status register

Interrupts

We have mentioned that the basic problem of input and output is that it is device-dependent and the time taken is beyond the computer's control. In programmed input/output all transfers of data to and from the CPU to peripherals are under the control of the program. The programmer then has to set the control flags (for example if the printer is busy) at the interface unit and test them. The CPU has to be programmed to look at the status flags of all the peripherals to find out which is ready (*software polling*). Since the computer must be managed in such a way that it can perform other tasks while waiting for input and output, the programmer's task is complicated. Ideally the details of input and output should not be the concern of the programmer and interrupts or an interrupt structure can do the job. An interrupt signals to the CPU that it must carry out the instruction currently being acted on and then stop working on that program and store the contents of all the registers in memory. Each interrupt has a unique set of instructions stored in its own area of memory, and servicing is instituted by hardware to maximise speed of execution. The program counter is set to the start of the interrupt routine and the CPU then carries out the servicing of the interrupt. When it has finished, the registers are restored from memory and the CPU continues with the program that was interrupted.

The system design must set priorities for interrupts, the priority being determined by how long the cause of interruption can wait before being serviced. The shorter the wait, known as the *crisis time*, the higher the interrupt's priority. If it is necessary to interrupt a low-priority interrupt because one of a higher priority has occurred, then the interrupt's register state must be stored, as well as the original register state before the first interrupt. There are usually no more than three, and sometimes fewer, levels of interrupt.

Character codes

The input and output of a computer comprise numbers and letters (alphanumerics); special characters such as punctuation marks, brackets and symbols; and control codes. The two codes most commonly used are the seven-bit ASCII (American Standard Code for Information Interchange) and the eight-bit EBCDIC (Extended Binary Coded Decimal Interchange Code). The ASCII code is usually made up to eight bits by appending a most significant bit (MSB); either a zero in all cases and signifying nothing, or a *parity* bit to act as a simple error check. The parity bit may be an even parity bit (that is the eight bits have an even number of 1s) or an odd parity bit (odd number of 1s), depending on the whim of the system designer. When the 8-bit code is sent serially it is transmitted with the MSB last so that b_0 is sent first and the parity bit, b_7, last.

Table 12.3 *The ASCII code*

$b_6 b_5 b_4$		0	16	32	48	64	80	96	112
$b_3 b_2 b_1 b_0$		000	001	010	011	100	101	110	111
0	0000	NUL	DLE	SP	0	@	P	`	p
1	0001	SOH	DC1	!	1	A	Q	a	q
2	0010	STX	DC2	"	2	B	R	b	r
3	0011	ETX	DC3	#	3	C	S	c	s
4	0100	EOT	DC4	$	4	D	T	d	t
5	0101	ENQ	NAK	%	5	E	U	e	u
6	0110	ACK	SYN	&	6	F	V	f	v
7	0111	BEL	ETB	'	7	G	W	g	w
8	1000	BS	CAN	(	8	H	X	h	x
9	1001	HT	EM	)	9	I	Y	i	y
10	1010	LT	SUB	*	:	J	Z	j	z
11	1011	VT	ESC	+	;	K	[	k	{
12	1100	FF	FS	,	<	L	\	l	¦
13	1101	CR	GS	-	=	M	]	m	}
14	1110	SO	RS	.	>	N	^	n	~
15	1111	SI	US	/	?	O	_	o	DEL

Table 12.3 shows the ASCII code (the bits are b_{0-6}). Thus the character '6' is encoded as $\times$0110110 in 8-bit binary ($\times$ is a zero, or parity bit) and is character 54 in decimal. For example in the word-processing program WordPerfect, the number 6 can be typed in by

pressing the 6 on the keyboard or entered as character 54 from character set 0 (the ASCII set) by typing <Ctrl>v0,54<Enter>. The codes numbered 0-32 are control codes (including SP, the 'space' character) and the rest are called printing characters.

Interfacing methods

The computer is connected to each interface unit by a port and signals are transmitted along wires connecting the port to the interface unit. The transmission can be two-way with two separate channels (known as *full duplex* or FDX), two-way using one channel (*half-duplex* or HDX) and one way (*simplex*). The transmission can be synchronous or asynchronous. In asynchronous serial transmission each character is sent with its own start and stop codes, while in synchronous serial transmission a stream of characters is sent, which is broken up into blocks by synchronisation code at the start and an end-of-message character at the end of the block. A low-priced asynchronous interface IC called the UART (Universal Asynchronous Receiver and Transmitter) can be obtained very cheaply (from £2 to £5). The receiver section converts input serial start, data, parity and stop bits to parallel data out and verifies correct transmission. The transmitter section converts parallel data to serial form and adds on the start, parity and stop bits automatically. The word length is selectable from 5 to 8 bits.

There are two commonly-used serial interface standards, the RS-232C (USA) and the V24 (CCITT, Europe), for connecting computers to peripherals and to *modems* (modulator-demodulator) used for telephonic data transmission. Both use 25-way connectors, but whereas the RS-232C standard defines connector-pin usage, the V24 standard refers to circuits. Although the cables have 25 wires in them, the data are transmitted and received serially on just two of them. The others are used for control and synchronisation signals.

Data loggers and other instruments are often interfaced to computers with the IEEE standard 488 bus, which uses a 24-way connector and one bus can support up to fifteen instruments. The cable length must be no more than 2 m per instrument, with a maximum length of 20 m.

Keyboards

Pressing a key on the standard QWERTY keyboard produces a binary code which is transmitted to the computer, which sends it to the CRT screen as a 'printed' character (known as *echoing*) and then interprets it within the context of the operating program. Control codes are sent by pressing the <Ctrl> key plus another or <Ctrl> + <Shift> + key. The keyboard/CRT combination is a VDU (Visual Display Unit). Commonly keys operate switches of the contacting kind with a spring return, and these gradually wear out. The most popular non-contacting switch is the Hall-effect switch in which the depression of a key causes a small permanent magnet to be brought near to a piece of semiconductor. The semiconductor carries a current in a direction normal to the magnet's motion and a voltage is produced at right angles to this to indicate depression of the key. The absence of contacts improves switch life, and there is less possibility of contact bounce.

The switches are wired to a matrix (see figure 12.6) which must be scanned to detect which switch has been closed. Suppose we use a 12 × 12 matrix and the switch on row

3, column 5 has been closed, thus connecting the row with the column. The rows can be scanned in turn by putting 0 V onto the row in time with a clock. When row 3 is scanned, column 5 will go to 0 V and be detected by the column decoder. Thus the ROM array will output the binary code at row 3, column 5. The scanning must be sufficiently rapid that fast keying is detected. If a maximum of ten keys can be depressed in a second, then an adequate scanning time would be 20 ms, which with 144 keys would imply a clock frequency of 144/0.02 = 7.2 kHz. In practice much faster clocks would be used. The keyboard's output is serial and it is normally connected to the computer by a five-pin DIN (*Deutsche Industrie Normal* – German industrial standard) connector.

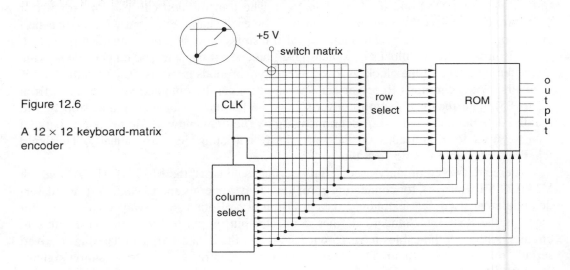

Figure 12.6

A 12 × 12 keyboard-matrix encoder

Printers

Printers have evolved very quickly in the last few years. The 9-pin dot-matrix printer was followed by the 24-pin printer, which in turn was superseded by the ink-jet, bubble-jet and laser printers. The dot-matrix print head comprises an array of needle-like wires with flat ends that strike a ribbon impregnated with wet ink. A character consists of an array of dots which can be printed or left blank, arranged in columns of nine dots in the case of the 9-pin printers. When a particular dot is to be printed the wire is pushed onto the ribbon by a solenoid. The matrix is shown in figure 12.7, where the letter 'f' is shown together with the 8-bit code for printing the dots in each column. The bit code for printing the character is readily derived from the matrix by writing a 0 for no dot and a 1 for a dot, starting with the LSB at the bottom and the MSB at the top. There are eleven columns for each letter, the first and last being blank normally, so that the letters do not run into each other. Preceding the dot matrix bytes is an 8-bit code giving information about the character so that it prints proportionally, or is double struck (for bold type) or has a 'descender' if necessary (for example, the letter 'g' requires dots to be printed in the descender row). The descender row uses the ninth pin. Individual characters are readily designed using the matrix and can be downloaded from a character file when needed.

The nine-pin dot-matrix printer was always much inferior to a typewriter in quality of print and gave a rather spotty appearance to printed characters. This could to some extent be overcome by printing half-spaced matrix columns and calling the result 'near-letter quality' (NLQ), though printing was thereby slowed. The 24-pin printer gave much better quality of lettering and could also be programmed easily, but was rapidly overtaken by ink-jet and laser printers offering very high print quality. Ink-jet printers work by projecting a spherical blob of ink towards a specified point on the paper. If the blob is small enough and of uniform size, the print quality will be good if the right type of paper is used. Unlike dot-matrix printers, ink-jet printers do not use ribbons to hold the ink but an ink reservoir, so the density of the printing remains constant. Unless ribbons are changed often, or the pressure of the needles is increased as the ink in the ribbon dries out or is used up, the printing will gradually fade, like that of a typewriter.

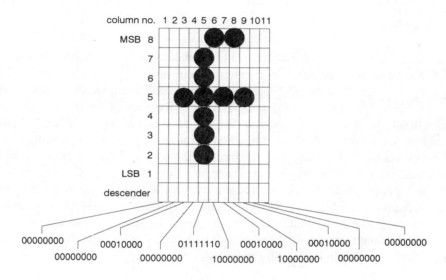

Figure 12.7 The dot matrix for 9-pin printers, showing the dots, and their 8-bit codes, for the letter 'f'

Laser printers contain a cylindrical drum coated with a photoconductive material. The photoconductive surface will hold an electric charge until exposed to light, so that as a laser diode is aimed at selected parts of the drum, the charge leaks away. The drum is rotated past a developer containing the ink powder, which is given the same polarity charge as the drum. The charged ink is repelled by the similarly charged areas of the drum and attracted to the uncharged areas that the laser has defined. The drum then passes over the paper which has an opposite charge to the ink, so that the ink powder is stuck to the paper. The paper then passes through a fuser which applies heat and pressure to the plastic ink and sticks it to the paper. The printing density therefore is reasonably constant with time. The printing resolution of a laser printer depends on the size of the light spot coming from the laser, but is normally about 40 μm, compared to 250 μm for 9-pin, and 90 μm for 24-pin, dot-matrix printers.

Dot-matrix colour printers use a four-colour ribbon and a program to 'mix' the colour specified by the computer. The printer can only print dots of the colours on the ribbon and these must be placed on the page individually, so that the eye averages out the coloured dots to give the illusion of a uniform shade. The problem is that there are not enough dots per unit area to prevent sudden jumps in colour shades that lead to visible bands on the printed page. Ink-jet printers 'mix' up to six colours to achieve the required shade (they still print separate dots of each colour like dot-matrix printers) and because the ink-blob is small, there is less of a jump from one shade to the next and better uniformity is achieved. Because there are so many dots to print, the printing times are very slow compared to black-and-white graphics (by a factor of 5-10). However, the colour printer market is in a state of rapid evolution, and there is little doubt that high-quality, low-cost, colour printers will soon be available, using other technologies than dot-matrix and ink-jet printing.

Magnetic discs

Magnetic discs have become one of the standard means of transferring information, data and programs from one computer to another. There are three common sizes of flexible ('floppy') disc: 8 inch (203 mm), $5\frac{1}{4}$ inch (133 mm) and $3\frac{1}{2}$ inch (89 mm) diameter. Rigid discs (or hard discs, or Winchesters) started off at a diameter of 14 inches (356 mm), but now quite small discs are being used as the bit density has gone up and diameters of 89 mm or less (down to 33 mm) have become common. Currently bit densities are in the order of 100 kbits/mm². The magnetic medium of the disc is a thin layer of ferrite powder which is stuck to plastic for floppies or aluminium for hard discs. The magnetic surface layer can be magnetised in a certain direction by a write head to form a 1 bit and in the opposite direction for a 0 bit. These magnetised bits can then be read by a either a special read head or the same head that was used to write. In flexible disc drives the disc is actually pushed into physical contact with the head via a slot cut into the plastic envelope, and so the discs (and heads) gradually wear out.

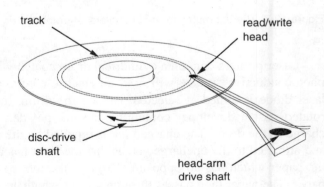

Figure 12.8

A magnetic disc and its read/write arm. The head size is exaggerated

Figure 12.8 shows the head positioned over the disc like the needle arm of an old record player. The arm can be driven by a stepper motor to give accurate positioning. Since the tracks are 0.3 mm wide or less on a flexible disc and perhaps 10 µm wide on a hard disc, positioning is very critical. The head and arm assembly is made as light and as rigid as possible for fast acceleration and motion. The coil in the head which detects the magnetic field of the bits during reading may be less than 100 µm across. Most discs are now double sided and require at least two heads, one for the top and one for the bottom. Hard discs usually have multiple heads for each side. On large machines a Winchester may contain five or more discs, each with a capacity of at least a Gbyte.

Figure 12.9 shows a $5\frac{1}{4}$-inch flexible disc in its sealed plastic envelope. The hole in the middle accommodates the drive shaft, which rotates the disc at about 300 r/min. A small hole next to the centre hole is used as an index mark, which is can be detected by a photocell and tells the disc controller where the start of the track is. The tracks are rings on the disc which are produced by the disc formatting software.

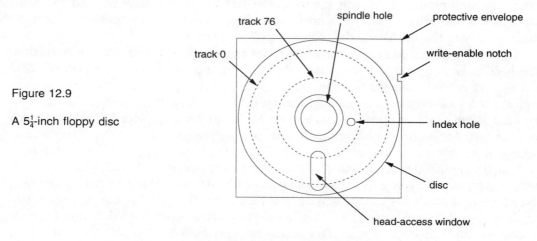

Figure 12.9

A $5\frac{1}{4}$-inch floppy disc

Figure 12.10 shows the original IBM 3740 disc format, which illustrates the principles. In this format one side of the disc is given 77 tracks, three of which store information about the disc or can be used as alternatives to bad sectors. The remaining 74 tracks are each formatted so that each starts with a 46-byte gap to give the read/write mechanism a breathing space, a 1-byte address and another 33-byte gap and then 26 sectors in which the data is written, followed by a 241-byte gap.

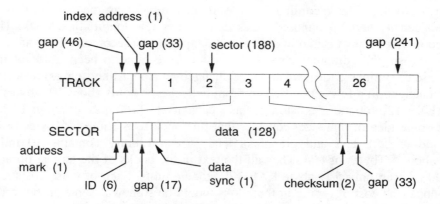

Figure 12.10 The original IBM 3740 floppy-disc format. The numbers in parentheses are the bytes used

Figure 12.10 shows also the arrangement within a sector, which starts with a 1-byte address, followed by a 6-byte track-and-sector identifier (ID), another gap of 17 bytes and a data synchronisation byte. The data is then written or read from the next 128 bytes, which terminate with a 2-byte checksum error detector and a final gap of 33 bytes. Thus in this format the data bytes total only $128 \times 26 = 3.33$ kbytes/track, or about 250 kbytes a side and 500 kbytes for a two-sided disc. Higher density formatting, with 512 bytes in a data cell instead of 128, is now the norm and gives 1.2 Mbytes for a two-sided $3\frac{1}{2}$-inch disc. These are kept in hard plastic containers, but are basically much the same as the older, less-well-protected, $5\frac{1}{4}$-inch floppies. The technology has now reached the point where 10 or 20 Mbytes can be stored on a floppy, but these would no longer be compatible with the 1.2-Mbyte disc technology.

Hard discs are sealed in containers filled with an inert gas to avoid dust contamination. The head does not need to contact the disc since the rigidity of the disc prevents disc wobble and possible damage by head contact. The gap between disc surface and head is critical for legible writing and reading, and must be maintained accurately, especially as the track widths on hard discs are smaller than on floppies and the speed of rotation is ten times as fast. Access times on discs are limited by the speed of rotation. On a flexible disc turning at 300 r/min, the average access time, when the head is in position over the right track, is $0.5 \times 60/300 = 0.1$ s, rather a long time. In addition if the head has to be moved there is a further delay of about 0.2 s. The data can be transferred at a relatively fast rate, since a single track holds about 7 kbytes in a 1.2-Mbyte disc and this can be read in one revolution, or 0.2 s, giving a data rate of 35 kbytes/s. Hard discs have data-transfer rates at least ten times as large. The average access time for hard discs is also much less than for flexible discs, 10-15 ms being typical.

Backing stores and archives

The cheapest storage medium is magnetic tape, where the cost per bit stored has always been low, but is now being dramatically reduced by recent developments in helically-scanned tapes. In these the recording is not along the length of the tape, but at an angle to it, almost vertically. The recording head is rotated rapidly as the tape moves and achieves a bit density about a hundred times that of in single longitudinal track. These tapes are used in video cameras and in digital-audio tape (DAT) recorders. They are much cheaper than the special computer tape decks which have hitherto been used for mass storage. The use of automatic retrieval and tape changeover (robot-line storage) means that tapes can be changed in a few seconds. The 8mm helical tape holds about 10 Gbytes and the Exabyte EXB-120 robotic mass storage has a capacity of 116 tapes, or about 1 Tbyte.

To have some idea of that size, consider that this book uses about 6 Mbytes of disc storage (including both text and graphics). The British Library contains 10 million volumes, or about 5 Tbytes of text. Thus all the text in the books of the largest library in Britain would take up only 5 Exabyte EXB-120 storage units. Each of these costs a few thousand pounds (the cost of typing in the books would be somewhat greater, but optical character recognition has the potential sharply to reduce these costs). Transmission over an optical-fibre network at 1 Gbit/s would enable remote stations to read a book in a few milliseconds, though access time would be ten seconds or so. The telephone network has too small a bandwidth for this task, the data rate being only about 1 kbyte/s.

CRT displays

The construction of the CRT is described elsewhere in this book and will not be considered here, rather we shall say a little about the display of information on the CRT monitor. Colour displays have a screen coated with red, green and blue phosphors arranged in groups of three (a colour dot) — one of each colour. Three separate electron beams can be used to excite these into emitting light, or one beam which is colour modulated. The colour dots are separated by about 0.3 mm, which thereby becomes the smallest feature that it is possible to show on the screen. In fact the screen resolution may be worse than this because of the way in which it is addressed and the associated display memory of the computer, but the graphics cards now used with desk-top computers can now give approximately this resolution. Take for example a screen which is 250 mm by 200 mm. If the colour-dot spacing is 0.3 mm, that means the screen contains a total of about 500 kdots. Let us further suppose that we modulate the beam to mix the colour required at each dot and that this requires 8 bits (256 colours). We thus require a total of 8×500 k = 4 Mbits, or 500 kbytes of video RAM to store the information necessary to drive the display. This amount of fast-access memory would have been prohibitively costly until recently.

The first colour displays were driven by a colour graphics adaptor (CGA), which had the relatively poor resolution of 320×200 pixels (a pixel is the smallest part of the screen that could be written to) and only four colours (that is 2-bit colour), but required only 16 kbytes of video RAM. Soon the enhanced graphics adaptor (EGA) came on the market offering 640×350 pixels and 16 (4 bits) colours, using 112 kbytes of RAM. The video graphics adaptor (VGA) of IBM had slightly better resolution with 640×480 pixels but the same 16 colours. The latest graphics adaptors are the super VGA (SVGA) with 800 $\times$ 600 pixels and the enhanced SVGA (ESVGA) with 1024×768. This is about the maximum number of pixels available with a small, cheap monitor.

Another problem with CRT displays is flicker. The eye retains an image for about 40 ms, the image persistence time, so the display must be refreshed at a faster rate than this to avoid perceived intensity changes. The screen is written to one horizontal line at a time, the whole screen requiring 600 lines to make up one frame, or complete screen in the case of the SVGA. In some cases the screen is split into one group of odd-numbered lines and another of even-numbered and these are refreshed alternately (known as interlacing). This has the effect of slowing down the frame-refreshing frequency ('frame rate') by a factor of two, so that a 70Hz vertical-scanning frequency results in a 35Hz frame rate. Though interlacing at 70 Hz is better than not interlacing at 35 Hz, it is not as good as a non-interlaced 70Hz frame rate. Flicker-free displays require a frame rate of at least 50 Hz.

12.1.5 Computer networks

Linking computers facilitates rapid transfer of data, but it also enables smaller, less-powerful computers to make use of the greater power of others, and provides access potentially to much greater storage. The link may be via the telephone system using a modem, in which case the distance between computers can be 20,000 km, or they may be linked by cables installed for the purpose. In most of the latter cases the link is no more than a few hundred metres long and the set of linked computers is called a local-area network (LAN), as opposed to the wide-area network (WAN) that uses public telephone lines or perhaps a microwave link. LANs soon gained widespread acceptance because they made better use of expensive resources — for example a laser colour printer can be connected to all the computers at a company site. LANs can of course be connected to WANs to give global interconnection possibilities.

Figure 12.11 shows a variety of network types. The circles (called nodes or stations) represent physical devices — printers, computers and so on, while the lines represent the transmission path, which is usually a cable in a LAN. The bus network of figure 12.11a has each station linked to a bus, which can be a single coaxial cable, down which the data must be serially transmitted. The Ethernet™ is an example of a bus network. Figure 12.11b shows a tree network, such as some parts of a telecommunications network might resemble, but which is not used in LANs. The star network of figure 12.11d is the configuration of a PABX telephone exchange (private automatic branch exchange). The ring network of figure 12.11c is used in the Cambridge ring LAN, which was one of the first put into use by M V Wilkes, to link computers in Cambridge. The network of figure 12.11e resembles some parts of the telephone system.

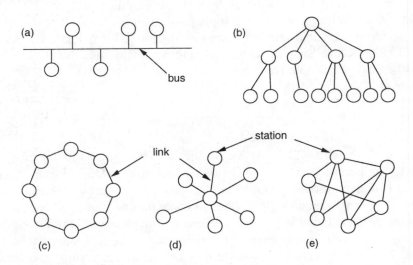

Figure 12.11 Network topologies: (a) bus (b) tree (c) ring (d) star (e) distributed

The critical points in a network depend on its topology. In a bus network, the bus is the critical point and when severed (for example, by unwittingly unplugging it) the network goes down. In the star network, the central node is vital, and in the tree network the nodes increase in importance the higher up the tree they are. Ring networks look most vulnerable, since all the stations receive and transmit all the data, and if any one breaks down, the network goes down too. This problem can be overcome by using bypass relays as figure 12.12 shows.

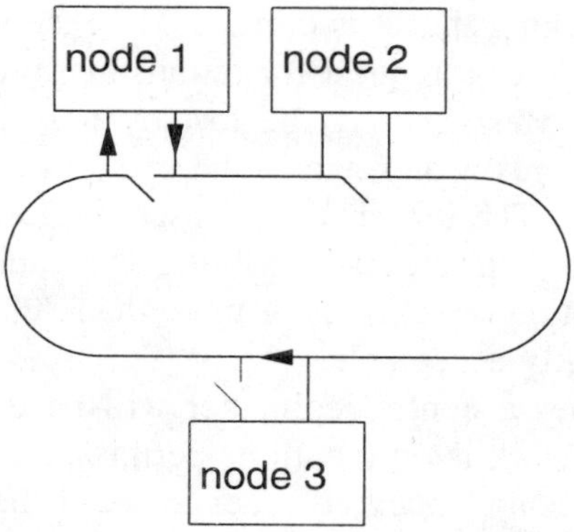

Figure 12.12

Ring LAN with bypass relay. Node 3
is down and has been bypassed

Access to the network

Access to the network must be controlled by a signalling procedure known as a *protocol*. Protocols are arranged in a hierarchy called the open-systems interconnect (OSI) model. In this the protocols are arranged in a hierarchy of levels or layers (see figure 12.13). The lowest level is the physical and the protocol for this layer would concern itself with connector layout, voltages, etc. The RS-232C serial interface specification is a physical-layer protocol. The next highest layer is the data-link layer in which the protocol must define the way data is moved from one station to another in the network. Examples are the high-level data-link control (HLDLC) and the binary synchronous communications (BISYNC) protocol, which define the format of message blocks. The third level is the network layer whose protocol defines how messages are broken into packets and routed from station to station (packet switching). The transport layer protocol defines the method of error detection and correction and the multiplexing methods used. The highest three layers are not concerned with the network itself, but with logging-in and out (session layer), character codes (presentation layer) and applications programs (applications layer).

Figure 12.13

The OSI protocol hierarchy. The bottom layer is the
physical interface and the top three have nothing to
do with the network proper

7	APPLICATION
6	PRESENTATION
5	SESSION
4	TRANSPORT
3	NETWORK
2	DATA LINK
1	PHYSICAL

When there is no control node, a station wishing to send a message first listens to the bus to see that it is not carrying traffic, and then places a packet of data on the bus containing the address to which it is sent and an error-checking code. The packet is received by all the other stations, but only the one with the right address opens it and reads its contents, checking that it is correct. The receiving station then sends an acknowledgment if all is well, or a request for retransmission if it is incorrectly sent. This is known as *handshaking*. Messages can be lost by noise or attenuation or when by mischance another station sends a message at the same time and there is a collision in which both messages are lost. The Ethernet LAN uses single-core, 5mm diameter, 50Ω coaxial cable with a maximum data rate of 10 Mbit/s. The cable attenuation is such that repeaters are necessary for station separations of more than 700 m. But for campus and local industrial use this is perfectly acceptable.

In some cases a control computer will be designated which is called the *host* and access to the network is only with its permission. The host computer looks at each of the other stations in some specified order to see if they have data for the network, a process called polling. The host then routes the data to its destination. In this way only one node can transmit data at any time and the host can determine the priority of the message or its source.

Ring LANs often use a system of channel sharing called *token passing*. A token is a distinctive pattern of bits that circulates round the ring when no traffic is carried. A node wishing to transmit must wait for the token before it can do so. When the token arrives, the node seizes it and puts its message packet onto the ring. The message is delivered and its receipt is confirmed, the token then being freed. The token cannot be used more than once, nor can it be held indefinitely, so that no node can appropriate the channel for too long. Token-ring networks offer a better service than bus networks when the level of traffic become high and collisions become more frequent, as figure 12.14 shows qualitatively.

Figure 12.14

Time delays versus traffic for Ethernet and token-ring networks. The Ethernet copes less well with high traffic volumes

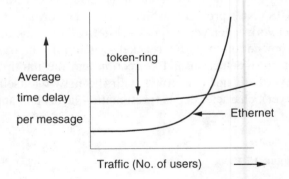

12.1.6 Programming languages

There have been, in the main, three areas of application for programming languages that have quite different characteristics:

> (1) scientific: complex algorithms needed
> (2) office and business: masses of data, simple algorithms
> (3) real-time control: for missiles, communications. Speed paramount

It seems unlikely that a single language could ever serve all three purposes equally well. Any language that does is likely to be unwieldy, possessing many facilities of little use most of the time, but which may lead into traps for the ignorant or unwary. Programming languages developed haphazardly at first, being poor relations to the hardware they ran on, but eventually they came to play a leading part in computing. Table 12.4 is a list of some of the more popular programming languages.

Originally computers were programmed by physically connecting various parts together. When the first stored-program computers came into existence, so did the art of programming on paper, but writing machine code in 1s and 0s was an exceedingly tiresome, time-consuming and error-prone activity. Wilkes wrote (of his first experience at programming the EDSAC I in 1949) 'the realization came over me with full force that a good part of the remainder of my life was going to be spent in finding errors in my own programs' — hardly a profitable use of time. The solution proposed by Wilkes in the first programming text[11] was to provide a 'pseudo code' which was relatively easy to learn and could be used with considerably fewer errors than were produced by writing in machine code. Each instruction of the pseudo-code was translated into machine code by an interpreter, which was itself a machine-code program. Computer users at Manchester had stayed with machine code using an alphabet of 32 symbols, based on the teleprinter code, to represent groups of five binary digits. By March 1954, Tony Brooker had produced a working version of the Mark I autocode system, which was a 'high-level' language that had many similarities to FORTRAN I.

Table 12.4 *Programming languages*

Language	Year	Application	Comment
FORTRAN	1957	Scientific	Numerous updated versions
ALGOL	1958	Scientific	No input/output handling
COBOL	1960	Data processing	Still very widely used
LISP	1960	List processing	Used in artificial intelligence
BASIC	1964	Scientific	Easy to learn. Not 'serious'
PL/1	1965	All purpose	Too big for convenience
Pascal	1970	Scientific, teaching	Developed for teaching
C	1972	Control, scientific	From Bell Labs for UNIX
Ada	1979	Scientific, control	Named after Byron's daughter
C++	1985	Control, scientific	For object-oriented programming
Java	1995	Internet	Object-oriented language also

Statements in high-level languages are no different in principle from pseudo-code and can be executed by means of an interpreter; it is, however, usually more efficient to translate all the program statements into machine code by means of a program called a compiler. Once a program is working, the compiled code can be stored and used on subsequent occasions without any need for recompiling, while interpreted programs must be re-interpreted each time they are run. Most modern high-level languages are compiled,

[11] *The Preparation of Programs for an Electronic Digital Computer* by Wilkes, Wheeler and Gill, Addison-Wesley (1953)

but there are still some occasions when interpreted programs might be desired.

In the early days the time penalty of interpreted programs was hardly noticeable since the programs spent virtually all their time in floating-point subroutines — there were no hardware implementations of floating-point arithmetic. However, as early as 1953 floating-point arithmetic circuitry became available (on the IBM 704), when it was realised that interpreting pseudo code was slower by far than running a machine-code program. The need for a compiled language then became urgent.

The first widely-used high-level language was FORTRAN I (FORmula TRANslation) developed by J W Backus at IBM in a surprisingly casual way ('we made [it] up ... as we went along'). The reason for this was that designing the language was felt to be the easy part of the job, while the hard part was designing the compiler. To some extent this was true, if you overlooked the consequences of bad language design, and in the event the FORTRAN I compiler was very efficient. FORTRAN I came out in 1957, and was superseded by FORTRAN II in 1958, FORTRAN IV in 1962, FORTRAN 77 in 1978, FORTRAN 90 in 1991 and FORTRAN 95 in 1996[12] (and it could be argued that PL/1 — Programming Language 1 — of 1965 was just a bloated version of FORTRAN). One might suppose that the later versions were simply fine tuning of the highly-successful original, but they are in fact — particularly FORTRAN 77 — almost different languages, made necessary by the competition of 'better' programming languages.

Table 12.5 *BNF definitions for numerical types*

\<unsigned integer\> ::=	\<digit\>	
	|	\<unsigned integer\>\<digit\>
\<integer\> ::=	+\<unsigned integer\>	
	|	−\<unsigned integer\>
	|	\<unsigned integer\>
\<decimal fraction\> ::=	.\<unsigned integer\>	
\<exponent part\> ::=	$10^{\text{\<integer\>}}$	
\<decimal number\> ::=	\<unsigned integer\>	
	|	\<decimal fraction\>
	|	\<unsigned integer\>\<decimal fraction\>
\<unsigned number\> ::=	\<decimal number\>	
	|	\<exponent part\>
	|	\<decimal number\>\<exponent part\>
\<number\> ::=	+\<unsigned number\>	
	|	−\<unsigned number\>
	|	\<unsigned number\>

The initial expectations of Backus and IBM for FORTRAN were probably exceeded and the language became the industry's standard for scientific computation. The latest versions, FORTRAN 90 and FORTRAN 95, are still by far the most widely used for

---

[12] The F programming language is a subset of FORTRAN 90 which has found some favour among teachers and users.

scientific computation using today's supercomputers. For business use, where there was much greater need for file and text-handling, COBOL was developed and it proved at least as successful, and if anything more durable, than FORTRAN. However, in the mid-1950s, FORTRAN was seen as a limited language, somewhat tied to IBM machines, which did not encourage the writing of well-structured programs. The perceived faults of FORTRAN led to the development of ALGOL (ALGOrithmic Language) as a second-generation language to replace it. The first specification, ALGOL-58, was described in a new formal syntax by Backus, but his description did not tally with that of P Naur, thereby indicating difficulties of describing what programming languages could do. Naur modified Backus's notation and the Backus-Naur Form (BNF) of notation came to light. BNF is now used to describe all programming languages. The complete description of ALGOL in BNF occupied only 15 pages and was issued as ALGOL-60 in 1960. Table 12.5 shows the definition of various types of number in BNF. The symbol ::= stands for 'is defined as' and stands for 'or', so an 'unsigned integer' is defined as a 'digit or an unsigned integer and a digit'. As one progresses down the list the definitions make use of preceding definitions until 'number' is defined.

In spite of all its advocates and its greatly-admired elegance, ALGOL failed to catch on, partly because IBM would not back it, and partly because it completely omitted the important field of input/output. Though not widely used, ALGOL soon became the standard by which other languages were judged ('a significant advance on most of its successors').

Of other languages there were legions, certainly several hundred and possibly over a thousand, of which Pascal, Ada and C have been prominent. Pascal originated as a language for teaching good programming techniques and became widespread in universities with some uptake by industry. Ada was the result of a US Department of Defense initiative when it came to the conclusion that no existing language would serve its need for embedded-computer control of missiles and missile systems. It was designed to promote the writing of programs in which most of the programmers' errors would be caught and pinpointed at compilation and not while running. However, it now appears that C (including C++) has become the most widely-used programming language in industry. C is a high-level language which has some of the ability of assembly language. C came out of the development of the UNIX operating system by Bell Labs; it is a small language that is, above all, fast. Much commercial software is written in C, as is the C compiler itself.

C has been criticised on the grounds that it permits sloppy programming and is not suited to proper software design. Modula2 was developed in an attempt to make a language which would not permit non-rigorous programming, and it succeeds, perhaps at the expense of flexibility. Nevertheless Modula2 has made some headway in teaching establishments, though it has so far failed to make a great impact on industry.

The development today of *object-oriented* programming resembles the development of structured programming twenty years ago in being fast evolving and fashionable. It is designed to enable programs readily to make use of other data classes defined by other programs. The end result should be the construction of programs from well-tested parts. The most widely used object-oriented programming language is C++ at present. More recently, Sun Microsystems have produced Java for use on the Internet. Java is a development of C++ which enables its users to create dynamic pages as well as download pages faster. Java and developments of it are certain to become important languages in the remainder of this century and well into the next.

It is probably true to say that the software run on a computer is more expensive than the hardware; certainly it is true of the machine on which this is written. Software also makes more people wealthy than hardware. Yet software started out as a poor relation — the important thing was making bigger, faster computers. What happened was that the hardware evolved at a different rate to software and the problems of hardware manufacture proved more amenable to solution than those of software. It must be the aim for programmers to produce software which is

- *portable*, that is can be run efficiently on any machine without any or much modification
- *maintainable*, that is it can readily be altered to take care of new circumstances by someone without a detailed knowledge of the program
- *well-documented*, so that other programmers can easily understand the way it works

and this can only be done within the constraints of the language used.

12.2 Microprocessors and microcontrollers

The first microprocessor to be made commercially was the 4-bit Intel 4004 of 1971 and though today 4-bit devices are no longer much used, they can still be found in toys, washing machines, heating-system controllers and a wide variety of household goods. By 1980 a variety of cheap 8-bit microprocessors, such as the Rockwell 6502 and the Intel 8080/8085 series had come on the market. Faster 8-bit microprocessors and microcontrollers followed with more and more facilities added to fill what were seen as gaps in a highly-competitive market. The 8-bit microcontroller is now replacing the 4-bit device, even in applications requiring relatively little processing power. 16-bit, 32-bit and even 64-bit devices have become available and are used mostly for specialised purposes such as PWM motor (mostly 16-bit microcontrollers) and robotic control. The robotics field in particular has urgent need of fast microcontrollers with large word sizes. Table 12.6 lists some of the devices, which is only a small sample of those available now and in the past.

The microprocessor is not a stand-alone computer, since it lacks memory and input/output control. These are the missing parts that the microcontroller supplies, making it more nearly a complete computer on a chip. Figure 12.15 shows a block diagram of the component parts of a microcontroller in which the component parts of a microprocessor are enclosed in a dashed rectangle. It can be seen that the microcontroller is a microprocessor with some additional on-chip features:

(1) A block of ROM to store program and data
(2) A small block of RAM
(3) Several ports for input and output

Table 12.6 *Some readily-available microprocessors and microcontrollers*

Device No.	Clock freq. (MHz)	Data size (bits)	Prog. Mem. (bytes)	Prog. Mem. Type	Internal RAM (bytes)	Add'ss'ble Ext. Mem. (bytes)	Parallel I/O	Serial I/O[A]	No. of Timers[B] (× bits)	No. of Interrupt Sources	Package Type[C]	Price (1-off)
80C51 types												
P80C31	12	8	0	–	128	64k	16	1 SCI	2 × 16	5	40-pin DIP	£3.00
P80C32	12	8	0	–	256	64k	24	0	3 × 16	7	40 pin DIP	£4.80
P80C51FA	12	8	0	–	256	64k	24	0	3 × 16	7	40-pin DIP	£5.60
P87C748EBPN	16	8	2k	OTP	64	64k	2	0	1 × 16	2	24-pin DIP	£7.00
P87C51	12	8	4k	OTP	128	64k	32	1 SCI	2 × 16	5	40-pin DIP	£12.00
P87C524EBFFA	16	8	16k	EPROM	512	64k	4	2	3 × 16	multi	40-pin CER	£35.00
PIC types												
PIC16C54RC/P	4	8	512†	OTP	32	–	12	0	1 w/dog	0	18-pin DIP	£3.40
PIC16C58-04P	4	8	2k†	OTP	73	–	12	1 SCI	1 w/dog	0	18-pin DIP	£4.00
PIC16C74-JW	20	8	4k‡	EPROM	192	–	33§	SPI/I²C	3	12	40-pin DIP	£21.00
PIC16C620-04P	4	8	512‡	OTP	80	–	13	0	1	4	18-pin DIP	£3.60
Other 8-bit types												
MC68HC11A1FN	8	8	512	E²PROM	256	64k	7	1 SPI / 1 SCI	1 × 16 / 1 w/dog	18	52-pin PLCC	£7.50
ST62E20F1/HWD	8	8	4k	EPROM	64	0	12–20§	4	1 × 8	any i/p	20-pin DIL	£33.00
MSM80C88A	5	8	0	–	0	1M	0	0	0	2	40-pin DIP	£15.00
16-bit types												
N80C186XL-20	20	16	0	–	0	1M	0	0	3 × 16	15	68-pin PLCC	£13.00
N80C188	10	16	0	–	0	1M	0	0	3 × 16	15	68-pin PLCC	£11.00
MC68HC001FN10	10	8/16	0	–	0	16M	0	1 SPI	0	10	68-pin PLCC	£13.00

Notes: [A] SCI = Serial Comms Interface I²C = Inter-Integrated Circuit bus interface [B] w/dog = watchdog [C] DIP = dual-in-line plastic, CER = ceramic. Most devices are available in several package styles, some using SMT, surface-mount technology † = 12-bit word ‡ = 14-bit word § = also has an 8-channel, 8-bit ADC

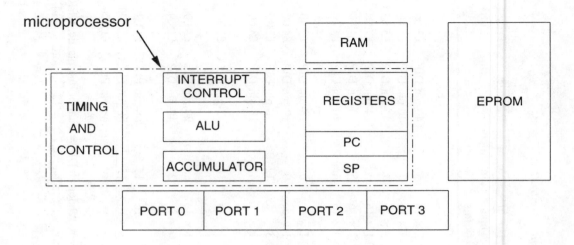

Figure 12.15   Block diagram of a microcontroller. IR, PC and SP stand for instruction register, program counter and stack pointer respectively

Microprocessors are for moving and processing as bytes relatively large amounts of data internally and to and from external sources. A microprocessor is really the CPU of a computer. A microcontroller, however, is as likely to be required to perform bit manipulations as much as byte, and must be able to accept and output control and timing signals. The memory stack in a microcontroller is usually much smaller than that of a microprocessor, while a microprocessor needs quite a lot of additional memory to perform its functions.

Both microcontrollers and microprocessors are programmed by writing code into a microcomputer which is then translated by a compiler into machine code that is stored in ROM. The programming, ROMing and debugging require a 'development system' for the particular device or device family. You can get microcontroller development systems which support high-level languages, but the code produced is less efficient. Software simulators are available that run on most microcomputers to help with program debugging. Emulators are also used, which carry out all the operations that the microcontroller can perform on external hardware. An emulator is the same as the microcontroller so far as external devices are concerned, while a simulator can only show what the contents of the RAM and ROM are at each stage in the program. In order to understand a microcontroller well enough to make it perform useful tasks there is no substitute for learning about all its features, which is why this chapter is full of detail.

12.2.1   The 8051 8-bit microcontrollers

The very-popular 8051 series of microcontrollers has dozens of variants, some of which are listed in table 12.6. The 8031 has no on-board ROM and is therefore the cheapest, the 8051 includes one-time programmable ROM (OTPROM) and the 8751 is for prototype production, since it has EPROM (erasable PROM) on board. For most general purposes the 8-bit microcontroller is the best choice and we shall be looking in some detail at one of the most popular, the Intel 8051. The 8-bit word gives a maximum resolution of 0.4% (1 in 256), which is enough for all but high-precision tasks and those requiring great speed. The 8051/8751 comes in a 40-pin DIP shown in figure 12.16, but the effective number of pins is increased to 64 by giving 24 of them dual functions.

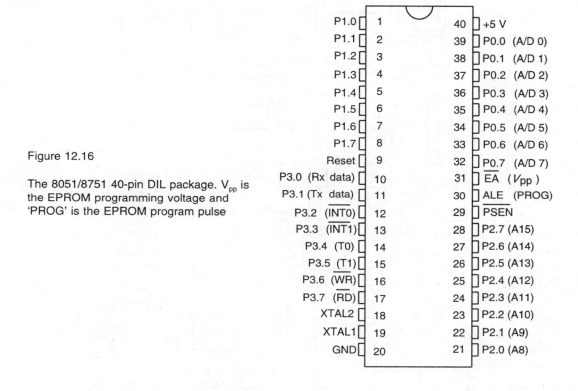

Figure 12.16

The 8051/8751 40-pin DIL package. V_{pp} is the EPROM programming voltage and 'PROG' is the EPROM program pulse

Figure 12.17 shows the architecture of the various parts of the 8051. Unlike computers, which can be programmed in high-level languages and require very little knowledge of the computer's organisation, microcontrollers require one to have considerable knowledge of their architecture. The most important parts of the microcontroller are the RAM, the special-function registers (SFRs) and the ports.

Figure 12.18 shows the layout of the RAM and the SFRs. To make the device do what we want requires us to know in more or less detail what is stored in this RAM and the SFRs. The RAM consists of 128 bytes (of 8 bits) arranged as four register banks, each containing 8 registers given the labels R0 to R7. Each RAM byte has a number associated with it which is its address, so one can either use a register's label, or its absolute RAM address. The register banks must be selected, so that the CPU knows in which bank R3, for example, is. On power up or reset, bank0 is selected; to select any other banks the appropriate settings must be made in the Program Status Word (PSW) SFR.

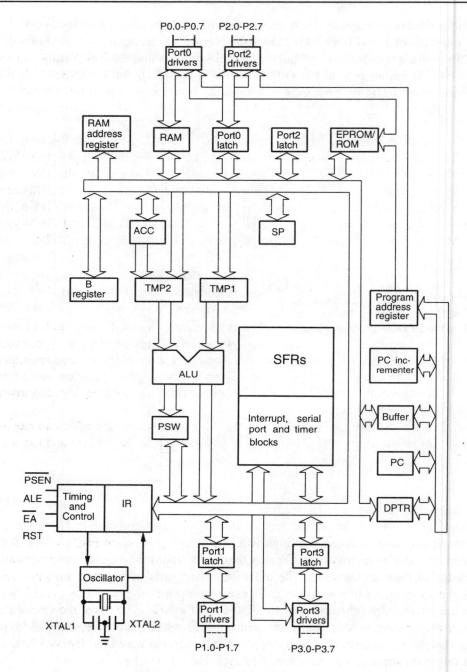

Figure 12.17 Architecture of an 8051 microcontroller

Above the register banks is an area of RAM (20h to 2Fh, inclusive[13]) which is bit addressable. This saves space when we want to store bits, as we do not have to use a whole byte to do so. In RAM locations from 30h to 7Fh is a general purpose area which can be used as a scratch-pad. The whole of the RAM area above 07h (register bank0) can be used as a memory stack to store variables, data, etc. It is best to put the stack above register bank3 and the bit-addressable bytes, that is above 2Fh, to avoid accidentally overwriting data.

RAM locations above 7Fh are reserved for the defined SFRs and can *only* be used when defined. Attempts to used undefined RAM above 7Fh will cause errors. We shall now look at these SFRs. ROM is used for storing code, RAM for data that can be altered as the program runs. The 8751 uses Harvard architecture where code and data are stored in different areas with the same address: the CPU decides which is which. If off-chip ROM is needed it will automatically be accessed by calls to addresses above 0FFF. If you wish all ROM calls can be made externally by connecting $\overline{E\,A}$ (pin 31) to GND.

12.2.2 The 8051 clock and the machine cycle time

The master clock can be set to any frequency between about 1 MHz and 16 MHz (or 12 MHz for some types). The user selects the master clock frequency by connecting a quartz crystal between pins 18 and 19 on the DIP, together with some auxiliary capacitors. This oscillator frequency determines the speed at which any operation is carried out. However, the machine cycle is actually 12 oscillator cycles, not one, as each instruction involves a number of smaller steps.

The machine cycle is the smallest time in which the microcontroller can execute the simplest instruction. If the instruction takes C machine cycles to execute and the oscillator frequency is f Hz, then the time taken is

$$t_{op} = \frac{12C}{f} \tag{12.1}$$

Thus if the oscillator frequency is 12 MHz, the time of execution is C μs. The 8051 can perform at most 1 million instructions per second (1 Mips). The operation mnemonics (op codes) are all listed under the appropriate headings in this chapter, together with the number of bytes taken in memory and the time for execution in machine cycles. With this information the space used in memory and the time taken to perform subroutines can be calculated. It is often essential that the precise time taken to do something is known, as well as whether a location in memory is close enough to be called by its label. This is called relative addressing and is discussed later.

---

[13] Hexadecimal (base 16) numbers will be written with an 'h' appended, decimal numbers with a 'd' appended and binary with a 'b' appended, thus 52h = 82d = 01010010b

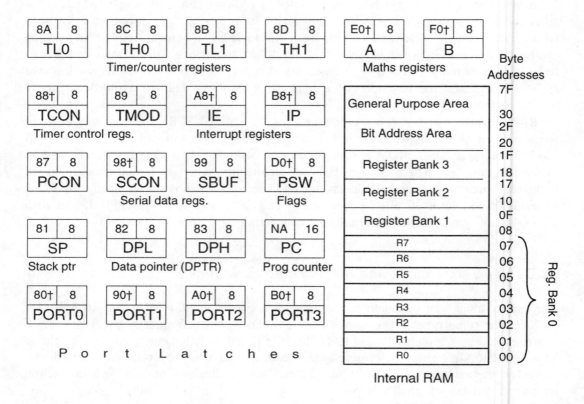

Figure 12.18 The layout of RAM and the SFRs: † = bit addressable

12.2.3 *The special-function registers (SFRs)*

The defined special-function registers are given in table 12.7, below. SFRs can be used in some operation codes by their names and in others by their addresses. Since any address in a program *must* start with a number, the addresses above, such as that of port3, B0, must be written as 0B0, with a leading zero. Assembly errors result if this is not done.

The program counter (PC) and the data pointer (DPTR)

These are the only 16-bit SFRs, all the others are 8-bit. Program bytes are fetched from memory address which is in the PC. These addresses can be in ROM with 16-bit addresses from 0000h to 0FFFh (on chip) or from 0000h to FFFFh off-chip or a mixture of the two. Any address above 0FFFh will result in an off-chip call. As soon as an instruction address is fetched from the PC it is incremented, so that it always points to the *next* instruction. Unlike all the other SFRs, the PC has no address. The DTPR register is made up of two 8-bit registers, DPH and DPL, which have individual addresses. They contain the addresses for both internal and external code and external data. The DPTR is under program control and can be specified by its name (for all 16 bits) or by the names of the high and low bytes (DPH and DPL). Its hexadecimal address is the two 8-bit bytes of DPH and DPL, 83h and 82h.

Table 12.7 *The 8051 SFRs*

Name	Function	RAM address
A	Accumulator	0E0h
B	Arithmetic	0F0h
DPL	Data Pointer Low byte; addressing external memory	82h
DPH	Data Pointer High byte; addressing external memory	83h
IE	Interrupt Enable control	0A8h
IP	Interrupt Priority	0B8h
P0	Port0 latch	80h
P1	Port1 latch	90h
P2	Port2 latch	0A0h
P3	Port3 latch	0B0h
PCON	Power down CONtrol, user flags	87h
PSW	Flags	0D0h
SBUF	Serial port data BUFfer	99h
SCON	Serial port CONtrol	98h
SP	Stack Pointer	81h
TCON	Timer/counter CONtrol	88h
TMOD	Timer/counter MODe	89h
TL0	Timer0 Low byte	8Ah
TL1	Timer1 Low byte	8Bh
TH0	Timer0 High byte	8Ch
TH1	Timer1 High byte	8Dh

The A and B arithmetic registers

A is the accumulator. It is used to perform addition, subtraction, multiplication and division (of integers) and also Boolean algebra on bits. B is used as an auxiliary register for multiplication and division only.

The program status word (PSW)

The PSW byte is set out below.

7	6	5	4	3	2	1	0
C	AC	F0	RS1	RS0	OV		P

The functions of the bits are as follows:

Bit address	Name	Function
PSW.7	C	Carry flag, used in arithmetic, in jumps and in Boolean operations
PSW.6	AC	Auxiliary carry flag, used only in BCD arithmetic
PSW.5	F0	User flag 0
PSW.4	RS1	Register bank select bit 1
PSW.3	RS2	Register bank select bit 2

RS1	RS2	Selects
0	0	bank0
0	1	bank1
1	0	bank2
1	1	bank3

Bit address	Name	Function
PSW.2	OV	Overflow, used in arithmetic
PSW.1		Not used
PSW.0	P	Parity flag for the accumulator. P = 1 indicates that the sum of the bits in A is odd

The stack pointer (SP)

The stack is a location in internal RAM which is used for fast transfer of data. The SP points to the location of the item on the top of the stack. When an item is put on the stack the SP is incremented *before* it is placed, and *after* it is removed the SP is decremented. The stack starts at RAM location 07h on powering up the chip or on reset, and can run up to top of RAM (7Fh). If the stack is expanded beyond 7Fh, disaster can result, since other memory gets overwritten including program data. The stack pointer can be set by the programmer and normally is put as high as possible in RAM, above 2Fh.

Counters and timers

Counters and timers are the distinguishing feature of microcontrollers. The 8051 has two counter/timers which are set up and controlled by the TMOD and TCON SFRs. Counting and timing can be done with software, but this may take too much memory, so two 16-bit up counters, T0 and T1, are provided. When counting internal clock pulses they are called timers and when counting external pulses they are called counters, but the action is the same. They are each divided into two 8-bit registers called TL0, TL1, TH0, TH1 for low and high bytes. The timers count machine cycles, that is the oscillator frequency divided by 12d. The timer control register, TCON, is shown below.

7	6	5	4	3	2	1	0
TF1	TR1	TF0	TR0	IE1	IT1	IE0	IT0

Bit address	Name	Purpose
TCON.7	TF1	Timer1 overflow flag. Set when timer1 overflows. Cleared when CPU vectors to the interrupt service routine.
TCON.6	TR1	Timer1 runs/stops when software sets this to 1/0
TCON.5	TF0	Timer0 overflow flag.
TCON.4	TR0	Timer0 run control bit.
TCON.3	IE1	External Interrupt1 flag. Set to 1 by external interrupt and cleared to 0 when interrupt is serviced.
TCON.2	IT1	Interrupt1 control bit. Set/cleared by software to 1/0 to specify falling edge/low level interrupt triggering.
TCON.1	IE0	As IE1, but for External Interrupt0.
TCON.0	IT0	AS IT1.

Timers have four modes of operation that can be set by bits 1 and 0 in the TMOD register, which is shown below.

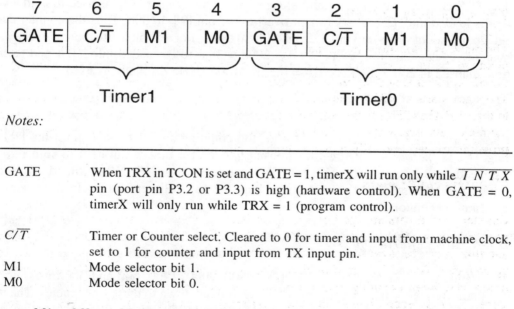

Notes:

GATE	When TRX in TCON is set and GATE = 1, timerX will run only while $\overline{INTX}$ pin (port pin P3.2 or P3.3) is high (hardware control). When GATE = 0, timerX will only run while TRX = 1 (program control).
$C/\overline{T}$	Timer or Counter select. Cleared to 0 for timer and input from machine clock, set to 1 for counter and input from TX input pin.
M1	Mode selector bit 1.
M0	Mode selector bit 0.

M1	M0	Mode	Action
0	0	0	13-bit timer
0	1	1	16-bit timer/counter
1	0	2	8-bit autoreload timer/counter
1	1	3	Timer0: TL0 is an 8-bit timer/counter controlled by the standard timer0 control bits, TH0 is an 8-bit timer controlled by timer1 control bits. Timer1: stopped

Each timer can operate independently of the other in modes 0,1 and 2 but not in mode 3.

Mode 0

The 13-bit counter mode. It uses THX as an 8-bit counter and TLX as a 5-bit counter. The pulse input (machine cycle) is divided by 32d in TLX, so that THX counts the oscillator frequency divided by 384d (12d × 32d). Thus a 12MHz oscillator frequency would become a timer frequency of 31.25 kHz. The timer flag is set whenever THX goes from FFh to 00h (256 counts), or in 8.2 ms for this case.

Mode 1

This works like mode 0, but uses 8 bits in TLX, so dividing the machine frequency by 256d. A 12MHz oscillator frequency divides down to $12 \times 10^6/12/256 = 3906$ Hz, so 256 counts of this would be 65.5 ms.

Mode 2

The TLX counter is used as a stand-alone 8-bit counter, while THX is used to hold a value that is loaded into TLX every time it overflows from FFh to 00h. Thus the counter can be set up to count from anything from 00h to FEh. For example, to count exactly 200 machine cycles requires 56d (= 256d − 200d = 38h) to be loaded into THX and then the time elapsed is $200 \times 12/12 \times 10^6 = 0.2$ ms if a 12MHz oscillator is used.

Mode 3

Choosing mode 3 for timer1 causes it to stop counting and place its control bit TR1 and flag TF1 in the service of timer0. Choosing mode3 for timer0 causes it to split into two independent 8-bit counters. TL0 sets TF0 and TH0 sets TF1. When timer0 is mode 3 timer1 can be used, but it cannot set its flag, but can still generate a clock or become a baud-rate generator.

Baud rate generation

The standard baud rates are 19200, 9600, 4800, 2400, 1200, 600 and 300 bits/s, and the oscillator frequency must be capable of being divided down to these frequencies. The usual crystal frequency chosen is 11.0592 MHz since 11059200/12 gives a machine cycle frequency of 921.6 kHz = 48 × 19200. Normally the serial-port mode1 is selected in the serial port control register, SCON (described later), then timer mode2 is selected for timer1 in the TMOD register, and the baud rate is

$$\text{Baud} = \frac{2^{\text{SMOD}} f_{\text{osc}}}{32 \times 12 \times (256 - \text{TH1})} \tag{12.2}$$

SMOD is bit 7 of the PCON register that can be set to 1 or cleared to 0. TH1 is the reload value of the high byte of timer1, which can be loaded by programming the TH1 register (RAM address 8Dh). One usually knows what baud rate is desired and really wants to know the value of the number to load into TH1, which is given by

$$TH1 = 256 - \frac{2^{SMOD} \times f_{osc}}{384 \times \text{baud rate}} \tag{12.3}$$

For example if the baud rate is to be 600, with SMOD = 1 and f_{osc} = 11.0592 MHz, equation 12.3 gives 160d, or A0h, as the number to be put into TH1. However, mode0 in SCON can be used in which case no timers need to be set up and the baud rate is just $f_{osc}/12$. For standard baud rates, this requires a very slow clock rate.

Interrupts

Interrupts force the program to call a subroutine located at some specified place in memory. Some interrupts are hardware generated to save time and are the only way in which real time control can be obtained. The interrupt enable (IE) SFR is shown below.

7	6	5	4	3	2	1	0
EA		ET2	ES	ET1	EX1	ET0	EX0

Bit address	Name	Function
IE.7	EA	Enables/disables all interrupts when EA = 1/0.
IE.6	–	Not used
IE.5	–	Not used
IE.4	ES	Enables/disables serial port interrupt when ES = 1/0.
IE.3	ET1	Enables/disables timer1 when ET1 = 1/0.
IE.2	EX1	Enables/disables external interrupt1 when EX1 = 1/0.
IE.1	ET0	As ET1 but for timer0.
IE.0	EX0	As EX1 but for external interrupt0.

There are five interrupts, three of which are generated automatically by internal operations: timer0 flag, timer1 flag and the serial port interrupt (RI or TI). Two interrupts are triggered by external signals with circuitry connected to $\overline{INT\ 0}$ and $\overline{INT\ 1}$ (that is the P3.2 and P3.3 pins). Of course the program can interrupt itself simply by setting the appropriate flag.

The programmer can assign priority to interrupts with the interrupt priority (IP) SFR and can enable them with the interrupt enable (IE) SFR. All interrupts can be blocked by these controls. When the interrupt is serviced by the routine placed by the programmer at the location specified by the chip manufacturers, the interrupt is cleared and the program

PC address is restored at the end of the interrupt subroutine with the RETI instruction.

Reset

Reset is the overriding interrupt: it cannot be blocked. In this case the PC is not saved and the reset causes a jump to ROM address 0000h and run from there. When high is applied to RST pin the reset is begun. When a low is then applied to RST pin, the internal registers have the values listed below.

SFR	Reset to	SFR	Reset to
A	00h	PSW	00h
B	00h	SP	07h
DPTR	0000h	SBUF	××h
IE	0××00000b	SCON	00h
IP	×××00000b	TCON	00h
P0	FFh	TMOD	00h
P1	FFh	TH0	00h
P2	FFh	TL0	00h
P3	FFh	TH1	00h
PCON	0×××××××b (HMOS)	TL1	00h
	0×××0000b (CHMOS)		

Note: × = undefined

Internal RAM is NOT affected by reset, but in power up it is in a random state. Register bank0 is selected after reset as the PSW contains 00h.

Interrupt priority

Each source of an interrupt (except reset) can be controlled by the IE SFR, but the order of servicing interrupts is set by the IP SFR. However, the IP SFR can only give priority when the appropriate bit is set to 1. Thus if several interrupts are priority 1, then the order of servicing is IE0 > TF0 > IE1 > TF1 > serial port (RI or TI). Priority can only be given to TI (or RI) if it is set to 1 in the IP SFR and all the others to 0. The IP SFR is shown below.

7	6	5	4	3	2	1	0
			PS	PT1	PX1	PT0	PX0

Bit address	Name	Priority to	Vector address
IP.7	–	Not used	–
IP.6	–	Not used	–
IP.5	–	Not used	–
IP.4	PS	Serial port	0003h
IP.3	PT1	Timer1	000Bh
IP.2	PX1	External interrupt1	0013h
IP.1	PT0	Timer2	001Bh
IP.0	PX0	External interrupt0	0023h

The interrupts must be serviced by instructions located in ROM at the addresses given in the table under the 'Vector address' heading. Interrupts that are not serviced for any reason must persist until they are, or they are lost.

Input/output pins and ports

The 8051 DIP has 40 pins, which is not enough. 32 of the pins are used by 4 ports, and 24 of these can be used for dual functions, effectively expanding the pin count to 64.

Pin	Alternative use	Pin	Alternative use
P0.0	Address/data bit 0	P3.0	Receive data
...	...	P3.1	Transmit data
...	...	P3.2	External interrupt0
P0.7	Address/data bit 7	P3.3	External interrupt1
P2.0	Address bit 8	P3.4	Timer0 output
...	...	P3.5	Timer1 output
...	...	P3.6	Write strobe
P2.7	Address bit 15	P3.7	Read strobe

Strobes are activated by 0s, not 1s. Other pins on the DIP are 9 (reset), 18 (XTAl1), 19 (XTAL2), 20 (GND), 29 (program store enable, $\overline{P\ S\ E\ N}$) 30 (address latch enable, ALE, or the EPROM program pulse), 31 (external enable, $\overline{E\ N}$, or the EPROM programming voltage), 40 (V_{CC} or V_{DD}, +5 V).

The ports cannot drive big loads and are limited to a little over 1 mA. All the port pins have latches and the associated latch circuitry depends on the alternative function (if any). Port1 has no dual function. Ports P0 and P2 are used for accessing 16-bit addresses of external memory, P0 the low bytes and P2 the high bytes. P3 has multiple alternative uses such as serial data input/output, external interrupts and timers. Serial data is input or output in conjunction with the serial data buffer, SBUF. External interrupt0 is controlled via bit TCON.1, external interrupt1 via bit TCON.3 and the external counters via TMOD. All the ports can be used as input/output ports instead of these alternatives.

Input and output of serial data

Serial data must use port P3. The serial port control register, SCON, controls data communications, while the PCON register controls the data rate. The data is held in SBUF, which comprises two separate registers, one for write and one for read. Both have the same address: 99h. The SCON register is shown below:

7	6	5	4	3	2	1	0
SM0	SM1	SM2	REN	TB8	RB8	TI	RI

Bit address	Name	Function
SCON.7	SM0	Serial port mode specifier. Set/cleared by program to 1/0 to select mode
SCON.6	SM1	Serial port mode specifier. As SM0

SM0	SM1	Mode	Action
0	0	0	Shift register, baud rate $= f_{\text{machine}}$
0	1	1	Variable baud rate set by timer1
1	0	2	Fixed baud rate. If SMOD = 0 it is 1/64d of f_{osc} and if SMOD = 1 it is 1/32d of f_{osc}
1	1	3	Variable baud rate set by timer1

Bit address	Name	Function
SCON.5	SM2	Multiprocessor communication mode
SCON.4	REN	Set/cleared by program to enable/disable reception
SCON.3	TB8	9th bit sent in modes 2 and 3. Set/cleared by program
SCON.2	RB8	9th bit received in modes 2 and 3. In mode 1, if SM = 2, it is the received stop bit. Not used in mode 0
SCON.1	TI	Tx interrupt flag. Set by hardware, cleared by program
SCON.0	RI	Rx interrupt flag. Set by hardware, cleared by program

Serial data communications are slow, so serial data flags in SCON are required to aid data flow. Input data cannot be under program control, of course, and must generate an interrupt when received. The TI and RI flags are ORed to generate an interrupt. The program must respond and clear the flag, unlike timer flags that are cleared automatically.

Data transmission and reception

Any data written to SBUF is transmitted. TI (SCON.1) is set to 1 after transmission to signal that SBUF is empty so that further data can be sent to it. Data will be received if the receive enable bit, REN (SCON.4) is set to 1 for all modes. For mode 1 only, the receiver interrupt flag, RI (SCON.0), must be cleared to 0 also. RI is set after data is received in all modes. Setting REN by program is the only way of limiting the receipt of unexpected data. Reception can begin in modes 1, 2 and 3 if RI is set when the serial stream of bits begins. RI must have been reset by the program before the last bit is received or the incoming data will be lost. Incoming data is not transferred to SBUF until the last data bit has been received. Then the previous transmission can be read from SBUF while new data is being received.

Mode bits can be set with bits M0 (SCON.7) and M1 (SCON.6). Virtually any submultiple of the machine frequency can be set for the baud rate. The SMOD bit in PCON must be set with timer 1 in modes 1, 2 or 3 for variable baud rates. Serial data modes 2 and 3 are used for communicating with second processors.

12.2.4 Moving data

An important part of effective use of the microcontroller is moving data around the chip efficiently. There are several different ways of achieving this, depending on what and where the data is and the manner in which is addressed.

Immediate addressing

In this mode the data itself is entered in the program operation code. Numbers that are to be read literally (that is as pure numbers) are entered by using the symbol # followed by the number subscripted by h for hexadecimal, or d for decimal numbers. # is an *immediate data* identifier. Data is moved by the MOV mnemonic followed by the destination address and then the source address, separated by a comma:

      MOV A,#12h
      MOV R1,#0F2h

Note that a number must begin with a digit (0 to 9), so if a hexadecimal number such as F2h is to be used, it must be prefixed by a zero. Unfortunately, omitting the # causes the compiler to assemble the number as an opcode: there is no way of trapping this error other than by using a simulator. This lack of safety in programming is a recurring problem for microprocessor programmers.

Register addressing

Some SFRs have mnemonics and can be called by them and they have addresses and can be called by them too. For example, here are several ways of MOVing data by immediate and register addressing:

Mnemonic	Comment
MOV A,#2Fh	;Copy the immediate data 2Fh into A. *(2B, 1C)*
MOV A,R1	;Copy the contents of register R1 into A. *(1B, 1C)*
MOV R1,A	;Copy the contents of A into register R1. *(1B, 1C)*
MOV R1,#0AFh	;Copy the immediate data AFh into register R1. *(1B, 1C)*
MOV DPTR,#0A2F1h	;Copy the immediate data A2F1h into the DPTR. *(3B, 2C)*

The numbers in parentheses after the comments respectively indicate the number of bytes of memory taken by the instruction and the number of machine cycles taken in execution. Comments must be preceded by a semicolon to be ignored by the assembler. The register bank that R1 is in must be specified by the appropriate bit settings in the PSW SFR (start up default setting is bank0).

Direct addressing

All 128 bytes of internal RAM and the SFRs can be addressed directly using the single-byte address assigned (see figure 22.4). The 128 RAM addresses go from 00h to 7Fh. But RAM address 00h to 1Fh are also the locations of the 32 working registers in the banks:

bank	register	address
0	R0	00h
...	...	...
0	R7	07h
1	R0	08h
...	...	...
1	R7	0Fh
2	R0	10h
...	...	...
2	R7	17h
3	R0	18h
...	...	...
3	R7	1Fh

On reset RS0 and RS1 in the PSW SFR are set to 0 so bank0 is selected and the SP is set to 07h. Thus as the stack is used, register space in banks 1, 2 and 3 gets used up in turn. The programmer must ensure that the SP is set high enough not to corrupt data put in the register banks during the program.

Data can be copied from one address to another by direct addressing:

```
MOV addr1,addr2    ;Copy the contents of addr1 to addr2. In register-transfer
                   ;notation, [addr1] ← [addr2]. (3B, 2C)
MOV A,addr1        ;[A] ← [addr1]. (2B, 1C)
MOV addr1,A        ;[addr1] ← [A]. (2B, 1C)
MOV addr1,#0F1h    ;[addr1] ← #0F1h. (3B, 2C)
MOV addr1,R2       ;[addr1] ← [R2]. (2B, 2C)
MOV R2,addr1       ;[R2] ← [addr1]. (2B, 2C)
```

Moves involving the accumulator, A, are always the fastest.

Indirect addressing

With indirect addressing the register holds the address that is used to store the data. The MOV command only indirectly addresses registers 0 and 1, the so-called *pointing registers*. Indirect addresses are denoted by a preceding @, for instance

MOV @Rp,#0D3h	;Copy the immediate data D3h into the address contained in ;Rp. Or, [[Rp]] ← 0D3h. *(2B, 1C)*
MOV @Rp,addr1	;[[Rp]] ← [addr1]. *(2B, 2C)*
MOV addr2,@Rp	;[addr2] ← [[Rp]]. *(2B, 2C)*
MOV @Rp,A	;[[Rp]] ← [A]. *(1B, 1C)*
MOV A,@Rp	;[A] ← [[Rp]]. *(1B, 1C)*

The address in Rp must be a RAM or a SFR address. Indirect addressing is always used whenever external RAM is used, but with the special mnemonic MOVX to remind us that a call on external RAM is taking place. Since external RAM can go to 64 kbytes, the DPTR is most often used to store the address:

MOVX A,@DPTR	;[A] ←[[DPTR]]. *(1B, 2C)*
MOVX @DPTR,A	;[[DPTR]] ← [A]. *(1B, 2C)*

Moving data from ROM

Data in ROM can be accessed by indirect addressing using the accumulator and the PC or DPTR. The mnemonic is MOVC:

MOVC A,@A+DPTR	;[A] ←[[A + DPTR]]. *(1B, 2C)*
MOVC A,@A+PC	;[A] ← [[A + PC + 1]]. 1 is added because the PC ;increments *before* the instruction executes. *(1B, 2C)*

The data pointer has separately-addressable bytes:

MOV DPTR,#0AF1Eh	;Put AF1Eh into the DPTR
MOV DPH,#0AFh	;Put AFh into the DPTR high byte
MOV DPL,#1Eh	;Put 1Eh into the DPTR low byte
MOV 83h,#0AFh	;Put AFh into the DPTR high byte
MOV 82h,#1Eh	;Put 1Eh into the DPTR low byte
MOVC A,@A+DPTR	;Copy the contents of ROM address AFC0h into A

Lines 2 & 3 and 4 & 5 above accomplish the same thing as line 1. The last line adds AF1Eh to A2h to form the address AFC0h. Suppose the PC contains B22Dh, then the code

MOV A,#0DEh	;[A] ← DEh
MOVC A,@A+PC	;[A] ← [[A+PC+1]]

would result in 3B0Ch being placed in A.

PUSH and POP

These are stack commands that PUSH data onto the top of the stack or POP (remove) it from the top of the stack. The stack pointer is incremented by one *before* a PUSH and decremented by one *after* a POP. For example:

PUSH addr1	;Increment the PC then copy the data in addr1 into the ;stack address specified by the PC. *(2B, 2C)*
POP addr2	;Copy the data at the internal RAM address specified by the ;stack pointer into addr2 then decrement PC. *(2B, 2C)*

The stack pointer resets to 07h, or is at 07h on start up. The first PUSH onto the stack therefore puts data in 08h, which is register R0 of bank1.

Examples:

MOV SP,#30h	;[SP] ← 30h. The SP is set to the lowest address in the ;general purpose area in RAM
MOV R0,#2Bh	;[R0] ← 2Bh
PUSH R0	;[SP] ← 31h. RAM address 31h now contains 2Bh
PUSH 00h	;[SP] ← 32h. 00h is the same as R0 if register bank0 is ;selected. Address 32h now contains 2Bh
POP 01h	;[SP] ← 31h. R1 contains 2Bh
POP 0A0h	;[SP] ← 30h. Port2 latch now contains 2Bh

'PUSH R0' is the same as 'PUSH 00h' only if register bank0 is selected. Direct addresses, not register names, must be used for most registers as the stack mnemonics have no means of telling which bank is in use. When the stack pointer reaches FFh it rolls over to 00h. RAM ends at 7Fh, so pushes above this result in errors. The stack pointer is usually set to 30h or higher. The stack pointer can be PUSHed to and POPped from the stack.

Exchanging data: XCH

The mnemonic XCH can be used to swap data between the accumulator and RAM addresses or registers, for instance

XCH A,R3	;[A] ← [R3], [R3] ← [A]. *(1B, 1C)*
XCH A,addr	;[A] ← [addr], [addr] ← [A]. *(2B, 1C)*
XCH A,@Rp	;[A] ← [[Rp]], [[Rp]] ← [A]. *(1B, 1C)*
XCHD A,@Rp	;Exchange the lower nibbles between A and the address in Rp. The higher order nibbles are unaffected. *(1B, 1C)*

Exchanges between A and any port location copy the data in port *pins* to A while data in A is copied to the *latch*. XCH is a very convenient way to save A without PUSHing or

POPping. When using XCHD, the upper nibbles of A and the exchanged address are unchanged.

12.2.5 Logical operations

Boolean algebra can be performed on both bits and bytes. Bit operations are very useful in control applications. Only the bit-addressable registers can be specified in bit operations. The following Boolean byte-level operations can be performed:

mnemonic	Boolean
ANL (ANd Logical)	AND
ORL (OR Logical)	OR
XRL (eXclusive oR Logical)	XOR

The operations are written with two operands separated by a comma. The destination of the operation is the first operand, which must be an address in RAM or the accumulator. Indirect addresses and immediate data cannot be destinations. For example:

ANL A,#10h ;AND each bit in A with the corresponding bit of the ;immediate number 10h, put the result in A. *(2B, 1C)*

ORL A,R1 ;OR each bit in A with the corresponding bits in R1 and put ;the result in A. *(1B, 1C)*

XRL A,addr ;Exclusive OR the bits in A and the bits in the RAM ;address and store the result in A. *(2B, 1C)*

ANL addr,#2Bh ;AND each bit in the RAM address with the corresponding ;bit of the immediate data 2Bh and put the result in the ;RAM address. *(3B, 2C)*

ORL A,@Rp ;OR each bit in A with the corresponding bit in the RAM ;address stored in Rp. Store the result in A. *(1B, 1C)*

The following are unary (they have only one operand) operations that can only be used on the accumulator:

CPL A ;ComPLement. NOT A. If A = B2h, CPL A makes A = ;4Dh. *(1B, 1C)*

CLR A ;CLeaR. Sets A to zero. *(1B, 1C)*

There are also byte manipulations that can only be carried out on the accumulator:

RL A ;Rotate the byte in A one bit to the left. The MSB becomes ;the LSB. *(1B, 1C)*

RLC A ;Rotate the byte in A and the carry bit to the left by one ;bit. The carry bit becomes the LSB and the MSB the ;carry. *(1B, 1C)*

```
        RR A            ;Rotate the byte in A one bit to the right. (1B, 1C)
        RRC A           ;Rotate the byte in A and the carry one bit to the right. The
                        ;carry becomes the MSB and the LSB the carry. (1B, 1C)
        SWAP A          ;Exchange low and high nibbles in A. (1B, 1C)
```

For example:

```
        CLR C           ;C = 0
        MOV A,#E6h      ;[A] ← 11100110b
        RL A            ;[A] ← 11001101b = CDh, C = 0
        RLC A           ;[A] ← 10011010b = 9Ah, C = 1
        RR A            ;[A] ← 01001101b = 8Dh, C = 1
        RRC A           ;[A] ← 10100110b = A6h, C = 1
        SWAP A          ;[A] ← 01101010b = 6Ah, C = 1
```

When the *destination* of a logical operation is the port address, the *latch* register not the pins is used both as the source of the data and as the destination of the result. Logical operations that use the port as a source but not a destination use the pins as source. For example, suppose port0 has its pins grounded (zero) and its latch contains FFh, then

```
        ANL P0,#0F0h    ;Reads FFh from the latch, ANDs it with F0h to give F0h
                        ;and sends this to the latch, which sets the lower latch bits
                        ;of port0 to zero, so turning off the lower output transistors
                        ;and pulling the lower pins up to 1
        ANL A,P0        ;Reads the port0 *pins* and not the port latch
```

Bit-level operations

These are useful for control. They can only be done with bit-addressable locations in RAM. The bytes of RAM that are bit addressable are from 20h to 2Fh, with bit address running from 00h to 7Fh. Thus the address of bit 4 of RAM address 28h is 44h.

byte address	bit address
20h	00h-07h
...	...
2Eh	70h-77h
2Fh	78h-7Fh

The SFRs do not all have bit addresses, those that do are

SFR	Byte addr	Bit addr
A	0E0h	0E0h-0E7h
B	0F0h	0F0h-0F7h
IE	0A8h	0A8h-0AFh
IP	0B8h	0B8h-0BFh
P0	80h	80h-87h
P1	90h	90h-97h
P2	0A0h	0A0h-0A7h
P3	0B0h	0B0h-0B7h
PSW	0D0h	0D0h-0D7h
TCON	88h	88h-8Fh
SCON	98h	98h-9Fh

When bit operations are used in the mnemonics, the assembler recognises hexadecimal addresses such as 9Fh as bit 7 of the timer control register, TCON. One can also write TCON.7 in the mnemonic and the assembler recognises this too. Using hexadecimal bit addresses makes a program more difficult for people to follow.

Bit-level Boolean operations are often done on the carry flag of the PSW, mnemonic C, which is a useful way of controlling the program according to the status of C.

ANL C,b	;AND C with addressed bit, b, put result in C. *(2B, 2C)*
ANL C,/b	;AND C and the complement of the addressed bit and put ;the result in C. b is unaffected. *(2B, 2C)*
ORL C,b	;OR the carry flag with bit b, store in C. *(2B, 2C)*
ORL C,/b	;OR C with the complement of b, store in C. *(2B, 2C)*
CPL C	;Invert C. *(1B, 1C)*
CPL b	;Invert b. *(2B, 1C)*
CLR C	;Set C to zero. *(1B, 1C)*
CLR b	;Set b to zero. *(2B, 1C)*
MOV C,b	;Copy b into the carry. *(2B, 1C)*
MOV b,C	;Copy the carry into b. *(2B, 2C)*
SETB C	;Set the carry flag to 1. *(1B, 1C)*
SETB b	;Set b to 1. *(2B, 1C)*

Other flags must be set using their bit addresses. As with byte operations the port bits used as destination are latches not pins. A port bit used only as a source is the pin. Latch bit logical operations are CLR, CPL, MOV, SETB only.

As an example, suppose we wish to double the contents of R0 and store the result in R1 (low byte) and R2 (high byte), the assembler instruction code for this is

MOV A,R0	;[A] ← [R0]
CLR C	;[C] ← 0, we must be sure C is 0
RLC A	;Doubles [R0], with carry
MOV R1,A	;Low byte in R1

```
CLR A           ;[A] ← 00h, clears A to receive carry
RLC A           ;The carry bit was bit 9, now bit 0 of A
MOV R2,A        ;[R2] ← [A], stores the carry (high byte) in R2
```

12.2.6 Arithmetic operations

There are 24 arithmetic operation codes of the following types:

```
INC A           ;Increment [A] by 1. (1B, 1C)
INC Rn          ;Increment [Rn] by 1. (1B, 1C)
INC addr        ;Increment the contents of RAM address by 1. (2B, 1C)
INC @Rp         ;Increment [[Rp]] by 1. (1B, 1C)
DEC A           ;Decrement [A] by 1. (1B, 1C)
...             ;etc. DEC is just like INC
ADD A,Rn        ;[A] ← [A] + [Rn]. C = 1 when [A] > FFh. (1B, 1C)
ADD A,addr      ;[A] ← [A] + [addr]. C  = 1 when [A] > FFh. (2B, 1C)
ADD A,#n        ;[A] ← [A] + #n. C = 1 when [A] > FFh. (2B, 1C)
ADD A,@Rp       ;[A] ← [A] + [[Rp]]. C = 1 when [A] > FFh. (1B, 1C)
ADDC A,Rn       ;[A] ← [A] + [Rn] + [C]. (1B, 1C)
...             ;etc. ADDC is like ADD, but with carry
SUBB A,Rn       ;[A] ← [A] – [Rn], with carry. (1B, 1C)
...             ;etc. SUBB is just like ADDC, except – for +
MUL AB          ;Multiply [A] by [B] and store in A. Low order byte is left
                ;in A and the high order in B. If the product > FFh, the
                ;overflow flag is set, otherwise it is cleared. The carry is
                ;always cleared. (1B, 4C)
DIV AB          ;Divide [A] by [B], store in A. The integer part of the
                ;quotient is stored in A, and the integer remainder in B.
                ;The carry and overflow flags are both cleared. If [B] = 0
                ;before division, [A] and [B] are both undefined, C = 0 and
                ;OV = 1. (1B, 4C)
DA A            ;Decimal adjust A. Can only be used after ADD or ADDC,
                ;not on its own. (1B, 1C)
```

There are several difficulties with arithmetic on the 8751. The first is that the numbers are of only 8 bits, so that accuracy is easily lost (and is at best ±0.4%). To maintain accuracy one must use two-byte arithmetic. In any case it is up to the programmer to take care of any carries and borrows. The flags affected are C, AC (auxiliary carry), OV (overflow), P (parity) all in the PSW. Another problem is that any arithmetic routine using division has to be protected by the programmer from inadvertent division by zero. The following is a list of the flag-affecting operations:

Mnemonic	Flags affected
ADD	C, AC, OV
ADDC	C, AC, OV
ANL C,addr	C
CJNE	C
CLR C	C = 0
CPL C	C = C'
DA A	C
DIV	C = 0, OV
MOV C,addr	C
MUL	C = 0, OV
ORL C,addr	C
RLC	C
RRC	C
SETB C	C = 1
SUBB	C, AC, OV

All flags are in the PSW. Flag C is set during borrowing. The parity flag is set at any time the 1s in A add up to odd, no matter what the operation.

Incrementing and decrementing

These instructions can be the best way of getting a desired result – simply keep INCing or DECing. No flags are affected.

INC A	;[A] ← [A] + 1. *(1B, 1C)*
INC Rn	;[Rn] ← [Rn] + 1. *(1B, 1C)*
INC @Rp	;[[Rp]] ← [[Rp]] + 1. *(1B, 1C)*
INC addr	;[addr] ← [addr] + 1. *(2B, 1C)*
INC DPTR	;[DPTR] ← [DPTR] + 1. *(1B, 2C)*
DEC A	;[A] ← [A] − 1. *(1B, 1C)*
...	;etc. DEC is like INC, *except* that the DPTR cannot be ;DECed

The contents overflow from FFh to 00h. The DPTR cannot be DECremented, but there is INC DPTR, which violates the principle of regularity – rules should be made without exceptions.

Addition

The first operand must be A, which is also used to store the result. C is set to 1 if there is a carry out of bit 7, the MSB, otherwise it is cleared to 0. AC set to 1 if there is a carry out of bit 3 (lower nibble), otherwise it too is cleared. OV is set to 1 if there is a carry out of bit 7, but not bit 6; or a carry out of bit 6 but not bit 7: in Boolean terms, OV = C7 XOR C6.

Unsigned or signed addition

The programmer can decide if the numbers used in the program are unsigned or signed. Signed numbers use bit 7 as a sign bit in the most significant byte of the group of bytes chosen by the programmer to represent the largest number needed. Bits 0-6 of the most significant byte and 0-7 of any other bytes represent the magnitude and bit 7 of the MSB the sign. If bit 7 is 1, the number is negative and if bit 7 is 0, it is positive. All negative numbers are in 2's complement form. A single-byte number can therefore range from −128d to +127d, and as 000d is positive, there are 128 negative and 128 positive numbers. The OV flag is not used for unsigned addition or unsigned subtraction. Unsigned addition uses the carry flag to show numbers > FFh, for example

> MOV A,#0D5h
> ADD A,#0A7h

The accumulator will contain 7Ch while C contains 1. The carry must be saved in a more significant byte.

Signed addition may be carried out in two ways: by adding like and unlike signed numbers. There is no way that addition of unlike signed numbers can exceed −128d or +127d. For example

$$-002d = 11111110b$$
$$\underline{+045d = 00101101b}$$
$$+043d = 00101011b$$

The carry, C = 1, and OV = 0 as there was a carry from bit 6, the next most significant. The carry is ignored as OV = 0. When positive numbers are added you can get overflow:

$$+046d = 00101110b$$
$$\underline{+099d = 01100011b}$$
$$+145d \quad 10010001b = -121d, \text{ wrong!}$$

But C = 0 and OV = 1, so corrective action can be taken, which is to complement the sign bit. If two positive numbers are added without exceeding +127d, then C = 0 and OV = 0. Adding two negative numbers that do not exceed −128d gives:

$$-033d = 11011111b$$
$$\underline{-044d = 11010100b}$$
$$-077d = 10110011b$$

C = 1 and OV = 0, so the result is correct. Adding two negative numbers that give too big a result:

$$-096d = 10100000b$$
$$\underline{-064d = 11000000b}$$
$$-160d \quad 01100000b = +096d$$

C = 1 and OV = 1, so the result can be corrected by complementing the sign bit. To summarise:

C	OV	Action
0	0	None
1	0	None
0	1	Complement the sign
1	1	Complement the sign

The only action required is to complement the sign when OV = 1.

Multi-byte signed numbers work similarly but a chain of carries forms. Only the most significant byte requires a sign bit and then all other bytes are treated like unsigned numbers. To use multi-byte numbers we must employ a carry:

ADDC A,#0D2h

Note that the least significant bytes are added with ADD and only higher bytes are added with ADDC.

Subtraction

There are two ways in which subtraction can be achieved: (1) by 2's complement addition and (2) by direct subtraction using SUBB. With unsigned numbers the carry flag, C, must be cleared to zero before subtracting or else it is included. For multi-byte subtractions C is cleared for the first byte but is included for subsequent bytes, as in

$$022d = 00010110b$$
$$\text{subtract} \quad \underline{101d = 01100101b}$$
$$-79d \quad 10110001b = 177d$$

The carry, C = 1 and OV = 0. The 2's complement of the result is 079d. Subtracting 22d from 101d gives C = 0, OV = 0 and the correct result. Signed subtraction is like signed addition and if OV = 1 the sign bit of the result must be complemented.

Multiplication and division

These operations use registers A and B and treat *all* numbers as *unsigned*. In multiplication the high order byte of the result is placed in B and the low order in A. If the result is > FFh, OV = 1 indicating that you must look for a high order byte in B. There cannot be overflow from the B register as FFh × FFh is the largest product whose value is FE01h

```
MOV A,#0A1h
MOV B,#0C2h
MUL AB          ;A = 02h, B = 7Ah, OV = 1
```

In division DIV AB means divide [A] by [B] and put the integer part of the quotient into A and the integer remainder in B. For example, if A = 0FEh (254d), B = 23h (35d), then DIV AB gives [A] = 07h (07d), [B] = 09h (09d). Before division OV = 0 unless B contains zero, when OV is set to 1. The carry is always reset.

Decimal arithmetic

Decimal numbers can be used if they are put in BCD form. BCD numbers occupy 4 bits, so two can be put in a register. Unfortunately the CPU can only add in binary and the result needs to be adjusted to give the right answer. The CPU can adjust if given the DA A command (decimal adjust A). Only addition can be performed on BCD numbers. For example:

```
MOV A,#38h
ADD A,#24h        ;[A] ← 5Ch
DA A              ;[A] ← 62d
ADDC A,#19h       ;[A] ← 7Bh, C = 0
DA A              ;[A] ← 81d
ADDC A,#23h       ;[A] ← A4h, C = 0
DA A              ;[A] ← 04d, C = 1
```

The result of adding 38h to 24h is 5Ch, which is 62d. Adding 23h to 81d gives 104d, so there is a carry of 1. BCD numbers can be used only with ADD and ADDC.

12.2.7 Jumps

Jumps give the microcontroller its real power: that of decision making. Instead of executing the program instructions in sequence, the microcontroller can be made to go unconditionally or conditionally to a different point in the program by means of *jumps* or calls on subroutines. There are jumps made according to the condition of certain specified bits, or bytes, and unconditional jumps. Jumps and calls are also classified according to their ranges, that is, how much the program counter has to be changed. There is a *relative* address range of +127d to −127d bytes from the instruction following the jump or call; an *absolute* range of anywhere on the same 2kbyte page as the instruction following; and a *long* range address of 0000h to FFFFh — anywhere in the memory. Relative addresses are identified by *labels* written to the left of the instruction mnemonic.

Bit jumps

The bit jumps are

```
JC rad       ;Jump to relative address if carry = 1. (2B, 2C)
JNC rad      ;Jump to rad if C = 0. (2B, 2C)
JB b,rad     ;Jump to rad if the addressable bit, b = 1. (3B, 2B)
JNB b,rad    ;Jump to rad if b = 0. (3B, 2C)
JBC b,rad    ;Jump to rad if the b = 1, then clear b to 0. (3B, 2C)
```

JBC is used for clearing flags. If a JBC mnemonic refers to a port bit, the state of the port latch SFR is changed, not cleared. The following example shows the function of JNC, JNB and JBC:

Label	Mnemonic	Comment
NEXT:	MOV R2,#32h	;[R2] ← 32h
	MOV A,R2	;[A] ← 32h, this initialises A
LOOP:	ADD A,R2	;[A] ← [A] + [R2], C = 0
	JNC LOOP	;Do LOOP until C = 1, then go on to next ;instruction
	MOV A,R2	;[A] ← 32h, initialising A
FLAG:	ADD A,R2	;[A] ← [A] + [R2]
	JNB 0D7h,FLAG	;The carry flag's bit address is D7h. The loop is ;repeated until C = 1
	JBC 0D7h,NEXT	;When C = 1, clear C to 0, go to NEXT and ;repeat the program

Using the JBC instruction instead of JB makes a CLR C instruction unnecessary. Labels cannot be words in the instruction set (*reserved* words). The label must be at an address within +127d and −128d of the next instruction to the jump. Jump ranges can be calculated by looking up the bytes used for each mnemonic. Normally the assembler should trap out-of-range addressing.

Example

In this example JNC is used to select different routines according to which byte is larger in 16-bit subtraction. The object is to subtract the smaller of two 16-bit numbers from the larger. We start off with one 16-bit number in R0 (high byte) and R1 (low byte) and the other in R2 (high byte and R3 (low byte) and the result is to be stored in R4 (high byte) and R5 (low byte). The whole subroutine is called by LCALL from the main program (LCALL is discussed in the next section).

SUBTR:	MOV A,R1	;The low byte in R1 is copied to A
	MOV B,R3	;The other low byte in R3 is copied to B
	CLR C	;Make sure C = 0
	SUBB A,B	;[A] ← [R1] − [R3], low byte subtracted
	MOV R4,A	;Low byte of result is stored in R4
	MOV A,R0	;High byte is copied into A
	MOV B,R2	;Other high byte copied to B
	JNC SUBT1	;Go to SUBT1 if C = 0, that is if [R1] > [R3]
	CLR C	
	INC B	;C = 1 if [R1] < [R3], so a carry of 1 is added to ;high byte that was in R2
SUBT1:	SUBB A,B	;[A] ← [R0] − [R2], high bytes subtracted

```
                        JNC HI_SAV    ;Go to HI_SAV if C = 0, that is if [R0] > [R2]
                        MOV A,R3      ;If C = 1, the order of subtraction must be
                                      ;reversed
                        MOV B,R1
                        CLR C
                        SUBB A,B      ;[A] ← [R3] − [R1], low byte subtracted but
                                      ;numbers are reversed
                        MOV R5,A      ;Store low byte in R5
                        MOV A,R2      ;High byte
                        MOV B,R0      ;Other high byte
                        JNC SUBT2     ;Go to SUBT2 if C = 0
                        CLR C
                        INC B         ;If C = 1, must increment high byte
        SUBT2           SUBB A,B      ;[A] ← [R2] − [R0], high byte subtracted but
                                      ;numbers are reversed
        HI_SAV:         MOV R3,A      ;high byte stored in R3
                        RET           ;Return to main program
```

Note that the assembler treats upper case and lower case letters alike; CLR, CLr, cLR, ClR, Clr, cLr, clR and clr are all read as CLR. Clarity of presentation should determine which is used.

Byte jumps

The byte jumps are:

```
    CJNE A,addr,rad     ;If [A] ≠ [addr], jump to relative address. If [A] < [addr], then C
                        ;:= 1, else C = 0. (3B, 2C)
    CJNE A,#n,rad       ;If [A] ≠ n, jump to relative address. C = 1 if [A] < n, else C =
                        ;0. (3B, 2C)
    CJNE Rn,#n,rad      ;If [Rn] ≠ n, jump to relative address. C = 1 if [Rn] < n, else C
                        ;:= 0. (3B, 2C)
    CJNE @Rp,#n,rad     ;If [[Rp]] ≠ n, jump to relative address. C = 1 if [[Rp]] < n, else
                        ;C = 0. (3B, 2C)
    DJNZ Rn,rad         ;[Rn] ← [Rn] − 1, then jump to rad if [Rn] ≠ 0. (2B, 2C)
    DJNZ addr,rad       ;[addr] ← [addr] − 1, jump to rad if [addr] ≠ 0. (3B, 2C)
    JZ rad              ;Jump to rad if [A] = 0, flags and A unaffected. (2B, 2C)
    JNZ rad             ;Jump to rad if [A] ≠ 0, flags and A unaffected. (2B, 2C)
```

For example:

```
                        MOV R0,#FFh
        WAIT:           DJNZ R0,WAIT          ;Does loop 254d times =
                                              ;508d machine cycles
                        MOV A,#00h            ;initialises A
        LONG_WAIT:      INC A
```

        CJNE A,#FFh,LONG_WAIT   ;Does loop 255d times =
                                 ;765d machine cycles

Byte jumps are economical and powerful instructions.

Unconditional jumps

And finally the unconditional jumps:

    JMP @A+DPTR   ;Jump to the address formed by adding [A] to [DPTR]. *(1B, 2C)*
    AJMP sad   ;Jump to the absolute short-range address. *(2B, 2C)*
    LJMP lad   ;Jump to the absolute long-range address. *(3B, 2C)*
    SJMP rad   ;Jump to relative address. *(2B, 2C)*
    NOP   ;Do nothing. This wastes time and can also be used to save
                   ;places in the program for later use. *(1B, 1C)*

An example of the use of NOP is

    CLR P1.0   ;Pin0 of port1 is driven low
    NOP   ;1 (machine) cycle elapsed
    NOP   ;2 cycles elapsed
    NOP   ;3 cycles elapsed
    SETB P1.0   ;Pin0 of port1 is driven high. 4 cycles elapsed.

The effect is to put a low pulse on pin P1.0 for exactly 4 machine cycles (48 oscillator cycles).

12.2.8 Calls and subroutines

Calls are made to subroutines which are located in short-range addresses (those on the same page as the instruction) or in long-range addresses (those anywhere in ROM). Interrupts require subroutines to be placed at the interrupt servicing address. For example, a program could be written which tested each port pin in turn to see what data had been received, a technique called polling. Polling takes time (a few ms) and the program could be made to respond immediately to the receipt of data at a port (within a few μs), by using an interrupt. Software interrupts can be written using *calls* to execute a subroutine. A call results in a jump to the address of the subroutine. When the subroutine has been executed the program resumes at the instruction immediately following the call. The PC is stored at the top of the RAM stack when a call is executed, requiring 2 bytes of RAM as the PC contains 16 bits. This is the location of the *next* instruction to be executed. When a *return* is encountered at the end of the subroutine the stack is automatically POPped twice to restore the PC. The subroutine return mnemonic is RET and the interrupt return instruction is RETI. The mnemonics for calls are:

    ACALL sad   ;Call the subroutine located at the short-range address which is

```
                              ;located on the same page as the instruction after ACALL. Push
                              ;the PC contents onto the stack. (2B, 2C)
        LCALL lad             ;Call the subroutine from anywhere in memory. Push the PC
                              ;contents onto the stack. (3B, 2C)
        RET                   ;Pop two bytes from the stack to the PC, returning the program
                              ;to the next instruction after ACALL or LCALL. (1B, 2C)
```

In the following example AJMP is used to skip the interrupt subroutine and start timer1.

```
        ORG 0000h             ;ORG is an assembler directive, meaning ORiG-
                              ;inate the code on the next line at the address in
                              ;ROM given. Here, it puts the first instruction of
                              ;the program at ROM address 0000h
        AJMP NEXT             ;Jump over interrupt subroutine
        ORG 001Bh             ;Put timer1 interrupt subroutine here
        CLR 8Eh               ;Stop timer1, reset TR1 = 0. (TR1 is timer1's
                              ;control bit whose bit address is 8Eh)
        RETI                  ;Return and enable interrupt structure
        ...                   ...
        ...                   ...
NEXT:   MOV 0A8h,#88h         ;Enable timer1 interrupt in the IE register. A8h is
                              ;the IE SFR address. 88h is 10001000 so that
                              ;bit7 and bit3 are set to 1. Bit7 is the
                              ;master interrupt enable and bit3 is the timer1
                              ;interrupt enable
        MOV 89h,#00h          ;Set timer operation mode0. 89 is the TMOD
                              ;SFR address
        MOV 8Bh,00h           ;Clear TL1 counter
        MOV 8Dh,00h           ;Clear TH1 counter
        SETB 8Eh              ;Timer1 control bit is bit5 in TCON, whose bit
                              ;address is 8Eh. Start timer1, set TR1 = 1
```

12.2.9 Look-up tables

Because arithmetic is very limited on an 8-bit microcontroller, it is better to compute the values of functions such as $\sin x$, $\ln x$, x^2 or $\sqrt{x}$, and put them into a table which can be looked up for the appropriate values. The table takes up some memory, but can be read much faster than the computation can be completed. The look-up table can be placed anywhere in ROM using the ORG directive and the bytes can be read in using the DB (define byte) directive. For example a table for $\sqrt{x}$ might begin

```
        ORG 2000h             ;Start table at ROM location 2000h
        DB 00h                ;ROM address 2000h contains 00d
        DB 01h                ;2001h contains 01d
```

```
DB 01h          ;2002h contains 01d
DB 02h          ;2003h contains 02d, √3d = 2d
DB 02h          ;2004h contains 02d
...             ...
DB 10h          ;20F0h contains 16d, √F0h = 16d
...             ...
DB 10h          ;2100h contains 16d
```

If the numbers are restricted to one byte, or 8 bits, looking up is straightforward. For example, if the square root of 09h is wanted, the DPTR is set to 2000h, which is the *base address* (the table begins at 2000h). The correct number is fetched with an indirect-address instruction such as MOVX A,@A+DPTR, if 09h (the *offset*) is first placed in A.

```
MOV DPTR,#2000h     ;2000h is the base address in external ROM
MOV A,#09h          ;09h is the number whose square root is wanted
MOVX A,@A+DPTR      ;A contains the desired square root
```

If greater precision is needed, we must increase the number of bytes in the square root to two and look them up separately as high and low bytes. Though more tedious, it does not differ much from the example given. The same is true for other 16-bit tables.

Programs are easier to write and to understand if names are used for constants. The assembler directive EQU can be used as in

```
EQU base_addr,#10F0h
MOV DPTR,#base_addr
EQU hibyte,#1000h
EQU lobyte,#1200h
```

This has been an exceedingly detailed section, but there is no middle way between the superficial discussion of microprocessors in general and the close examination of a particular model. Only by carefully studying the instruction set and facilities of a specific device can one begin to put it to work usefully. The 8051 series has much in common with other microcontrollers, so that a working knowledge of it will assist in using all of them.

12.3 Interfacing

Microcontrollers must send and receive signals from the external environment: that is their purpose. But most data comes in a form which is incompatible with the format required by the microcontroller, and many devices to be controlled either require data in a different form or have power or current sourcing or sinking requirements that are beyond a micro-controller's limited capacity. Interfacing devices are required to overcome these problems and so enable the microcontroller to perform its functions. Many ICs have been specifically designed for interfacing with particular microcontroller families, while other more simple devices have general applicability.

12.3.1 Transferring data

Data is transferred between a microcontroller and its peripherals via a collection of wires called a bus, which is divided into three subsections: the address bus, the data bus and the control bus. The address bus conveys the identity or number of the device to or from which data is transferred, and the control bus signals the type of data to be transferred and also provides synchronisation. The address bus has as many lines as there are bits in the memory address, usually 16 for 8-bit microcontrollers. The number of locations that can be addressed is then $2^{16} = 65536$. The data bus is the defining bus for microcontrollers: an n-bit microcontroller has n lines in its data bus, so the 8051 microcontrollers have eight data lines. The control bus will be smaller, normally with four lines for an 8-bit micro-controller, and a total of only 16 control codes. Different microcontrollers have different ways of signalling and each has to be learnt as the need arises. Figure 12.19 shows the set up with the number of lines in each bus written over them with an oblique line through the bus.

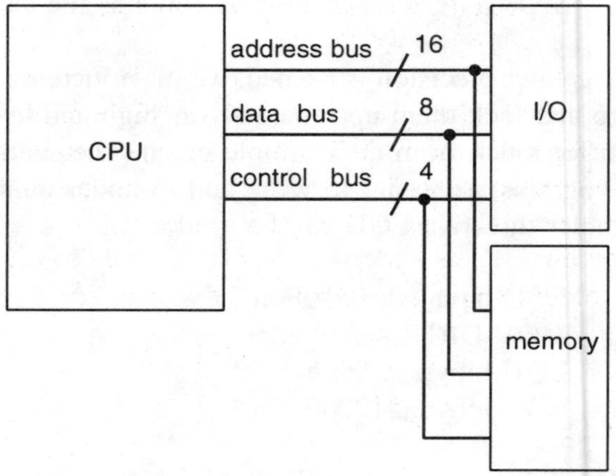

Figure 12.19

Memory, data and control buses in 8-bit microcontrollers

Data may be written to or read from an external device according to the state of the write and strobe (an enabling pulse) signals. When the write signal is HIGH (logic 1) data is *read* and when it is LOW (logic 0) data is written, so that write is active low. This is indicated by placing an asterisk after it: write*. The timing diagram for a write instruction might look like figure 12.20 which shows the address lines, data lines and write* and strobe* (which is also active low) signals.

Figure 12.20

Timing diagram for a write operation

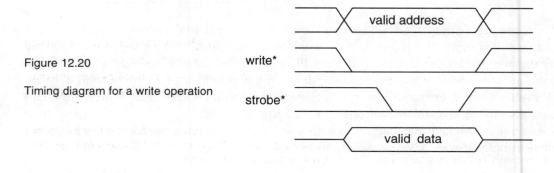

In figure 12.20 neither the horizontal time axis nor the vertical voltage axis is indicated, which is the custom. The three kinds of transition shown in figure 12.21 mean: firstly (figure 12.21a) that the address data changes from one defined state to another at the time indicated by the crossover; secondly (figure 12.21b) that the write* and strobe* lines change from whatever state they were in to a logical 0 at the instant indicated; thirdly (figure 12.21c) that the state changes from whatever it was to a logical 1; and fourthly (figure 12.21d) that the data lines change from an undefined state to a defined state at the times indicated.

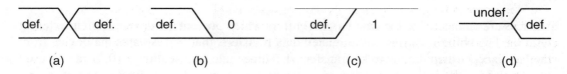

| (a) | (b) | (c) | (d) |

Figure 12.21   (a) Transition from one defined state to another (b) Transition from a defined state to logical 0 (c) Transition from a defined state to logical 1 (d) Transition from an undefined state to a defined state

A similar timing diagram to that of figure 12.20 can be constructed for the process of reading data but the write* signal is then set to logical 1. For a write cycle the master[14] microcontroller puts the address to be written to on the address bus and the data to be transferred on the data bus. It then sets write* to logical zero and when all the signals have settled down it sets strobe* to zero (it is said to *assert* strobe*). After a fixed length of time strobe* is released to 1 and the master stops driving the data bus.

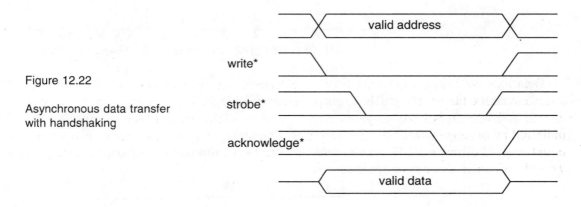

Figure 12.22

Asynchronous data transfer with handshaking

The problem is that the process of transferring data from the master CPU to the slave peripheral and vice versa takes time and that time will vary from one slave device to another, so how long must strobe* be asserted? Obviously, as long as it takes the slowest slave to accept the data. This means that all data transfers must run at the pace of the

---

[14] 'Master' here means the device which initiates an action. 'Slave' means a device which is read from or written to by the master device. Devices which are written to are sometimes called receivers and devices which are read from are sometimes called transmitters. Masters and slaves can do either or both.

slowest slave, which is clearly not a good thing as it will slow down the operation considerably: there is no point in spending 1 ms on an operation that need only take 10 ns. The solution is to communicate from the slave to the master so that the master knows that the data has been received. This form of acknowledgement is called *handshaking*, and the data transfer is called asynchronous, in contrast to that just described which is termed synchronous. Thus an additional signal, acknowledge*, is required and the timing diagram for the write operation is that of figure 12.22.

12.3.2 *Data buffering*

Devices are connected to the data bus via buffers which prevent the device from interfering with the bus while it carries out unrelated data transfers, that is it isolates the device from the bus except when data is to be transferred. Three-state (see section 9.10.3) can be used for the purpose of isolation as they present a high impedance to the bus when not in use. Figure 12.23 shows an octal three-state CMOS buffer in the 74 series of logic circuits, the 74HCT240, in 20pin DIL form.

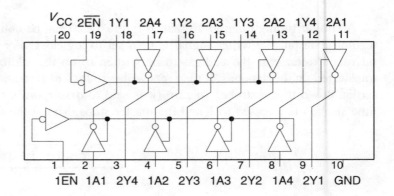

Figure 12.23

The 74HCT240 octal three-state buffer

The truth table for the buffer is given below in table 12.7

Table 12.7 *The truth table for the 74241 three-state buffer*

1EN	1A	1Y	2EN	2A	2Y
1	1	1	1	1	1
1	0	0	1	0	0
0	1	Z	0	1	Z
0	0	Z	0	0	Z

Notes: Z = high-impedance state. The 74240 buffer inverts the data; the 74241 does not.

When EN is HIGH the data is transferred. When EN is LOW the outputs are high impedance regardless of the states of the inputs. For example, if we wish to connect a hex-adecimal keypad to the bus we must connect the four data lines to one side of the octal buffer and also the keypad status line which tells whether a key has been pressed. This leaves three input lines spare which can be connected to ground (it is always unwise to leave unconnected inputs floating). Figure 12.24 shows these connections. Since the grounded inputs will read as zeros they are made the MSBs (pins 2A2-4) and the status line is the LSB of the upper nibble (2A1). The two enable pins are connected together.

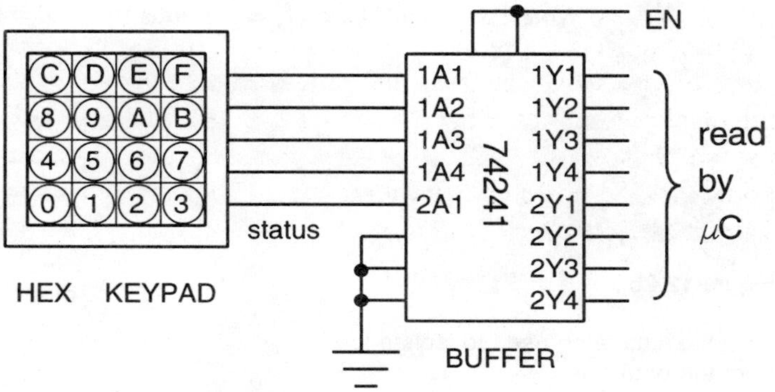

Figure 12.24

Hex keypad and data buffer

12.3.3 Latches

Ports 1, 2 and 3 of the 8051 microcontroller have output buffers which can drive up to four TTL loads each and port 0 can drive eight. But microcontrollers cannot supply much current to drive loads and one way around this problem is to use a latch to source or sink current. Generally speaking latches can sink more current than they can source. The 74 series of ICs contains a number of latches such as the 74HCT373 which is a three-state octal D-type latch, capable of supplying ±35 mA at each pin. The data pins of the latch can be written to and the data transferred to the output pins, thus either driving them HIGH (sourcing current) or LOW (sinking current). Two control pins are used on the 74373 latch, the latch enable (EN) and the output control (OC*). Its truth table is that of table 12.7, which shows that the OC* is normally LOW and when HIGH it causes the latch to present a high impedance to the bus. Maintaining both control pins at LOW keeps the data on the outputs. With OC* LOW and EN HIGH the data on the input pins is transferred to the output pins.

Figure 12.25 shows the latch connected to one load which takes current (the LED) and one which supplies current. The latch will dissipate less power when sinking current and this is to be preferred if several loads are to be activated simultaneously. The resistor R should be about 150 Ω if $V_{CC} = 5$ V, to restrict the LED current to about 20 mA. If the current into the latch pins is to be no more than 35 mA, then R_L should be not less than 130 Ω if V_{CC} is 5 V. This current is enough to drive many relays and so large loads can be accommodated.

Table 12.7 *Truth table for the 74373 three-state octal latch*

OC*	EN	Input	Output
0	1	1	1
0	1	0	0
0	0	×	Q_p
1	×	×	Z

Notes: × = don't care Q_p = previous value Z = high-impedance state.

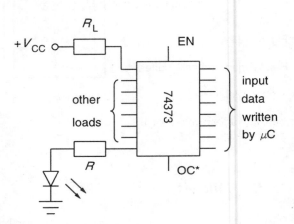

Figure 12.25

An octal data latch used to isolate loads from the data bus

12.3.4 Interfacing the asynchronous bus

When several devices are connected to the microcontroller each is given an identifying number which is its address so that it can be written to or read from as required. An asynchronous bus requires two signals to be generated by external logic: a valid* and an acknowledge* signal. The valid* signal is generated when the address is the correct one and strobe* is asserted (that is LOW). The acknowledge* signal is generated when the data has been successfully read from or written to the device whose address is valid. If the address of the device is entered via a hexadecimal keypad then an 8-bit comparator (sometimes called an equality detector) will serve to generate the valid* signal. The 74HCT688 is an 8-bit device which will serve the purpose. Figure 12.26a shows the circuit for this. Normally each device on a bus will be attached to an equality detector which has the address entered, perhaps by switches as in figure 12.26b.

The timing of acknowledge* depends on the operational details of the devices being addressed. Figure 12.27 shows a timing diagram with delayed acknowledge*. In some cases it is necessary to delay assertion of acknowledge* and for this purpose an 8-bit shift register could be used to produce a delay varying from 1-8 clock cycles.

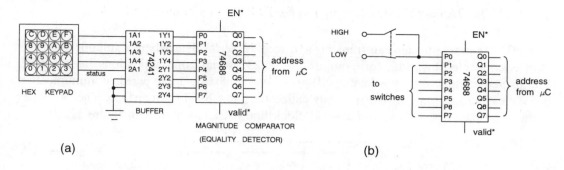

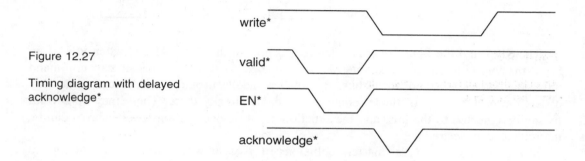

(a)

(b)

Figure 12.26 Using an equality detector to generate a valid* signal (a) Address entered by keypad (b) Address entered by switches

Figure 12.27

Timing diagram with delayed acknowledge*

Inverters can also be used (as well as other logic gates) to produce reasonably accurate unclocked delay times. The disadvantage of using gates, as in figure 12.28a, is that the delay is variable depending on the device family, the supply voltage and manufacturing tolerances. Thus a double inversion using a standard 74HCT04 CMOS IC can have a delaying effect of from 50 ns up to 120 ns. If a reasonably constant and definite delay is wanted, the circuit of figure 12.28b will work well. There are two inputs on this shift-register chip, either can be used. The delay at each output, Q_n, is $n/f_c + \Delta T$, where f_c is the clock frequency and ΔT is the propagation delay from clock input to gate output.

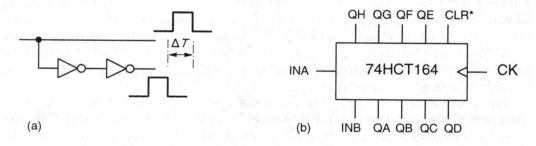

(a)

(b)

Figure 12.28 Delay circuits using (a) Gates (inverters here) (b) An 8-bit shift register

12.3.5 The Inter-IC (I^2C) bus interface

The 8051 microcontroller can transmit and receive serial data when a suitable interface chip is connected to port3, for example the RS232 serial communications interface IC. Philips Semiconductors have developed special on-chip hardware to permit communication between ICs using only a two-wire bus called the Inter-IC (I^2C) interface. The two lines are a serial data line (SDA) and a serial clock line (SCL) as shown in figure 12.29.

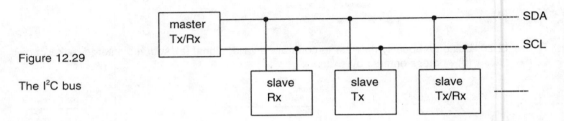

Figure 12.29

The I^2C bus

The SDA and the SDL are connected to V_{CC} and so the bus is HIGH when idle. If an 8051 master with the I^2C bus interface is attached to the two lines it can command numerous slaves attached to the bus. Removal or addition of further devices in no way upsets the performance of those remaining or those already there. There may be several masters connected to the bus, and the functioning of devices as masters or slaves can be interchanged if desired.

In a system with several masters collisions may occur when two devices attempt to initiate an action simultaneously. These collisions can be dealt with by software to avoid loss of data. We shall only consider systems with one master device which is an 8051 microcontroller.

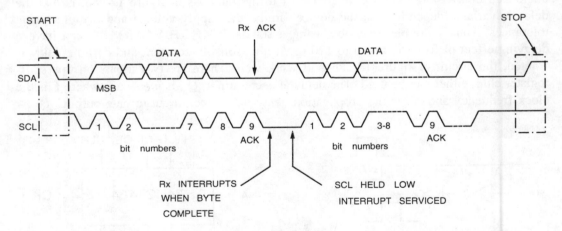

Figure 12.30 Timing diagram for the I^2C bus showing STOP and START conditions, and acknowledge (ACK) signal

Data transfers take place serially at a rate of one bit per clock cycle. The START of transfer is recognised by the condition shown in figure 12.30, where the data line goes from HIGH to LOW while the clock line is HIGH. End of transfer, the STOP condition, is signalled by the data line moving from LOW to HIGH while the clock line is HIGH. Only the master can generate STOPs and STARTs. Once the start condition is recognised, the data line is recognised as busy until the stop signal is received. After each byte of eight bits is transferred (in the order MSB first and LSB last) an acknowledge bit is sent. A receiver can delay data transfers with a WAIT signal generated by holding the SCL LOW. If the master does not receive an acknowledge signal after starting data transfer (perhaps because the device is already busy) then it must retransmit the data after a program-determined delay.

However, prior to any attempt at data transmission, the master must transmit the unique address of the device to be accessed. The address is a seven-bit number followed by a read/write* (or receive/transmit*) bit. This must be acknowledged by the addressed device before data can be sent or received. The address byte is compared by each device on the bus using an equality comparator and this sends the acknowledgement. Figure 12.31 shows the complete data transferral cycle. It is not strictly necessary for the master to signal STOP if it wishes to continue data tranfer with a different device, it merely sends out another address byte and data transfer, continuing for as long as needed before sending a STOP and freeing the bus. This is probably slightly more efficient than using STOPs after each device use is terminated, but in multimaster systems it would be unwise not to terminate each device transaction with STOP.

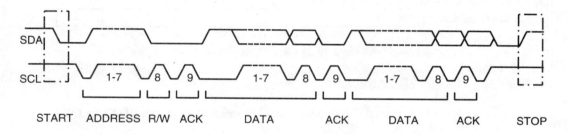

Figure 12.31 A complete data transfer on the I²C bus

Memory addressing on the I²C bus requires more than just the device address, it needs the word sub-address also. The memory chip has a word address register which is automatically incremented by one with each byte of data written. The sub-address must be sent after the device address so that the byte at that address (or a series of bytes starting from there) can be transferred. The byte sub-address is an 8-bit word, not a 7-bit word like the device address. Figure 12.32 shows the bit sequence in a frame from the master when sub-addressing is used.

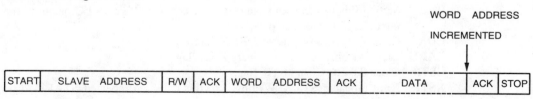

Figure 12.32 Using word addresses with master transmitting to slave device

Microcontroller hardware for the I²C bus

An 8051-series microcontroller (for example the 83C751) requires some special registers for the use of the I²C bus, and some of the other hardware has changed functioning. The SCL and SDA lines are attached to pins P0.0 and P0.1 of port0 respectively, which have open collector (actually drain for the device is MOS) outputs. Three SFRs control the I²C port: the I²C control SFR, called I2CON; the I²C configuration SFR, called I2CFG; and the I²C data SFR, called I2DAT. The layouts of these SFRs are shown in figure 12.33.

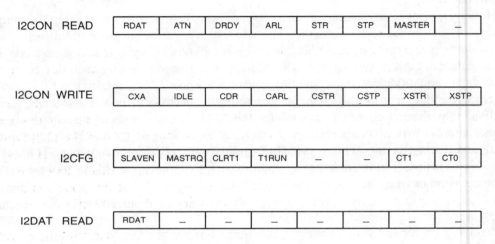

I2CON READ							
RDAT	ATN	DRDY	ARL	STR	STP	MASTER	–

I2CON WRITE							
CXA	IDLE	CDR	CARL	CSTR	CSTP	XSTR	XSTP

I2CFG							
SLAVEN	MASTRQ	CLRT1	T1RUN	–	–	CT1	CT0

I2DAT READ							
RDAT	–	–	–	–	–	–	–

Figure 12.33 The I²C registers

Notes: <u>I2CON READ register</u>

RDAT Received DATa bit. Content is identical to that in RDAT in I2DAT register.

ATN ATteNtion flag, set when any of DRDY, ARL, STR or STP flags are set.

DRDY Data ReaDY flag.

ARL ARbitration Loss flag.

STR STaRt flag.

STP SToP flag.

MASTER Flag set when device is a master.

<u>I2CON WRITE register</u>

CXA Clear Xmit Active. Writing 1 to CXA clears the transmit-active state.

IDLE Setting this to 1 causes a slave device to idle and ignore I²C bus until the next START is detected.

CDR Clear Data Ready. Clears DRDY flag.

CARL Clear ARbitration Loss. Clears ARL flag.

CSTR Clear STaRt. Clears STR flag.

XSTR Xmit repeated STaRt. Writing 1 to this causes a repeated START to be sent. Only for masters.

XSTP Xmit SToP. Sends a STOP. Xmit active state is set.

The I2CFG register

SLAVEN	Writing 1 to this enables slave functions of I²C interface.
MASTRQ	Request for bus control as master.
CLTR1	Clear timer1 interrupt flag. Always read as 0.
T1RUN	Writing 1 to this lets timer1 run. Will only run inside frames and is cleared by SCL toggling, START and STOP. Writing 0 to this stops and clears timer1.
CT1,CT0	Bits programmed according to frequency of crystal oscillator, f_{osc}.

I2DAT register

RDAT Received DATa bit. Taken from SDA on every rising edge of SCL.
XDAT Xmitted DATa bit. Determines the next bit to be transmitted.

In I²C operation timer1 is used to generate the I²C clock, that is SCL, and is not available for normal use. The clock rate for SCL is made independent of the actual microcontroller's oscillator frequency by setting bits CT0 and CT1 in the I2CFG SFR. Timer1 is clocked at the machine cycle frequency, $f_M = f_{osc}/12$, and the SCL is set to toggle when bit 4 of timer1 inverts (toggles). The preloaded values in the two LSBs of the timer vary the number of counts needed to cause bit 4 to toggle. What this means is that the SCL clock frequency is maintained reasonably constant at about 100 kHz for oscillator frequencies varying from 10-16 MHz.

Timer1 also looks for bus hang-ups by timing out the bus after 1024 machine cycles, a number which cannot be varied by the user. Timing out occurs if SCL has not changed state within a frame for this length of time (which is about 100 SCL clock cycles). A service routine is required for the interrupt generated by the timing out.

Of the five interrupts in the 83C751 microcontroller, two are used for the I²C bus. The IE2 flag of the interrupt enable (IE) register is used for to enable I²C interrupts which occur at START, STOP and when a new data bit appears on the bus. The service routine for the interrupt must be located at 023h. The timer1 overflow interrupt is enabled by the ETI flag and the interrupt service routine for this must be located at 01Bh.

12.4 16-bit and 32-bit microprocessors

The 8-bit microcontroller and microprocessor is inadequate for tasks which require relatively large amounts of data to be handled or processed, or for making extensive computations. In fact some industry estimates put the growth rate (*not* the absolute numbers sold) in 16-bit microcontroller sales as double that of the 8-bit device's. There is no possibility of going into the subject of 16-bit and 32-bit microcontrollers in any detail, so this section contains only a few remarks about one of the most popular 32-bit microcontrollers, the Motorola 68020, a development of the 16-bit 68000 (the 6800 series are 8-bit devices). A detailed exposition would take many more pages than we can afford. The manufacturer's data books are hundreds of pages long and must be consulted for operational details. The 68020 is one of a family of microprocessors developed and manufactured by Motorola, which comes in a square 114-pin package shown in figure 12.34.

The 68020 is more powerful than the 8051 microcontroller and has a much larger instruction set, besides being more expensive. It can also handle vastly more data at a greater speed, and address huge amounts of external memory, but it has only eight 32-bit data registers and eight 32-bit address registers on board.

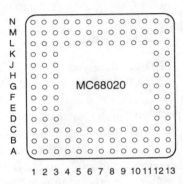

Figure 12.34

68020 pin configurations shown from pin side of package

Figure 12.35 shows the block diagram for the device, from which we see that additional ICs must be bought and connected to the microcontroller to perform whatever function is required. Since the applications are specific and therefore restricted, this makes more sense than providing lots of costly on-board hardware most of which will be unused. There are for example no on-board timers or serial communications ports.

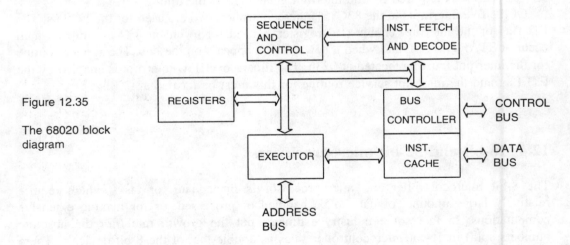

Figure 12.35

The 68020 block diagram

The registers available on the 68030 are shown in figure 12.36, which shows that the data registers can be used as 8-bit, 16-bit and 32-bit registers. Data can be processed as bytes (8-bits), words (16-bit) or as longwords (32-bits) by using extensions to the instructions: INST.B, INST.W or INST.L respectively, where INST is the instruction mnemonic. If no extension is used 16-bit (word) data is assumed.

The address registers can be split into 16-bit and 32-bit addresses. One of the address registers (A7) can be used as a user-controlled stack pointer (USP). In addition to these sixteen registers there is a 32-bit stack pointer (SP) register (split into two 16-bit words) and a 32-bit program counter (PC). The 8-bit code-condition register (CCR) contains five user-settable flags denoted C (bit 0), V, Z, N and X (bit 4).

Figure 12.36

The 68020 registers

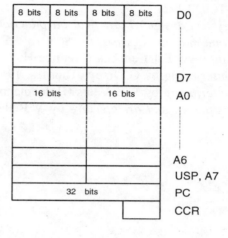

Instructions are of the two-operand variety:

INST source address,destination address

which is self explanatory, or of the one-operand kind:

INST address

in which the operation is carried out on the data at the specified address and the result is stored in that address. Addressing can be done in any of fourteen different modes in the 68020. An instruction cache is used to speed up data transfer so that the relatively slow clock frequency is not a handicap when handling 32-bit longwords. A floating-point coprocessor can be purchased to handle floating-point arithmetic operations and yields considerable computing power. Only when a great deal of data needs processing should the 32-bit microprocessor be used: it is not a cheap general-purpose device.

Suggestions for further reading

The first places to look for information on microprocessors and microcomputers are the manufacturers' data books, which are indispensable for the use of any of their products. It takes some time before any sensible choice from the numerous variants in each family can be made. The following provide some useful additional help:

Application notes for 80C51-base microcontrollers by Philips Semiconductors (1993)

80C51-based 8-bit microcontrollers: data handbook by Philips Semiconductors (1994)

Programming and interfacing the 8051 microcontroller by S Yerelan and A Ahluwalia (Addison-Wesley 1995)

Microprocessor systems by R J Mitchell (Macmillan 1995)

Digital and microprocessor engineering by S J Cahill (Ellis Horwood, 2nd ed., 1993)

Microcontrollers, architecture, implementation and programming by K J Hintz and D Tabak (McGraw-Hill, 1992)

Microprocessor system design, a practical introduction by M J Spinks (Newnes, 1992)

Microcomputers and microprocessors, the 8080, 8085 and Z80 programming, interfacing and troubleshooting by J E Uffenbeck (Prentice-Hall, 2nd ed., 1991)

Microprocessor interfacing by G Dixey (Thornes, 1991)

Microprocessor interfacing by R E Vears (Heinemann-Newnes, 1990)

Computer systems design and architecture by V P Heuring and H F Jordan (Benjamin-Cummings 1997)

13 Measurements, electrical transducers and EMC

FROM TIME to time students and practitioners of electronic engineering have to make measurements of some property of a physical system. Usually the property measured is converted into an electrical signal by means of a transducer, a device to convert energy from one form into another. Often the readings obtained are subjected to no critical analysis as it is obvious that their variability is small, or the range and magnitude of the data is entirely within expectations, given the sort of equipment used for the job. Nevertheless all engineers should be aware of the limitations in their apparatus and need the ability to perform simple statistical analysis of data.

The accuracy of electronic equipment is often determined by the type of transducer used to collect physical data from the outside world, and so more than an inkling of the types of electrical transducers available — and their performance as well — is necessary.

Electronic equipment is also subject to and the cause of electromagnetic interference (EMI) which can in extreme cases cause malfunction. In other cases less severe interference can degrade performance, though still within acceptable limits. The fitness of electrical or electronic equipment to function in its environment without suffering from or causing excessive EMI is called electromagnetic compatibility (EMC), and is now the subject of considerable and demanding EU legislation.

We will start by considering electrical standards and then the analysis of experimental errors; the following section deals with transducers, and the chapter ends with a description and discussion of EMC.

13.1 Standards

Physical properties such as force, pressure, density, energy, thermal conductivity etc. have SI (Système Internationale) units which are derived from a number of base units, namely mass (the kilogramme, kg), length (the metre, m), time (the second, s), temperature (the kelvin, K), electrical current (the ampere, A), amount of substance (the mole, mol) and luminous intensity (the candela, cd). Thus force is given in terms of the derived SI unit the newton (N) which is 1 $kgms^{-2}$ and has *dimensions* of mass (M), length (L) and (time)$^{-2}$ (T^{-2}). The order and style of writing units like this is very important, since 'm' also stands for milli (10^{-3}). Thus 1 ms means 1 millisecond and not 1 meter-second, which has to be written as 1 sm to avoid any confusion; likewise a kelvin-metre must be written Km and not mK, which is 10^{-3} K. When dealing with powers of subunits, the whole of the subunit is taken to be raised to the power indicated, thus 1 mm^2 means 1 $(mm)^2$ and so 10^{-6} m^2, not 1 $m(m^2)$ or 10^{-3} m^2. If in any doubt one must use brackets or other means of expressing the quantities unambiguously.

National standards laboratories maintain standards which are then used to calibrate secondary standards that can be used in other laboratories with somewhat reduced accuracy. In the UK this is the National Physical Laboratory, Teddington, which is always keen to discuss measurements with any interested parties in industry, commerce or academia. All measurements of a physical property in the UK should be traceable to an NPL standard.

The kilogramme unit of mass is defined by a cylinder of platinum kept at Sèvres in France. Various copies of this standard are sent for comparison periodically. Careful weighing on a balance can achieve an accuracy of about 1 in 10^9 but routine weighings in air without buoyancy corrections are no better than 1 in 1000. With proper correction for the buoying of the weights and the weighed matter by air, accuracies of about 1 in 10^6 can be achieved in ordinary laboratories with good balances.

The unit of length, the metre, is now based on the speed of light in a vacuum, and is the distance travelled by light in $1/299\,792\,458 = 3.335\,640\,952$ ns. It is relatively easy to translate this definition into marks on a ruler of some kind which can be used in the field with an accuracy of about 1 in 1000. Most surveying, however, is done by timing pulses of laser light to a much greater accuracy than this, say 1 in 10^5.

The unit of time is now defined by the frequency of radiation coming from electronic transitions between well-defined states in various atoms, for example the Cs-133 isotope or hydrogen. The standard is used to send out time signals by radio which are accurate to about 1 part in 10^{12}, and these are used in laboratories to calibrate counters and timers. Frequency is certainly the most accurate measurement that can be made routinely outside of calibration and standards laboratories, perhaps to 1 in 10^9.

The electrical base unit is the ampere which is defined in terms of the force between two current-carrying wires. Measuring this force requires a special current balance and accuracy is low compared to other measurements, about 1 in 10^6. The secondary unit of resistance, the ohm (Ω) can be derived from the reactance of a standard air capacitor, whose capacitance can be calculated from its geometry to give an accuracy of about 1 in 10^7. The other secondary electrical unit is the volt (V), which is based on the Josephson effect. When two superconductors are separated by a thin insulating layer and are immersed in an alternating electromagnetic field of frequency, f, the voltage across the Josephson junction as it is called, is a multiple of $hf/2q$, where h is Planck's constant and q the magnitude of the electronic charge. The accuracy is about 1 in 10^8. This standard is used to calibrate zener diodes and bandgap references kept at a constant temperature, which are used as secondary standards with an accuracy of about 1 in 10^6.

Temperature is defined in terms of the triple point of water at 273.16 K. The accuracy for realising this temperature is about 0.01 K. The practical temperature scale is realised by using platinum resistance thermometry between 13.8044 K (the triple point of hydrogen) and 630.755°C (the freezing point of antimony), where $T_{°C} = T_K - 273.15$. A Pt/Pt-10%Rh thermocouple is used between 630.755°C and 1064.43°C (the freezing point of gold). Planck's radiation law is used with an optical pyrometer above this temperature. The practical scale can be realised from about 1-10 mK up to the gold point, but the absolute scale only to about 5-10 parts in 10^5.

13.1.1 The accuracy of laboratory instruments

It is essential before using any laboratory instrument to read the manufacturer's instruction book, many of which are extremely well-written and full of pertinent information about accuracy and how to get the best out of the equipment. A huge range of digital meters is now available to measure almost any physical property. Table 13.1 gives a few examples gleaned from popular suppliers' catalogues.

Table 13.1 *A selection of digital instruments*

Instrument	Model	Function/range[A]	Accuracy[B]	Price[C]
Oscilloscope	Goldstar	5 mV/cm to 5 V/cm	3%	
	OS-9020P	0-20 MHz		£280
	OS-9100P	0-100 MHz	3%	£600
Multimeter	Fluke 45	DC voltage	0.025% + 2	£650
		AC voltage		
		50-10000 Hz	0.2% + 10	
		DC amps	0.05% + 2	
		AC amps		
		50-10000 Hz	0.5% + 10	
Multimeter	Wavetek T100B	DC voltage	0.5% + 1	£70
Counter[D]	Black Star			
	Nova 2400	10 Hz - 2.4 GHz	0.001% + 1	£300
LCR bridge[E]	Prism 6401	R: 10^{-3} - 10^8 Ω	0.25%	£750
		L: 10^{-7} - 10^4 H	0.25%	
		C: 10^{-13} - 10^{-2} F	0.25%	
		Q: 0.1 - 99	0.25%	
LCR bridge[E]	RS	R: 2 kΩ - 2 MΩ	0.8% + 1	£85
		L: 2-500 mH	2% + 2	
		C: 20 nF - 200 μF	1% + 1	
LCR bridge[F]	Seesure	R: 10 Ω - 1 MΩ	2%	£38
		L: 100 μH - 10 H	2%	
		C: 1 nF - 100 μF	2%	
Balance	Sartorius PT610	0-610 g	±0.2 g	£350
Water tester	Hanna	pH 0-14	±0.2	£125
		conductivity 0-0.2 S/m	±4 mS/m	
		temperature 0-60°C	±1°C	
		ORP ±1 V	±5 mV	
Thermometer/	Digitron	−60°C - +590°C	0.1% + 1	£600
Anemometer/	Mistral AF200	1.5 - 58 m/s	±0.2 m/s	
Manometer		0-200 kPa	±500 Pa	
Thermometer/	RS	0°C - 50°C	±1°C	£25
Hygrometer		0-100% RH	±6%	
Barometer/	RS	794-1050 mbar	±5 mbar	£50
Thermometer/		−5°C to +55°C	±1°C	
Clock/calender				

Notes to table 13.1:   [A] Range for stated accuracy, not necessarily the full range of the instrument [B] For up to 1 year after calibration. 0.1% + 1 means 0.1% of reading plus least significant digit [C] For single quantities [D] Accuracy depends on sampling time [E] LCR bridges typically operate at 100 Hz and 1 kHz fixed frequencies [F] No digital display

These examples are chosen as typical of the type of instrument in the general price range given; calibration instruments are specifically excluded. When used over the frequency ranges given the instruments listed in table 13.1 will be accurate to the degree stated provided they have been recalibrated within the last year. Most suppliers and all manufacturers now provide a calibration traceable back to national standards (NAMAS calibration), or at lower cost one which uses reputable calibration instruments.

Inexpensive (£200 and above) DC voltage and current calibration sources are available which have typical accuracies of 0.1-0.2% of full-scale reading over a few decades, starting from about 10 mV or 10 mA full scale. Inaccurate measurements — assuming the instrument is operating within specification — have many origins, but by far the most frequent causes are

(1) Failure to take account of the input impedance of the instrument
(2) Operation outside the frequency range of the instrument

Input impedances are often overlooked as a source of inaccuracy: the average laboratory oscilloscope has an input impedance of 1 MΩ in parallel with 20 pF, but using a metre of coaxial lead will add another 100 pF or so to this and reduces the input impedance to 10 kΩ at 120 kHz. Inexpensive digital multimeters have input impedances around 10 MΩ in parallel with 50 pF, which means that the source resistance must be less than 10 kΩ for 0.1% DC voltage accuracy. The frequency range is often only a few tens of kHz.

Waveforms

Another cause of confusion is a misapprehension of the nature of the waveform being measured, or a failure to realise what the meter reading means for a given waveform. Of course, any unknown or suspect waveform must be looked at on an oscilloscope. Digital multimeters are increasingly made to give true rms readings for any waveform within the limitations of the instrument, but many are average-reading instruments which require correction to rms values. Even the 'true' rms meters are usually restricted to a fairly small *crest factor*, which is defined as the ratio of the peak to the rms value.

Example 13.1

What are the crest factors of (a) a sine wave (b) a square wave (c) a triangular wave and (d) a half-wave rectified sinewave?
    This amounts to finding the rms values in terms of peak values.
(a) In section 1.7.1 we calculated the rms value of a sinewave and found it was the peak value divided by $\sqrt{2}$, so the crest factor of a sinewave is $\sqrt{2} = 1.414$.

(b) The rms value of a square wave is the same as the peak value, so the crest factor is 1.
(c) The rms value of a triangular wave is $1/\sqrt{3}$ times the peak value and so its crest factor is $\sqrt{3} = 1.732$.
(d) A half-wave rectified sinusoidal voltage must have half the heating effect of a full sinewave, since it omits every other half cycle, that is

$$P_{1/2} = \frac{P_{RMS}}{2} = \frac{1}{2}\frac{V_p^2}{2} = \frac{V_p^2}{4} = V_{RMS}^2$$

assuming a notional 1Ω load. Thus $V_{RMS} = V_p/2$ and its crest factor is therefore 2. Pulsed or spiky waveforms will have high crest factors (see problem 13.1).

If the meter is average reading it first rectifies the AC waveform and then finds its average value. This is converted by the meter to an rms value, assuming the input to be sinusoidal. If the input is not sinusoidal the meter reading must be corrected to true rms as in the next example.

Example 13.2

What is the correction factor to convert the reading of an average-reading voltmeter to true rms values when the input is (a) a sinewave (b) a square wave (c) a triangular wave and (d) a half-wave rectified sinewave?

(a) The average value of a rectified sinewave is

$$V_{AV} = \frac{\int_0^{T/2} V_p \sin \omega t \, dt}{T/2} = \left[\frac{V_p \cos \omega t}{\omega T/2}\right]_{T/2}^0 = \frac{V_p[1 - (-1)]}{\pi} = \frac{2V_p}{\pi}$$

and the rms value is $V_p/\sqrt{2}$, so the ratio is

$$\frac{V_{RMS}}{V_{AV}} = \frac{V_p/\sqrt{2}}{2V_p/\pi} = \frac{\pi}{2\sqrt{2}} = 1.111$$

But we know that an average-reading voltmeter gives a true rms reading for sinewaves, that is the correction factor is 1, and *all* waveforms are multiplied internally by 1.111 before their 'rms' value is displayed.
(b) The ratio of rms to average value for a square wave is 1, but the internal correction factor of 1.111 means that the displayed value is too high; the true rms value is actually $1/1.111 = 0.9$ times the displayed value.
(c) The average value of a triangular waveform is $0.5V_p$ and its rms value is $V_p/\sqrt{3}$, giving

$$\frac{V_{RMS}}{V_{AV}} = \frac{V_p/\sqrt{3}}{0.5V_p} = \frac{2}{\sqrt{3}} = 1.1547$$

Thus the displayed value will be too low and must be multiplied by $1.1547/1.111 = 1.039$.
(d) The average value of a half-wave rectified sinewave is V_p/π, half that of a full-wave

rectified sinewave. The rms value we have already found to be $V_p/2$ so the ratio is

$$\frac{V_{RMS}}{V_{AV}} = \frac{V_p/2}{V_p/\pi} = \frac{\pi}{2} = 1.571$$

The meter will read low and the correction factor is $1.571/1.111 = 1.414$.

Resolution, accuracy and readability

The resolution of an instrument is the smallest change that one can see in its reading. In the case of digital instruments this is ±LSB. In the case of an instrument with a scale such as a moving-coil meter or oscilloscope it is $\pm\frac{1}{2}$(smallest division). Thus if a $4\frac{1}{2}$-digit meter reads 123.45 mV, we can see that it has a resolution of 0.01 mV or 10 μV or $1/20\,000 = 0.005\%$ of FS. Oscilloscopes typically have a grid displayed on the screen with a maximum vertical or horizontal range of 8 squares (1 square = 1 cm), subdivided into 5, so the resolution is $0.5 \div (8 \times 5) = 1.25\%$ of FS. Both of these are likely to be substantially smaller than the accuracy, which is taken from the maker's specification, assuming the instrument has been calibrated in the last year. The Fluke 45 meter in table 13.1 has a 5-digit display, so its resolution is $1/100\,000 = 0.001\%$, but its accuracy is only 0.025% + 2 LSDs. A reading of 456.78 mV can therefore indicate a true value anywhere from 456.65 to 456.91 mV. An oscilloscope reading of 6.5 cms at 200 mV/cm is 1.3 V_{pp} = $1.3/2\sqrt{2} = 460$ mV$_{rms}$. The resolution is ±0.1 cm = ±20 mV, but the accuracy is 3% of the reading or ±39 mV. In many cases it is possible to set up an oscilloscope with a calibrator so that its accuracy is the same as its resolution, but it has to be checked very frequently to maintain this.

Readability refers to the ease with which a reading can be written down into a notebook. Oscilloscopes are difficult to read and mistakes in recording readings on them are extremely frequent, quite apart from a tendency to overlook the conversion from peak-to-peak to rms values. Provided the reading is steady, digital displays are the easiest to read and lead to the fewest errors in transcription to notebooks. Moving-coil instruments are better than oscilloscopes in their readability, but are markedly inferior to digital meters. They are becoming more and more obsolete, and though many panel meters are moving-coil types they are used for semi-quantitative or just qualitative work. They are useful when readings are varying somewhat — as in a fuel gauge or speedometer — and give a visual idea of the rate of change in a quantity.

Significant figures

An often-overlooked fact is that observers, however much we try to automate data collection, are ultimately human and therefore prone to err. One way in which error can be reduced is to record only those figures in the data that are truly significant and leave out the rest. For example if the voltage showing on the meter is 4.56384 V, but you know that the last four digits vary randomly[1], then there is no need to write down more than 4.563 V for that reading, recording no more than two of the random digits (and even this

---

[1] By 'randomly' we mean that any of the digits from 0-9 appears to be as likely to occur as any other.

is too many unless a very large number of readings is taken). And if the data is known to require an accuracy of 1%, then writing down 4.56 V should suffice. Not only does the recording of the additional data waste time and money, it also leads to a greater probability that some of the significant digits will be wrongly written down. It is an observable fact that long strings of digits are difficult to transcribe with accuracy unless time and trouble are taken: yet an all-too-common and reprehensible practice is to write down all the digits displayed without thinking. Computer output should also be truncated at two random digits or at one digit more than the accuracy required strictly warrants.

Business and commerce abound with absurd examples of the over-recording of useless digits: what on earth is the point of reporting sales of £12,183,256 in 1996 for example? In these cases the accuracy cannot be better than about 1% and in any event most people will only take in the first two digits; the sales should have been stated as £12.2 million. Presumably someone spent as much time recording the last five useless digits as he did the first three, but if the likelihood of error is equal for all the digits, more care could have been devoted to the really significant ones, were these alone recorded.

It is good engineering practice to write down numbers with implied accuracy: 12.2 mA implies an accuracy of ±0.1 mA at best, 12.20 mA implies at best ±0.01 mA and so on. This is especially important when final results are presented after analysis.

13.1.2 Experimental errors

If one makes one reading with a meter, properly used and maintained, then one can only assume the error is whatever the maker says it should be. In many cases, however, one makes a series of measurements of a single, supposedly constant quantity, and one finds the readings vary somewhat.

Table 13.2 *Hydrophone readings in mV at 3 kHz*

Observation No.	Reading	Running mean
1	407	—
2	374	392
3	418	400
4	415	404
5	397	402
6	409	403
7	372	399
8	363	394
9	374	392
10	387	392
11	402	393
12	414	394
13	392	394
14	418	396
15	427	398
16	424	400

For example consider the data in table 13.2, which are voltages taken from an apparatus repeatedly set up before and disassembled after each reading of the same quantity (in this case the output of a hydrophone at 3 kHz).

Note that the last two digits in the data presented appear to be random: the first is not as it is either 3 or 4 and no other digit. The variability of the data is of unknown origin: the measuring instruments were well maintained and would not be expected to produce much variability. Every effort was made to keep conditions constant, but try as one might, the data are inherently variable. We could then ask questions such as: What is the 'true' value of the hydrophone voltage? How accurate is this value? What are the chances that a single reading will be out by a given amount? Though individually the observations are somewhat variable their running mean is reasonably constant and varies rather slowly, settling down at about 400 mV. We should be right in thinking that taking a lot more reading will reduce the variation in the running mean still more. Thus the mean or average of the readings is the best estimate we can make as to the 'true' value and is given by

$$\mu = \frac{1}{n} \sum_{j=1}^{n} x_j \qquad (13.1)$$

The best measure of the variability of the data is the *variance*, which is given by

$$s_n^2 = \frac{1}{n} \sum_{j=1}^{n} (\mu - x_j)^2 \qquad (13.2)$$

If we carry out the computation for the data of table 13.2 we find $s_n = 19.8$, where s is the *standard deviation* and is the square root of the variance. Because the number of observations is limited, there is a small correction that is usually applied to the standard deviation which makes it the best estimate of the standard deviation of an infinite number of observations, σ. This estimate of the true standard deviation, denoted s_{n-1}, is given by

$$s_{n-1} = s_n \sqrt{\frac{n}{n-1}} = 19.8 \sqrt{\frac{16}{15}} = 20.4 \qquad (13.3)$$

The difference is so slight that it can be ignored, but it should be applied to samples with less than about ten observations.

If the errors in the readings are truly random then the readings will follow a Gaussian or normal distribution, for which, in an infinitely large sample, 68% of the readings will be within one standard deviation of the mean. In our example, we should therefore expect $0.68 \times 16 = 11$ observations to lie between $400 - 20$ and $400 + 20$, or between 380 and 420. Actually ten do. Table 13.3 gives a few probabilities for a normal distribution as a function of $m = y/\sigma$, where $y = |x - \mu|$. $P(m)$ is the proportion of the readings that are expected to lie within m standard deviations of the mean.

We can see now that the chances of an observation lying more than a few standard deviations from the mean is very small, since 95.5% lie within $\pm 2\sigma$ and 99.7% within $\pm 3\sigma$. If we consider our data, this means that 95.5% of the observations, that is 15 of them ($0.955 \times 16 = 15.3$), should lie between $400 - 2 \times 20.4 = 359$ mV and $400 + 2 \times 20.4 =$

441 mV. In fact all of them lie within these limits (though only just).

The expected range of our data, if it is truly normally distributed with mean 400 mV and $\sigma = 20$ mV, can be calculated by finding the value of m for which

$$P(m) = 1 - 0.5/n = 1 - 0.5/16 = 0.969$$

We can look this up in table 13.3 and find $m = y/\sigma = 2.16$ (on interpolation). Hence

$$y = |x - \mu| = 2.16\sigma = 2.16 \times 20 = 43$$

Thus the expected range is from 357 to 443 mV.

Table 13.3 *Normal probabilities*

m	0.1	0.2	0.3	0.4	0.5	0.6	0.7	0.8	0.9	1.0
$P(m)$	0.08	0.16	0.24	0.31	0.38	0.45	0.52	0.58	0.63	0.68

m	1.2	1.4	1.6	1.8	2.0	2.2	2.4	2.6	2.8	3.0
$P(m)$	0.77	0.84	0.89	0.93	0.955	0.972	0.984	0.991	0.995	0.997

Note: For $m > 3$, $P(m) \approx 1 - (0.8/m)\exp(-0.5m^2)$

The accuracy of the mean

The standard deviation and the mean of the data given in table 13.2 are not expected to be very different from the true values and we can use both with reasonable confidence. But we might well ask: How likely is the true mean to be within such and such a value of the experimental mean? We can imagine that as the number of observations increases the mean will settle down more and more closely to the true value. It turns out that the accuracy of the experimental data improves in proportion to $1/\sqrt{n}$ and that the *standard error* of the mean is given by

$$S_n = \frac{s_n}{\sqrt{n}} = \frac{20}{\sqrt{16}} = 5 \tag{13.4}$$

We can then use table 13.3, substituting S_n for σ, to say what the probability that a given value is the true mean. For example, how likely is it that the true mean is 411 mV? This value is 11 mV higher than the mean of the data, that is $11/5 = 2.2$ standard errors from the mean, for which we find $P(m) = 0.972$. This means that there is a 0.972 likelihood that the mean will be less than 411 mV and only a $1 - 0.972 = 0.028$ chance that the mean will be 411 mV or more. Some might consider this probability slight.

To summarise: the standard deviation tells us how likely it is that a single value is the true value of the quantity we are measuring; the standard error how close the mean of a number of observations is likely to be to the true mean. The standard error of the mean decreases as the inverse square root of the number of observations: to increase the accuracy tenfold we must increase the number of observations a hundredfold.

Example 13.3

The data given below are temperature differences in K recorded in a laboratory apparatus. What are the sample's mean and standard deviation? What is the standard error of the mean? How likely are readings of 2.39 K and 1.12 K? How probable is it that the true mean is 1.85 K? What range of values would be expected?

ΔT(K)	2.05	1.95	1.86	1.51	1.20	1.59	1.83	1.79	1.42	1.31
ΔT(K)	1.61	1.45	1.29	1.47	1.77	1.52	1.58	1.60	1.91	2.11

Using equation 13.1 we find the mean to be 1.64 K, and equation 13.3 gives the best estimate of the standard deviation as 0.258 K. The standard error of the mean from equation 13.4 is 0.056 K. A reading of 2.39 K is 0.75 K higher than the mean, that is 0.75/0.258 = 2.9σ. Table 13.3 gives $P(m) = 0.996$ (interpolated) for $m = 2.9$, which means the probability of the reading is $1 - 0.996 = 0.004$, only 0.4%, rather unlikely. A reading of 1.12 K is 0.52 K below the mean and that is 0.52/0.258 = 2σ, giving $P(m) = 0.955$, so the probability is 0.045, 4.5%; though not large, it means such a reading will arise on average once in 22 observations.

Equation 13.4 gives the standard error of the mean as 0.056 K, so a mean of 1.85 K is 0.21 K higher than the sample's mean, which is 0.21/0.056 = 3.75S_n, a value that is not even listed in table 13.3. In fact for this we find $P(m)$ from the approximation given below the table to be 0.99981, so the probability that the true mean is 1.85 K is only 0.019% — a vanishingly small chance.

There are 20 observations so we first need to find the value of m for which $P(m)$ is $1 - 1/40 = 0.975$. Interpolation of table 13.3 leads to $m = 2.25$. Thus the limits are within $\pm2.25\sigma$ of the mean, that is from 1.06 K to 2.22 K. The data actually range from 1.20 K to 2.11 K, which suggests that they are distributed rather differently from an ideal Gaussian. Further analysis should be undertaken, or better further observations.

It is a worthwhile exercise to round the data off to two figures and recalculate the mean, the best estimate of the mean and standard deviation. One finds little difference when taking all three figures.

Combining different estimates of the mean

Sometimes we are given estimates of a quantity from different sources with different errors associated with them. How should we combine them so as to make use of all the data? For example, we are told by one observer that his estimate of a resistance is $181.51 \pm 0.38 \, \Omega$, while another observer of the same resistance says it is $181.65 \, \Omega \pm 0.34\%$. How can we combine both estimates so as to give a better estimate of the true mean, with a smaller error than either?

The first thing to notice is that the difference in the means is less than the errors of either; they are said to be *consistent*. Calling the first mean R_1 and its error δR_1, and the second mean R_2 and its error δR_2, the best estimate of the mean is

$$R = \frac{R_1(\delta R_2)^2 + R_2(\delta R_1)^2}{(\delta R_1)^2 + (\delta R_2)^2}$$

The errors must be put in the same form: let us use percentages, that is $\delta R_1 = 0.21\%$ and $\delta R_2 = 0.34\%$. The formula above gives us $R = 181.55 \ \Omega$.

The best estimate of the combined error is given by

$$\frac{1}{\delta R} = \sqrt{\left(\frac{1}{(\delta R_1)^2} + \frac{1}{(\delta R_2)^2}\right)}$$

which gives $\delta R = 0.18\%$, and $R = 181.55 \pm 0.33 \ \Omega$. The reduction in the error is not very large.

13.1.3 *When there are several known sources of error*

Quite often one calculates a quantity such as power or current from measurements made of other quantities such as voltage or resistance. In these cases the errors of each of the contributory measurements must be combined (assuming we know them) to give an estimate of the error in the desired quantity. As an example suppose we have measured the voltage across and resistance of a resistor and we want to know the power developed in it. Let us say the voltage is found to be 6.43 V and the resistance 3.04 Ω, and that the instruments we are using have known accuracies of 0.1% (voltmeter) and 0.15% (ohmmeter). The power is calculated as

$$P = V^2/R = 6.43^2/3.04 = 13.60 \ \text{W}$$

Let the error in V be δV and that in R be δR, then we can write

$$P = \frac{(V \pm \delta V)^2}{R \pm \delta R} = \frac{V^2(1 \pm \delta V/V)^2}{R(1 \pm \delta R/R)} = P_0 \frac{(1 \pm \epsilon_V)^2}{1 \pm \epsilon_R}$$

where P_0 is the true power and ϵ_V and ϵ_R are the relative errors in V and R respectively. If the errors are small we can approximate

$$P \approx P_0[(1 \pm 2\epsilon_V)(1 \mp \epsilon_R)] \approx P_0(1 \pm 2\epsilon_V \mp \epsilon_R)$$

Assuming the worst outcome the errors add and $P = P_0[1 \pm (2\epsilon_V + \epsilon_R)]$. Then the error in the power is twice that in the voltage (= 0.2%) plus that in the resistance (0.15%), for a total of 0.35% or 0.048 W, that is $P = 13.60 \pm 0.05$ W, rounding the error slightly.

This is, however, a conservative view and might be the best view where safety, say, is concerned. But the error is probably going to be less than this, and is in fact the r.m.s. error, given by

$$\epsilon = \sqrt{(2\epsilon_V)^2 + (\epsilon_R)^2} = \sqrt{(0.2)^2 + (0.15)^2} = 0.25\%$$

and then $P = 13.600 \pm 0.034$ W.

In general we say that if $x = F(a,b,c)$ and $f_a = \partial F/\partial a$ etc. then

$$\delta x = \sqrt{(f_a\delta a)^2 + (f_b\delta b)^2 + (f_c\delta c)^2} \qquad (13.5)$$

where δx, δa etc. are now the actual, not the relative, errors. Notice that if one term such as $f_b\delta b$ is much greater than any of the others (say a factor of three or more), then it dominates the overall error and renders the others negligible.

Example 13.4

The output voltage of a differential amplifier is given by

$$V_o = \frac{-R_2 V_1}{R_1} + \left(1 + \frac{R_2}{R_1}\right)\left(\frac{R_4 V_2}{R_3 + R_4}\right)$$

If the resistances are all 5% tolerance and $R_1 = 4.7$ kΩ, $R_2 = 100$ kΩ, $R_3 = 10$ kΩ and $R_4 = 220$ kΩ, what is the output voltage error when $V_1 = V_2 = 0.1$ V?

The error in the resistances are $\delta R_1 = 0.235$ kΩ, $\delta R_2 = 5$ kΩ, $\delta R_3 = 0.5$ kΩ and $\delta R_4 = 11$ kΩ. Then there are four error terms:

$$\frac{\partial V_o}{\partial R_1}\delta R_1 = \left(\frac{R_2 V_1}{R_1^2} - \frac{R_2 R_4 V_2}{R_1^2(R_3 + R_4)}\right)\delta R_1 = 4.63 \text{ mV}$$

$$\frac{\partial V_o}{\partial R_2}\delta R_2 = \left(\frac{-V_1}{R_1} + \frac{R_4 V_2}{R_1(R_3 + R_4)}\right)\delta R_2 = -4.63 \text{ mV}$$

$$\frac{\partial V_o}{\partial R_3}\delta R_3 = \left(1 + \frac{R_2}{R_1}\right)\left(\frac{-R_4 V_2}{(R_3 + R_4)^2}\right)\delta R_3 = -4.63 \text{ mV}$$

$$\frac{\partial V_o}{\partial R_4}\delta R_4 = \left(1 + \frac{R_2}{R_1}\right)\left(\frac{R_3 V_2}{(R_3 + R_4)^2}\right)\delta R_4 = 4.63 \text{ mV}$$

The amplifier is well designed because all the sources of error have equal impact. The overall error in V_o is

$$\delta V_o = \sqrt{4.63^2 + (-4.63)^2 + 4.63^2 + (-4.63)^2} = 9.26 \text{ mV}$$

13.2 Transducers

Transducers turn one form of energy into another, but those considered here will all be electrical transducers that either turn electrical energy into some other form of energy or vice versa. We exclude from the category devices which operate with large amounts of power. Though vital parts of many systems, transducers tend to be the poor relations of both electronics and computers, but without well-designed and well-utilised transducers the system as a whole may suffer from impaired performance, or at worst be rendered useless. Because they often have to work in uncontrolled or hostile environments, transducers are usually ruggedly made to withstand extremes of temperature, pressure or mechanical abuse etc. We shall not be concerned with this aspect of transducers but only with their performance as electrical components, albeit of a special kind.

There are two basic types of electrical transducer, shown diagrammatically in figure 13.1, the modulators and the generators. The modulators need an electrical input in addition to the primary source input. Some transducers can be made to work in reciprocal mode: that is they can take an electrical input and turn it into some other form: piezoelectric transducers are an example of this; they can turn electrical energy into mechanical vibrations (sound) or vice versa.

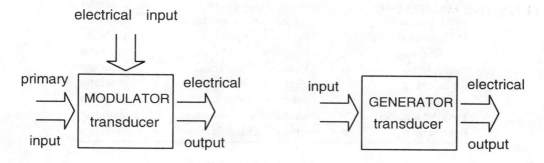

Figure 13.1 The two basic types of electrical transducer

Table 13.4 lists some electrical transducers — probably you will be able to think of many more — but it covers a wide range, including several different ways of energy conversion. The sensing or transducing element can made from a variety of materials even when the same kind of energy is converted to electricity. A great deal of ingenuity has gone into devising transducers and improving the efficiency of transduction. To take one example, acoustical-electrical transduction has utilised electrostriction, piezoelectricity, magnetostriction, electromagnetic induction and optical fibres to effect the conversion. We shall discuss in turn some of the more common electrical transducers.

13.2.1 Temperature transducers

The transducers that must be immersed in or have contact with the medium whose temperature is measured often have to work in extreme conditions. The sensing element is usually sheathed to provide mechanical and chemical protection. If fast response times are required the element must be left bare and replaced when it fails. Remote-sensing thermometers are becoming more popular because they can have rapid responses and do not require contact with the object measured. Optical fibres have also extended the range of applications for remote temperature sensing. The contacting types of transducer include metal resistance elements (especially platinum), thermistors, diodes and thermocouples. The platinum resistance element is rather bulky which limits the response time, but it is an excellent means of calibrating the other types.

Table 13.4 *Electrical transducers*

Transducer	Conversion	Type
radio or television aerial	electromagnetic → electrical	generator
resistance thermometer	thermal → electrical	modulator
pyrometer	thermal → electrical	generator
PIN diode	optical → electrical	modulator
laser diode	electrical → optical	generator
thermocouple	thermal → electrical	generator
Peltier heat pump	electrical → thermal	generator
strain gauge	mechanical → electrical	modulator
Hall-effect fluxmeter	magnetic → electrical	modulator
coil fluxmeter	magnetic → electrical	generator
microphone	acoustic → electrical	generator
loudspeaker	electrical → acoustic	generator
solar cell	optical → electrical	generator
position encoder	mechanical → electrical	modulator
lightmeter	optical → electrical	modulator

Thermocouples

These are the most popular industrial temperature transducers and rely on the Seebeck effect in which a voltage is produced when two junctions between dissimilar metals are held at different temperatures (see figure 13.2).

The voltage produced at the terminals is proportional to the temperature difference between the junctions. In figure 13.2, T_1 is the cold or reference junction (usually the ice point, 0°C) and T_2 the temperature to be measured:

$$V_{\Delta T} = S_{AB}(T_2 - T_1) \;\Rightarrow\; T_2 = V_{\Delta T}/S_{AB} \quad \text{(in °C)}$$

where $T_1 = 0°C$ and S_{AB} is the Seebeck coefficient (sometimes called the thermoelectric power) of the junction formed from A and B. Seebeck coefficients are rather small, a few tens of μV/K implying that accurate voltage measurement is necessary even for modest accuracy. The quoted accuracy of commercial thermocouples is about ±0.0075T where T is the Celsius temperature: calibration must be undertaken to improve on this figure.

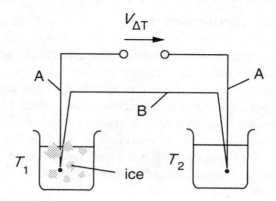

Figure 13.2

A thermocouple circuit

The inconvenience of maintaining a cold reference junction has led to the manufacture of electronic cold-junction compensation which is now provided with almost all digital thermocouple instruments. However, their accuracy is seldom better than ±1°C, even over restricted temperature ranges. Temperatures near room temperature are best measured with mercury-in-glass thermometers, if possible. These can easily be obtained with an accuracy of 0.2°C or better; they are inexpensive and easy to read and use.

Because any metal-to-metal contact will produce a thermal e.m.f. one cannot just attach copper wire to a thermocouple to connect it to a voltmeter, but must use one of the metals of the junction as in figure 13.2, or more usually a compensating cable which is designed to eliminate spurious thermal voltages. Compensating cables must be of the correct type for the materials used. Cable connections must also be made with the correct type of connector to avoid spurious e.m.f.s Table 13.5 lists some of the more common types of thermocouple. Commercial thermocouples have been given letters to designate the type, such as B, J, K, N, R, S and T.

Table 13.5 *Standard industrial thermocouples*

Junction metals	Type	S^A	Range^B (°C)	Comments
platinum-30% rhodium/				high T, dear, S low
platinum-6% rhodium	B	11†	100 to 1820	(zero at near 20°C)
iron/constantan (Fe/Cu-Ni)	J	53	−300 to 1200	obsolescent
chromel/alumel (Cr-Ni/Al-Ni)	K	41	−330 to 1370	rugged, commonest
nicrosil/nisil (Ni-Cr-Si/Ni-Si)	N	28	−270 to 1300	improved type K
platinum-13% rhodium/platinum	R	19†	−150 to 1760	low S, high T, dear
platinum-10% rhodium/platinum	S	12†	−150 to 1760	low S, high T, dear
copper/copper-nickel	T	43	−200 to 400	5ms response time

Notes: [A] Seebeck coefficient near room temperature in μV/K, except † at 1400°C.
[B] for tabulated e.m.f. in IEC 584 and other sources such as Kaye and Laby's *Tables of physical and chemical constants.*

Resistance thermometers

These are nearly always made with platinum for the sensor because it can be used at high temperatures without degradation and can achieve much greater accuracy (with care ±0.1°C) than thermocouple thermometers. Only its relatively large size (30 mm × 4 mm × 1 mm is standard) and slow response time (about 10 s for the unclad element) prevents its greater use. The construction shown in figure 13.3 has laser-trimmable links for the automatic adjustment of the resistance to 100 Ω at 0°C. The large area of platinum on the lower left is a fine-trimming pad which can be cut down the centre to give great control of the final resistance (normally to ±0.1% or better).

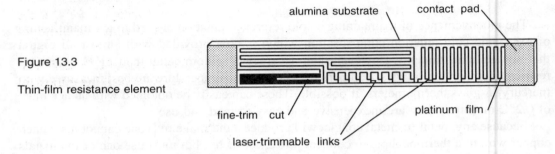

Figure 13.3

Thin-film resistance element

The temperature coefficient of the resistance is about 0.4%/K, which means that to obtain a resolution of ±0.1 K requires a resistance measurement with a resolution of 0.04 Ω in 100 Ω at 0°C, a requirement well within the capabilities of modern methods of instrumentation when a four-wire measurement is used. Self-heating should be guarded against when the sensor is not in good thermal contact with a large mass. In still air self-heating causes a rise of 200 K/W, so to keep the rise of temperature down to 0.1 K requires a current of about 2 mA or less.

Thermistors

Thermistors utilise the non-linear change of resistance with temperature of semiconductors. As a result the thermistor resistance change with temperature is given by

$$R_T = A \exp(B/T) \tag{13.6}$$

where A and B are constants which must usually be found for each thermistor by the user, because of manufacturing variations, and T is the temperature in K. Unfortunately the calibration has to be repeated from time to time as drift occurs. Usually a quick check with some other thermometer is all that is needed to see if recalibration must be carried out. Recalibration can be done fairly rapidly by using standard thermometers or two fixed points. The thermistors most often used have negative temperature coefficients (NTC thermistors), for which the constant, B, in equation 13.6 is positive. But, by suitable doping of the material used in the element, positive temperature coefficients (PTC) can be obtained, for which B is negative. Series-connected PTC thermistors can be used for over-current protection of transformers for example. The miniature bead type of thermistor can have a rapid response time (< 100 ms) but is rather fragile and they are not particularly cheap (about £3-£4 each). The next example gives some idea of the resistance changes in thermistors.

Example 13.5

A bead thermistor has a resistance of 10 kΩ at 20°C and 4 kΩ at 44°C. If its resistance obeys equation 13.6 what will its relative sensitivity (relative change in resistance/°C) be at 30°C? If the resistance is monitored with a constant-current source, what is its maximum value given that the self-heating should be < 1 mW, and the working temperature range is from 25°C to 35°C? How small a temperature change can be detected at 25°C if the resistance of the bead is measured by measuring the voltage across it with a voltmeter of 1 mV resolution? (Take $T_K = T_{°C} + 273.15$ K)

Using equation 13.6 we find from the given data $A = 0.05513$ Ω and $B = 3549.6$ K. The relative change of resistance with temperature is found by taking logarithms of equation 13.6 and differentiating the result:

$$\ln R = \ln A + \frac{B}{T} \quad \Rightarrow \quad s = \frac{1}{R}\frac{dR}{dT} = \frac{-B}{T^2}$$

Substituting for B and $T = 25°C = 298.15$ K leads to $s = -0.03993$ Ω/Ω/°C = −3.993%/K, in other words the resistance falls by almost 4% for every degree rise in temperature at 25°C. This is a ten-fold larger change than that of a resistance thermometer.

The self-heating will be most at the lowest temperature since the resistance will be largest there. Thus at 25°C where the resistance is 8.163 kΩ, $I^2R = 1$ mW, so that $I = 0.35$ mA. The voltmeter can detect changes of 1 mV and the voltage across the bead is $0.35 \times 8.163 = 2.857$ V, giving a relative resistance resolution of $0.001/2.857 = 3.5 \times 10^{-4}$. The smallest temperature change that can be detected is therefore

$$s\delta T = 0.35 \times 10^{-3} \quad \Rightarrow \quad \delta T = 0.35 \times 10^{-3}/0.03993 = 8.8 \text{ mK}$$

We see that thermistors can be used to detect very small changes in temperature. They are, however, not very linear, which restricts each device to a fairly restricted range of temperatures. Linearising methods are available, but they usually produce a loss in sensitivity. Some ASICs (application-specific ICs) have been designed for use with thermistor instruments.

13.2.2 *Displacement transducers*

This term is used for devices which measure changes in length of the range from a few microns up to about a metre or so. Other kinds of transducer also measure length, but in this range the choice is relatively restricted. There are three main types: resistive or potentiometric, capacitive and inductive. The resistive type uses a potentiometer wiper attached to the item being measured and will not be further discussed beyond saying that it is cheap, simple and has at best an accuracy of about 0.1% in the range from 0.1 m to 10 m.

Capacitive displacement transducers have one plate of a variable capacitor attached to the moving object. Figure 13.4 shows one type, which has three plates to improve accuracy, can be used for large displacements and gives a linear output.

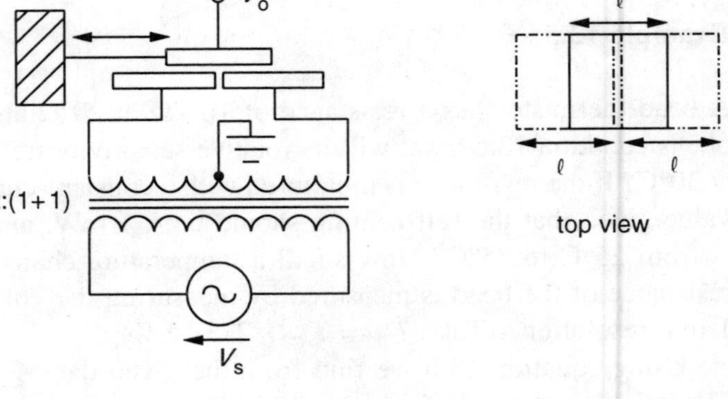

Figure 13.4

A displacement transducer made from a 3-plate capacitor

$2:(1+1)$

top view

The supply frequency is usually a few kHz and feeds a transformer with a centre-tapped secondary. All three plates have the same dimensions and the gap between the two bottom plates is minimal. If the transformer's turns are correctly adjusted to give zero output when the moving plate covers equal areas of the two halves of the two bottom plates, then the output is zero. When the top plate moves from the zero position to the left by a small amount, δl, the capacitance of the left-hand capacitor changes by δC to $C + \delta C$ while that of the right-hand capacitor changes to $C - \delta C$.

The equivalent circuit for the transducer is that of figure 13.5a which reduces to that of figure 13.5b, so that the output is

$$V_o = \frac{V_s}{2}\frac{\delta C}{C} = \frac{V_s}{2}\frac{\delta l}{l/2} \Rightarrow \frac{V_o}{V_s} = \frac{\delta l}{l}$$

A phase-sensitive detector is needed to give the direction of the displacement. The transducer is relatively easy to make and the transformer adjustment is also not very difficult as it is merely a matter of taking off a few turns from one of the secondaries.

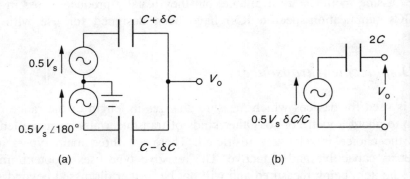

Figure 13.5 Equivalent circuits for the displacement transducer of figure 13.4

The inductive type of displacement transducer frequently encountered is the linear variable differential transformer or LVDT, shown in figure 13.6. They are quite difficult to make and are usually purchased from specialist manufacturers at prices starting around £50. They can be AC or DC energised and have a relatively small range of movement, typically between ±1 mm for the smallest to ±50 mm for the larger (and more costly) ones. The ferromagnetic rod which moves with the object must be carefully positioned in the winding assembly so that it does not touch the sides and moves smoothly to give a linear output. The split secondary coils are in series opposition so that when the rod is located in the centre the output voltage is zero. A PSD is needed to give the direction of travel.

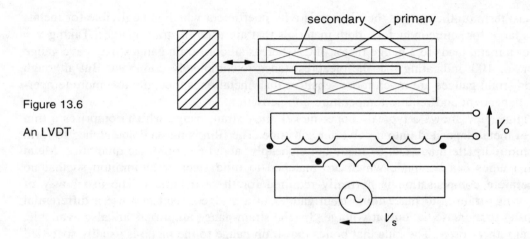

Figure 13.6

An LVDT

13.2.3 Strain gauges

Strain gauges are a sort of displacement transducer since they respond to changes in length, but they are often used for measuring force, pressure or acceleration — quantities which can be derived from a change in length. The strain gauge is a thin film resistor which is glued to the object which is undergoing strain, so that its resistance changes with the length of the resistor.

Consider a thin-film resistor with cross-sectional area, A, length, l, and resistivity, ρ. Its resistance is

$$R = \frac{l\rho}{A} \;\Rightarrow\; \ln R = \ln l + \ln \rho - \ln A$$

so that for small changes

$$\frac{\delta R}{R} = \frac{\delta l}{l} - \frac{\delta A}{A} + \frac{\delta \rho}{\rho}$$

The cross-sectional area and the resistivity will both change when the length changes. The change in area is related to the change in length by Poisson's ratio, ν,

$$\frac{\delta A}{A} = \frac{-2\nu\delta l}{l}$$

If the volume is unchanged in the deformation, $v = 0.5$, but for many metals v is between 0.3 and 0.4, implying some increase in volume with increase in length. Thus the relative change in resistance is given by

$$\frac{\delta R}{R} = \frac{\delta l}{l}\left(1 + 2v + \frac{\delta \rho / \rho}{\delta l / l}\right)$$

whence the *gauge factor* is derived

$$GF = \frac{\delta R / R}{\delta l / l} = 1 + 2v + \frac{\delta \rho / \rho}{\delta l / l}$$

The last term on the right is the piezoresistivity coefficient which is negligible for metals but is large for semiconductors, both materials that are used in strain gauges. Taking $v = 0.5$ for a metal leads to a gauge factor of 2, whereas silicon strain gauges may have gauge factors of 100, indicating that the piezoresistance is completely dominant. But although silicon strain gauges are much more sensitive than metal ones they are also more temperature dependent and require temperature compensation.

Figure 13.7 shows a typical copper-nickel alloy strain gauge which comprises a thin film of metal deposited onto a polyester substrate. The film is masked and etched to form the required pattern and can be made quite cheaply: about £2 for single quantities. Metal strain gauges can be made which are matched to mild steel or aluminium so that no temperature compensation is normally required for these materials. The usual way of measuring strain is to place four strain gauges in a bridge circuit and use a differential amplifier to increase the output voltage. Special strain-gauge amplifiers are also available, though rather costly. The glue that bonds the strain gauge to the metal is usually specified by the manufacturer as it is essential to measure the right strain.

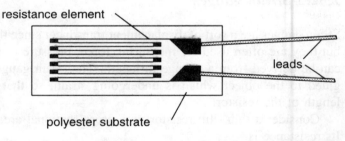

Figure 13.7

A copper-nickel strain gauge

resistance element

leads

polyester substrate

Specially arranged strain gauges on metal substrates are used to measure force in what is often called a load cell. These are supplied with electrical connectors and have very good linearity and overall accuracy (±0.02% typically), but they are expensive (c. £200). They can be used, for example, in electronic weighbridges for loads ranging from 1 kg to 10 tonnes.

Relatively inexpensive static pressure transducers (c. £20) are now made from small silicon diaphragms (about 3 mm × 3 mm) which have ion-implanted piezoresistors put into them in a bridge formation. They operate over pressure ranges of from 0-1 psi (pounds per square inch, 1 psi = 6.9 kPa = 69 mbar = 0.068 atmospheres) to 0-30 psi, that is up to about two atmospheres. The electrical supply required is about 10 V DC and their responsivities are about 1 µV/Pa.

13.2.4 Piezoelectric transducers

These are almost always made from lead zirconate-titanate or $Pb(Zr,Ti)O_3$, hence its usual acronym PZT. Strictly speaking the material is an electrostrictive ferroelectric, but after exposure to a large electric field (3 MV/m) during poling the material behaves as if it were piezoelectric. PZT acquires a surface charge in response to pressure ($\equiv$ deformation) and this can be detected with a suitable amplifier. Alternatively, if a charge is placed on its surface, its thickness will change in proportion. Though not particularly good at sensing DC pressures, PZT is very useful where alternating or varying pressures are concerned. Figure 13.8a shows a typical PZT plate element excited by a voltage, V_s, which causes the silvered faces to move in and out as shown by the arrows. The response is greatest at the mechanical resonant frequency of the PZT plate and so by adjusting the plate dimensions the frequency of maximum response can be varied.

The equivalent circuit of a PZT transmitting transducer is that of figure 13.8b, where the mechanical resonant frequency is the same as the electrical, $\omega_0 = 1/\sqrt{(LC)}$. The static capacitance is C_0 which is usually much greater than C. The power dissipated in the device and the power sent out as sound waves are both given by I^2R where I is the current in the resonant branch, given by V_s/R at resonance.

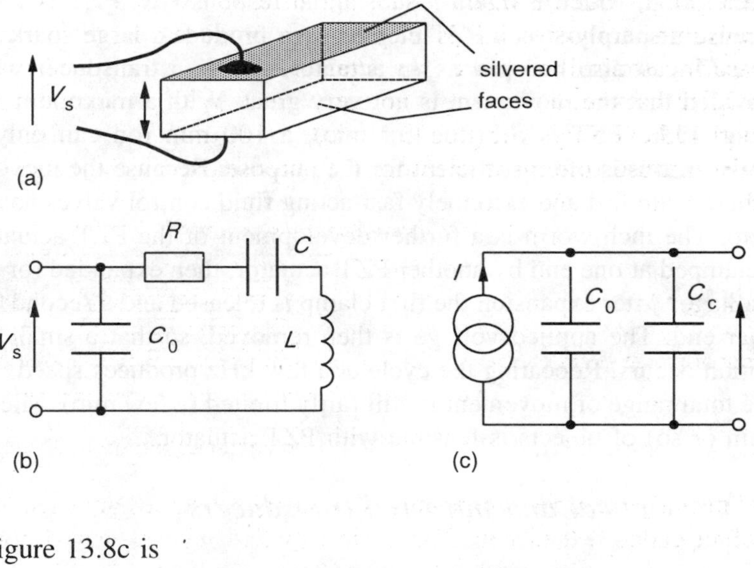

Figure 13.8

(a) A PZT transducer element and its equivalent circuits
(b) As a transmitter
(c) As a receiver

The current source in figure 13.8c is

$$I = k\frac{dP}{dt} = AD\frac{dP}{dt}$$

where D is the piezoelectric strain coefficient, about 0.5 nC/N for PZT and A the area of one silvered face. The output voltage then becomes

$$V_o = \frac{1}{C_T}\int I dt = \frac{ADP}{C_T}$$

where $C_T = C_0 + C_c$. We can calculate C_0 from the element dimensions and relative permittivity

$$C_0 = \frac{\epsilon_0 \epsilon_r A}{l}$$

where $\epsilon_r \approx 1000$ for PZT and so $\epsilon_0 \epsilon_r \approx 10$ nF/m. Taking C_c to be negligible (it often is not!), then we find

$$V_o = \frac{DlP}{\epsilon_0 \epsilon_r} \approx 0.05lP$$

Thus for a PZT element 1 mm thick, we find an output voltage of 50 μV/Pa, or 5 V/bar, a fairly large responsivity. PZT makes sensitive ultrasonic transmitters and receivers, and is widely used in medicine, sonar, non-destructive testing, cleaning and cutting.

When used as an accelerometer the effective pressure is ρla where ρ = density of PZT ≈ 5000 kg/m^3 and a = acceleration. Then the output voltage becomes

$$V_o \approx 0.05 \times 5000 \, l^2 a = 250 l^2 a$$

an element of thickness 10 mm will have an output voltage of about 25 mV per m/s^2 of acceleration, which is again a substantial responsivity. PZT is also used as a gas lighter because a sharply-struck PZT element can produce a large spark.

PZT can also be used as an actuator, that is a transducer which causes movement, provided that the movement is not very great. With a maximum allowable strain of only about 0.1% (PZT is a brittle ceramic), a 100 mm rod can only move about 0.1 mm; however, this is often sufficient for the purpose. Because the movement is very rapid PZT actuators are fast and extremely fast-acting fluid control valves have been developed using them. The inch worm is a further development of the PZT actuator in which a PZT rod is clamped at one end by another PZT actuator, then expanded (or contracted) by applying a voltage. After expansion the first clamp is released and a second clamp-actuator fixes the other end. The applied voltage is then removed, so that a small amount of translational motion occurs. Repeating the cycle at a few kHz produces speeds of the order of 1 mm/s. The total range of movement is still fairly limited (a few mm). Micropositioning (to within a nm or so) of objects is feasible with PZT actuators.

13.2.5 *Optical and infra-red transducers*

We have discussed LEDs, PIN diodes, laser diodes and solar cells in section 2.5, which dealt with optical diodes. A number of other transducers are used to convert light to electricity, including photoresistors, pyroelectric sensors and transducers which essentially rely on the temperature rise produced by electromagnetic radiation; the latter have been discussed in section 13.2.1 as temperature transducers.

Photoresistors use a thin film of cadmium sulphide (CdS) to detect incoming light centred on the wavelength 530 nm, near the bandgap cut-off wavelength of 505 nm. The resistance falls from about 1 MΩ in the dark to 10 kΩ at 10 lux (roughly equivalent to a room lit by one candle) and 0.5 kΩ at 1 klux (a room lit by a 100W tungsten filament lamp). They are used in lightmeters and automatic light switches and have a response time of about 50 ms.

Pyroelectric detectors are used in many passive infra-red (PIR) intruder alarms and respond effectively to heat radiated from the body. This heat is detected between wavelengths of 7 μm and 14 μm, which eliminates most solar emission. The pyroelectric element is once more PZT which has an appreciable pyroelectric coefficient, defined by

$$\Pi = \frac{1}{A}\frac{dQ}{dT} \tag{13.7}$$

Unfortunately the response is zero at constant heat flux and so the device will only detect moving heat sources unless the radiation is interrupted by a mechanical chopper. This produces an increase in surface temperature, δT, which decays as $\exp(-t/\tau)$ in time and transforms to the frequency domain as

$$\delta T = \frac{\beta W}{1 + j\omega\tau} \tag{13.8}$$

where β is a constant and W is the radiant intensity at the detector surface in W/m^2. The output is then maximal when the chopping angular frequency is $1/\tau$, and if $\tau \approx 100$ ms, the best chopping frequency is about 1-2 Hz. Chopping is not necessary for intruder sensors. Substituting $\delta T = \delta Q/\Pi A$ from equation 13.7 into equation 13.8 gives

$$\delta Q = \frac{\beta\Pi A W}{\sqrt{1 + \omega^2\tau^2}} \approx \beta\Pi A W$$

And so when this charge appears across the capacitance of the element, the induced voltage is

$$V_{\mathrm{o}} = \frac{\delta Q}{C} \approx \frac{\beta\Pi A W}{C}$$

For a small (linear dimensions a few mm) PZT detector the output voltage is about 3 kV/W, that is $\beta\Pi/C = 3000$ and $V_{\mathrm{o}} = 3000AW$.

Exposed skin emits a surprisingly large amount of radiant heat, depending on conditions, in the order of 10-100 W/m^2, so that a human face emits about 1 W. At 5 m from a detector with a concentrator of area, $A = 10^{-3}$ m^2, it will produce a heat flux of about

$$W = \frac{1}{2\pi r^2} = 6 \text{ mW/m}^2 \quad \Rightarrow \quad WA = 6 \ \mu\text{W}$$

which will give an output of about 20 mV. The detector requires a high-input-impedance amplifier as close to it as possible to reduce cable capacitance. Intruder alarm sensors have built-in FET amplifiers.

13.3 Electromagnetic compatibility

Electrical equipment of all kinds produces electromagnetic disturbances which may propagate through the air or through wiring to other electrical equipment, which will be disturbed by it. The former equipment is the source of electromagnetic interference (EMI) and the latter is the susceptor. Both source and susceptor are regulated by law within the countries of the European Union (one of the most recent being Directives No. 89/336/EEC and 92/31/EEC of January 1996), and the European standards on this subject would themselves fill several books. But the import of the legislation is that all electrical equipment — appliances and tools, and all electrical parts of larger assemblies — must be capable of operating in an EMI environment as well as conforming with EU standards on electromagnetic emissions. This fitness for service is termed electromagnetic compatibility (EMC). Equipment which is deemed to have passed the EMC requirements of the EU can be labelled with a CE sign. This does not necessarily mean that the device will emit little or insignificant electromagnetic radiation, nor that it cannot be susceptible to electromagnetic radiation, only that these are within certain limits. Nor does a CE sticker imply that the device bearing it has been tested, only that the manufacturer is sure it will pass such a test: if it does not then legal redress can be obtained. It can be an expensive matter to consider EMC only after a product is designed; when EMC is borne in mind at product inception the cost of achieving conformity with the law can be greatly reduced.

13.3.1 Sources of EMI

Sources of EMI can be many and various, the commonest ten are listed below:

1.   Power semiconductors which switch at mains frequencies to produce wire-conducted EMI. The voltage spikes can be very large and have significant harmonics at frequencies up to beyond 100 kHz.
2.   Household appliance motors and portable power tools.
3.   Brushes of DC motors produce arcing and rapid current reversals which give rise to broadband EMI.
4.   Thermostatically-controlled heaters and driers produce conducted broadband EMI which depends on the type of current control.
5.   Fluorescent lamps produce conducted and radiated EMI peaking around 1 MHz.
6.   Desk-top computers can cause EMI at their clock frequencies. Older models which are enclosed in unscreened, non-conducting plastic cases are particularly bad.
7.   Other electronic equipment within a screened enclosure, especially multiplexers, timers and flip-flops. Because of its proximity to susceptors, and because it is probably within a screened enclosure, its influence though possibly considerable can easily be overlooked. Digital pulses with sharp edges cause interference at $f = 1/2\pi\tau_r$, where τ_r is the rise time.
8.   Motor vehicles. Not only ignition systems, which are usually adequately suppressed, but also alternators and other electrical equipment.
9.   Radio and TV transmitters. Powerful sources only over narrow frequency ranges, though equipment near a large transmitter may need special protection.

10. Atmospheric sources such as lightning and extraterrestrial emissions which are especially enhanced during times of heightened sunspot activity. These are in the tens of MHz range.

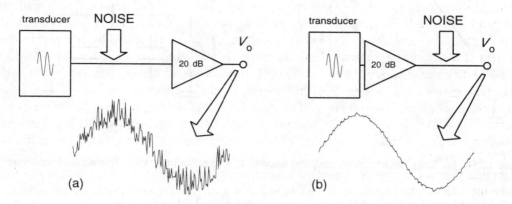

Figure 13.9 The effect of preamplifier positioning (a) far from and (b) close to the transducer

EMI can be stopped at its source, or during transmission, or at the susceptor; but often the only choice is the last of these three. Sometimes very simple measures will suffice, such as moving an amplifier from one end of a cable to the other as shown in figure 13.9a, where the noise which is coupling into the cable is amplified with the signal. By moving the preamplifier down the cable to the transducer (figure 13.9b), the noise introduced through the cable is left unamplified, while the desired signal is amplified. In the example shown the amplifier gain is only 20 dB, but the effect of positioning is very obvious.

Apart from placing components where the effects of noise may be reduced, the chief ways of reducing EMI in order of expense are grounding (cheapest), shielding and filtering (most costly). Grounding is closely bound up with conductor shielding.

13.3.2 Conductor shielding: capacitive coupling

EMI can be coupled into a susceptor circuit either capacitively or inductively; these are sometimes called electric-field coupling and magnetic-field coupling respectively. Consider the two wires shown in figure 13.10a, which are connected to ground via stray capacitances, C_1 and C_2, and to each other by capacitance C_{12}. The susceptible conductor, numbered 2, is also grounded via resistance, R_2. Any voltage source connected to source conductor, number 1, will produce a noise voltage across R_2 and C_2 as shown in the equivalent circuit of figure 13.10b. From this figure we can deduce the noise voltage to be

$$\mathbf{E_n} = \frac{j\omega C_{12}R_2\mathbf{E_s}}{1 + j\omega R_2(C_{12} + C_2)} \tag{13.9}$$

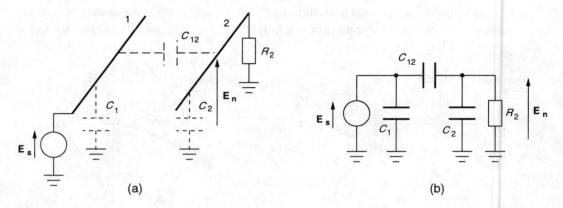

Figure 13.10 (a) Stray capacitances coupling source, 2, and susceptor, 1 (b) The equivalent circuit

The inter-conductor coupling capacitance can be calculated from

$$C_{12} \approx \frac{\epsilon_0 \epsilon_r \pi l}{\ln(2x/d)}$$

where l is the length of the adjacent conductors, x their separation and d their diameter. With air as the coupling medium, $\epsilon_r = 1$ and C_{12} is at most 40 pF/m when the conductors are in contact. Thus equation 13.9 can be reduced to

$$E_n \approx \omega R_2 C_{12} E_s \qquad (13.10)$$

at low frequencies. If R_2 is large or the frequency is high, then equation 13.9 reduces to

$$E_n \approx \frac{C_{12} E_s}{C_{12} + C_2} \qquad (13.11)$$

that is a capacitive divider circuit with noise that is independent of frequency. At high frequencies, given comparable magnitudes for C_{12} and C_2, it can be seen that the noise voltage is comparable to the source voltage.

Capacitive coupling is greatly reduced by screening the conductors as shown in figure 13.11a, where the susceptor is screened by a cylindrical metal shield earthed at one end. Additional capacitances, C_{1S} and C_{2S}, are introduced between shield and source and shield and susceptor conductors. The unscreened length of susceptor conductor has stray capacitances, C'_{12} and C'_2 which are much smaller than they would have been for the unscreened conductor.

Figure 13.11b shows the equivalent circuit for the screened conductor, and this can be reduced to that of figure 13.11c: identical to that of figure 13.10b apart from the additional screen capacitances. Thus equations 13.10 and 13.11 for the noise voltage induced become

$$E_n \approx \omega C'_{12} R_2 E_s$$

at low frequencies and

$$E_n \approx \frac{C'_{12}E_s}{C'_{12} + C_{2S} + C'_2} \approx \frac{C'_{12}E_s}{C_{2S}}$$

at high. Since $C'_{12} \ll C_{12}$ and $C_{2S} \gg C'_{12}$ the noise voltage is much attenuated.

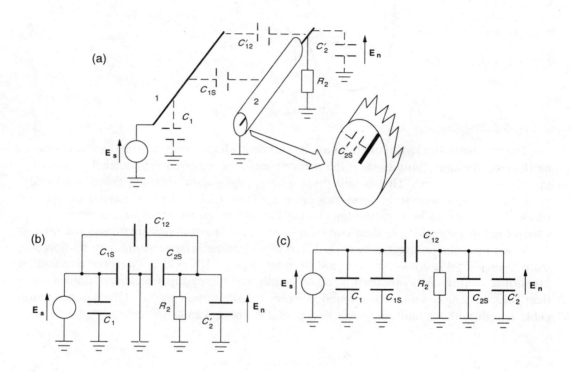

Figure 13.11 (a) Coupling between source conductor, 1, and shielded susceptor conductor, 2 (b) The equivalent circuit (c) The redrawn equivalent circuit

13.3.3 Conductor shielding: inductive coupling

Inductive (or magnetic-field) coupling occurs when magnetic flux from a source links with a conductor as in figure 13.12.

The induced e.m.f. is

$$E_n = \omega B A \cos\theta$$

where θ is the angle between **B** and the normal to A. By grounding both ends of the susceptor conductor as in figure 13.13a the ground loop area is maximum and so is the noise voltage.

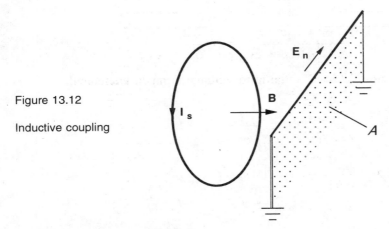

Figure 13.12

Inductive coupling

The magnetic flux will penetrate a non-magnetic shield conductor and so screening is ineffective; the only thing to do if the source cannot be removed or modified is to reduce the ground loop's area. This is best done with a screen conductor grounded at one end, preferably the one nearest the conductor ground. Figure 13.13b shows the arrangement, and figure 13.13c shows how grounding the far end of the screen introduces a ground loop, which though less effective than that of figure 13.13a, nevertheless increases the noise.

Some idea of the effectiveness of different grounding arrangements may be obtained from figure 13.14, in which the test frequency was 100 kHz. The effectiveness of grounding declines with increasing cable length and increasing frequency. Cables more than 0.1λ in length should be grounded every 0.05λ, so that at 10 MHz the maximum cable length without multiple grounding points is about 3 m.

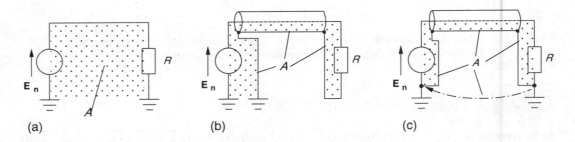

(a)   (b)   (c)

Figure 13.13 Ground loop areas for differing grounding points

The twisted pair shown in figure 13.14c is as effective as the coaxial screened cable of figure 13.14d when the twists are made small (about 50/m) and the frequency is 100 kHz or less. Screened coaxial cable is effective at up to 200 MHz but braided screen types begin to become porous — particularly if the braiding gets pulled apart — above this. Continuously screened cable is, however, available to overcome this problem. A more serious concern is the increasing loss in coaxial cables at high frequencies.

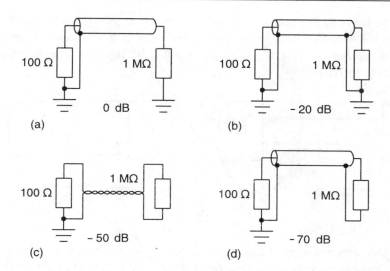

Figure 13.14 The effect of differing screening and grounding arrangements on inductive coupling

Grounding

Good grounding is necessary to minimise inductive coupling. There are three ways of making ground connections: the serial ground (sometimes called common ground) shown in figure 13.15a, the parallel or separate ground shown in figure 13.15b, and the multipoint ground to a highly-conductive ground plane shown in figure 13.15c.

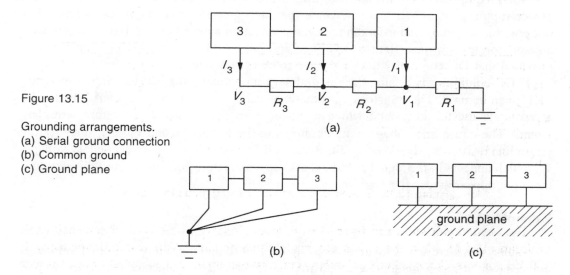

Figure 13.15

Grounding arrangements.
(a) Serial ground connection
(b) Common ground
(c) Ground plane

The cheapest and worst form of grounding is the serial ground because the currents to ground are summed, that is $V_1 = (I_1 + I_2 + I_3)R_3$ and $V_2 = V_1 + (I_2 + I_3)R_2$ and $V_3 = V_2 + I_3R_3$. The currents are noise currents and the resistances are the ground-connection resistances. Serial grounds can be used when the grounded parts are not much affected by noise or if the noise is minimal, but parallel connection should be used when this is not the case, making sure that the ground connection is of low resistance. If grounding braid is used on an equipment case for example, it must be tightly connected with bolts and locking washers to bare metal.

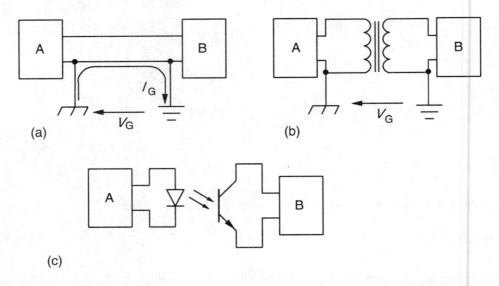

Figure 13.16 A ground loop and its removal (a) with an isolation transformer (b) by opto-isolation

Noisy equipment and any drawing high current such as motors, heaters, relays and power supplies, should be given separate (parallel) ground connections. Sometimes it is not possible to remove an offending ground connection as in figure 13.16a, in which case an isolation transformer (figure 13.16b) or an opto-isolator (figure 13.16c) can be used, provided that DC continuity does not have to be maintained.

If DC continuity is required a longitudinal transformer, such as that shown in figure 13.17, can be used. The transformer, sometimes called a common-mode choke, is usually a ferromagnetic toroid around which identical turns of the signal and return wires are wound. The return line voltage is V_1, as shown in figure 13.17b, and $V_1 \approx V_G$ since R_1, the return line resistance, should be small. Because the turns ratio is 1:1 the secondary voltage, V_2, which appears in the signal circuit, is also approximately equal to V_G and this does not therefore appear across the load, R_L.

Example 13.6

Two subsystems are wired together as in figure 13.17b and a ground-to-ground voltage is induced which has a magnitude of 35 mV at 15 kHz. The connected wires are wound round a ferrite toroid to form a neutralising, longitudinal transformer with equal inductances of 5 mH on primary and secondary. If the line resistances are 1.3 Ω and the load resistance is 50 Ω what is the noise voltage in the load? If the signal voltage at 15 kHz is 60 mV, what is the signal-to-noise ratio in dB?

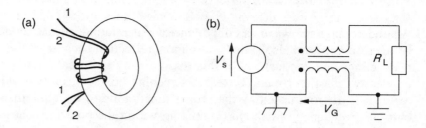

Figure 13.17

The longitudinal transformer
(a) Physical appearance
(b) Circuit diagram
(c) Equivalent circuit

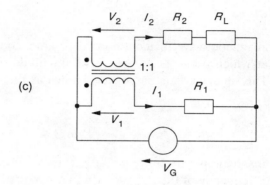

We must assume that the transformer is ideal, making $V_1 = V_2$ in figure 13.17c. By Kirchhoff's voltage law, the upper mesh of figure 13.17c produces

$$V_1 + I_1R_1 = V_2 + I_2(R_2 + R_L)$$

Substituting $V_1 = V_2$ and $R_1 = R_2 = 1.3\ \Omega$, $R_L = 50\ \Omega$ gives $I_1 = 39.5I_2$. Then I_1 is found by using Kirchhoff's voltage law on the lower mesh of figure 13.17c:

$$\mathbf{V_G} = \mathbf{V_1} + \mathbf{I_1}R_1 = \mathbf{I_1}(j\omega L + R_1)$$

The transformer primary reactance, $\omega L = 471\ \Omega$, far larger than R_1, so the magnitude of $\mathbf{I_1}$ is

$$I_1 = \frac{V_G}{\omega L} = \frac{35 \times 10^{-3}}{471} = 74\ \mu A$$

Thus $I_2 = 74/39.5 = 1.9\ \mu A$, giving a noise voltage of $I_2R_L = 1.9 \times 50 = 95\ \mu V$. The signal-to-noise ratio is

$$SNR = 20\log_{10}(S/N) = 20\log\left(\frac{60 \times 10^{-3}}{95 \times 10^{-6}}\right) = 56\ dB$$

The original SNR was 4.7 dB, some 51 dB less. The noise voltage could be further reduced by reducing the line resistance, or putting more turns on the transformer.

13.3.4 Sheet metal shields

An electromagnetic wave can only penetrate a certain depth into conducting media before being absorbed, and also suffers substantial reflection because of the impedance mismatch between air and conductor. For this reason many electrical equipments are placed inside conductive plastic or metal boxes. A screening box serves a twofold purpose: it prevents external EMI from upsetting the proper functioning of the equipment inside, and it prevents electromagnetic radiation originating within the box from causing EMI outside.

Absorption losses

A plane electromagnetic wave incident normally onto a conducting surface will penetrate to a depth, δ, at which point its electric field amplitude is reduced to 37% (1/e) of its surface value. This characteristic parameter, known as the *skin depth*, is given by

$$\delta = \sqrt{\frac{2}{\mu\sigma\omega}}$$

where μ is the magnetic permeability (= $\mu_0\mu_r$), σ is the electrical conductivity in S/m and ω is the angular frequency of the radiation. In non-magnetic materials we can take μ to be μ_0 (= $4\pi \times 10^{-7}$ H/m), and for copper shields at room temperature $\sigma = 55$ MS/m, giving a skin depth formula for these which can be written

$$\delta_{Cu} = 68f^{-1/2} \text{ mm} \tag{13.12}$$

We can express the absorption in a given thickness of metal, x, in dB using

$$A_{dB}(x) = -20\log_{10}[\exp(-x/\delta)] = 8.686x/\delta \tag{13.13}$$

Example 13.7

A copper EMI screen is 1 mm thick. Calculate the absorption in dB for a plane e.m. wave of frequency (a) 1 kHz and (b) 1 MHz. Repeat for an aluminium shield of the same thickness. At what frequency will the absorption be 60 dB for each metal?
(The conductivity of aluminium is 60% of that of copper.)
    The skin depths calculated from equation 13.12 are for copper (a) 2.15 mm at 1 kHz and (b) 0.068 mm at 1 MHz. The skin depth goes as $\sigma^{-1/2}$ so for aluminium the skin depth is increased by a factor of $1/\sqrt{0.6} = 1.29$ giving skin depths of 2.77 mm at 1 kHz and 0.088 mm at 1 MHz. Substitution into equation 13.13 gives the absorption in copper as (a) 4 dB and (b) 128 dB and in aluminium as (a) 3.1 dB and (b) 98 dB.
    Equation 13.13 gives, for an absorption of 60 dB,

$$\delta = \frac{8.686x}{60} = \frac{8.686 \times 1}{60} = 0.145 \text{ mm}$$

Equation 13.12 then leads to

$$f_{Cu} = (68/\delta)^2 = (68/0.145)^2 = 220 \text{ kHz}$$

The frequency will be increased by a factor of $1/0.6 = 1.67$ for aluminium, to 367 kHz.

In practice the reflection losses at 1 kHz will be much greater than the absorption, so the shields are not as bad as it appears at this frequency.

Reflection of plane e.m. waves from plane conductors

How much electromagnetic energy passes through the surface of a conductor depends on the relative impedances of the electromagnetic wave, Z_w, and the conductor, Z_c. For a plane wave in air the wave impedance is the free-space impedance, Z_0, which is given by

$$Z_0 = \sqrt{\frac{\mu_0}{\epsilon_0}} = 377 \ \Omega$$

The surface impedance of a conductor is given by

$$Z_c = \sqrt{\frac{\omega\mu}{\sigma}} = \sqrt{\frac{2\pi f\mu_0}{\sigma}}$$

for a non-magnetic metal. Inserting $\sigma = 55$ MS/m for copper into this equation yields

$$Z_{Cu} = 0.38f^{1/2} \ \mu\Omega$$

Thus $Z_{Cu} \ll Z_w$ and the transmission coefficient is

$$T_{Cu} = \frac{4Z_{Cu}Z_w}{(Z_{Cu} + Z_w)^2} \approx \frac{4Z_{Cu}}{Z_w} = 4 \times 10^{-9}f^{1/2}$$

The reflection loss in dB is $-20\log T$, or

$$R_{Cu} = -20\log T_{Cu} = -20\log(4 \times 10^{-9}) - 10\log f$$

$$= 168 - 10\log f \tag{13.14}$$

At low frequencies the reflection loss is very much larger than the absorption loss, which can virtually be ignored for thin shields. Note that the reflection losses go down with increasing frequency while the absorption losses go up, resulting in minimal losses at a shield thickness equal to the skin depth.

The effects of near-field radiation

Below a distance from the source which is roughly $\lambda/2\pi$ the radiation cannot be considered a plane wave. (Note that $\lambda/2\pi$ is nearly 1000 km at 50 Hz.) This is near-field radiation and has different characteristics to far-field or plane-wave radiation. The near-field wave impedance depends on the ratio of the electric and magnetic fields, *E/H*. The near-field radiation from an electric source such as a dipole antenna is high impedance and has a high E/H ratio. Thus electric sources have high wave impedances in the near field and the transmission coefficient is even smaller than for a plane wave and even more radiation is therefore reflected from a conducting shield.

If the source is a low-impedance or magnetic source, such as a loop antenna, the wave impedance is lower in the near field than in the far, and the reflection losses are reduced. The wave impedance of a magnetic point source is

$$Z_M = \mu_0 \omega r$$

And thus the reflection losses of a conducting shield are

$$R = -20\log\left(\frac{4Z_c}{Z_M}\right) = 20\log\left(\frac{Z_M}{4Z_c}\right) = 20\log\left(\frac{\mu_0 \omega r}{4\sqrt{\mu\omega/\sigma}}\right)$$

Substituting the previously-used values for copper leads to

$$R_M(\mathrm{Cu}) = 14.3 + 10\log r^2 f \quad \mathrm{dB} \tag{13.15}$$

Thus if $r = 1$ m and $f = 50$ Hz, $R = 31.3$ dB, compared to the plane-wave value from equation 13.14 of 151 dB.

For shields less than about 0.1δ in thickness the reflection losses are approximately $10\log r^2 f$, that is about 14 dB less than that given in equation 13.15, and copper shields are ineffective. Iron shields perform better against low-frequency (below about 100 kHz) magnetic sources because of their increased permeability but they must be fairly thick (up to 5 mm) to achieve large EMI reductions.

Shield integrity

It is impossible to maintain a complete shield around a component or sub-assembly because it must be connected to other parts of the equipment by wires, and it must have a cover of some sort through which access is gained. These are weak points in the screen and in most cases more EMI comes through them than through the shield itself, especially at high frequencies.

Breaks in the shield act as attenuating waveguides and the leakage depends on the largest dimension of the hole rather than its area, provided this is larger than about 0.01λ. If the large dimension of the slot is L the cut-off frequency is

$$f_c = \frac{c}{\lambda_c} = \frac{c}{2L} \tag{13.16}$$

The cut-off frequency is actually near the frequency of maximum penetration, but below this frequency the radiation is attenuated. For $f \ll f_c$ the attenuation is

$$A_{dB} = \frac{30d}{L} \quad (\text{in dB}) \tag{13.17}$$

where d is the distance travelled in the waveguide by the wave.

Example 13.8

An EMI shield is made of copper 1 mm thick, which is spot welded every 25 mm to join the top to the sides as shown in figure 13.18, with a seam overlap of 20 mm. What is the cut-off frequency, the absorption and reflection of the shield, and the attenuation of the seam at 200 MHz?

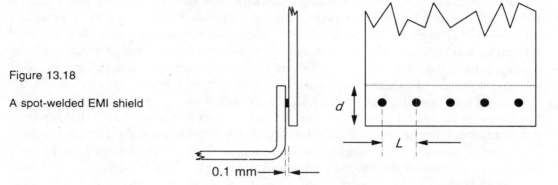

Figure 13.18

A spot-welded EMI shield

0.1 mm

The welded seam forms a waveguide of width, $L = 25$ mm, height 0.1 mm and length (in the direction of propagation), $d = 20$ mm. The entry slot of the waveguide is 0.1 mm × 25 mm, so that the largest dimension of the slot is 25 mm. The cut-off frequency is therefore, according to equation 13.16,

$$f_c = c/2L = 3 \times 10^8/2 \times 25 \times 10^{-3} = 6 \text{ GHz}$$

200 MHz is well below the cut-off frequency and therefore the waveguide attenuation, using equation 13.17, is

$$A_{dB} = \frac{30d}{L} = \frac{30 \times 20 \times 10^{-3}}{25 \times 10^{-3}} = 24 \text{ dB}$$

This is a rather poor figure and means the shield's efficacy has been seriously impaired. By halving the distance between the weld spots the attenuation can be increased to 48 dB. It is interesting to note that if the seam overlap, d, had been 10 mm the attenuation would have been only 12 dB.

The reflection losses from the shield are $168 - 10\log(2 \times 10^8) = 85$ dB. The skin depth from equation 13.12 is 4.8 μm = 0.0048 mm, so the attenuation by equation 13.13 is

$$A_{dB} = 8.686 \times \frac{1}{0.0048} = 1810 \text{ dB}$$

The EMI at 200 MHz can only enter via the welding seam.

A poorly-designed lid on an EMI shield will let in virtually everything at high frequencies. Screw holes also need careful design to avoid high-frequency EMI. Braided screens on coaxial cables must not be pulled apart to leave holes or they will leak at high frequencies. Terminations should be made properly to BNC connectors to maintain the shield integrity all the way to the outer shield of the equipment. The braid should not be twisted but forced into all-round uniform contact with the connector outer screen.

13.3.5 Power lines

The EMI which gets onto wires connecting or supplying power to sub-assemblies can be reduced by filtering. It may be necessary to make a separate power supply for an especially sensitive unit or even as a last resort to use batteries to eliminate all possible power-line EMI. It is difficult and expensive to make a low-noise wide-band amplifier running from a mains supply, but relatively straightforward if a battery supply is used. Power-line filters must be designed to absorb EMI rather than reflect it, in other words lossy filters are needed. RL, RC and RLC rather than LC filters with a low Q-factor should be used. A ferrite bead with a metal film on its outside makes a reasonably-good, lossy, cheap, power-line filter. Small ferrite beads or hollow cylinders can be purchased for a few pence each. If the use of ferrite beads comes as an afterthought, split cylinders can be used to clip around the offending wire, but the cost is much higher.

The characteristic impedance, Z_0, of power leads should be made as low as possible since an instantaneous current change of δI becomes a noise voltage in the line of $Z_0 \delta I$. This impedance is dependent on the relative positions of the leads. In figure 13.19 are shown three common configurations. Figure 13.19a shows two copper conductors placed side by side as on a printed-circuit board (PCB). This has a characteristic impedance given by

$$Z_0 = \frac{120}{\sqrt{\epsilon_r}} \ln\left[\pi(x/b + 1)\right] \ \Omega$$

If $b \gg x$, and both are much greater than the conductor thickness, then Z_0 is a minimum. For typical PCB material $\epsilon_r = 4$ and $Z_0 = 69 \ \Omega$.

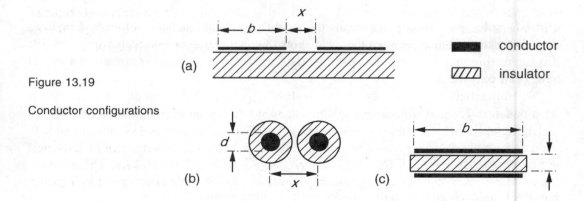

Figure 13.19

Conductor configurations

Figure 13.19b shows two round wires placed in close proximity side by side for which $Z_0 = 120/\sqrt{\epsilon_r}$, where ϵ_r is the relative permittivity of the insulation, assuming the wires are touching and that $x > 2d$, which is usually the case. Typically $\varepsilon_r = 3.5$ for PVC, the usual insulator for ordinary wires, and then $Z_0 = 64\ \Omega$, a moderate value.

Figure 13.19c shows the configuration with the lowest impedance for which

$$Z_0 = \frac{377x}{b\sqrt{\epsilon_r}}$$

If $\varepsilon_r = 2.3$ (polythene) and $x/b = 0.05$ (say 0.25 mm thick polythene and 5 mm wide conductors) then $Z_0 = 12.5\ \Omega$.

Conductor shape is also important when the frequencies are high, because of the skin effect. The current in a conductor is effectively confined to a surface layer whose thickness is equal to the skin depth. Thus the effective resistance of conductors for AC is increased and is given by

$$R_{AC} = \frac{\text{Area}}{\text{perimeter} \times \delta} \times R_{DC}$$

The area is the cross-sectional area normal to the direction of current flow. For round wires of radius, r, the AC resistance is

$$R_{AC} = \frac{r}{2\delta} \times R_{DC}$$

And for thin rectangular conductors, thickness t, it is

$$R_{AC} = \frac{t}{2\delta} \times R_{DC}$$

Flat rectangular conductors make better use of material than round wires.

13.3.6 Components

The components used in filters or any other circuit are not ideal. Capacitors in particular have leads which have an inductance of about 1 µH/m, causing self resonance, sometimes at a lower frequency than is desirable. For example a 5 nF capacitor with leads 20 mm long will have an inductance of about 40 nH and a self-resonant frequency of

$$f_{sr} = \frac{1}{2\pi\sqrt{(LC)}} = 11\ \text{MHz}$$

This is not a very high frequency, yet the capacitor cannot be used above about $0.4f_{sr}$ or 4 MHz. By keeping the leads as short as possible their inductance is minimised and self resonance occurs at higher frequencies. Electrolytic capacitors have substantial inductance and so operate over limited frequency ranges, usually below about 10 kHz. The addition of parallel polycarbonate or polystyrene capacitors of smaller value is sufficient to improve the high-frequency performance.

For line filtering at frequencies higher than about 100 MHz a feed-through capacitor (see figure 13.20a) can be used with the EMI shield as shown in figure 13.20b. Feed-through capacitors have their outer case and one half of the capacitor plates joined together and this is bolted to the screen (figure 13.20c). Their inductance is thus very small and their self-resonant frequencies are in the GHz region. Figure 13.20d shows the special circuit symbol for a feed-through capacitor.

Ferrite beads for slipping onto wires are designed to produce large eddy-current losses at high frequencies and effectively act as resistances, thereby absorbing high-frequency EMI. At low frequencies they present a small reactance and pass signal and power frequencies readily.

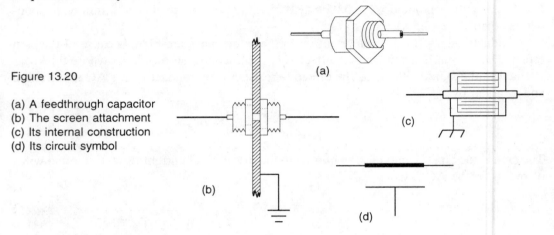

Figure 13.20

(a) A feedthrough capacitor
(b) The screen attachment
(c) Its internal construction
(d) Its circuit symbol

Example 13.9

A copper wire of circular cross-section has a diameter of 0.9 mm, a characteristic imped-ance of 75 Ω, a length of 100 mm and a capacitance of 100 pF/m. It connects to a feed-through capacitor of value 1 nF. What is the resonant frequency of the circuit if the load resistance is effectively infinite? What is the circuit's Q-factor at resonance? If the noise voltage is 5 mV, what is the noise voltage in the load at the resonant frequency? If a ferrite bead of effective resistance 100 Ω at the resonant frequency is slipped over the wire, what is then the load noise voltage?

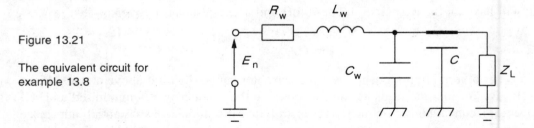

Figure 13.21

The equivalent circuit for example 13.8

The equivalent circuit is shown in figure 13.21, in which the resistance and inductance are due to the wire and nothing else. The wire's inductance is calculated from its characteristic impedance and capacitance using

$$Z_0 = \sqrt{\frac{L_w}{C_w}} \quad \Rightarrow \quad L_w = Z_0^2 C_w = 75^2 \times 10^{-10} = 0.56 \ \mu H/m$$

The wire is 100 mm long so $C_w = 10$ pF, which is negligible compared to that of the parallel feed-through capacitor. The wire's inductance is 56 nH, so the resonant frequency is

$$\omega_0^2 = \frac{1}{LC} = \frac{1}{56 \times 10^{-9} \times 1 \times 10^{-9}} \quad \Rightarrow \quad \omega_0 = 1.34 \times 10^8 \ rad/s$$

which is 21.3 MHz. At this frequency the skin depth in copper is

$$\delta = \frac{68}{\sqrt{21.3 \times 10^6}} = 0.0147 \ mm$$

Because the skin depth is small compared to the wire's radius, the effective resistance of the wire is given by

$$R_w = \frac{l}{\sigma A_{eff}} = \frac{l}{\sigma 2\pi r \delta} = \frac{0.1}{55 \times 10^6 \times 2\pi \times 0.45 \times 10^{-3} \times 0.0147 \times 10^{-3}} = 44 \ m\Omega$$

where A_{eff} is the effective area of the wire carrying current at 21.3 MHz, $2\pi r \delta$.

The wire forms a series RL circuit which connects to the feed-through capacitor and load in parallel. The load impedance is high enough to be ignored compared to the reactance of the feed-through capacitor. The Q-factor of the wire is

$$Q = \frac{\omega_0 L_w}{R_w} = \frac{1.34 \times 10^8 \times 56 \times 10^{-9}}{44 \times 10^{-3}} = 170$$

The voltage appearing across the reactive components is the noise voltage, E_n, magnified by the Q-factor, or $QE_n = 170 \times 5 = 850$ mV. Since the load is in parallel with the feed-through capacitor, this is the noise voltage in the load also.

Putting a ferrite bead on the wire will increase the effective series resistance from 44 mΩ to 100 Ω and then Q and QE_n are reduced by a factor of $0.044/100 = 4.4 \times 10^{-4}$. The noise voltage across the load thus decreases to $850 \times 4.4 \times 10^{-4} = 0.37$ mV.

Suggestions for further reading

Experimental measurements: precision, error and truth by N C Barford (Wiley 2nd ed. 1985)
Handbook of modern sensors: physics, design and applications by J Fraden (American Institute of Physics/OUP 1996)
Sensors and transducers by M J Usher and D A Keating (Macmillan 2nd ed. 1996)
Electrical and electronic measurement and testing by W Bolton (Longman 1993)
Principles of measurement systems by J P Bentley (Longman 1995)

Principles of engineering instrumentation by D C Ramsay (Arnold, 1996)
Electronic noise and low noise design by P J Fish (Macmillan 1993)
Electrical interference and protection by E Thornton (Ellis-Horwood 1990)

Problems

1 Determine the mean and standard deviation of the data below, which are repeated observations of the same quantity using the same instruments. What is the standard error of the mean? What is the probability that a reading of 214.5 will be obtained? How likely is it that the true mean is 210.0? How many more readings are required to reduce the standard error of the mean to 0.2? Plot a histogram of the data (divide the range into five equal intervals) and compare it to an ideal Gaussian distribution of the same mean and standard deviation. Comment.

211.3, 210.2, 210.0, 211.2, 212.2, 212.9, 211.2, 210.2, 209.8, 210.6, 210.9, 209.7

[210.85, 0.98, 0.28, 10^{-4}, 1.5×10^{-3}, 12]

2 Examine figure P13.2 and use the range and number of data points[2] to estimate the mean and standard deviation of the values indicated by dots. How probable is it that a reading will be −1 mV? In what length of time will a reading of 20 mV be obtained on average?

[10 mV, 3 mV, 1.3×10^{-4}, 720 μs]

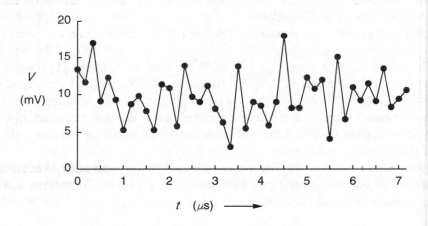

Figure P13.2

3 Two voltages were obtained by different instruments of different accuracies. One voltage is given as 32.1 ± 1.0 mV and the other as 91.5 mV ± 2%. What is the error in the sum and in the difference between the voltages, expressed as a percentage?
[1.7% and 3.5%]

---

[2] The data was generated by the Box-Müller method described in Appendix 3.

4 An r.m.s. current is estimated from the heat produced in a resistance in a time of 50 ms. If the error in the heat measurement is 1%, the error in the time measurement is 1 ms and the error in the resistance measurement is 0.8%, what is the most likely error in the current estimate? If the resistance measurement were error free, what would the change in the most likely error in the current be? If the time taken for the measurement were to be doubled, what would the most likely error be? *[1.2%, −0.07%, 0.8%]*

5 A resistance is measured by two different methods and the results are given as 2.057 ± 0.024 kΩ and 2.071 kΩ ± 1.55%, what is the best estimate of the true value? What is the error of the estimate? *[2.062 kΩ ± 0.93%]*

6 The electrical power derived from a wind turbine is proportional to the cube of the wind speed, and is supplied to a grid at constant voltage. How accurately must the speed be determined if the error in the calculated current is to be less than 10%?

7 A coaxial cable has a braided outer conductor which is 0.3 mm thick and has an effective slot length of 1.8 mm and an effective slot width of 0.06 mm. What is the frequency, f_c, at which the waveguide transmission is greatest? Below what frequency is the waveguide transmission negligible? Find the attenuation in the waveguide at 10 GHz. Repeat these calculations for a coax whose braid is maltreated so that the slot length is 5 mm and the width 1 mm. *[83 GHz, 1.7 GHz, 6.7 dB; 30 GHz, 600 MHz, 2.4 dB]*

8 Calculate the frequency at which the skin depth in copper is 0.3 mm at 300 K. What is the attenuation in this thickness of copper at this frequency? What are the reflection losses at this frequency? Repeat the calculations for aluminium and for steel. (The conductivity of steel is 4.8 MS/m and $\mu_r = 500$.) How much will the figures for copper be affected by a temperature increase of 50 K? *[48.5 kHz, 8.7 dB, 121 dB. 80.9 kHz, 8.7 dB, 117 dB; 1.17 kHz, 8.7 dB, 99.7 dB]*

9 Two conductors of circular cross-section and diameter 0.5 mm are placed as close together as possible along their lengths of 75 mm. If the insulation on the wires is made of PVC 0.3 mm thick with $\varepsilon_r = 2.3$, what is the capacitance between the conductors? The peak-to-peak voltage across one of the conductors is 100 mV and the capacitances to ground are negligible. At what frequency will the rms noise voltage in the other conductor be 10 mV if its resistance to ground is 1.2 kΩ. What will the rms noise voltage be at this frequency if each conductor has a capacitance to ground of 15 pF?
[4.12 pF, 9.1 MHz, 6.1 mV]

10 The susceptor in problem 13.9 is screened by a coaxial conductor over all but 5 mm of its length. Calculate the capacitance between screen and inner conductor, assuming the screen is of negligible thickness, using

$$C_{2S} = \frac{2\pi\varepsilon_r\varepsilon_0 l}{\ln(d_2/d_1)}$$

where d_1 is the inner conductor diameter and d_2 the screen diameter. What is the noise-coupling capacitance between the conductors now? What is the maximum r.m.s. noise voltage? What is the noise half-power frequency?
[11.4 pF, 0.275 pF, 0.833 mV, 10.5 MHz]

11 Show that absorption and reflection losses are a minimum for a sheet conductor when its thickness is the same as the skin depth.

12 A sensitive piece of equipment suffers from EMI emanating at 50 Hz from a low-impedance source 0.5 m away. A copper shield 0.3 mm thick is suggested as a means of reducing the EMI from its unshielded magnitude of 65 mV. Is this a 'thin' shield? What will the induced voltage be? If the shield were made from iron of the same thickness whose conductivity is 8 MS/m and $\mu_r = 600$, what will the induced voltage be? How thick would shields of copper and iron have to be to reduce the EMI to 1 mV? Is it better to shield the source or the susceptor? *[18.6 mV, 48.5 mV, 11.5 mm, 4.3 mm]*

13 A copper power line on a PCB is 0.2 mm thick, 100 mm long and 3 mm wide. A ground plane is on the other side of the PCB which is 1.5 mm thick. If the power line's inductance is 0.12 μH, what is its self-resonant frequency? What is the AC resistance of the line at this frequency? What is the Q-factor of the line at the self-resonant frequency? If the noise voltage at this frequency is 1 mV, what is the induced noise voltage in the power line? (Take $\varepsilon_r = 4$ for the PCB and $C = \varepsilon A/t$. Assume σ_{Cu} to be 58 MS/m.)
[173 MHz, 57 mΩ, 2268, 2.268 V]

14 A thermistor obeys equation 13.6 with $A = 0.04$ Ω and $B = 3500$ K, and has a thermal resistance of 100 K/W. It is used in still air to measure temperatures of 25°C and 0°C. What must the sensing current be if the error in measuring these temperatures is to be less than 10 mK? (Take 0°C to be 273.15 K.) *[< 141 μA, < 82.5 μA]*

Appendix 1: Logic symbols

THE SYMBOLS used in this work are non-standard, nevertheless they are those that are almost exclusively used in data books, journal papers, industrial specifications and so on. The standards for logic symbols which are increasingly demanded by governmental and other official bodies worldwide (not just in the EU) are those set out in BS 3939, section 21 and ANSI/IEEE standard 91 (1984) in the USA. For simple combinational logic circuits containing a few ANDs, NORs, NOTs etc. the symbols used in this book are more readily interpreted than the standard ones, but the standard forms are probably superior to the ad hoc boxes used for all the other digital circuits: counters, timers, multiplexers etc.

Figure A1.1 shows the logic symbols for the five common logic gates. The logic output negation is readily understood, but the identical shapes of the boxes means that the symbols (called 'general qualifying symbols' see table A1.1) inside must be read with care, especially as their meanings, except for AND, are not self-evident.

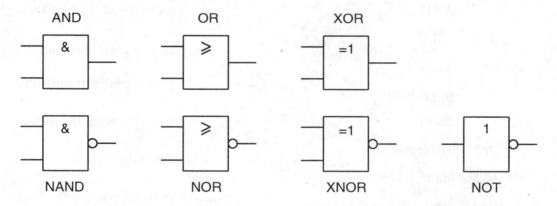

Figure A1.1 ANSI/IEEE symbols for the five common logic gates

The symbols for other logic circuits are based upon an input/output box as in figure A1.2, but with a common block added on top comprising a rectangle with two pieces cut out of the lower corners. This common block acts identically on all the inputs in the divisions in the lower rectangle. The general qualifying symbol is written in the centre top of the common block. The input lines are drawn on the left-hand side and the output lines on the right. The common block can have outputs as well as inputs, these being a function of the outputs in the lower rectangle. Input and output qualifying symbols can be used on the input and output lines. Some of these are given in table A1.1 along with a selection of general qualifying symbols.

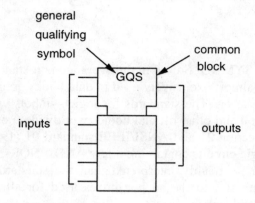

Figure A1.2

Use of common block in logic symbols

Table A1.1 *(a) General qualifying symbols (b) I/O qualifying symbols*

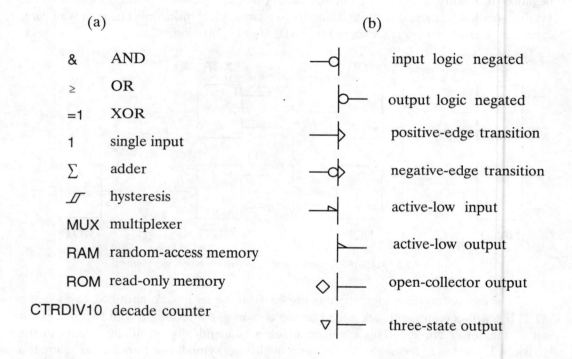

(a)

&	AND
≥	OR
=1	XOR
1	single input
Σ	adder
⎍	hysteresis
MUX	multiplexer
RAM	random-access memory
ROM	read-only memory
CTRDIV10	decade counter

(b)

input logic negated

output logic negated

positive-edge transition

negative-edge transition

active-low input

active-low output

open-collector output

three-state output

The use of some of these symbols in indicated in figure A1.3 which shows an arithmetic logic unit (ALU) and function generator which has an input of two four-bit words (A0-4 and B0-4). The four-bit output word is F0-4. The bracketed numbers in the lower rectangle indicate the significance of the bits, ranging from (1) the LSB through to (8) the MSB. Cn is the carry in for addition and the inverted carry in for subtraction. CP* and CG* are the propagate and generate active low outputs for cascaded operation when words of more than four bits are used. The common-block select inputs S0-2 determine the function to be performed on the inputs according to table A1.2.

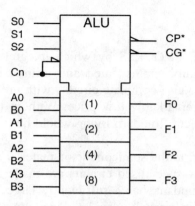

Figure A1.3

Logic symbol for the 74LS381 ALU/function generator

Table A1.2 *Function table for the 74LS381*

S2	S1	S0	function	
0	0	0	CLEAR	
0	0	1	B minus A	arithmetic operations
0	1	0	A minus B	
0	1	1	A plus B	
1	0	0	$A \oplus B$	Boolean operations
1	0	1	$A + B$	
1	1	0	$A \cdot B$	
1	1	1	PRESET	

Appendix 2: IC fabrication

THE PROCESS by which integrated circuits are manufactured is very complex, highly sophisticated and utilises many different techniques; yet it manages to produce complete circuits with millions of components at a remarkably high yield and at a remarkably low price. Without this process the permeation of electronics into every aspect of modern living would be far less and many products would be unthinkable without it.

Silicon is the semiconductor material used for the overwhelming majority of ICs — only a few specialised circuits are made from gallium arsenide (GaAs) — because it is cheap, abundant, has a sufficiently large bandgap (just over 1 eV), can be produced in large single crystals, can be doped both n-type and p-type, and has an oxide (called variously silica, silicon dioxide or SiO_2) which can be used as a surface barrier against impurities. The first stage of the process is producing thin, single-crystal slices (or wafers) of silicon of a specific resistivity and conductivity type.

A2.1 Wafer production

Silicon is obtained from quartz sand, which is almost pure silica. The sand is heated with carbon to form silicon of relatively low purity — about 99% — according to the chemical equation

$$2SiO_2(s) + 3C(s) \rightarrow Si(s) + 3CO(g)\uparrow + SiO(g)\uparrow$$

The impure silicon is powdered and reacted at high temperature with hydrogen chloride gas to form trichlorosilane gas:

$$Si(s) + 3HCl(g) \rightarrow SiHCl_3(g)\uparrow + H_2(g)\uparrow$$

The gaseous trichlorosilane is led away from the reaction chamber and condensed to form a liquid, which can be repeatedly distilled to give a high-purity, electronic-grade chemical. Electronic-grade, polycrystalline silicon is produced by reducing this very pure trichlorosilane with hydrogen at high temperatures once more:

$$SiHCl_3(g) + H_2(g) \rightarrow Si(s) + 3HCl(g)\uparrow$$

Polycrystalline material is useless for device fabrication and must be turned into single-crystal material by melting it in a special type of furnace called a Czochralski crystal grower. The materials used in the hot parts of the furnace must also be of high purity so that no inadvertent doping of the silicon occurs. Since silicon has a melting point of 1410°C, the melt can be contained in a quartz (SiO_2) vessel contained within a carbon susceptor that is heated by an RF coil. Doped silicon (n-type or p-type) can be added to the original material to make a product whose resistivity is whatever is wanted (typically 0.1 Ωm). A seed crystal of the desired crystallographic orientation is lowered into the surface of the melt and slowly withdrawn to form rod of material about 150 mm in diameter and about 3 m long. By counter-rotating both melt and seed crystal temperature and impurity fluctuations are kept to a minimum. Figure A2.1 shows a diagram of part of a Czochralski puller, or crystal grower, which is enclosed within a gas-tight container filled with helium or some other inert gas. The diameter of the silicon boule (the solidified single crystal material) is monitored and kept constant by controlling the withdrawal rate.

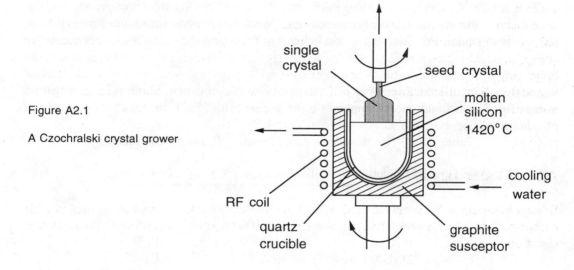

Figure A2.1

A Czochralski crystal grower

After the crystal has cooled down it is ground to a constant diameter and an orientation flat is ground along its length, so that it is readily aligned during subsequent processing. Then it is sliced by a diamond saw into wafers and its resistivity is checked using a special surface resistance probe. Variations in resistivity are usually kept to within ±10% of the nominal value, say from 0.09 to 0.11 Ωm if the target is 0.1 Ωm. Some segregation of dopant always occurred and not all the boule will be within the target range. Any material that is outside the range can be sent back for reprocessing.

Slicing damages the surface of the wafer very considerably and this damage is removed from one side by carefully polishing with a very fine abrasive paste such as diamond dust. A highly polished mirror finish is obtained and the number of defects/m^2 can be ascertained by etching the silicon with a mixture of nitric and hydrofluoric acids (Sirtl etch). The wafer thickness and parallelism are also checked to see if they are within specification. The wafers are usually made about 0.6 mm thick so that breakages are minimised while as many wafers as possible are obtained from the crystal. Since it takes about five minutes to cut one wafer and the saw blade is about 0.4 mm thick, the cutting up of an entire ingot takes about a week if only one saw is used.

A2.2 Oxidation, photolithography and diffusion

The finished wafer can now be oxidised prior to diffusion. Oxidation of the silicon surface produces a barrier against impurities which can be selectively removed from desired parts by hydrofluoric acid (a solution of hydrogen fluoride in water)[1]. By chance the oxide is strongly adherent to the underlying silicon (in most cases oxides are not) and this is a very important reason for the original choice of silicon for IC manufacture.

The process of oxidation can be carried out slowly using dry oxygen by simply exposing the silicon to an atmosphere containing oxygen at a fairly high temperature: 1000°C - 1200°C. Dry-grown oxide films are of a better quality than wet-grown, but the time taken to grow a thick layer is considerable. Wet-grown oxide is produced by exposing the wafer to steam. Provided the oxide layer is not too thin the oxide growth is parabolic

$$x^2 = Bt$$

where x is the oxide thickness and B is the parabolic rate constant. Table A2.1 gives some values for B as a function of temperature for wafers with (111) surfaces[2].

Table A2.1 *The parabolic rate constant, B, for (111) silicon*

Type of oxidation	Temperature (°C)	B ($\mu m^2/h$)
wet	1200	0.72
wet	1100	0.51
wet	1000	0.29
dry	1200	0.045
dry	1100	0.027
dry	1000	0.012

Thus at 1100°C in wet oxygen it will take about two hours to grow a 1 μm thick oxide layer, but in this time in dry oxygen the layer will be only 0.23 μm thick. For diffusion masking wet oxygen and 1-2 μm thick layers are used and thin dry-grown oxide is used for MOS gate insulation for example. Devices can also be isolated from each other by oxide layers rather than reverse-biased p-n junctions.

Oxidation uses up silicon at a rate of 0.44 μm for each μm of oxide grown. This produces patterns on the surface after diffusion, since not all the oxide is removed, but only selected areas where diffusion is wanted. The pattern so produced is used to align the various masks that are used to create the diffusion patterns and hence the devices in the silicon.

---

[1] Hydrofluoric acid (HF) is *very dangerous* since it can permeate the tissues unnoticed — it causes very little irritation — and after several hours begins to destroy the underlying bone, at which point pain is felt. The antidote is several injections of calcium gluconate into the areas near the affected bone. When using HF always wear gloves and assume that any spillage in the working area is HF. It can readily be diluted to a safe level with water and mopped up.

[2] (111) are the Miller indices of the plane which denote the crystallographic orientation. Most wafers are either (111) or (100) orientation.

After oxidation the wafers are coated with a thin film of photosensitive material known as *photoresist* (and often as KPR — Kodak Photo-Resist). When this is exposed to ultra-violet light through a mask, only some parts are exposed which become polymerised and insoluble in an appropriate solvent[3]. The wavelength of UV radiation is about 100-200 nm, which means that the edge of any line between exposed and unexposed resist will be ragged to approximately this extent. Thus if a channel is cut 1 μm wide in the oxide it will have a variable width of 0.8-1.2 μm or so, which is probably not acceptable. When linewidths become less than about 2 μm wide, UV radiation cannot be used. Electron beams have very small wavelengths and can be used with special resists to do the same job as UV light, though the process is slower. An advantage of electron-beam lithography is that no mask is used: the beam is merely deflected onto the areas to be exposed. Linewidths for electron-beam lithography can be 0.1 μm or less.

Figure A2.2a-g shows the photolithographic process (photolithography means writing by light on stone, literally). The polymerised resist is then removed to leave a wafer surface that is partly oxidised and partly bare. The bare silicon surface is exposed to dopants (electrically-active impurities) that can be diffused into the surface to a closely controlled depth (up to 10 μm) as in figure A2.2h.

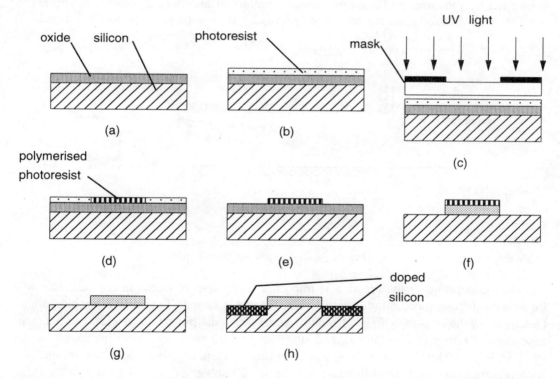

Figure A2.2 (a) Oxidised wafer (b) Photoresisted wafer (c) Masking and exposure to UV light
(d) Polymerised resist (e) Removal of unpolymerised resist (f) Removal of exposed oxide
(g) Removal of polymerised resist (h) Exposure to dopants and diffusion

---

[3] This is a negative resist. A positive resist becomes more easily removed after exposure to UV.

Alternatively, for very small dopant depths (1-3 µm), ion implantation can be used, the oxide this time acting as a barrier to the incoming energetic ion beam, so preventing doping. Ion implantation requires expensive equipment (as do many process in IC production) but is capable of very precise control of dopant concentrations.

A2.3 Epitaxy

The production of certain devices is made easier if a thin complementarily-doped layer of silicon is deposited on the original wafer. Thus if the original wafer were p-type, one would deposit an n-type layer on the surface. The layer so deposited adopts the same crystallographic orientation as the substrate (the original wafer) and is called an *epitaxial* layer (strictly it ought to be called an epitactic layer, but the former usage is now universal) from the Greek meaning 'arranged upon'. There are many ways of accomplishing epitaxy, but the oldest is to use chemical vapour deposition (CVD) which involves heating up the substrate in an atmosphere containing silicon species which are reduced by hydrogen or decompose to silicon. If the rate of deposition is carefully controlled the epitaxial layers are virtually free from defects. The resistivity and dopant type can be controlled by judicious additions of minute amounts of appropriate molecular species in the gas that is passed over the wafers. Figure A2.3 shows the arrangement often used.

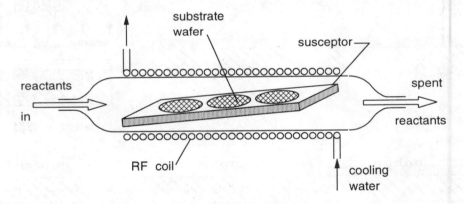

Figure A2.3 An epitaxial reactor

The temperature of the reaction is important since dopant atoms in the substrate will be able to diffuse more rapidly as the temperature is raised. Typically tetrachlorosilane ($SiCl_4$) or trichlorosilane ($SiHCl_3$) require hydrogen reduction at 1100°-1200°C and will take about 10 mins to deposit a layer 5 µm thick. Silane prediluted with hydrogen can be used at 950°-1000°C. Silane is a gas at room temperature, but tetrachlorosilane and trichlorosilane are both liquids and some form of bubbler is needed to introduce these species into the hydrogen stream prior to entering the reactor. Close control of temperature and flow rates is necessary if uniform layers are to be produced. To dope the layers one generally uses prediluted cylinders of hydrogen and gaseous dopant compounds such as diborane, B_2H_6 (p-type layers), phosphine, PH_3 or arsine, AsH_3 (for n-type layers). The thickness and resistivity are controllable to ±10% fairly readily. All the subsequent processing to form working devices takes place within the epi-layer; the substrate is little more than a means of handling the epi-layer.

A2.4 Making a BJT

As an example of device production by the processes described above, consider the making of an npn BJT for which a buried layer is required to keep down the series resistance of the collector. The sequence of steps is shown in figure A2.4, which is a cross-sectional sequence with a much-exaggerated, and not-to-scale, vertical scale.

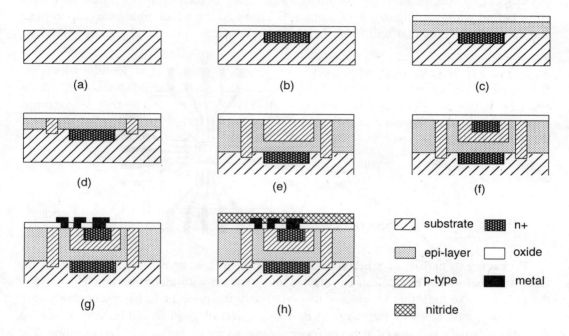

Figure A2.4 Stages in the making of a BJT (a) Substrate (p-type) (b) Buried layer (n+) (c) Epi-layer (n-type) (d) Isolation (p-type) (e) Base (p-type) (f) Emitter (n+) (g) Contact metal (h) Nitride passivation

The n-type buried layer is shown in figure A2.4b with the oxide layer on top after diffusion. The steps in the silicon and the oxide layers are not shown in the diagrams, which are highly schematic. The buried layer is produced by the oxidation, photolithography and diffusion on a p-type substrate (figure A2.4a), as described in section A2.3, and this procedure is followed for all the subsequent diffusions too. The substrate is boron-doped to about 0.1 Ωm and the buried layer is antimony-doped very heavily n-type (denoted n+) to about 10 μΩm resistivity. An epi-layer, doped n-type with phosphorus to give a resistivity of about 0.1 Ωm, is grown on top of the buried layer (figure A2.4c). The transistor is isolated from the rest of the circuit by a boron diffusion (figure A2.4d), which goes right through the epi-layer to the substrate, and then the p-type base is formed by a p-type boron diffusion (figure A2.4e).

The emitter diffusion (figure A2.4f) is very critical as it determines the base width and hence the beta of the transistor, it is usually done with phosphorus at a relatively low temperature to achieve fine control of the diffusion depth as well as to prevent too much diffusion from the buried layer. Contact windows are etched after this and aluminium (or other metal) is evaporated over the wafer and then etched to form the interconnections between the components of the circuit (figure A2.4g). A passivation and scratch-protection layer of silicon nitride or silicon oxynitride is deposited over the whole wafer by CVD (figure A2.4h) at the end of wafer processing. Bonding pad windows are etched through the passivation so that each chip can be connected to the pins of its package.

After wafer processing the individual circuits are run through electrical tests and devices out of specification are identified. The wafer is then broken up into individual chips before wire bonding to a lead frame (see figure A2.5), which holds the chip in place and contains the pins that will form the electrical contacts in the final packaged device.

Figure A2.5

A chip mounted in a lead frame prior to wire bonding

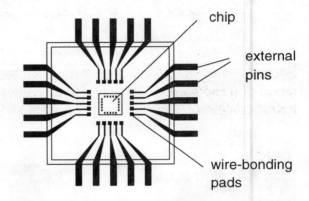

Packaging in plastic, or a hermetically-sealed metal can, or metal-ceramic box, is the final stage before the device can be retested and stamped with identification. The yield of the process is so high that in spite of its complexity the price of ICs is exceedingly low. If the failure rate at every step were to be 1% the overall yield would be very small. As the defect density is reduced it becomes economic to make larger and larger chips. The first ICs were only about 2-3 mm across, whereas now 20×20 mm is used routinely when required.

Suggestions for further reading

There are many books on this topic, but the interested reader might like to try

Integrated circuit design, fabrication and test by P Shepherd (Macmillan 1996)
Integrated circuit engineering by L J Herbst (Oxford University Press, 1996)
VLSI design techniques for analog and digital circuits by R L Geiger, P E Allen and N R Strader (McGraw-Hill, 1990)

Appendix 3: Generating normally-distributed variables

I N SEVERAL PLACES in this work there has been a need to generate normally-distributed variables, either for the purpose of illustrating errors, or to make noisy waveforms where the rms noise must be known. This can be done using random numbers and the Gaussian distribution function, which is generally rather tedious if reasonable accuracy is required. A quicker way is to use the Box-Müller method. This has as its starting point two random variables, x and y, which range from 0 to 1 and then using the formulas

$$u = \sigma\sqrt{-2\ln x} \times \cos(2\pi y) + \mu$$

$$v = \sigma\sqrt{-2\ln x} \times \sin(2\pi y) + \mu$$

to generate two variables, u and v, which are independent and normally-distributed with means of μ and standard deviations of σ. The formulas are readily adapted to use with a pocket calculator.

Index